Lecture Notes in Mathematics 1510

P. P. Kulish (Ed.)

Quantum Groups

Proceedings of Workshops held in the Euler International Mathematical Institute, Leningrad, Fall 1990

Springer-Verlag
Berlin Heidelberg New York
London Paris Tokyo
Hong Kong Barcelona
Budapest

Editor

Petr P. Kulish
Euler International Mathematical Institute
Fontanka 27, 191011, St. Petersburg, Russia

Mathematics Subject Classification (1991): 16W30, 17B37, 57M25, 81R50

ISBN 3-540-55305-3 Springer-Verlag Berlin Heidelberg New York
ISBN 0-387-55305-3 Springer-Verlag New York Berlin Heidelberg

Printed in Germany

Typesetting: Camera ready by authors
Printing and binding: Druckhaus Beltz, Hemsbach/Bergstr.
46/3140-543210 - Printed on acid-free paper

FOREWORD

This is the first volume of the Proceedings of the Euler International Mathematical Institute in St.Petersburg. Springer-Verlag has kindly agreed to incorporate these Proceedings in the format of the Lecture Notes in Mathematics.

Due to a variety of circumstances the first Semester of the Institute, in the fall of 1990, was not fully fledged. Instead of a combined research effort it was organized in the form of short workshops. So this volume contains the results of individual authors, brought to the attention of the participants, rather than the outcome of joint work carried out within the walls of the Institute. This will be changed as soon as the Institute begins to work in full strength.

I hope that the results obtained at the Euler Institute will interest the international mathematical community, so that our publications will become an important part of the Lecture Notes in Mathematics. I would like to thank Springer-Verlag for their invaluable help and good will.

Director of EIMI L. Faddeev

PREFACE

The Euler International Mathematical Institute in Leningrad was founded with the aim of enhancing contacts of Soviet and foreign mathematicians and to create opportunities for them to meet during two or three month periods to discuss current problems and carry out research within the mathematical topic chosen for this period.

The first semester (Fall 1990) was devoted to quantum groups, in view of the growing interest to this subject manifested by mathematicians of various specializations. Organizational questions combined with the desire to bring together as many participants as possible determined the structure of the first semester which was divided into three workshops:

1. Quantum groups, deformation theory and representation theory (15–28 October 1990);
2. Quantum groups, symmetries of dynamical systems and conformal field theory (12–25 November 1990);
3. Quantum groups, low-dimensional topology and link invariants (2-15 December 1990).

Each workshop was attended by approximately 15 Soviet mathematicians and 15 colleagues from Europe, Japan and the U.S.. All participants gave one-hour talks most of which are included in these Proceedings. According to the topical division, the contributions are grouped into three sections whose titles coincide with the titles of the workshops. The talks were delivered in the mornings, and in the afternoons some informal seminars were devoted mainly to detailed discussions of the problems mentioned in the talks. In particular, during the first workshop there was a seminar on unsolved problems in quantum groups. Some of these problems are published in Chapter 4 of these Proceedings. Many interesting problems are enounced also in the contributed papers; indeed one of the papers carries the title "On some unsolved problems in quantum group theory".

From the topics of the workshops and the titles of the contributions it is clear that quantum groups and quantum Lie algebras draw the attention of mathematicians from quite different fields, including topology, algebra and analysis. These concepts are also actively exploited in modern theoretical physics. It may be interesting to note that almost all conferences on mathematical and theoretical physics, from high energy to solid state physics, planned for this summer (1991) include a section on quantum groups.

The organizers of the first semester of the Euler International Mathematical Institute hope that the contributions of actively working experts in quantum groups collected in these Proceedings will influence the rapid development of this field.

Leningrad
September 1991

P.P.Kulish

TABLE OF CONTENTS

II. Quantum Groups, Symmetries of Dynamical Systems and Conformal Field Theory

III. Quantum Groups, Low-Dimensional Topology and Link Invariants

IV. Open Problems in Quantum Group Theory

ON SOME UNSOLVED PROBLEMS IN QUANTUM GROUP THEORY

V.G.DRINFELD

Physical & Technical Institute of Low Temperatures, Kharkov

In this paper k denotes a field of characteristic 0. "Algebra" means "algebra over k" unless it is clear that an algebra over some other ring is considered.

1. Quantization of Lie bialgebras

The definition of quantization of Lie bialgebras can be found in §6 of [1].

Question 1.1. *Can every Lie bialgebra be quantized?*

It is known (cf. [1], §9) that if a certain cohomology group $Der^2\mathfrak{g}$ associated to a Lie bialgebra $\mathfrak{g}$ is zero then $\mathfrak{g}$ can be quantized.

Theorem. *Every Lie bialgebra* $\mathfrak{g}$ *can be quantized modulo* h^4.

If $\mathfrak{g}$ is a Lie bialgebra then by a quantization of $\mathfrak{g}$ modulo h^n I mean a Hopf algebra deformation A of the universal enveloping algebra $U\mathfrak{g}$ over $k[h]/(h^n)$ with a mapping $\delta : A \to A\otimes A$ such that the restriction to $\mathfrak{g}$ of the reduction of $\delta \bmod h$ is cocommutator map $\mathfrak{g} \to \mathfrak{g}\otimes\mathfrak{g}$ and the following identities hold:

$$\Delta - \Delta' = h\Delta, \delta' = -\delta, \delta(ab) = \Delta(a)\delta(b) + \delta(a)\Delta'(b)$$
$$Alt \circ (\delta\otimes id)\circ\delta = 0, (\Delta\otimes id)\circ\delta = (id\otimes\delta)\circ\Delta + \sigma_{23}\circ(\delta\otimes id)\circ\Delta$$

Here $\Delta' = \sigma\circ\Delta$, $\delta' = \sigma\circ\delta$ where $\sigma : A\otimes A \to A\otimes A$ interchanges the tensor factors, Alt is the canonical map $A\otimes A\otimes A \to \Lambda^3 A$ and $\sigma_{23} : A\otimes A\otimes A \to A\otimes A\otimes A$ is defined by $\sigma_{23} = id\otimes\sigma$. Note that if B is a quantization of $\mathfrak{g}$ in the sense of [1] then B/h^nB is a quantization $\bmod h^n$.

Let me describe a quantization of $\mathfrak{g} \bmod h^3$ which will be denoted by $A_3(\mathfrak{g})$. Choose a basis $e_i \in \mathfrak{g}$ and denote by c_{ij}^k and f_i^{jk} the structure constants of $\mathfrak{g}$, i.e., $[e_i, e_j] = c_{ij}^k e_k$ and the cocommutator maps e_i to $f_i^{jk} e_j\otimes e_k$. $A_3(\mathfrak{g})$ is the algebra over $k[h]/(h^3)$ with generators x_i and defining relations

$$x_i x_j - x_j x_i = c_{ij}^k x_k + h^2(f_i^{lm} f_j^{pq} c_{lp}^r \{x_m, x_q, x_r\}/12 + a d_i^s d_j^t c_{st}^k x_k) \tag{1.1}$$

Here $\{x,y,z\}$ denotes $(xyz + xzy + yxz + yzx + zxy + zyx)/6$, $d_i^s = f_i^{jk} c_{jk}^s$, $a \in k$ is arbitrary. Δ and δ given by

$$\Delta(x_i) = x_i\otimes 1 + 1\otimes x_i + h f_i^{jk} x_j\otimes x_k/2 - h^2 f_i^{jk} f_j^{lm}(\{x_k, x_l\}\otimes x_m + x_m\otimes\{x_k, x_l\})/12, \tag{1.2}$$

$$\delta(x_i) = f_i^{jk} x_j \otimes x_k - h^2 (f_i^{rs} f_r^{tu} f_t^{vw} \{x_u, x_v\} \otimes \{x_s, x_w\}/12 + b d_j^l d_k^m f_i^{jk} x_l \otimes x_m) \quad (1.3)$$

Here $b = a + 1/576$ and $\{x, y\}$ denotes $(xy + yx)/2$. Up to isomorphism $A_3(\mathfrak{g})$ does not depend on a (the isomorphism has the form $x_i \to x_i + h^2 \beta d_i^j d_j^l x_l, \beta \in k$).

The above formulae for $A_3(\mathfrak{g})$ were obtained by a straightforward and mysterious computation. The first mystery is that the cohomological obstruction to quatization mod h^3 is zero. The second one is the role of d_i^j in (1.1), (1.3) (d_i^j is nothing but the matrix of the derivation $D : \mathfrak{g} \to \mathfrak{g}$ defined in §8 of [1] which is so to say the classical analog of the square of the antipode).

Note that $A_3(\mathfrak{g})$ is an *even* quantization, i.e., there is a k-linear map $\tau : A_3(\mathfrak{g}) \to A_3(\mathfrak{g})$ such that $\tau(ab) = \tau(a)\tau(b), \tau(ha) = -h\tau(a), \tau^2 = id, (\tau \otimes \tau) \circ \delta = \delta \circ \tau$ and τ is a coalgebra antihomomorphism (τ acts identically on the generators $x_i \in A_3(\mathfrak{g})$). From this and the oddness of 3 one can deduce that $A_3(\mathfrak{g})$ can be lifted uniquely to an even quantization mod h^4 denoted $A_4(\mathfrak{g})$. I have not computed $A_4(\mathfrak{g})$ explicitly.

Note that 1) $A_3(\mathfrak{g})$ is defined by Eqs. (1.1)-(1.3) whose right-hand sides are obtained from c_{ij}^k, f_k^{ij} and x_r by using acyclic tensor calculus[1], 2) the fact that Eqs. (1.1)-(1.3) define a quantization mod h^3 can be obtained by using only formal rules of acyclic tensor calculus and the identities

$$c_{ij}^k + c_{ji}^k = 0 \quad (1.4)$$

$$f_k^{ij} + f_k^{ji} = 0 \quad (1.5)$$

$$c_{ij}^r c_{rk}^l + c_{jk}^r c_{ri}^l + c_{ki}^r c_{rj}^l = 0 \quad (1.6)$$

$$f_r^{ij} f_l^{rk} + f_r^{jk} f_l^{ri} + f_r^{ki} f_l^{rj} = 0 \quad (1.7)$$

$$c_{rs}^k f_k^{ij} + c_{tr}^j f_s^{it} + c_{ts}^i f_r^{jt} - c_{tr}^i f_s^{jt} - c_{ts}^j f_r^{it} = 0 \quad (1.8)$$

which mean that $\mathfrak{g}$ is Lie bialgebra. A quantization with properties 1), 2) will be called *universal*. $A_4(\mathfrak{g})$ is a universal quantization mod h^4.

Question 1.2. *Does there exist a universal quantization for Lie bialgebras?*

I do not know whether universal quantization mod h^n is unique up to isomorphism but it can be shown that if *some* universal quantization mod h^n can be lifted to a universal quantization mod h^{n+1} then it is true for *all* universal quantizations mod h^n. Therefore it is easy to see that for a concrete n the problem of existence of universal quantization mod h^n can be solved algorithmically. It would be interesting to solve it for $n = 5$.

Remark. Formulae (1.1) - (1.3) are homogeneous in the following sense: if f_k^{ij} is replaced by λf_k^{ij} and h by $\lambda^{-1} h$ then the right-hand sides of (1.1) and (1.2) remain the same while the right-hand side of (1.3) is multiplicated by λ. In this situation let us say that the universal quantization is homogeneous. It is easy to show that if there is a universal quantization mod h^n then it can be chosen to be homogeneous.

[1] Acyclicity means,e.g., that we are not allowed to form the expression $c_{ir}^s c_{js}^t c_{kt}^r$ where r, s, t form a cycle in the oriented graph whose vertexes are indices i, r, j, s, k, t and an arrow from one index to another means that the first one is a lower index of some tensor and the second one is an upper index of the same tensor. Acyclic expressions make sense if $\dim \mathfrak{g} = \infty$ and moreover if $\mathfrak{g}$ is an object of an arbitrary k-additive tensor category [2] while nonacyclic expressions make sense in rigid tensor category.

2. Does there exist a quantized Campbell-Hausdorff series?

First consider the usual Campbell-Hausdorff series $F(x,y) = ln(e^x e^y) = x + y + [x,y]/2 + [x,[x,y]]/12 + \ldots$. Suppose we have a Lie algebra $\mathfrak{p}$ with a fixed basis. Denote by f_k^{ij} the structure constants of $\mathfrak{p}$. Let us write $x, y \in \mathfrak{p}$ as $x = (x_1, \ldots, x_n), y = (y_1, \ldots y_n)$. Then $F(x,y) = (F_1, \ldots F_n)$ where

$$F_k = x_k + y_k + f_k^{ij} x_i y_j/2 + f_k^{ij} f_j^{rs} x_i x_r y_s/12 + \ldots$$

$F(x,y)$ has the following properties:

a) $F_k(x,y)$ is expressed in terms of x_i, y_i and f_t^{rs} by means of acyclic tensor calculus with rational coefficients (cf. footnote 1);

b) $F_k = x_k + y_k + f_k^{ij} x_i y_j/2+$ terms of degree greater than 2;

c) F satisfies the associativity law

$$F(F(x,y),z) = F(x,F(y,z)); \tag{2.1}$$

to be more precise (2.1) is a formal consequence of (1.5) and the Jacobi identity (1.7)

Until now we tacitly assumed that x_i, y_i, z_i were *commuting* variables. Now suppose that the only relations between x_i, y_i, z_i are the following ones: $[x_i, y_j] = 0$, $[x_i, z_j] = 0$, $[y_i, z_j] = 0$ for all i, j.

Question 2.1. *In this situation does there exist a formal series $F(x,y)$ which satisfies properties a)-c)?*

A positive answer to question 1.2 implies a positive answer to question 2.1. To construct $F(x,y)$ one can proceed as follows. Consider the Lie coalgebra $\mathfrak{a}$ with structure constants f_k^{ij}. Then the free Lie algebra $\mathfrak{g}$ generated by the vector space $\mathfrak{a}$ has a natural Lie bialgebra structure. We suppose that a universal quantization for Lie bialgebras exists. Choose it to be homogeneous (cf. the remark at the end of section 1). Apply this quantization to $\mathfrak{g}$. We obtain a Hopf algebra which is a topologically free associative algebra over $k[[h]]$ with generators $x_1, x_2, \ldots, x_n$. Then $\Delta(x_i) = \widetilde{F}_1(x_1 \otimes 1, \ldots, x_n \otimes 1, 1 \otimes x_1, \ldots 1 \otimes x_n)$ for some formal series $\widetilde{F}_1(x,y)$ with coefficients in $\mathbb{Q}[[h]]$. Put $\widetilde{F} = (\widetilde{F}_1, \ldots, \widetilde{F}_n)$. Then $\widetilde{F}$ has the required properties a), c) and instead of b) we have $\widetilde{F}_k = x_k + y_k + +h f_k^{ij} x_i y_j/2+$ terms of degree greater than 2. Besides, $\widetilde{F}$ remains unchanged if f_k^{ij} is replaced by λf_k^{ij} and h by $\lambda^{-1} h$. Therefore, we can obtain the desired F by putting $h = 1$.

3. "Naive" quantization of solutions to the classical Yang-Baxter equation

Question 3.1. *Given an associative algebra A and a solution $r \in A \otimes A$ to the classical Yang-Baxter equation*

$$[r^{12}, r^{13}] + [r^{12}, r^{23}] + [r^{13}, r^{23}] = 0 \tag{3.1}$$

does there always exist a formal series $R = R(h) = 1 + hr + \sum_{n=2}^{\infty} R_{(n)} h^n$ satisfying the quantum Yang-Baxter equation $R^{12}R^{13}R^{23} = R^{23}R^{13}R^{23}$?

Question 3.2. *Does there exist a universal solution to the above quantization problem?*

Just as in section 1 "universality" means that 1) the coordinates $R^{ij}_{(n)}$ of the tensors $R_{(n)} \in A \otimes A$ are expressed by means of acyclic tensor calculus in terms r^{ij} (the coordinates of r) and m^k_{ij} (the structure constants of A), 2) the Yang-Baxter equality for R is a formal consequence of (3.1) and the associativity identity $m^r_{ij} m^l_{rk} = m^l_{ir} m^r_{jk}$.

If r satisfies (3.1) and the "unitarity" condition $r^{21} = -r$ then it is natural to look for an R such that $R^{21} = R^{-1}$. Let us call this "unitary quantization problem". This problem always has a solution (cf. [3], [4]), but if r depends continuously on a parameter s and the rank of the tensor r is not constant then the R constructed in [3], [4] does not seem to be continuous in s.

Question 3.3. *Does there exist a universal solution to the unitary quantization problem?*

4. "Sophisticated" quantization of solutions to the classical Yang-Baxter equation

If $r \in \mathfrak{g} \otimes \mathfrak{g}$ where $\mathfrak{g}$ is not an associative algebra but a Lie algebra then (3.1) makes sense but the quantization problems discussed in section 3 do not make sense. However if the tensor $t = r^{21} + r$ is $\mathfrak{g}$-invariant[2] there is a good definition of quantization of r. According to [5], [6] there is a "quasi-triangular quasi-Hopf algebra" $(A, \Delta, \varepsilon, \Phi, R)$ over $k[[h]]$ canonically associated to $(\mathfrak{g}, t)$, namely $A = (U\mathfrak{g})[[h]], \Delta : A \to A \otimes A$ is the usual comultiplication, $\varepsilon : A \to k[[h]]$ is the usual counit, $R = e^{ht/2}, \Phi = \varphi(ht^{12}, ht^{23})$ where $\varphi(X, Y)$ is a formal power series over $\mathbb{Q}$ in non-commuting variables X, Y satisfying Eqs.(2.12)-(2.14) from [6] and the equation $\varphi(0,0) = 1$. It is proved in [6] that φ exists and $(A, \Delta, \varepsilon, \Phi, R)$ does not depend on φ up to "twisting". The quantization problem is as follows: for given $\mathfrak{g}$ and r find an $F \in (U\mathfrak{g} \otimes U\mathfrak{g})[[h]]$ such that $F \equiv 1 - rh/2 \bmod h^2$, $(\varepsilon \otimes id)(F) = 1 = (id \otimes \varepsilon)(F)$ and the quasi-triangular quasi-Hopf algebra $(A, \widetilde{\Delta}, \varepsilon, \widetilde{\Phi}, \widetilde{R})$ obtained by twisting $(A, \Delta, \varepsilon, \Phi, R)$ with F is a Hopf algebra. By definition, this means that $\widetilde{\Phi} = 1$, i.e., $F^{23} \cdot (id \otimes \Delta)(F) \cdot \Phi \cdot (\Delta \otimes id)(F^{-1}) \cdot (F^{12})^{-1} = 1$. In this case $\widetilde{R} = F^{21} \cdot R \cdot F^{-1}$ satisfies the quantum Yang-Baxter equation and $\widetilde{R} \equiv 1 + rh \bmod h^2$.

All the questions and remarks from section 3 make sense also for the quantization problem discussed in this section. Of course a negative answer to a "naive" question 3.1, 3.2 or 3.3 would imply a negative answer to the corresponding "sophisticated" question.

5. Other quantization problems

Analogs of questions 1.1 and 1.2 can also be formulated for other quantization problems such as quantization of coboundary Lie bialgebras (cf. [1], §§4, 10), quantization of quasi-Lie bialgebras (cf. [5], §2) or quantization of quasitriangular and coboundary quasi-Lie bialgebras[3] (cf. [5], §3). The existence of universal quantization is proved

---

[2] t is always invariant with respect to the smallest Lie subalgebra $\mathfrak{g}_r \subset \mathfrak{g}$ such that $r \in \mathfrak{g}_r \otimes \mathfrak{g}_r$, because (3.1) implies that $[r^{12} + r^{13}, t^{23}] = 0, [t^{12}, r^{13} + r^{23}] = 0$.

[3] By a quasitriangular (resp. coboundary) quasi-Lie bialgebra I mean a Lie algebra $\mathfrak{g}$ with a fixed invariant symmetric (resp. skew-symmetric) element of $\mathfrak{g} \otimes \mathfrak{g}$ (resp. $\mathfrak{g} \otimes \mathfrak{g} \otimes \mathfrak{g}$)

for both quasitriangular and coboundary quasi-Lie bialgebras (cf. [6] and §3 from [5]), but in coboundary case the proof is much worse than in the quasitriangular one (in the latter case the proof is more constructive, the quantization is canonical up to twisting, and there is a connection with conformal field theory via Knizhnik-Zamolodchikov equations).

6. An isolated quantization problem

Let $\mathfrak{p}$ be the set of formal series $\sum_{i=-\infty}^{n} f_i D^i$ where $f_i = f_i(x)$ are C^∞ functions on $\mathbf{R}/\mathbf{Z}$ and $D = d/dx$. $\mathfrak{p}$ is an associative algebra and therefore a Lie algebra. There is an invariant scalar product on $\mathfrak{p}$ given by $(\sum_i f_i D^i, \sum_j D^j g_j) = \sum_i (f_i, g_{-i-1})$ where (f_i, g_{-i-1}) is the L^2 scalar product. Denote by $\mathfrak{p}_1$ (resp. $\mathfrak{p}_2$) the set of all elements $\sum_i f_i D^i \in \mathfrak{p}$ such that $f_i = 0$ for $i < 0$ (resp. $i \geq 0$). Then $(\mathfrak{p}, \mathfrak{p}_1, \mathfrak{p}_2)$ is almost a Manin triple in the sense of §3 from [1] (it is not a Manin triple because the natural map $\mathfrak{p}_2 \to \mathfrak{p}_1^*$ is not surjective since not every continuous linear functional on $C^\infty(\mathbf{R}/\mathbf{Z})$ is of the form $u \to (u, f), f \in C^\infty(\mathbf{R}/\mathbf{Z})$). So we almost have a Lie bialgebra structure on $\mathfrak{p}_1$.

Question 6.1. (T.G.Khovanova) *Can $\mathfrak{p}_1$ be quantized in some sense?*

7. Quantization of some "classical" constructions

Suppose that a finite group H acts on a Lie bialgebra $\mathfrak{g}$. Then $\mathfrak{g}^H = \{a \in \mathfrak{g} | ha = a$ for all $h \in H\}$ is a subalgebra of $\mathfrak{g}$, while $\mathfrak{g}_H$ (the biggest quotient space of $\mathfrak{g}$ on which H acts trivially) is a quotient coalgebra of $\mathfrak{g}$. Using the natural bijection $\mathfrak{g}^H \to \mathfrak{g}_H$ one obtains a Lie coalgebra structure on $\mathfrak{g}^H$. In other words the cocommutator $\varphi : \mathfrak{g}^H \to \mathfrak{g}^H \otimes \mathfrak{g}^H$ is the composition of the commutator $\mathfrak{g}^H \to \mathfrak{g} \otimes \mathfrak{g}$ and the natural projection $\mathfrak{g} \otimes \mathfrak{g} \to \mathfrak{g}^H \otimes \mathfrak{g}^H$. Therefore φ is a 1-cocycle and $\mathfrak{g}^H$ is a Lie bialgebra. I prefer to denote it $\mathfrak{g}_H^H$ since in general it is neither a subbialgebra nor a quotient bialgebra.

Question 7.1. *Does there exist a quantum analog of the construction of $\mathfrak{g}_H^H$?*

Suppose H acts on a Poisson-Lie group G. Denote by $\mathfrak{g}$ the Lie bialgebra corresponding to G. Then $\mathfrak{g}^H$ corresponds to the group $G^H = (g \in G | hg = g$ for all $h \in H)$ with a Poisson bracket on it which can be defined *without* using the group structure on G. In order to do it one has to apply the following general construction to the algebra of functions on G. Suppose a finite group H acts on a Poisson algebra A. Let B be the biggest quotient algebra of A with respect to multiplication on which H acts trivially. Then B has a natural Poisson structure. Indeed, $B = A/A \cdot V = A^H/(V \cdot V)^H$ where $V \subset A$ is the vector space generated by elements of the form $ha - a, a \in A$. On the other hand $(V \cdot V)^H$ is a Poisson ideal in A^H.

Question 7.2. *Does there exist a quantum analog of this construction?*

Of course, if H acts on an associative algebra A one can try to put $B = A^H/(V \cdot V)^H$ or may be $B = A^H/I$ where I is the ideal of A^H generated by W^H and $W = \{xy + yx | x, y \in V\}$. Perhaps the crucial test for a good definition is the following one: if A is a flat algebra over $L = k[h]/(h^n)$ and A/hA is the coordinate ring of a smooth variety then B should be flat over L.

The following examples show the importance of Question 7.1. First of all a simple Lie algebra of type B, C, F or G with the usual bialgebra structure (cf. [1], §3) can be obtained as $\mathfrak{g}_H^H$ where $\mathfrak{g}$ is a simple Lie algebra of type A, D or E with the usual bialgebra structure and H is generated by automorphism of $\mathfrak{g}$ induced by a nontrivial automorphism of the Dynkin diagram. Of course twisted affine Kac-Moody algebras with the usual bialgebra structure can be obtained from non-twisted ones by applying the $\mathfrak{g}_H^H$ construction. The following two examples are due respectively to M.Nazarov and G.I.Olshansky. Let $\mathfrak{a}$ be the superalgebra $\mathfrak{g}(n|n)$; define a Lie superalgebra structure on $\mathfrak{g} = \mathfrak{a}[\lambda]$ by means of the simplest rational solution to the classical Yang-Baxter equation (cf. Example 3.3 from [1]). The involution $\sigma : \mathfrak{a} \to \mathfrak{a}$ which sends (a_{ij}) to $(a_{2n+1-i,2n+1-j})$ has the remarkable property that $(\sigma(x), \sigma(y)) = -(x, y)$ where (x, y) denotes the invariant scalar product on $\mathfrak{a}$. Therefore the involution $\tau : \mathfrak{g} \to \mathfrak{g}$ given by $\tau(a\lambda^n) = \sigma(a)(-\lambda)^n$ is a superbialgebra automorphism, and we can consider $\mathfrak{g}_H^H$ where H is generated by τ. We can also introduce a Lie superbialgebra structure on $\mathfrak{a}$ itself as in Example 3.2 from [1]. Then σ is a superbialgebra automorphism, and we can consider $\mathfrak{a}_G^G$ where G is generated by σ. The Lie superalgebra $\mathfrak{a}_G^G$is coboundary but not quasitriangular in the sense of §4 from [1]. As a Lie superalgebra $\mathfrak{a}_G^G$ is isomorphic to the "queer" superalgebra $\mathfrak{q}(n)$. Probably Nazarov and Olshansky have quantized the above bialgebras $\mathfrak{g}_H^H$ and $\mathfrak{a}_G^G$ by means of R-matrices (I am not quite sure of it because as far as I know they did not prove that their deformations were really flat). They have also considered another involution σ whose invariants form the Palamodov superalgebra $\mathfrak{p}(n)$.

8. Quantized Kac-Moody algebras

Let $\mathfrak{g}$ be a Kac-Moody algebra with an invariant scalar product. Recall that there are two versions of quantized Kac-Moody algebras: Jimbo's $U_q\mathfrak{g}$ (cf [7]) and $U_h\mathfrak{g}$(cf. Example 6.2 from [1]). The main difference between them is that $U_h\mathfrak{g}$ is complete in the h-adic topology.

Lusztig proved [8] that if λ is an integer dominant weight then the irreducible $\mathfrak{g}$-module $L(\lambda)$ with the highest weight λ can be deformed to a module over $U_q\mathfrak{g}$. More precisely for generic q the dimensions of the weight spaces of the irreducible $U_q\mathfrak{g}$-module $L(\lambda)$ with highest weight λ are the same as for $q = 1$ and therefore $L_q(\lambda)$ is a deformation of $L(\lambda)$.

Question 8.1. *What is the situation for an arbitrary λ?*

If $\dim \mathfrak{g} < \infty$ then this question is trivial because according to Proposition 4.3 from [9] there is an algebra isomorphism $U_h\mathfrak{g} \to (U\mathfrak{g})[[h]]$ whose restriction to the Cartan subalgebra is identical. In general there is no reason to expect that $U_h\mathfrak{g}$ is isomorphic to $(U\mathfrak{g})[[h]]$ as an algebra (this is certainly wrong if $\mathfrak{g}$ is affine; moreover, in this case $U_h\mathfrak{g}/h^3U_h\mathfrak{g}$ is not isomorphic to $(U\mathfrak{g})[h]/h^3(U\mathfrak{g})[h]$ as an algebra[4]). But to obtain an answer to Question 8.1 it would be enough to prove a weaker statement. The point is that each element of a Verma module over $U\mathfrak{g}$ is annihilated by some $I_\beta, \beta \in \Gamma_+$, where

[4] It is easy to deduce from (1.1) that a quantization mod h^3 of a Lie bialgebra $\mathfrak{g}$ is a trivial deformation of $U\mathfrak{g}$ in the sense of algebras iff $\beta \cup \beta = 0$ where $\beta \in H^1(\mathfrak{g}, \mathfrak{g}\otimes\mathfrak{g})$ is the class of the cocommutator $\mathfrak{g} \to \mathfrak{g}\otimes\mathfrak{g}$ and $\cup$ denotes the pairing $H^1(\mathfrak{g}, \mathfrak{g}\otimes\mathfrak{g}) \otimes H^1(\mathfrak{g}, \mathfrak{g}\otimes\mathfrak{g}) \to H^2(\mathfrak{g}, Sym^3\mathfrak{g})$ induced by the map $(\mathfrak{g}\otimes\mathfrak{g})\otimes(\mathfrak{g}\otimes\mathfrak{g}) \to Sym^3\mathfrak{g}$ given by $(a\otimes b)\otimes(c\otimes d) \mapsto [a, c]\cdot b\cdot d$

Γ_+ is the semigroup generated by simple roots and I_β is the left ideal of $U\mathfrak{g}$ generated by the elements of $U\mathfrak{g}$ on which the Cartan algebra acts according to some $\gamma \in \beta + \Gamma_+$. Therefore, the actions of $U\mathfrak{g}$ on a Verma module can be continued to an action of the completion $\hat{U}\mathfrak{g}$ of $U\mathfrak{g}$ with respect to the topology whose base of neighborhoods of zero is formed by all the I_β. We can also introduce $\hat{U}_h\mathfrak{g}$ by the first completing $U_h\mathfrak{g}/h^r U_h\mathfrak{g}$ with respect to a similar topology and then considering the inverse limit with respect to r. A positive answer to Question 8.1 would follow from a positive answer to the following question.

Question 8.2. *Does there exist an algebra isomorphism $\hat{U}_h\mathfrak{g} \to (\hat{U}\mathfrak{g})[[h]]$ whose restriction to the Cartan algebra is identical?*

Recall that a Kac-Moody algebra $\mathfrak{g}$ with invariant scalar product corresponds to a Cartan matrix $A = (A_{ij})$ with the following properties: a) symmetrizability, i.e., $d_i A_{ij} = d_j A_{ji}$ for some $d_i \neq 0$, b) $A_{ii} = 2$, c) if $i \neq j$ then $A_{ij} \in \mathbb{Z}, A_{ij} \leq 0$. However to an *arbitrary* symmetrizable matrix A Kac associated in [10] a Lie algebra $\mathfrak{g}(A)$ which is a quotient of a Lie algebra $\tilde{\mathfrak{g}}(A)$ with very simple defining relations by some mysterious ideal (if A satisfies conditions b) and c) then Gabber and Kac proved that this ideal is generated by Serre's relations). Note that the definition of a Lie bialgebra structure on $\mathfrak{g}(A)$ given in Example 3.2 from [1] makes sense for an arbitrary symmetrizable A.

Question 8.3. (I.M.Gelfand) *How to define a quantum analog of $\mathfrak{g}(A)$ for an arbitrary symmetrizable A? How does the "size" of this quantum algebra depend on A (or rather on $B = (B_{ij})$ where $B_{ij} = exp(hA_{ij})$)?*

The second question makes sense also for algebras $\mathfrak{g}(A)$ themselves. I.M.Gelfand thinks it is quite analogous to the question of how the "size" of an irreducible representation of a Kac-Moody algebra with highest weight λ depends on λ.

9. Set-theoretical solutions to the quantum Yang-Baxter equation

The quantum Yang-Baxter equation

$$R^{12}R^{13}R^{23} = R^{23}R^{13}R^{12} \tag{9.1}$$

makes sense even if R is not a linear operator $V \otimes V \to V \otimes V$ but a map $X \times X \to X \times X$ where X is a set. Of course to each set-theoretical solution to (9.1) there corresponds a solution in the usual sense (apply the functor $X \mapsto$ the free module generated by X). Maybe it would be interesting to study set-theoretical solutions to (9.1) Note that a mapping $X \times X \to X \times X$ can be considered as a pair of operations on X and (9.1) as curious identities for these operations. The only thing I know about set-theoretical solutions to (9.1) is the following couple of examples.

Example 1 (V.V.Lyubashenko). If $R(x, y) = (S(x), T(y))$ then R satisfies (9.1) iff $ST = TS$. Of course such an R satisfies the "unitarity" condition $R^{21} = R^{-1}$ iff $T = S^{-1}$.

Example 2 (B.B.Venkov). If $R(x, y) = (x, x \circ y)$ for some operation $\circ$ on X then (9.1) is equivalent to the following distributivity identity:

$$x \circ (y \circ z) = (x \circ y) \circ (x \circ z) \tag{9.2}$$

Such an R never satisfies the "unitarity" condition except the case $R = id$. The identity (9.2) is satisfied, e.g., if X is subset of a group G invariant with respect to conjugation and $x \circ y = xyx^{-1}$,or if X is a root system and $x \circ y = s_x(y)$ where s_x is the reflection corresponding to x.

References

[1] Drinfeld V.G., *Quantum groups*, Proc. ICM-86 (Berkeley) **1** (1987), 798-820.

[2] Deligne P. and Milne J., *Tannakian categories*, Lect. Notes Math. **900** (1982), 101-228.

[3] Drinfeld V.G., *On constant, quasiclassical solutions of the quantum Yang-Baxter equation*, Sov. Math. Dokl. **28** (1983), 667-671.

[4] Moreno C. et Valero L., *Produits star invariants et équation de Yang-Baxter quantique constante*, Dans les Actes des Journées Relativistes (24-29 avril 1990, Aussois, France).

[5] Drinfeld V.G., *Quasi-Hopf algebras*, Algebra Anal. **1** no. 6 (1989), 114-148. (in Russian)

[6] Drinfeld V.G., *On quasitriangular quasi-Hopf algebras and a group closely connected with* $Gal(\bar{\mathbb{Q}}/\mathbb{Q}$, Algebra Anal. **2** no. 4 (1990), 149-181. (in Russian)

[7] Jimbo M., *A q−difference analogue of $U(\mathfrak{g})$ and the Yang-Baxter equation*, Lett.Math.Phys. **10** (1985), 63-69.

[8] Lusztig G., *Quantum deformations of certain simple modules over enveloping algebras*, Adv. Math. **70** (1988), 237-249.

[9] Drinfeld V.G., *On almost cocommutative Hopf algebras*, Leningrad Math. J. **1** no. 2 (1990), 321-342.

[10] Kac V.G., *Infinite dimensional Lie algebras*, Birkhäuser, Boston a.o., 1983.

PHYSICAL & TECHNICAL INSTITUTE OF LOW TEMPERATURES, LENIN PROSPECT 47, KHARKOV, 310086, USSR

QUANTUM SYMMETRY

Murray Gerstenhaber[(1)], Anthony Giaquinto[(2)] and Samuel D. Schack[(3)]

[(1)] University of Pennsylvania, [(2)] University of Michigan
[(3)] State University of New York at Buffalo

In this paper we lay the foundation for studying quantum groups as part of algebraic deformation theory by introducing the quantum symmetric group and the concept of quantum symmetry. These point the way from infinitesimal methods – the cohomology theory of bialgebras – through algebraic deformation theory to quantization. We show how the general principle that a quantum group should be viewed as a space of quantum symmetric elements apply to $\mathbf{M}(n) = \mathcal{O}(M(n))$ and $\mathbf{SL}(n) = \mathcal{O}(Sl(n))$, and derive from this the formulas for the "preferred presentation" of their quantum linear spaces. One striking aspect of this approach is that deformations and quantizations are produced directly from particular solutions (modulo invariants) of the *classical* Yang-Baxter equation without constructing quantum Yang-Baxter operators.

This paper was presented to the conference on Quantum Groups inaugurating the Euler International Mathematical Institute in Leningrad, October 14-28, 1990. We are grateful for the committed efforts of Academician L.D.Faddeev, Professor P.P.Kulish, and many others, as well as to the generosity of LOMI. We thank all who made this extraordinary gathering possible. Portions of this paper were announced in [18], which contained a serious error, corrected herein. Other portions were developed initially for the case $n = 2$ in the dissertation [19] of the second author, to whom are due, amongst other things, the formulas of section 13, which gave birth to the general theory of which they are now a consequence.

1. Introduction

If G is an algebraic group or monoid then $\mathcal{O}(G)$ will denote the bialgebra of polynomial functions on G. In particular, G may be $M(n)$, the $n \times n$ matrices, $GL(n)$, or $SL(n)$ in which case we denote their corresponding bialgebras by $\mathbf{M}(n)$, $\mathbf{GL}(n)$, and $\mathbf{SL}(n)$, respectively. The coefficient ring (tacitly always $\mathbb{R}$ or $\mathbb{C}$) is denoted k.

In the usual description of $\mathbf{M}_q(2) = \mathcal{O}_q(M(2))$ (cf., e.g., [8]), one observes that the polynomial ring $k[x_{11}, x_{12}, x_{21}, x_{22}]$ is a bialgebra with $\Delta x_{ij} = \sum_l x_{il} \otimes x_{lj}$. A "deformed" multiplication depending on a parameter q is then introduced by the Faddeev-Reshitikhin-Takhtajan (FRT) construction for a quantum Yang-Baxter matrix: Writing $x_{11} = a, x_{12} = b, x_{21} = c, x_{22} = d$, one sets $ab = qba, ac = qca, bd = qdb, cd = qdc$, $bc = cb$, and $ad - da = (q - q^{-1})bc$. The "original" comultiplication on the generators is then extended to all of the "deformed" polynomial ring by requiring it to be an algebra morphism. It is not obvious, however, that the dimensions of the homogeneous parts of the new graded ring are the same as that of the original polynomial ring; its Hilbert series might be different. In the language of algebraic deformation theory, there is no natural identification of the underling space of the new algebra with the old. We show, with the simplest of counterexamples, that the FRT construction need not be a deformation, even starting with a quantum Yang-Baxter matrix; the Hilbert series of the "deformed" algebra may be different from that of the old. Moreover, the new comultiplication is not identical with the old on elements of $k[a,b,c,d]$ whose degree is > 1. Using the concept of quantum symmetry, we exhibit all $\mathbf{M}_q(n)$ and $\mathbf{SL}_q(n)$ as deformations in the sense of [11]. The bialgebra cohomology of [18] shows that if G is a reductive Lie Group, then every bialgebra deformation of the ring G of polynomial functions on G is equivalent to a *preferred deformation* in which the coproduct is strictly unchanged (this being meaningful now that the underlying vector spaces of the new and old algebras are identified). We present the explicit preferred deformation of the quantum linear space of $\mathbf{M}_q(n)$ and will present the full bialgebra formulas for $\mathbf{M}_q(2)$ in a subsequent paper. We show also the relation between the preferred standard deformation of $\mathbf{M}(n)$ and the quantization of the universal enveloping algebra $U(\mathfrak{sl}(n))$. One of the main problems in quantum groups, from our point of view, is to start with an infinitesimal deformation of G and from it to construct an explicit quantization. This is not always possible since there may be obstructions. The earlier summary of results [18] gave an explicit exponential formula which, as was kindly pointed out by A.A.Kirillov and L.Takhtajan, was correct on generators but not on arbitrary elements of $\mathbf{M}(n)$. In effect, we complete that result here, showing how to quantize starting with an infinitesimal satisfying certain axioms. We also complete another result by showing how to describe, in an intrinsic way meaningful for all reductive groups, the quantum Yang-Baxter matrix giving the "standard" quantization. Using the bialgebra cohomology and "diagonal" deformations, we show that $\mathbf{SL}_q(n)$ is naturally contained in a family of deformations whose dimension is $1 + \dim(\bigwedge^2 \mathfrak{h})$, where $\mathfrak{h}$ is a Cartan subalgebra. This is the Sudbery family (cf. [31]), but written in this fashion, one knows what to expect for all semisimple G. The larger family of deformations of $\mathbf{M}(n)$ has $1+\dim(\bigwedge^2 M(n))$ parameters. This tells what to expect for arbitrary reductive G. The Dipper–Donkin group occurs in this family, distinguished by the fact that its infinitesimal has square 0. Finally, we consider deformations of $\mathbf{SL}(n)$ which are at the "boundary" of the deformation space. This is closely related to the construction of "universal deformation formulas" and the breaking of symmetry, which are discussed briefly. We show that these previously unknown boundary deformations appear in a family parametrized by $\bigwedge^2 \mathfrak{h}^\infty$, where $\mathfrak{h}^\infty$ is the subspace of $\mathfrak{h}$ orthogonal to the highest root. (One does not get one parameter more because the basic deformation in this family is a "jump" deformation.) This is only a preliminary paper. We do not discuss quantization over $\mathbf{Z}$ or in finite characteristics although the basic ideas remain meaningful. We hope, however, to have laid some very necessary foundations. Notations: The coefficient ring is denoted by k, but the same symbol is also used as a running index; there should be no confusion. We use V to denote a vector space whose dimension is always n.

2. Hochschild Cohomology

Let A be an associative algebra over a commutative unital ring k and M be an A-bimodule. A Hochschild m-cochain with values in M is a k-multilinear map

$$F^m : \underbrace{A \times \cdots \times A}_{\text{m times}} \longrightarrow M,$$

or equivalently a k-linear map still denoted $F^m : A^{\otimes m} \to M$ (where $\otimes$ will always mean $\otimes_k$). The k-module of these is denoted $C_h^m(A, M)$ or simply C_h^m. One sets $C_h^0(A, M) = M$. The Hochschild coboundary $\delta_h = \delta_h^A : C_h^m \to C_h^{m+1}$ is defined by

$$\begin{aligned}(\delta_h F^m)(a_1, \ldots, a_{m+1}) =& a_1 F^m(a_2, \ldots, a_{m+1}) \\ &+ \sum_{i=1}^{m}(-1)^i F^m(a_1, \ldots, a_i a_{i+1}, \ldots a_{m+1}) \\ &+ (-1)^{m+1} F^m(a_1, \ldots, a_m) a_{m+1}\end{aligned}$$

(If $m = 0$ and $F = u \in M$ then $\delta_h F(a) = au - ua$.) One has $\delta_h \delta_h = 0$. Denoting $\ker(\delta_h)$ by Z_h^m (the m-cocycles) and $\operatorname{im}(\delta_h(C_h^{m-1}))$ by B_h^m (the m-coboundaries), one sets $Z_h^m / B_h^m = H_h^m(A, M)$, the mth *Hochschild cohomology group* (actually k-module) of A with coefficients in M. There are many ways to obtain the same groups; the groups are intrinsic, but different approaches may be useful in different contexts.

When M is A itself, $H_h^*(A, A)$ has a very rich structure first fully described in [10]. There is an associative *cup product* on the direct sum $C_h^*(A, A) = \bigoplus_{m=0}^{\infty} C_h^m(A, A)$ defined by

$$(F^m \smile G^p)(a_1, \ldots, a_m, b_1, \ldots, b_p) = F(a_1, \ldots, a_m) G(b_1, \ldots, b_p).$$

(We may drop sub and superscripts when they are evident.) One has

$$\delta_h(F^m \smile G^p) = (\delta_h F^m) \smile G^p + (-1)^m F^m \smile \delta_h G^p$$

as a consequence of which the cup product is well-defined on cohomology classes. It is graded commutative on $H_h^*(A, A)$: if $\xi^m \in H_h^m$ and $\eta^p \in H_h^p$, then

$$\xi^m \smile \eta^p = (-1)^{mp} \eta^p \smile \xi^m.$$

There is a composition of cochains: For $1 \leq i \leq m$ define $F^m \circ_i G^p \in C_h^{m+p-1}(A, A)$ by

$$F^m \circ_i G^p(a_1, \ldots, a_{m+p-1}) = F(a_1, \ldots, a_{i-1}, G(a_i, \ldots, a_{i+p-1}), a_{i+p}, \ldots, a_{p+m-1}).$$

Set

$$F \circ G = \sum_{i=1}^{m} (-1)^{(m-1)(p-1)} F \circ_i G,$$

and define the graded commutator with respect to the "reduced" dimensions $m-1$ and $p-1$ to be

$$[F^m, G^p] = F^m \circ G^p - (-1)^{(m-1)(p-1)} G^p \circ F^m = (-1)^{(m-1)(p-1)} [G^p, F^m].$$

If $K \in C_h^r$ we then have the *graded Jacobi identity*

$$(-1)^{(m-1)(r-1)}[F^m,[G^p,K^r]]+(-1)^{(p-1)(m-1)}[G^p,[K^r,F^m]] + (-1)^{(r-1)(p-1)}[K^r,[F^m,G^p]]=0.$$

That is, $[-,-]$ yields a *graded Lie algebra.* Moreover,

$$[Z_h^m, Z_h^p] \subseteq Z_h^{m+p-1} \text{ and } [Z_h^m, B_h^p] \subseteq B_h^{m+p-1},$$

so the graded Lie multiplication induces one on $H_h^*(A,A)$. If ξ^m, η^p, ζ^r are cohomology classes of dimensions m,p,r, respectively, then $\smile$ and $[-,-]$ are related by

$$[\xi^m \smile \eta^p, \zeta^r] = [\xi^m, \zeta^r] \smile \eta^p + (-1)^{m(r-1)}\xi^m \smile [\eta^p, \zeta^r],$$

i.e., $[-,\zeta^r]$ is a graded derivation of $\smile$. Note also that if $\mu : A \otimes A \to A$ is the multiplication, then $\delta_h F^m = -[F,\mu] = (-1)^{m-1}[\mu, F]$. Further, if $F \in Z_h^m(A,A)$ with m *even* then $F \circ F = (1/2)[F,F]$ is in Z_h^{2m-1}. Thus $F \mapsto F \circ F$ induces a quadratic map $H_h^m \to H_h^{2m-1}$ on cohomology. If A is an algebra and $F,G \in Z_h^1(A,A)$, i.e., if F,G are derivations, then $[F,G]_G$ is just the usual commutator $FG-GF$ and is again a derivation. If A is a commutative algebra, then $Z_h^1(A,A) = H_h^1(A,A)$. Because another bracket product will appear, this one will henceforth be denoted $[-,-]_G$.

Historical Note. The concept of a graded Lie algebra was first introduced in [10] and later rediscovered in many contexts, resulting in frequent and varied misattribution. A graded k-algebra $\Psi = \coprod_{i\geq 0}\Psi_i$ which, like $H^*(A,A)$, simultaneously carries graded commutative and graded Lie structures linked as in $H^*(A,A)$ has been called both a *G-algebra* [17] and a *graded Poisson algebra.*

The foregoing can be dualized: Suppose that A is a coalgebra with comultiplication $\Delta : A \to A \otimes A$, which we write in the Sweedler notation as $\Delta a = \sum a_{(1)} \otimes a_{(2)}$, and suppose that M is an A-(bi)comodule with structure maps $\Delta_L : M \to A \otimes M$ and $\Delta_R : M \to M \otimes A$. We let $C_c^m(M,A)$ be the set of k-linear maps $F^m : M \to A^{\otimes m}$ and define $\delta_c = \delta_c^A : C_c^m \to C_c^{m+1}$ by

$$\delta_c F^m = (\mathrm{id}_A \otimes F)\Delta_L + (\sum_{i=1}^{m}(-1)^i \,\mathrm{id}_A^{\otimes i-1} \otimes \Delta \otimes \mathrm{id}_A^{\otimes m-i})F + (-1)^{m+1}(F \otimes \mathrm{id}_A)\Delta_R.$$

One has $\delta_c\delta_c = 0$ and therefore can define $H_c^m(M,A)$ exactly as algebra cohomology was defined. When $M = A$, then analogous to $\smile$ for algebras we can define $F \frown G = (F \otimes G)\Delta$, but we shall not need this here. Composition of cochains does not involve the (co)algebra structure, so the similar ("dual") definitions apply. As in the algebra case, the bracket on the coderivations Z_c^1 is the usual commutator.

For any vector space V, we denote by $\bigwedge^m V$ its mth exterior power, and by $\bigwedge V$ its exterior algebra. This is associative and graded commutative. If V is a Lie algebra $\mathfrak{g}$ then $\bigwedge \mathfrak{g}$ is a G-algebra with the *Schouten bracket* $[-,-]_S$ defined by

$$[a_1 \wedge \cdots \wedge a_m, b_1 \wedge \cdots b_p]_S =$$

$$(-1)^{(m-1)(p-1)} \sum (-1)^{i+j}[a_i,b_j] \wedge a_1 \wedge \cdots \wedge \widehat{a}_i \wedge \cdots \wedge a_m \wedge b_1 \wedge \cdots \wedge \widehat{b}_j \wedge \cdots \wedge b_p.$$

In fact $\bigwedge \mathfrak{g}$ is a universal G-algebra in the following sense: Consider the pair (Ψ_0, Ψ_1) of an arbitrary G-algebra Ψ. These have the following properties:

1. Ψ_0 is a commutative algebra and Ψ_1 is a symmetric bimodule.
2. Ψ_1 is a Lie algebra and Ψ_0 is a Lie module.
3. If $\lambda, \lambda' \in \Psi_1$ and $a, b \in \Psi_0$, then

$$\begin{aligned}
[\lambda, ab] &= [\lambda, a]b + a[\lambda, b], \\
[b\lambda, a] &= b[\lambda, a], \\
[\lambda, a\lambda'] &= [\lambda, a]\lambda' + a[\lambda, \lambda'].
\end{aligned}$$

Pairs (Ψ_0, Ψ_1) form a category $\mathcal{C}$ and G-alg $\to \mathcal{C}$ is a functor where $\Psi \mapsto (\Psi_0, \Psi_1)$. This functor has a left adjoint, namely

$$(\Psi_0, \Psi_1) \mapsto \bigwedge\nolimits_{\Psi_0} \Psi_1$$

with Schouten bracket. The adjunction gives a universal map

$$\bigwedge\nolimits_{\Psi_0} \Psi_1 \longrightarrow \Psi$$

for any G-algebra Ψ. In particular, for any algebra there is a canonical map $\bigwedge_{H^0} H^1 \to H^*$ sending $\xi_1 \wedge \cdots \wedge \xi_m$ to $\xi_1 \smile \cdots \smile \xi_m$ where the ξ_i are classes of derivations. When A is commutative, we have $\mathrm{Der}(A) = Z^1_h(A, A) = H^1_h(A, A)$ and $A = H^0_h(A, A)$ itself. There are few algebras with non-trivial Hochschild cohomology where that cohomology can be explicitly computed. However, for commutative smooth algebras, i.e., those like the ring of polynomial functions on a smooth affine algebraic variety (and its localizations), we have (cf. [20])

Theorem 2.1 (Hochschild-Kostant-Rosenberg). *If A is a commutative smooth k-algebra then the canonical mapping*

$$\bigwedge\nolimits_A \mathit{Der}(A) \longrightarrow H^*_h(A, A)$$

is a G-algebra isomorphism. If M is a symmetric A-bimodule, i.e., if $am = ma$ for all $a \in A, m \in M$ then there is an isomorphism

$$(\bigwedge_A \mathit{Der}(A)) \otimes_A M \longrightarrow H^*_h(A, M). \blacksquare$$

The cohomology of a commutative smooth algebra therefore consists of the *skew multiderivations* – A cochain F is skew if $F\sigma = (-1)^\sigma F$ for every permutation σ and it is a multiderivation if it is a derivation of each individual argument. Another feature of the cohomology of such algebras is that distinct cocycles are not cohomologous. Also $H^m_h(A, A) = 0$ for m greater than the transcendence degree of A over k (its geometric dimension), whereas, for an arbitrary finite-dimensional commutative algebra over a field there may be infinitely many non-zero Hochschild cohomology groups. These assertions follow from the Hodge decomposition of [15].

3. Separability and Coseparability

Let $B \to A$ be a k-algebra map and $f : M \to N$ be a morphism of left (or right or two-sided) A-modules. Then f is called *B-split* if it splits when viewed as a (left, right, or two-sided) B-module morphism, i.e. there is a B-module morphism $g : N \to M$ such that $fgf = f$. Equivalently, if $\ker(f)$, the kernel, and the image $f(M)$ have B-module complements in M and N. When k is a field, f is always k-split. Dually, if $B \to A$ is a coalgebra map then an A-split (left, right or two-sided) B-comodule map is one which splits when viewed as an A-comodule map. A k-algebra S is called *separable* (over k) if the following equivalent conditions hold:

1. Every morphism $f : M \to N$ of left (or right) S-modules which is k-split is also A-split. (In particular, if k is a field, then every left S-module morphism is split.)
2. The multiplication map $\mu : S \otimes S \to S$ is S-split when viewed as a morphism of S-bimodules. (Here $S \otimes S$ is an S-bimodule by $s(s' \otimes s'') = ss' \otimes s''$ and $(s' \otimes s'')s = s' \otimes s''s$.)
3. $H^1_h(S, M) = 0$ for all S-bimodules M.
4. $H^m_h(S, M) = 0$ for all S-bimodules M and all $m > 0$.

There are other equivalent conditions, most useful of which is the existence of a "separability idempotent", but the foregoing are the ones which dualize most easily to coalgebras. A k-coalgebra A satisfying any of the equivalent dual conditions is called *coseparable.* A separable algebra which is projective as a k-module is always finitely generated over k (cf. [33]). An algebra A over a field k is separable if and only if it is finite-dimensional and $\bar{k} \otimes A$ is semisimple, i.e., a finite direct sum of total matrix algebras. A coseparable coalgebra need not be finitely generated (even if k is a field); any direct sum of coseparable algebras is coseparable.

If $\mathcal{V}$ is any affine algebraic variety, then we denote by $\mathcal{O}(\mathcal{V})$ the k-algebra of polynomial functions on $\mathcal{V}$. Since the algebra of functions on $\mathcal{V} \times \mathcal{W}$ is $\mathcal{O}(\mathcal{V}) \otimes \mathcal{O}(\mathcal{W})$, an associative multiplication on $\mathcal{V}$ which is also a morphism of varieties induces a coassociative comultiplication $\Delta : \mathcal{O}(\mathcal{V}) \to \mathcal{O}(\mathcal{V}) \otimes \mathcal{O}(\mathcal{V})$. Since $\mathcal{O}(\mathcal{V})$ is from the start a commutative algebra with unit, it is almost a bialgebra, possibly lacking only the counit map $\varepsilon : \mathcal{O}(\mathcal{V}) \to k$, but if $\mathcal{V}$ has a unit element e for multiplication, then ε is just evaluation at e. In particular, if $M(n)$ is the algebra of all $n \times n$ matrices over k, then $\mathrm{M}(n)$, which is just the polynomial ring $k[x_{ij}], i, j = 1, \ldots, n$ in n^2 variables is a bialgebra. Here x_{ij} is the function whose value on the matrix unit e_{km} is $\delta_{ik}\delta_{jm}$, and $\varepsilon x_{ij} = \delta_{ij}$. If G is an affine algebraic group then $\mathcal{O}(G)$ is a Hopf algebra; the inverse map of G induces the antipode map of $\mathcal{O}(G)$. The following is practically the definition of "reductive":

Theorem 3.1. *The coalgebras* $\mathrm{M}(n)$ *and* $\mathcal{O}(G)$ *for any reductive* G *are coseparable.* ∎

Returning to algebras, suppose that A contains a subalgebra B (or that we have a given morphism $B \to A$) and that M is an A-bimodule. A Hochschild cochain $F \in C^m_h(A, M)$ is *normalized* if

$$F(a_1, \ldots, a_m) = 0 \text{ whenever any } a_i \in k$$

(where k is viewed as a subring of A). It is *B-relative* if

$$F(ba_1,\dots,a_m) = bF(a_1,\dots,a_m),$$
$$F(a_1,\dots,a_ib,a_{i+1},\dots,a_m) = F(a_1,\dots,a_i,ba_{i+1},\dots a_m), \text{ and}$$
$$F(a_1,\dots,a_mb) = F(a_1,\dots,a_m)b,$$

for all $b \in B$. If it is both, then $F(a_1,\dots,a_m) = 0$ whenever any $a_i \in B$. The normalized B-relative cochains with the usual Hochschild coboundary operator form a subcomplex $\overline{C_h}^{\bullet}(A,B;M)$ of the Hochschild cochain complex $C_h^{\bullet}(A,M)$. The inclusion therefore induces a morphism

$$\overline{H_h}^{*}(A,B;M) \longrightarrow H_h^*(A,M).$$

(Note: We will always denote a complex $\cdots \to C^m \to C^{m+1} \to \cdots$ by a symbol like $C^{\bullet}$, while a direct sum of groups like $\bigoplus H_h^m$ will be denoted by H_h^*, indicating that there is no boundary or coboundary map.)

Theorem 3.2 ([14]). *If B is a k-separable algebra and $B \to A$ is an algebra map, then the inclusion $\bar{C}_h^{\bullet} \hookrightarrow C_h^{\bullet}$ induces an isomorphism*

$$\overline{H_h}^{*}(A,B;M) \longrightarrow H_h^*(A,M) \blacksquare$$

The proof is elementary from the viewpoint of resolutions (cf. [14]) but the theorem (which implies, amongst other things, that simplicial cohomology is a special case of Hochschild cohomology) was long overlooked. Note that for B we can always take k itself. The theorem then gives the classical result that every cocycle is cohomologous to a normalized one, and that Hochschild cohomology can be computed using the subcomplex of normalized cochains. The dual theorem applies to the coalgebra cohomology $H_c^*(-,B)$ when A is coseparable and $B \to A$ is a coalgebra map.

4. Bialgebra Cohomology

Let A be a k-module which is simultaneously an algebra and a coalgebra with multiplication $\mu : A \otimes A \to A$, unit $\eta : k \to A$, comultiplication $\Delta : A \to A \otimes A$, and counit $\varepsilon : A \to k$. Then A is a *bialgebra* if Δ and ε are algebra maps or, equivalently, μ and η are coalgebra maps. Thus, for example $\Delta\mu = (\mu \otimes \mu)(23)(\Delta \otimes \Delta)$ where (23) interchanges the middle two factors in $A \otimes A \otimes A \otimes A$. The bialgebra A is a *Hopf algebra* if in addition it has an *antipode* $S : A \to A$, i.e. a k-linear map such that $A \xrightarrow{\Delta} A \otimes A \xrightarrow{S \otimes \mathrm{id}} A \otimes A \xrightarrow{\mu} A$ is identical with $A \xrightarrow{\varepsilon} k \xrightarrow{\eta} A$, and the same with $S \otimes \mathrm{id}$ replaced by $\mathrm{id} \otimes S$. It follows that S is an antihomomorphism for both Δ and μ. As mentioned in earlier $\mathbf{M}(n)$ is a bialgebra with $\Delta x_{ij} = \sum x_{ik} \otimes x_{kj}$, and $\mathcal{O}(G)$ for an affine algebraic group is a Hopf algebra, the antipode being induced by the inverse map $G \to G$. The category of left modules over a bialgebra A has an "internal" tensor product: If M and N are two left A-modules then $M \overline{\otimes} N$ is $M \otimes N$ with left module action given by $a(m \otimes n) = (\Delta a)(m \otimes n) = \sum a_{(1)} m \otimes a_{(2)} n$ where $\Delta a = \sum a_{(1)} \otimes a_{(2)}$. Writing $\Delta_r : A \to A^{\otimes r}$ for iterated comultiplication (note $\Delta_0 = \varepsilon$, $\Delta_1 = \mathrm{id}$, $\Delta_2 = \Delta$), $M_1 \otimes \dots \otimes M_r$ becomes a left A module $M_1 \overline{\otimes} \dots \overline{\otimes} M_r$ by $a(m_1 \otimes \dots \otimes m_r) = (\Delta_r a)(m_1 \otimes \dots \otimes m_r)$. The same remarks hold for right and two-sided modules. Note that $(\Delta a)(m \otimes n)$ can be written as $(\mu_L \otimes \mu_L)(23)(\Delta \otimes \mathrm{id}_M \otimes \mathrm{id}_N)(a \otimes m \otimes n)$ where μ_L is the left module structure map $A \otimes M \to M$ or $A \otimes N \to N$ depending on the context. Dually, if M and N are left A comodules with $\Delta_L : M \to A \otimes M$ and similarly for N, then $M \otimes N$ becomes an A-comodule $M \underline{\otimes} N$ with structure map given by the composite

$$M\underline{\otimes}N \overset{\Delta_L\otimes\Delta_L}{\to} A \otimes M \otimes A \otimes N \overset{(23)}{\to} A \otimes A \otimes M \otimes N \overset{\mu\otimes\mathrm{id}\otimes\mathrm{id}}{\to} A \otimes (M\underline{\otimes}N)$$

The bialgebra cohomology $H_b^*(A, B)$ is defined for ordered pairs (A, B) of bialgebras satisfying the properties:

1. B is an A-bimodule such that $\Delta : B \to B\overline{\otimes}B$ is an A-bimodule map while the module actions $A \otimes B \to B$ and $B \otimes A \to B$ are coalgebra maps, and
2. A is a B-bicomodule such that $\mu : A\underline{\otimes}A \to A$ is a B-bicomodule map while the comodule coactions $A \to B \otimes A$ and $A \to A \otimes B$ are algebra maps.

Note that any bialgebra morphism $f : A \to B$ makes (A, B) a bialgebra pair. For applications to deformation theory we use the pair (A, A) and id: $A \to A$. Define $C_b^{m,p}(A, B)$ or briefly $C_b^{m,p}$ to be $\mathrm{Hom}_k(A^{\underline{\otimes} m}, B^{\overline{\otimes} p})$. (N.B. In [18] we called this group $C_b^{p,m}$.) The Hochschild coboundary $\delta_h : C_b^{m,p} \to C_b^{m+1,p}$ and coalgebra coboundary $\delta_c : C_b^{m,p} \to C_b^{m,p+1}$ then commute. (There is a conceptual reason for this, as explained in [18] – we may view the double complex $C_b^{\bullet\bullet}$ as Hom, in an appropriate category of the bar resolution of A into the cobar resolution of B.) We define the total complex C_b^* by setting

$$C_b^m = \bigoplus_{i+j=m+1} C_b^{i,j} \text{ (note the indices)}$$

with total coboundary δ_b given by $\delta_b \mid_{C_b^{m,p}} = \delta_h + (-1)^m \delta_c$. The resulting cohomology groups are denoted $H_b^*(A, B)$. The lowest cohomology group is then $H_h^{-1}(A, B) = k$. Their relationship to Singer's groups (cf. [30]) is discussed in [18].

For the deformation theory we shall need primarily the "truncated" complex $\widehat{C}_b^{\bullet\bullet}$ obtained from $C_b^{\bullet\bullet}$ by replacing those $C_b^{m,n}$ with either $m = 0$ or $n = 0$ by 0. The corresponding groups are denoted $\widehat{H}_b^*$. Note that $\widehat{H}_b^0(A, A) = 0$ and there are no 1-coboundaries, so for any A we have $H_b^1 = Z_b^1$; this is the module of maps $f : A \to A$ which are simultaneously derivations and coderivations, i.e. "infinitesimal automorphisms" of the bialgebra structure of A. We show below for $A = \mathbf{M}(n)$ they can be identified with the elements of $M(n)$ itself and for $A = \mathcal{O}(G)$, (a reductive G) they are the elements of the Lie algebra, $\mathfrak{g}$, of G. As yet, we have not found a nontrivial graded Lie structure on $H_b^*(A, B)$ or $\widehat{H}_b^*(A, B)$. However, the first and third authors have, with Stasheff, found a nontrivial graded commutative, associative product whose meaning is still a mystery.

Suppose now that our bialgebra A is $\mathbf{M}(n)$. Its generators as a polynomial algebra are x_{ij}, $i, j = 1, \ldots, n$ where x_{ij} is the function on $M(n)$ whose value at the matrix unit e_{km} is $\delta_{ij}\delta_{km}$. Let A_d be the linear subspace of $A = \mathrm{M}(n)$ spanned by all monomials of total degree d, so $A = k + A_1 + A_2 + \ldots$. Then $A_1 = M(n)^*$, A is the symmetric algebra SA_1 on A_1 and $A_d = S^d A_1$, the d-th symmetric power of A_1. Of course, $A_1^* = \mathrm{Hom}_k(A, k) = M(n)^{**}$ is just $M(n)$, with dual basis $x_{ij}^* = e_{ij}$. Each A_d is a subcoalgebra. Since $\dim A_1 < \infty$ we have also $(\bigwedge A_1)^* = \bigwedge A_1^*$.

We wish to compute $H_b^*(A,A)$ and $\widehat{H}_b^*(A,A)$ for $A = \mathbf{M}(n)$ or $A = \mathcal{O}(G)$, G a reductive group. First, coseparability implies that the columns, $C_b^{m,\bullet} = C_c^\bullet(A^{\otimes m}, A)$, of the defining double complex are exact, so the total cohomology is that of the augmenting row which is a subcomplex of the Hochschild cochain complex $C_h^\bullet(A,k)$. By the theorem of Hochschild-Kostant-Rosenberg, $\bigwedge_A \mathrm{Der}(A,A) \otimes_A k \to H_h^\bullet(A,k)$ is an isomorphism. Now, there is a natural isomorphism

$$\bigwedge\nolimits_A \mathrm{Der}(A,A) \otimes_A k \cong \bigwedge \mathrm{Der}(A,k)$$

and any derivation $A \to k$ is completely determined by the images of the elements of degree one, A_1, which can be arbitrary elements of k, so

$$\bigwedge \mathrm{Der}(A,k) = \bigwedge \mathrm{Hom}(A_1,k) = \bigwedge A_1^*.$$

This can naturally be identified with $\bigwedge \mathfrak{g}$ where $\mathfrak{g}$ is the Lie algebra of G when $A = \mathcal{O}(G)$ for a reductive group and $\mathfrak{g} = M(n)$ when $A = \mathbf{M}(n)$. Thus, for example, when $A = \mathbf{M}(n)$ the matrix unit $e_{ij} \in M(n)$ corresponds to the derivation $\varepsilon \circ \partial/\partial x_{ij} : A \to k$.

The bialgebra cohomology groups $H_b^m(A,A)$ therefore consist of those elements of $\bigwedge^m \mathfrak{g}$ which are coalgebra cocycles. If $\gamma = x_1 \wedge \cdots \wedge x_m \in \bigwedge^m \mathfrak{g}$ then, as shown in [18] the coalgebra coboundary $\delta_c(\gamma) : A^{\otimes m} \to A$ may be described as

$$\mu_m \circ (X_1 \wedge \cdots \wedge X_m - X_1' \wedge \cdots \wedge X_m') \tag{1}$$

where X_i and X_i' correspond to the left and right invariant vector fields on G associated to the x_i and μ_m is the iterated multiplication map $A^{\otimes m} \to A$. The coalgebra cocycles are those elements $\gamma \in \bigwedge \mathfrak{g}$ that satisfy $[g,\gamma]_S = 0$ for all $g \in \mathfrak{g}$. By definition, these are the *invariants*, $(\bigwedge^m \mathfrak{g})^{\mathfrak{g}}$, of $\bigwedge^m \mathfrak{g}$ and so we have the following theorem which is proved but misstated in [18].

Theorem 4.1. *For $A = \mathbf{M}(n)$ or $\mathcal{O}(G)$ for reductive G the natural maps*

$$(\bigwedge\nolimits^m \mathfrak{g})^{\mathfrak{g}} \longrightarrow H_b^m(A,A)$$

$$\bigwedge\nolimits^m \mathfrak{g}/(\bigwedge\nolimits^m \mathfrak{g})^{\mathfrak{g}} \longrightarrow \widehat{H}_b^m(A,A)$$

are isomorphisms. Moreover, every class in $\widehat{H}_b^m(A,A)$ is represented by $(\mu_m \circ \delta_c\gamma, 0, \cdots, 0)$ for a unique γ in $\bigwedge^m \mathfrak{g}/(\bigwedge^m \mathfrak{g})^{\mathfrak{g}}$.■

There are no invariants in $\bigwedge^r M(n)$ for $r = 1, 2$ and there is a unique one (up to scalar multiple) for $r = 3$. The same is true for every reductive algebra $\mathfrak{g}$. Combining observations, the elements of $\bigwedge^2 \mathfrak{g} = \widehat{H}_b^2(A,A)$ may be viewed as the coalgebra cocycles among the skew biderivations of A into k, i.e. among the skew linear maps $A \otimes A \to k$ which are algebra derivations in each argument. For $M(2)$ and $\mathfrak{sl}(2)$ the invariant in $\bigwedge^3 \mathfrak{g}$ is $(e_{11} - e_{12}) \wedge e_{12} \wedge e_{21}$. In general, the invariants do not form an ideal for the exterior multiplication in $\bigwedge \mathfrak{g}$ but do for the graded Lie product so $\bigwedge \mathfrak{g}/(\bigwedge \mathfrak{g})^{\mathfrak{g}}$ is again a graded Lie algebra. This gives, in a special case, a graded Lie bracket for $H_b^*(A,A)$ and $\widehat{H}_b^*(A,A)$. The graded commutative product mentioned earlier is trivial in this case.

5. Algebraic deformation theory

Let A be an associative algebra over a ring k (here $\mathbb{R}$ or $\mathbb{C}$), and let $\mu : A\otimes A \to A$ be the multiplication. Denote the ring of formal power series in t over k by $k[\![t]\!]$ and the $k[\![t]\!]$ module of formal power series over A by $A[\![t]\!]$. (This is generally larger than $A\otimes k[\![t]\!]$.) A *deformation* of A is an associative $k[\![t]\!]$-bilinear multiplication $\mu_t : A[\![t]\!]\otimes_{k[\![t]\!]} A[\![t]\!] \to A[\![t]\!]$ which can be written in the form $\mu_t = \mu + t\mu_1 + t^2\mu_2 + \cdots$ where the μ_i are k-linear maps $A\otimes A \to A$ extended to be $k[\![t]\!]$-bilinear. (Here $\mu = \mu_0$. There is also an ambiguity in the use of both t and integers as subscripts, but this should cause no confusion.) We denote $A[\![t]\!]$ together with this μ_t simply by A_t. The associativity of μ_t can be written as $\mu_t \circ \mu_t = 0$, i.e.,

$$\sum t^m \sum_{i+j=m} \mu_i \circ \mu_j = 0.$$

Because μ and all μ_i are of dimension 2, we have

$$\mu \circ \mu_m + \mu_m \circ \mu = [\mu, \mu_m]_G = -\delta_h \mu_m.$$

Therefore

$$\sum_{\substack{i+j=m\\ i,j>0}} \mu_i \circ \mu_j = \delta_h \mu_m, \text{ all } m.$$

This implies in particular that $\delta_h\mu_1 = 0$, i.e., μ_1 must be a cocycle in the Hochschild theory, and that $\mu_1 \circ \mu_1 = \delta_h \mu_2$, i.e., $\mu_1 \circ \mu_1$ must be a coboundary. Of course, μ_t can be μ itself, i.e., we may have all $\mu_i = 0$ for $i > 0$, in which case we have the "null" deformation. (We are considering here only the deformation theory of associative algebras, cf. the original paper [11]. Nijenhuis and Richardson have developed the Lie theory in a parallel way, cf. [28]. There μ_1 is a 2-cocycle in the Chevalley-Eilenberg theory.) Deformations μ_t and $\nu_t = \mu + t\nu_1 + t^2\nu_2 + \dots$ are *equivalent* if there is a $k[\![t]\!]$-module map $f_t : A[\![t]\!] \to A[\![t]\!]$ of the form $f_t = \mathrm{id}+tf_1+t^2f_2+\dots$ the f_i being linear maps $A \to A$ (extended to be $k[\![t]\!]$-linear) such that $\nu_t(a,b) = f_t^{-1}\mu_t(f_ta, f_tb)$. Thus f_t is an algebra isomorphism $\nu_t \to \mu_t$. This implies that $\nu_1 = \mu_1 + \delta_h f_1$. We may therefore view the cohomology class $\overline{\mu_1}$ as the "infinitesimal" of the class of deformations equivalent to μ_t. Assuming (as we always do here) that A is unital, every cocycle is cohomologous to a normalized one, so every deformation is equivalent to one in which the old unit element still serves as the unit. (In fact, any separable subalgebra can be preserved.) In particular, any deformation of a unital algebra is unital. It is an easy exercise to show that an invertible element remains invertible when an algebra is deformed, although the inverse generally changes. If the infinitesimal μ_1 is a coboundary, then μ_t is equivalent to a ν_t with $\nu_1 = 0$. It is easy by induction to prove

Theorem 5.1. *If A is an algebra with $H^2_h(A,A) = 0$ (e.g., separable) then every deformation of A is equivalent to the null deformation.* ■

A deformation equivalent to the null deformation is called ***trivial***. An algebra for which all deformations are trivial is ***rigid***, or more precisely, ***analytically rigid***, cf. [17]. (It may happen that A is finite dimensional over a field and rigid, but that $H^2_h(A,A) \neq 0$. There are examples in the associative theory in all finite characteristics [14] and in the Lie theory [29] but so far no associative example is known in characteristic zero.)

Suppose now that $\mu_1 \in Z^2_h(A,A)$ or, equivalently, as one can easily check, that the multiplication in $A[t]/t^2$ given by $\mu + t\mu_1$ is associative. The necessary and sufficient condition that there exist a bilinear $\mu_2 : A \times A \to A$ such that $A[t]/t^3$ with multiplication $\mu + t\mu_1 + t^2\mu_2$ is associative is that the *obstruction* to μ_1, namely the class of the cocycle $\mu_1 \circ \mu_1 \in Z^3_h(A,A)$ must vanish. If we can divide by 2 then this equals $\frac{1}{2}[\mu_1, \mu_1]_G$, but in every case we have a quadratic obstruction map $ob : H^2_h(A,A) \to H^3_h(A,A)$ carrying $\overline{\mu_1}$ to $\overline{\mu_1 \circ \mu_1}$. If we have found μ_2, then there is an obstruction, again in $H^3_h(A,A)$ to the existence of μ_3, and so forth. (See [11] for details.) For a Noetherian ring we conjecture that there will be only a finite number of obstructions. In the next two sections we extend the algebraic deformation theory to bialgebras where in principle one again encounters a sequence of obstructions to deformation when an infinitesimal is given. However, for $\mathbf{M}(n)$ and $\mathbf{SL}(n)$ the only obstruction which we have actually encountered (in the present incomplete state of the theory) is the first.

It is important for some purposes to know how the cohomology groups of an algebra A behave under a deformation μ_t. We call $f \in Z^m_h(A,A)$ *liftable* or *extendable* if there is a power series $f_t = f + tf_1 + t^2 f_2 \ldots$ with $f_i \in C^m_h(A,A)$ such that $[f_t, \mu_t]_G = 0$. (Recall that the Hochschild coboundary operator is $-[-, \mu_t]_G$. In particular, $[f,\mu_1]_G + [f_1, \mu]_G = 0$, so $[f, \mu_1]_G$ must be a coboundary. The cohomology class of $[f,\mu_1]_G$ in $H^{m+1}_h(A,A)$ is therefore the *primary obstruction* to lifting f. If one passes the primary obstruction, then there is another, again in $H^{m+1}_h(A,A)$, and so forth. Suppose now that k is a field. In order to remain over a field, we extend the coefficients of A_t from $k[\![t]\!]$ to its fraction field $k((t))$, i.e., we now allow finitely many negative powers of t. Some cocycles $f \in Z^m_h(A,A)$ may now be liftable to coboundaries; we call these *jump cocycles.* (Every $f \in B^m_h(A,A)$ is liftable to a coboundary, for if $f = -[g,\mu]_G$ then we can lift it to $f_t = -[g, \mu_t]_G$.) We then have Coffee's Theorem (cf. [12]):

Theorem 5.2. *The cohomology of the deformed algebra A_t is given by*

$$H^m_h(A_t, A_t) = (\textit{liftable } m\textit{-cocycles})/(\textit{jump } m\textit{-cocycles}). \blacksquare$$

Therefore, cohomology can only drop. For simplicity, we assume henceforth that k has characteristic zero. Then we also have

Theorem 5.3. *The jump cocycles of $Z^m_h(A,A)$ are precisely those appearing as obstruction to lifting elements of $Z^{m-1}_h(A,A)$.* ■

This yields the

Corollary 5.4 ([17]). *If A has an Euler-Poincaré characteristic*

$$\chi = \sum_{i=0} (-1)^i \dim H^i_h(A,A)$$

then χ is preserved under deformation. ■

A *jump deformation* intuitively is one where at time $t = 0$ we have an initial algebra A_0 but for all sufficiently small $t \neq 0$ the algebras A_t are isomorphic. To formalize this, we introduce a second variable u and require that $A_{(1+u)t}$ be isomorphic to A_t over $k((t,u))$. From [12] we then have

Theorem 5.5. *If $\mu_t = \mu + t\mu_1 + \dots$ is a jump deformation of A then μ_1 is a jump cocycle.* ∎

A jump deformation therefore "destroys its own infinitesimal" μ_1. This μ_1 must then be the obstruction to lifting some derivation. In particular, we have

Corollary 5.6 ([17]). *If $H_h^1(A, A) = 0$ then A has no jump deformations.*

There are analogous results for finite characteristics. In characteristic zero, which we are assuming, every derivation ϕ is the infinitesimal of a formal family of automorphisms $\exp(t\phi)$. In some cases this may be viewed as a genuine Lie group of automorphisms, e.g., when $A = C^\infty(\mathcal{M})$ where $\mathcal{M}$ is a compact C^∞ manifold. (To extend to the analytic case we need presheaves of algebras, cf. [16].)

Historically, algebraic deformation theory began as an analogue of that of complex manifolds introduced by [9] and elaborated by Kodaira and Spencer. (Cf. the bibliography in [17].) Its application to quantization and other areas is due to Lichnerowicz (cf. [23]). The formal aspects of the algebraic deformation theory have now subsumed those of the analytic theory ([17]), but issues of convergence remain.

6. Universal deformation formulas

Let $\mathfrak{g}$ be a Lie algebra and for any associative k-algebra A, let $\mathrm{Der}(A)$ denote its Lie algebra of derivations $A \to A$. A *universal deformation formula* based on $\mathfrak{g}$ is an element of a certain complex which produces from each Lie algebra map $\mathfrak{g} \to \mathrm{Der}(A)$ a formal deformation of A. The universal enveloping algebra of $\mathfrak{g}$ will be denoted throughout by $U\mathfrak{g}$. The tensor algebra $\bigotimes U\mathfrak{g}$ over $U\mathfrak{g}$ can be made into a complex with coboundary operator $\delta : \bigotimes^m U\mathfrak{g} \to \bigotimes^{m+1} U\mathfrak{g}$ defined so: For $m = 0$, set $\delta = 0$ and for $m = 1$ define $\delta : U\mathfrak{g} \to U\mathfrak{g} \otimes U\mathfrak{g}$ by $\delta 1 = 1 \otimes 1, \delta a = 0$ for all $a \in \mathfrak{g}$, and

$$\delta(a_1 a_2 \cdots a_m) = -\sum a_{i_1} \cdots a_{i_r} \otimes a_{j_1} \cdots a_{j_s},$$

where the sum is over all partitions of $\{1, \dots, m,\}$ into two disjoint subsets $I = \{i_1 < \dots < i_r\}$ and $J = \{j_1 < \dots < j_s\}$. One then extends δ linearly to all of $U\mathfrak{g}$. (For $m = 1$ no such partition is possible, whence $\delta a = 0$.) On higher tensor powers of $U\mathfrak{g}$, δ is defined inductively by setting

$$\delta(\alpha^m \otimes \beta^p) = \delta\alpha^m \otimes \beta^p + (-1)^m \alpha^m \otimes \delta\beta^p,$$

where $\alpha^m \in \bigotimes^m U\mathfrak{g}, \beta^p \in \bigotimes^p U\mathfrak{g}$. Note that one must establish the consistency of the formula by showing that if $a, b \in \mathfrak{g}$ and $[a, b] = c$, then $\delta[a, b] = \delta(ab) - \delta(ba) = \delta c = 0$, but that is evident. On the other hand, if we fix a basis $x_1, \dots, x_r$ for $\mathfrak{g}$, then the theorem of Poincaré-Birkhoff-Witt asserts that the elements $x_1^{m_1} x_2^{m_2} \cdots x_r^{m_r}$ form a basis for $U\mathfrak{g}$, and the actual multiplication in $\mathfrak{g}$ never enters into the definition of δ. To compute the cohomology one may therefore assume $\mathfrak{g}$ to be abelian. In that case, however, if $\mathfrak{g}^*$ is the dual vector space to $\mathfrak{g}$ with dual basis $x_1^*, \dots, x_r^*$ and A is the polynomial ring $k[\mathfrak{g}^*]$, then we can map $\bigotimes U\mathfrak{g}$ into the Hochschild cochain complex $C_h^*(A, A)$ by sending x_i to $\partial/\partial x_i^*$ and $\otimes$ to $\smile$. If we complete $U\mathfrak{g}$ to allow formal power series $\sum c_{i_1 \dots i_r} x_1^{m_1} \cdots x_r^{m_r}$ then this becomes an isomorphism of complexes. This will not change the cohomology of $\bigotimes U\mathfrak{g}$ because δ is homogeneous. We may therefore apply the theorem of Hochschild-Kostant-Rosenberg to conclude that there is an isomorphism

$$H^*(\bigotimes U\mathfrak{g}) \longrightarrow H_h^*(k[\mathfrak{g}^*], k[\mathfrak{g}^*]) = \bigwedge \mathfrak{g}.$$

Moreover, since we are assuming that the characteristic is zero, for any vector space V the epimorhphism $\bigotimes^m V \to \bigwedge^m V$ has a natural splitting sending $v_1 \wedge \cdots \wedge v_m$ to $\frac{1}{m!}\sum(\mathrm{sgn}(\sigma))v_{\sigma_1} \wedge \cdots \wedge v_{\sigma_m}$, where the sum is over σ in the symmetric group S_m. Viewing $\bigwedge^m \mathfrak{g}$ in this way as a subspace of $\bigotimes^m \mathfrak{g}$, which in turn is contained in $\bigotimes^m U\mathfrak{g}$, we have

Theorem 6.1. *The elements of $H^*(\bigotimes U\mathfrak{g})$ are uniquely represented by the classes of the elements of $\bigwedge \mathfrak{g}$.* ∎

While the cohomology of $\bigotimes U\mathfrak{g}$ did not depend on the structure of the Lie algebra $\mathfrak{g}$, the composition in the complex, which we now define, does. It is given by the following three rules (in which we denote composition simply by juxtaposition):

(*i*) For elements in $U\mathfrak{g}$, composition is just their usual product.
(*ii*) If $a \in \mathfrak{g}$ then $a(\alpha \otimes \beta) = a\alpha \otimes \beta + \alpha \otimes a\beta$. Every element of $\mathfrak{g}$ is thus a derivation for the tensor product multiplication, so evaluation of an expression like $(a_1 \cdots a_r)(\alpha \otimes \beta)$ is given by Leibniz' rule.
(*iii*) If $\alpha_1, \ldots, \alpha_m \in U\mathfrak{g}$ and $\beta \in \bigotimes^p U\mathfrak{g}$ then

$$(\alpha_1 \otimes \cdots \otimes \alpha_m)\beta = \sum_{i=1}^{m} (-1)^{(i-1)(m-1)} \alpha_1 \otimes \cdots \alpha_{i-1} \otimes \alpha_i\beta \otimes \cdots \otimes \alpha_p.$$

If $\mathfrak{g}$ were a Lie algebra of derivations of some algebra A then these would be precisely the rules for composition in the subcomplex of $C^*(A, A)$ generated by the elements of $\mathfrak{g}$ under cup product (in place of tensor product) and the composition introduced in section 2. It follows that if $\alpha \in \bigotimes^m U\mathfrak{g}$, $\beta \in \otimes^n U\mathfrak{g}$ then setting

$$[\alpha, \beta] = \alpha\beta - (-1)^{(m-1)(n-1)}\beta\alpha$$

defines a graded Lie multiplication in which we have

$$\delta(\alpha) = -[\alpha, 1 \otimes 1].$$

Now, this bracket induces a graded Lie multiplication on $H^*(\bigotimes U\mathfrak{g})$. The isomorphism $H^*(\bigotimes U\mathfrak{g}) \to \bigwedge \mathfrak{g}$ is then a G-algebra map — in particular, it carries the bracket on $H^*(\bigotimes U\mathfrak{g})$ to the Schouten bracket on the exterior algebra of the Lie algebra $\mathfrak{g}$. On the other hand, the splitting map $\bigwedge \mathfrak{g} \to \bigotimes U\mathfrak{g}$, using which we represented the cohomology classes by cocycles, is not a Lie algebra map: if $\alpha, \beta \in \bigwedge \mathfrak{g}$ then $[\alpha, \beta]$ does not lie in $\bigwedge \mathfrak{g}$, but it is a cocycle, and its class is represented by the element $[\alpha, \beta]_S$ of $\bigwedge \mathfrak{g}$. We can now define a *universal deformation formula* or "udf" based on $\mathfrak{g}$ to be an element of $\bigotimes^2 U\mathfrak{g}[[t]]$ of the form

$$\bar{\mu}_t = 1 \otimes 1 + t\bar{\mu}_1 + t^2\bar{\mu}_2 + \ldots$$

where $\mu_i \in \bigotimes^2 U\mathfrak{g}$ (Here we have restored the composition symbol "o" to avoid confusion.) Equivalently,

$$\delta(\bar{\mu_1}) = 0 \quad \text{and} \quad \sum_{i=1}^{m-1} \bar{\mu_i} \circ \mu_{\bar{m}-i} = \delta(\bar{\mu_m}) \quad \text{for } m > 0.$$

In particular, $\bar{\mu_1}$ must be in $\mathfrak{g} \otimes \mathfrak{g}$. Equivalence of udf's is defined like that of deformations. We view $\bar{\mu_1}$ as the infinitesimal of the specific udf $\bar{\mu_t}$, and its cohomology class, which is represented by its skew part, as the infinitesimal of the class of equivalent udf's. If $\bar{\mu_t} \circ \bar{\mu_t}$ is not necessarily 0 but is an invariant of $\mathfrak{g}$, i.e., if all its coefficients are invariants, then $\bar{\mu_t}$ is a *quasi-universal deformation formula.* If A is a associative k-algebra then any Lie morphism $\mathfrak{g} \to \mathrm{Der}(A)$ induces a morphism $\bigotimes U\mathfrak{g} \to C^\bullet(A, A)$ carrying $\otimes$ to $\smile$ and preserving composition. As $1 \otimes 1$ is carried to $\mathrm{id}_A \smile \mathrm{id}_A$ which is just multiplication in A, this is also a morphism of complexes. We therefore have

Theorem 6.2. *If $\bar{\mu_t}$ is a universal deformation formula based on $\mathfrak{g}$ then any morphism $\mathfrak{g} \to \mathrm{Der}(A)$ induces a deformation of A.* ■

The first example is that where $\mathfrak{g}$ is the two-dimensional abelian Lie algebra with generators ϕ and ψ, and

$$\bar{\mu_t} = \exp t(\phi \otimes \psi) = 1 \otimes 1 + t\phi \otimes \psi + \frac{t^2}{2}\phi^2 \otimes \psi^2 + \dots,$$

cf. [12]. (In the special case where $\phi = \partial/\partial x, \psi = \partial/\partial y$ it was known earlier by Moyal, [25].) A "quasi-exponential" formula for the solvable two dimensional $\mathfrak{g}$ spanned by H, X with $[H, X] = \lambda X$ is given in [4]:

$$\bar{\mu_t} = 1 \otimes 1 + H \otimes X + \frac{t^2}{2}H(H+1) \otimes X^2 + \frac{t^3}{3!}H(H+\lambda)(H+2\lambda) \otimes X^3 \dots .$$

We shall see that $H \wedge X$ is the infinitesimal of the jump deformation of $\mathbf{SL}(2)$ to a quantum group which lies on the boundary of the deformation space. This suggests that, for Hopf algebras like $\mathcal{O}(G)$ with G reductive, those deformations which lie on the boundary of the deformation space are associated with udf's. This is true by explicit calculation for all $\mathbf{SL}(n)$. However, even in the foregoing "easy" case associated with $\mathbf{SL}(2)$ we do not yet have a closed form for the udf with infinitesimal $H \wedge X$, although that can be calculated, in principle, from the fact that it is equivalent to the one given. Now when a Lie group G with Lie algebra $\mathfrak{g}$ operates on an algebra A it induces a Lie morphism $\mathfrak{g} \to \mathrm{Der}(A)$. For example, if $\mathcal{M}$ is a C^∞ manifold, then any Lie group G of diffeomorphisms of $\mathcal{M}$ also acts on $A = C^\infty(\mathcal{M})$ and thereby induces a morphism $\mathfrak{g} \to \mathrm{Der}(A)$. If we have a udf based on $\mathfrak{g}$ and μ_t is the induced deformation of A, then the images of some of the elements of $\mathfrak{g}$ may not be liftable to derivations of A_t. As a result, not all of the automorphisms in G may be liftable to automorphisms of A_t. Universal deformation formulas may thus give rise to a form of "spontaneous symmetry breaking" : the very existence of symmetries points a path to their collapse! The quasi-exponential formula which we have just given generally breaks symmetry because the image of H generally is not liftable. Since the algebra generated by H

and X is contained in every simple Lie algebra, the presence of a simple Lie group of automorphisms therefore generally induces spontaneous symmetry breaking. Quasi-universal deformation formulas will also induce deformations when only the zero element of A is invariant under all of G. For example, suppose that G is not compact and operates on an open manifold, inducing an automorphism of the algebra of smooth functions vanishing at infinity. Every deformation of $\mathcal{O}(G)$ (G reductive) to a quantum group should give rise to a quasi-udf, but we have been unable to write down any of the formulas.

7. Deformations of Bialgebras

Let A be a k-algebra with multiplication μ and comultiplication Δ. A deformation A_t of A is a $k[\![t]\!]$-bilinear multiplication $\mu_t = \mu + t\mu_1 + t^2\mu_2 + \dots$ and comultiplication $\Delta_t = \Delta + t\Delta_1 + t^2\Delta_2 + \dots$ with which $A[\![t]\!]$ is again a bialgebra. The existence of units and counits is preserved by deformation (cf. [18]). If A is a Hopf algebra then so is A_t, for,as mentioned earlier, the antipode S is just a two-sided inverse to the identity map of A in the "convolution" multiplication of maps $f : A \to A$ defined by $f * g = \mu(f \otimes g)\Delta$. (The unit element for this product is $\eta\varepsilon$.) A deformation of A induces one of this algebra, and S remains invertible.

If (μ_t, Δ_t) is a deformation of A as bialgebra, then in particular μ_t defines an algebra deformation, so $\delta_h\mu_1 = 0$, and Δ_t defines coalgebra deformation, so $\delta_c\Delta_1 = 0$. The compatibility between μ_t and Δ_t, namely $\Delta_t\mu_t = \mu_t \otimes \mu_t(23)\Delta_t \otimes \Delta_t$, is equivalent to

$$\sum_{i'+i''+j'+j''=m} (\mu_{i'} \otimes \mu_{i''})(23)(\Delta_{j'} \otimes \Delta_{j''}) - \sum_{i+j=m} \Delta_i\mu_j = 0. \tag{2}$$

Those terms in which one index is m and the others are zero may be rearranged to give

$$[(\mu_m \otimes \mu + \mu \otimes \mu_m)(23)(\Delta \otimes \Delta) - \Delta \otimes \mu_m] + [(\mu \otimes \mu)(23)(\Delta_m \otimes \Delta + \Delta \otimes \Delta_m) - \Delta_m \otimes \mu].$$

Since μ_m and Δ_m are 2-cochains, the first term is just $\delta_c\mu_m$ and the second is $\delta_h\Delta_m$. (Recall the definition of the internal tensor product!) For $m = 1$, the sum vanishes so $(\mu_1, \Delta_1) \in C^{1,2} \otimes C^{2,1} = \widehat{C}^2$ is a 2-cocycle of the total complex $\widehat{C}^\bullet$. As with algebras, an equivalent deformation $({\mu'}_t, {\Delta'}_t)$ (definition evident) has an "infinitesimal" $({\mu'}_1, {\Delta'}_1)$ cohomologous to (μ_1, Δ_1). Also as with algebras, if $(\mu_1, \Delta_1) \in \widehat{Z}_b^2$ is given then there is a primary obstruction in $\widehat{H}_b^3$ to finding a (μ_2, Δ_2) such that we have a bialgebra mod t^3. This obstruction is the class of the 3-cocycle $(\mu_1 \circ \mu_1, \mu_1 \times \Delta_1, \Delta_1 \circ \Delta_1)$ where

$$\begin{aligned}\mu_1 \times \Delta_1 =& (\mu_1 \otimes \mu_1)(23)(\Delta \otimes \Delta) + (\mu \otimes \mu)(23)(\Delta_1 \otimes \Delta_1) + \\ &+ (\mu_1 \otimes \mu + \mu \otimes \mu_1)(23)(\Delta_1 \otimes \Delta + \Delta \otimes \Delta_1) - \Delta_1\mu_1,\end{aligned}$$

namely, the negative of the sum of all the terms in (2) with $m = 2$, but no index equal to 2. Passing this obstruction leads to another, and so forth. As mentioned earlier, we are unable, so far, to exhibit a graded Lie structure on $\widehat{H}_b^3$ in which $(\mu_1 \circ \mu_1, \mu_1 \times \Delta_1, \Delta_1 \circ \Delta_1)$ is just $(1/2)[(\mu_1, \Delta_1), (\mu_1, \Delta_1)]$, but when $A = \mathrm{M}(n)$ or $\mathcal{O}(G)$ for G reductive, it is easy to check that the graded Lie structure on $\bigwedge M(n)/(\bigwedge M(n))^{M(n)}$ or $\bigwedge \mathfrak{g}/(\bigwedge \mathfrak{g})^{\mathfrak{g}}$, respectively, is what we seek.

When A is coseparable, the vanishing of $H_c^n(A,A)$ for all $n>0$ (in particular of H_c^2) implies that every deformation is equivalent to one in which $\Delta_t = \Delta$, i.e., the coalgebra structure is strictly unchanged, but exhibiting such a deformation may be difficult. To make plain what this means we introduce some definitions which may be at variance with present usage but at least are precise. A *quantum group* will be a bialgebra A_t which is a deformation in the foregoing sense of a bialgebra of the form $\mathbf{M}(n)$ or $\mathcal{O}(G)$ where G is an affine algebraic group (although properly a deformation of $\mathbf{M}(n)$ is only a "quantum monoid"). A quantum universal enveloping algebra is similarly a deformation of the universal enveloping algebra. Note that it is not apparent that $\mathbf{M}_q(n)$, with its usual presentation, is a deformation (a quantum group, in our sense). The problem is that there is no obvious identification of its underlying vector space with that of $\mathbf{M}(n)$, enabling us to write its multiplication as a power series beginning with that of $\mathbf{M}(n)$. Also note that while Δ is unchanged on the *generators* x_{ij} it *is* function of q. For example, when $n=2$, $(\Delta x_{12})^2$ by definition is $(\Delta x_{12})^2 = (x_{11}\otimes x_{12} + x_{12}\otimes x_{22})^2 = x_{11}^2\otimes x_{12}^2 + x_{12}^2\otimes x_{22}^2 + x_{11}x_{12}\otimes x_{12}x_{22} + x_{12}x_{11}\otimes x_{22}x_{12} = x_{11}^2\otimes x_{12}^2 + x_{12}^2\otimes x_{22}^2 + (1+q^2)x_{11}x_{12}\otimes x_{12}x_{22}$.

If we view q as a formal power series in t with $q(0)=1$, (the natural choice, as it happens, being $(1-\sin t)/\cos t = \cos t/(1+\sin t)$) then fortunately the underlying $k[\![t]\!]$-module of $\mathbf{M}_q(2)$ (where coefficients are now extended to $k[\![t]\!]$) *can* be identified with that of $\mathbf{M}(2)[[t]]$. That, however, is not obvious here and generally not true under the (ambiguous) common definition of quantum group. The Faddeev-Reshitikhin-Takhtajan (FRT) construction of a "quantum group" usually starts with a "quantum Yang-Baxter" (qYB) matrix R defined as follows: Let V be a vector space over k and $R\in \mathrm{Aut}(V\otimes V)$. Then we can extend R to a vector space automorphism of $V\otimes V\otimes V$ in 3 ways: define R_{12}, R_{13}, R_{23} defined by letting R_{ij} be the identity on the tensor factor with the omitted index; R is *quantum Yang-Baxter* matrix if $R_{12}R_{13}R_{23} = R_{23}R_{13}R_{12}$. (If $\dim V = n$ and a basis is chosen then R is $n^2\times n^2$ and the R_{ij} are $n^3\times n^3$.) Now let $X=(x_{ij})$ be viewed as a matrix with coefficients in the non-commutative polynomial ring $k\langle x_{ij}\rangle$, set $X_1 = X\otimes \mathrm{I}$ and $X_2 = \mathrm{I}\otimes X$, where $\otimes$ here denotes the *Kronecker product.* (If $A=(a_{ij})$ and $B=(b_{kl})$ are matrices with coefficients in a non-commutative ring then the Kronecker product $A\otimes B$ is a matrix whose rows are indexed by the double index (i,k), columns by (j,l), and $(A\otimes B)_{(i,k),(j,l)} = a_{ij}b_{kl}$. Because of the non-commutativity, $A\otimes B\neq B\otimes A$ in general. One can give the rows and columns of $A\otimes B$ a single index by ordering lexicographically.) Setting $R(X_1X_2) = (X_2X_1)R$ introduces certain quadratic relations on the x_{ij}, and $k\langle x_{ij}\rangle$ modulo these relations is usually called a "deformation" or "quantization" of $k[x_{ij}]$. If we view $k[x_{ij}]$ as $\mathbf{M}(n)$, then extending the original comultiplication on the x_{ij} to be an algebra morphism gives the "deformation" or "quantization" of the bialgebra $\mathbf{M}(n)$ to one defined by R. The construction can fail to be a genuine deformation because, although the relations $RX_1X_2 = X_2X_1R$ certainly define an associative graded ring, it may be impossible to identify its underlying vector space, in a way which preserves the grading, with that of the original polynomial ring $k[x_{ij}]$.

As an example of the failure let $n=2$ and for simplicity write $x_{11}=a$, $x_{12}=b$,

$x_{21} = c$, $x_{22} = d$, so that

$$X_1X_2 = \begin{pmatrix} aa & ab & ba & bb \\ ac & ad & bc & bd \\ ca & cb & da & db \\ cc & cd & dc & dd \end{pmatrix},$$

$$X_2X_1 = \begin{pmatrix} aa & ba & ab & bb \\ ca & da & cb & db \\ ac & bc & ad & bd \\ cc & dc & cd & dd \end{pmatrix}.$$

Now any diagonal $n^2 \otimes n^2$ matrix is qYB. Take $R = \operatorname{diag}(\lambda, \mu, \nu, \sigma)$. Then the comutation relations imply, amongst others, that $\lambda ab = ba\mu = \mu ba$ (since μ is in the ground field) and $\lambda ba = ab\nu$, so unless $\lambda^2 = \mu\nu$ we have $ab = ba = 0$ and there are too few elements of degree 2 in the "deformed ring". We *do* have a bialgebra, but the underlying vector space is a proper quotient of the original and certainly not a deformation in our sense.

If $A = \mathcal{O}(G)$ with G reductive, or $\mathbf{M}(n)$ then by a *preferred deformation* of A we shall mean a formal deformation (!) in which the comultiplication is strictly unchanged. A *preferred presentation* of a given quantum group is an isomorphism with a preferred deformation. Similarly if $\mathfrak{g}$ is semi-simple Lie algebra and $A = U\mathfrak{g}$ (which is rigid as an algebra) then a preferred deformation is one in which only the comultiplication changes, and preferred presentation is defined similarly.

Our cohomology theory implies that if G is reductive Lie group or $\mathfrak{g}$ a semi simple Lie algebra, then any deformation of $\mathcal{O}(G)$ or $U\mathfrak{g}$ has a preferred presentation. The problem is to exhibit it, which we do here in the cases of $\mathbf{M}(n)$ and $\mathbf{SL}(n)$.

How do we "integrate", i.e. construct a global deformation from a $\gamma \in \bigwedge^2 \mathfrak{g}$ where $\mathfrak{g}$ is the Lie algebra of G or $\gamma \in \bigwedge^2 M(n)$, assuming that integration is possible? The actual infinitesimal of the deformation is a cocycle cohomologous to $\delta_c\gamma$. As noted in [18], since $\gamma \in \operatorname{Hom}_k(A \otimes A, k)$ we can also consider it as a cochain for the coalgebra $A \otimes A$. So considered, its coboundary $\delta_c^{A\otimes A}\gamma$ lies in $\operatorname{Hom}_k(A \otimes A, A \otimes A)$ and we have $\delta_c\gamma = \mu\delta_c^{A\otimes A}\gamma$. Now $\delta_c^{A\otimes A}\gamma$ is a coderivation of $A \otimes A$, so $\exp(t \cdot \delta_c^{A\otimes A}\gamma)$ is an automorphism of the coalgebra structure and $\mu\exp(t\widehat{\delta_c}\gamma)$ is a multiplication on $A[\![t]\!]$ which is a morphism of the coalgebra structure. It is not associative on all of $A[\![t]\!]$ but does define a new product on the generators which we should like to extend to all of $A[\![t]\!]$, if possible.

Suppose now that $V = k^n$ (column vectors of length n), so $\operatorname{End}V = M(n)$. For $f, g \in \operatorname{End}V$. Then $f \wedge g = \frac{1}{2}(f \otimes g - f \otimes g) \in \bigwedge^2 \operatorname{End}V \subset \otimes^2\operatorname{End}V = \operatorname{End}(V \otimes V)$. As before, set $(f\wedge g)_{12} = (f \wedge g)\otimes \mathrm{I} = \frac{1}{2}(f\otimes g\otimes \mathrm{I} - g\otimes f\otimes \mathrm{I})$, $(f\otimes g)_{13} = \frac{1}{2}(f\otimes \mathrm{I}\otimes g - g\otimes \mathrm{I}\otimes f)$ and $(f \wedge g)_{23} = \frac{1}{2}(\mathrm{I} \otimes f \otimes g - \mathrm{I} \otimes g \otimes f)$ as elements of $\operatorname{End}(V \otimes V \otimes V)$. It is easy to check for $\gamma \in \bigwedge^2 M(n)$ the graded Lie product

$$[\gamma, \gamma]_S = 2 \cdot \{[\gamma_{12}, \gamma_{13}] + [\gamma_{12}, \gamma_{23}] + [\gamma_{13}, \gamma_{23}]\}.$$

This must be an invariant if $\delta_c\gamma$ is to serve as the infinitesimal of a deformation of $\mathbf{M}(n)$. A matrix $R' \in M(n)^{\otimes 2}$ is called *classical Yang-Baxter* (cYB) if $[R'_{12}, R'_{13}] + [R'_{12}, R'_{23}] +$

$[R'_{13}, R'_{23}] = 0$. Thus, solutions in $\bigwedge^2 M(n)$ of the classical Yang-Baxter equation (modulo invariants) correspond to infinitesimal bialgebra deformations of $\mathbf{M}(n)$. If $R = 1 + tR' + t^2R'' + \ldots$ is qYB then R' is cYB, but it is not known if every cYB matrix R' can serve as the linear term of such an R. However, we have the following trivial

Theorem 7.1. *If R' is classical Yang-Baxter and $(R')^3 = 0$ then $\exp(tR')$ is quantum Yang-Baxter.* ∎

Suppose now that $\mathfrak{g}$ is a semisimple Lie algebra (which we can always view as an algebra of matrices). In $\bigwedge^2 \mathfrak{g}$ there is a well-known "standard" γ, henceforth denoted γ_q, with $[\gamma_q, \gamma_q]_S$ an invariant, described as follows: Choose a Cartan subalgebra $\mathfrak{h}$ and a system of positive roots, relative to which $\mathfrak{g}$ then has a decomposition $\mathfrak{n}_- \oplus \mathfrak{h} \oplus \mathfrak{n}_+$, where $\mathfrak{n}_+$ is spanned by the positive root vectors and $\mathfrak{n}_-$ by the negative ones. For each positive root α choose an X_α in the root space L_α and let $X_{-\alpha}$ be the unique element of $L_{-\alpha}$ such that $\kappa(X_\alpha, X_{-\alpha}) = 1$, where κ is the Killing form. Then $\gamma_q = \sum X_\alpha \wedge X_{-\alpha}$. This γ_q can also be described as follows: Using κ we can identify $\mathfrak{g}$ with $\mathfrak{g}^*$ and hence $\mathfrak{g} \otimes \mathfrak{g}$ with $\mathfrak{g} \otimes \mathfrak{g}^* = \mathrm{End}\mathfrak{g}$. Under this identification, $\gamma_q = \frac{1}{2} \sum (X_\alpha \wedge X_{-\alpha} - X_{-\alpha} \wedge X_\alpha)$ is identified with that element of $\mathrm{End}\mathfrak{g}$ whose restriction to $\mathfrak{n}_+$ is $+1 \cdot \mathrm{Id}$, to $\mathfrak{n}_-$ is $-1 \cdot \mathrm{Id}$, and to $\mathfrak{h}$ is 0. For example, in $\mathfrak{sl}(n)$ we have $\gamma_q = \sum_{i<j} e_{ij} \wedge e_{ji}$, which may, of course, also be viewed as an element of $\mathfrak{gl}(n)$ or $M(n)$. Clearly, if $h \in \mathfrak{h}$ then $[h, \gamma_q]_S = 0$, so if $\omega \in \bigwedge^2 \mathfrak{h}$ and $\gamma' = \gamma_q + \omega$ then $[\gamma', \gamma']_S = [\gamma_q, \gamma_q]_S$. Denoting the rank of $\mathfrak{g}$ ($= \dim \mathfrak{h}$) by r, the set of all $c\gamma_q + \omega$ with $c \in k$ and $\omega \in \bigwedge^2 \mathfrak{h}$ is thus a linear space $\mathcal{T}$ of dimension $\binom{r}{2} + 1$ and $\gamma' \in \mathcal{T}$ implies that $[\gamma', \gamma']_S$ is an invariant in $\bigwedge^3 \mathfrak{g}$ (which has only one invariant up to a constant multiple).

We now broaden our notion of equivalence of (infinitesimal) deformations. If G is a Lie group and $\mathfrak{g}$ its Lie algebra, than G operates in a nontrivial way on the space $\bigwedge^2 \mathfrak{g}$ of infinitesimal deformations by conjugation. We call two infinitesimals in the same orbit equivalent. Clearly if $\bar{\gamma}$ is in the closure of the union of the conjugates of $\mathcal{T}$, i.e. in $\overline{G\mathcal{T}}$, then $[\bar{\gamma}, \bar{\gamma}]_S$ is an invariant. After choosing a Cartan subalgebra $\mathfrak{h}$ and a set of positive roots, we can exhibit one such $\bar{\gamma}$, henceforth denoted γ_∞, so: Let α be the highest positive root vector and set $\gamma_\infty = [X_\alpha, \gamma_q]_S$. Since $[X_\alpha, [X_\alpha, \gamma_q]_S]_S = 0$ and $[\gamma_q, \gamma_q]_S$ is invariant we have

$$[\gamma_q, \gamma_q]_S = \exp(t \cdot \mathrm{ad}X_\alpha) \cdot [\gamma_q, \gamma_q]_S = [\gamma_q + t[X_\alpha, \gamma_q]_S, \gamma_q + t[X_\alpha, \gamma_q]_S]_S$$

so $[\gamma_\infty, \gamma_\infty]_S$ is actually 0, i.e. γ_∞ is cYB. For $\mathfrak{sl}(n)$, $\gamma_\infty = -\{(e_{11} - e_{nn}) \wedge e_{1n} + 2\sum_{j=2}^{n-1} e_{1j} \wedge e_{jn}\}$; if $n = 2$ it is just $(e_{11} - e_{22}) \wedge e_{12}$. Note that $\gamma_\infty^3 = 0$ and therefore $\exp(t\gamma_\infty)$ is quantum Yang-Baxter.

A significant open question is whether every γ with $[\gamma, \gamma]_S$ an invariant is actually integrable. That is, for $\mathbf{M}(n)$ or $\mathcal{O}(G)$ for G reductive is the primary obstruction the only one. Also, is $\overline{G\mathcal{T}}$ the space of all γ with $[\gamma, \gamma]_S$ an invariant?

8. The Quantum Symmetric Group

The *quantum symmetric group* qS_m is the abstract group with generators $\tau_1, \ldots, \tau_{m-1}$ and relations

1. $\tau_i \tau_j = \tau_j \tau_i$ if $|i - j| > 1$, and
2. $\tau_i^2 = 1$, all i.

While the defining relations for qS_m involve no parameters, the representations of this group that we introduce will. Also, qS_m is a "Coxeter group" (so named by Tits, for the history cf. [2]).For $m = 2$, the quantum symmetric group is just the ordinary symmetric group S_2; for $m = 3$ it is the infinite dihedral group. Axioms 1. and

3. $\tau_i\tau_{i+1}\tau_i = \tau_{i+1}\tau_i\tau_{i+1}$, $i = 1, \ldots, m-2$

give the braid group. With all three we have the Artin presentation of the ordinary symmetric group S_m, so there is an epimorphism $qS_m \to S_m$.

If V is any vector space over k then S_m operates on $V^{\otimes m}$ by permutation of the tensor factors. Suppose that $\dim V < \infty$ and choose $\gamma \in \bigwedge^2 \operatorname{End}V$. We shall define a representation $S_m(\gamma)_L$ (the "left representation") of qS_m on $V[\![t]\!]^{\otimes m}$ for which (since t ultimately is a real or complex parameter) we write simply $V^{\otimes m}$. Since $\bigwedge^2 \operatorname{End}V \subset \otimes^2 \operatorname{End}V = \operatorname{End}(V \otimes V)$ we may form $Q = Q(t) = \exp(t\gamma)$ and view it as an element of $GL(V \otimes V)$. Define $(12)^L_t = Q^{-1}(12)^L Q \in GL(V \otimes V)$ where $(12)^L$ interchanges tensor factors in $V \otimes V$.

Since $(12)^L\gamma = -\gamma(12)^L$ we have $(12)^L Q = Q^{-1}(12)^L$, so $(12)^L_t = Q^{-2}(12)^L = (12)^L Q^2$. Clearly $((12)^L_t)^2 = \mathrm{I} \otimes \mathrm{I}$ where $\mathrm{I} = \mathrm{id}_V$. Define $(i, i+1)^L_t : V^{\otimes m} \to V^{\otimes m}$ for $i = 1, \ldots, m-1$ to be $\mathrm{I}^{\otimes i-1} \otimes (12)^L_t \otimes \mathrm{I}^{m-i-1}$, i.e., it is $(12)^L_t$ acting in the i-th and $i+1$-st tensor factors. These $(i, i+1)^L_t$ (which depend on γ) are the *quantum interchanges*. The map $\tau_i \to (i, i+1)^L_t$ defines $S_m(\gamma)_L$ as a subgroup of $GL(V^{\otimes m})$, or, equivalently, realizes $V^{\otimes m}$ as a left qS_m-module. We may write simply qS_m for $S_m(\gamma)$ when γ is understood. Setting $\gamma = 0$ gives the standard representation of S_m on $V^{\otimes m}$ described above.

An $\alpha \in V^{\otimes m}$ is *quantum symmetric* if it is invariant under $S_m(\gamma)_L$. The subspace of the ordinary symmetric elements (those invariant under S_m) can be identified with $S^m V$, the m-th symmetric power of V, but there are generally fewer quantum symmetric elements than ordinary symmetric ones. We will denote the spaces of quantum and ordinary symmetric elements in $V^{\otimes m}$ by $\operatorname{sym}_q V^{\otimes m}$ and $\operatorname{sym} V^{\otimes m}$, respectively and say that there are *enough* quantum symmetric elements when the dimensions of these two spaces coincide for all m. Their direct sums over all m will be denoted by $\operatorname{sym}_q(k\langle V\rangle)$ and $\operatorname{sym}(k\langle V\rangle)$; these are the quantum and ordinary symmetric elements of the tensor algebra of V. For any finite dimensional vector space W there is an isomorphism $GL(W) \to GL(W^*)$ (where W^* is the dual of W) sending T to $(T^*)^{-1}$ which we denote by T^{-*}. We therefore also have a representation of qS_m on $(V^*)^{\otimes m}$. To distinguish, we denote that on $V^{\otimes m}$ by $S_m(\gamma)_L$ or just by $qS_{m\,L}$ and that on $(V^*)^{\otimes m}$ by $qS_{m\,R}$, and similarly write $(12)^L_t$, $(12)^R_t$, where $(12)^R_t = ((12)^L_t)^{-*}$. Define a representation of $S_m(\gamma)$ on $V^{\otimes m} \otimes (V^*)^{\otimes m}$ by diagonal operation: $(i, i+1)_t = (i, i+1)^L_t \otimes (i, i+1)^R_t$.

The representation $S_m(\gamma)_L$ may collapse to the standard representation of S_m. For this to happen we must have $((i, i+1)^L_t \cdot (i+1, i+2)^L_t)^3 = 1$ for all i, for which it is sufficient that $((12)^L_t(23)^L_t)^3 = 1$. To see what this means, note that $\gamma \in \operatorname{End}(V^{\otimes 2})$ and let $\gamma_{12}, \gamma_{13}, \gamma_{23}$ again denote the extensions to $\gamma \in \operatorname{End}(V^{\otimes 2})$ in which γ operates on the indicated tensor factors. Then $(12)^L\gamma_{13} = \gamma_{23}(12)^L$, $(12)^L\gamma_{12} = -\gamma_{12}(12)^L$, and so forth for the other indices. Also, we have $(12)^L_t = Q_{12}^{-2}(12)^L$. Using these rules, one finds that $((12)^L_t(23)^L_t)^3 = 1$ if and only if $\{(\exp(t\gamma))(12)(\exp(t\gamma))(13)\}^3 = \mathrm{I} \otimes \mathrm{I} \otimes \mathrm{I}$, i.e. we must have that

$$Q_{12}^{-2}Q_{13}^{-2}Q_{23}^{-2}Q_{12}^{2}Q_{13}^{2}Q_{23}^{2}[(12)(23)]^3 = \mathrm{I} \otimes \mathrm{I} \otimes \mathrm{I}.$$

The bracketed factor is $I \otimes I \otimes I$, so we have

Theorem 8.1. *The representations $S_m(\gamma)$, $S_{mL}(\gamma)$ and $S_{mR}(\gamma)$ collapse to the standard representations of S_m if and only if $Q_{12}^{-2}Q_{13}^{-2}Q_{23}^{-2} = Q_{23}^{-2}Q_{13}^{-2}Q_{12}^{-2}$, i.e. if and only if Q^{-2} is quantum Yang-Baxter.* ∎

As observed earlier, γ_∞ is classical Yang-Baxter and $\gamma_\infty^3 = 0$, which implies that $\exp(t\gamma_\infty) = Q$ is quantum Yang-Baxter so the theorem applies to γ_∞.

A bilinear form $(-|-)$ on V induces a form, still denoted $(-|-)$ on $V^{\otimes m}$ by $(v_1 \otimes \ldots \otimes v_m \mid \omega_1 \otimes \ldots \otimes \omega_m) = (v_1 \mid \omega_1)\ldots(v_m \mid \omega_m)$. If the original form is symmetric and non-degenerate then so is the induced one. We call γ *skew* relative to $(-|-)$ on V if it is skew relative to the induced form on $V \otimes V$. In that case Q is orthogonal relative to the form and therefore so is every element of $S_m(\gamma)_L$. It follows (in characteristic 0) that $V^{\otimes m}$ is completely reducible or semisimple as a $S_m(\gamma)_L$-module.

If A is a bialgebra and $\gamma \in \operatorname{Hom}(A^{\otimes 2}, k)$ then $\delta_c^{A\otimes A}\gamma$ is a "coinner coderivation" and $\exp(t \cdot \delta_c^{A\otimes A}\gamma) = T$ is an automorphism of the coalgebra structure of $A \otimes A$. Consequently, μT is a coalgebra map $A \otimes A \to A$. However, it does not give an associative multiplication. Suppose now that $A = \mathrm{M}(n)$, so $\widehat{H}_b^2(A, A) = \bigwedge^2 M(n)$. (Recall that elements of the latter may be viewed as skew biderivations of A into k, i.e. as skew linear maps $\gamma : A \otimes A \to k$ which are algebra derivations as functions of each argument; in this way $\bigwedge^2 M(n) \subset \operatorname{Hom}(A \otimes A, k)$.) Setting $M(n)^* = A_1$ we have, as earlier, $\mathrm{M}(n) = k + A_1 + S^2A_1 + S^3A_1 + \ldots$. We wish to study the restriction of T to $A_1 \otimes A_1$ where it becomes a linear map $A_1 \otimes A_1 \to A_1 \otimes A_1$.

To compute T, set $V = k^{n*}$,and let the standard basis of $V^* = k^n$ be $e_1, \ldots, e_n$ and denote the dual basis of $V = k^{n*}$ by $u_1(= e_1^*), \ldots, u_n(= e_n^*)$. Then

$$M(n) = \operatorname{End} k^n = k^n \otimes k^{n*} = V^* \otimes V$$

has basis $\{e_{ij} = e_i \otimes u_j\}$ where $(e_i \otimes u_j)e_k = e_i\langle u_j, e_k\rangle = \delta_{jk}e_i$. The dual basis in $A_1 = M(n)^* = k^{n*} \otimes k^n = V \otimes V^*$ is $\{x_{ij} = e_{ij}^* = u_i \otimes e_j\}$.

Note that the tensor algebra $\otimes A_1$ is just the non-commutative polynomial ring $k\langle A_1\rangle = k\langle X\rangle = k\langle x_{ij}\rangle$, where X is the matrix of the x_{ij}'s. Viewing the x_{ij} as variables it is meaningful to multiply $X \otimes X$ by an $n^2 \times n^2$ matrix with coefficients in k. Now

$$\begin{aligned}
&\delta_c^{A\otimes A}(e_{ij} \otimes e_{kl})(x_{mp} \otimes x_{rs}) \\
&= \sum_{\lambda,\mu}[x_{m\lambda} \otimes x_{r\mu}\langle e_{ij} \otimes e_{kl}, x_{\lambda p} \otimes x_{\mu s}\rangle - \langle e_{ij} \otimes e_{kl}, x_{m\lambda} \otimes x_{r\mu}\rangle x_{\lambda p} \otimes x_{\mu s}] \\
&= \delta_{jp}\delta_{ls}x_{mi} \otimes x_{rk} - \delta_{mi}\delta_{rk}x_{jp} \otimes x_{ls}
\end{aligned}$$

(Recall that $x_{ij} = e_{ij}^*$.) Extending $\partial/\partial x_{ij} = \partial_{ij}$, to a differential operator on $k\langle X\rangle$, this biderivation is

$$\delta_c^{A\otimes A}(e_{ij} \otimes e_{kl}) = \sum_{m,r} x_{mi}\partial_{mj} \otimes x_{rk}\partial_{rl} - \sum_{p,s} x_{jp}\partial_{ip} \otimes x_{ls}\partial_{ks}.$$

As an operator on the Kronecker product matrix $X \otimes X$, the left summand replaces every element in the (j, l) column by the corresponding entry in the (i, k) column, all other columns being replaced by 0. This is just $(X \otimes X)(e_{ij} \otimes e_{kl})$. Similarly, the right sum is just $(e_{ij} \otimes e_{kl})(X \otimes X)$. Therefore, viewing $\gamma \in \bigwedge^2 M(n) \subset \otimes^2 M(n)$ as an $n^2 \times n^2$ matrix with constant entries, we have

$$(\delta_c^{A\otimes A}\gamma)(X \otimes X) = (X \otimes X)\gamma - \gamma(X \otimes X).$$

Since $\exp(t\gamma) = Q(t) = Q$, we have

$$T(X \otimes X) = \exp(t \cdot \delta_c^{A\otimes A}\gamma)(X \otimes X) = Q^{-1}(X \otimes X)Q.$$

We now wish to compute $T^{-1}(12)T$ where $(12) : A_1 \otimes A_1 \to A_1 \otimes A_1$ is given by $(12)(x_{ij} \otimes x_{kl}) = x_{kl} \otimes x_{ij}$. This requires two observations, each easily verified. First, if S is the linear transformation defined by $S(X \otimes X) = P^{-1}(X \otimes X)P$ for some invertible scalar matrix P, then the composite TS is given by

$$TS(X \otimes X) = P^{-1}Q^{-1}(X \otimes X)QP.$$

(Note the reversal of order!) Second, $(12)(X \otimes X) = E(X \otimes X)E$ where $E = \sum e_{ij} \otimes e_{ji}$. Note that $E = E^{-1}$ and that, since $\gamma \in \bigwedge^2 M(n)$ we have $E\gamma E = (12)\gamma = -\gamma$, which implies that $EQE = Q^{-1}$. Thus

$$\begin{aligned}(T^{-1}(12)T)(X \otimes X) &= Q^{-1}EQ(X \otimes X)Q^{-1}EQ \\ &= Q^{-1}EQEE(X \otimes X)EEQ^{-1}EQ \\ &= Q^{-2}[(12)(X \otimes X)]Q^2\end{aligned}$$

Now $V^{\otimes 2}$ has basis $\{u_i \otimes u_j\}$ and $V^{*\otimes 2}$ has dual basis $\{u_i^* \otimes u_j^* = e_i \otimes e_j\}$. Moreover, $A_1^{\otimes 2} = V^{\otimes 2} \otimes V^{*\otimes 2}$. The entry $x_{ij} \otimes x_{kl}$ in the $((i,k),(j,l))$ place of $X \otimes X$ may be viewed as $(u_i \otimes u_k) \otimes (u_j^* \otimes u_l^*) \in V^{\otimes 2} \otimes V^{*\otimes 2}$. Denoting left and right multiplication by Q by L_Q and R_Q respectively, we have

$$(12)_t = (12)_t^L \otimes (12)_t^R = (L_Q^{-2} \otimes R_Q^2)((12)^L \otimes (12)^R).$$

But

$$((12)^L \otimes (12)^R)((u_i \otimes u_k) \otimes (u_j^* \otimes u_l^*)) = (u_k \otimes u_i) \otimes (u_l^* \otimes u_j^*) = (x_{kl} \otimes x_{ij}) = (12)x_{ij} \otimes x_{kl}.$$

Therefore $T^{-1}(12)T = (12)_t$ and we have a commutative diagram

$$\begin{array}{ccccc} A_1 \otimes A_1 & \xrightarrow{T} & A_1 \otimes A_1 & \xrightarrow{\mu} & S^2 A_1 \\ {\scriptstyle (12)_t}\downarrow & & \downarrow{\scriptstyle (12)} & & \downarrow{\scriptstyle \mathrm{id}} \\ A_1 \otimes A_1 & \xrightarrow{T} & A_1 \otimes A_1 & \xrightarrow{\mu} & S^2 A_1 \end{array}$$

in which the right square commutes since A is commutative.

9. The Quantized Enveloping Algebra

To prove that the infinitesimal $\gamma_q = \sum_{i<j} e_{ij} \wedge e_{ji}$ can be "integrated" to a genuine deformation we shall need a quantization of the universal enveloping algebra $U\mathfrak{sl}(n)$ of $\mathfrak{sl}(n)$. Note that $\mathfrak{sl}(n)$, and hence $U\mathfrak{sl}(n)$ operates on every $V^{\otimes m}$. Since $\dim V = n$, the commutant of this action is the group algebra kS_m if $m \leq n$ or a quotient of it if $m > n$. In the latter case, if $V^{\otimes m}$ is decomposed as an S_m-module the simple summands which appear are those corresponding to partitions of m into n or fewer parts. For each n we have associated to $\gamma_q = \sum_{i<j} e_{ij} \wedge e_{ji} \in \bigwedge^2 M(n)$ the quantum symmetric group $S_m(\gamma_q)_L$.

To define the related quantization of $U\mathfrak{sl}(n)$ we begin with $n = 2$. A *semiderivation* (sometimes called a skew derivation or σ-derivation) of an algebra B belonging to an automorphism σ of B is a linear map $D : B \to B$ such that $D(ab) = Da\cdot b+\sigma a\cdot Db$ for all $a, b \in B$. Alternatively, view B as a B-bimodule ${}_\sigma B_1$ with actions $b\cdot x\cdot b' = \sigma(b)xb'$; then a σ-derivation is simply an ordinary derivation $B \to {}_\sigma B_1$. The relationship between σ-derivations and quantization of $U\mathfrak{sl}(n)$ was first described by Montgomery and Smith [26]. Much of this section adapts and extends their work to our present setting. However, the crucial fact that the quantized $U\mathfrak{sl}(n)$ is the commutant of the quantum symmetric group is entirely new. (Jimbo [22] has proved that his quantized enveloping algebra is the commutant of the Hecke algebra.)

For $n = 2$ we denote the basis of V by u, v. An elementary calculation shows that for $\gamma = \gamma_q = e_{12} \wedge e_{21}$ we have

$$(12)_t(u\otimes u, u\otimes v, v\otimes u, v\otimes v) = (u\otimes u, -\sin t\, u\otimes v+\cos t\, v\otimes u, \sin t\, v\otimes u+\cos t\, u\otimes v, v\otimes v).$$

Therefore $u \otimes v + (-\sin t\, u \otimes v + \cos t\, v \otimes u) = (1 - \sin t)u \otimes v + \cos t\, v \otimes u$ is quantum symmetric, i.e. invariant under $(12)_t$, so $qu \otimes v + v \otimes u$ is quantum symmetric, where $q = \cos t/(1 + \sin t) = (1 - \sin t)/\cos t$.

Define $V^{\otimes m}(i,j)$ to be the subspace of $V^{\otimes m}$ consisting of all α which are bihomogeneous of total degree i in u and j in v (so $i + j = m$), and define σ on $\alpha \in V^{\otimes m}(i,j)$ by $\sigma\alpha = q^{i-j}\alpha$. Define X_q, Y_q to be semiderivations of the tensor algebra $\bigotimes V$ ($= k\langle V\rangle = k\langle x,y\rangle$, the non-commutative polynomial algebra) belonging to this σ with $X_q u = 0$, $X_q v = u$ and $Y_q u = v$, $Y_q v = 0$. Also $H_q = qX_qY_q - q^{-1}Y_qX_q$ is a semiderivation belonging to σ^2.

Theorem 9.1. *For every m, X_q and Y_q commute with the operation of $S_m(\gamma_q)_L$ on $V^{\otimes m}$.*

Proof. It is sufficient to show that they commute with all $(i,i+1)_t$, and in fact it is easy to see that it is sufficient to consider the case $m = 2$ and $(12)_t$. Now $X_q(12)_t v \otimes v = X_q v \otimes v = u \otimes v + q^{-1} v \otimes u$ which is symmetric, so $X_q(12)_t u \otimes v = X_q(-\sin tu \otimes v + \cos tv \otimes u) = (-q \sin t + \cos t)u \otimes u$, while $(12)_t X_q u \otimes v = (12)_t qu \otimes u = qu \otimes u$. But $-q\sin t + \cos t = q$. Similar arguments hold for $v \otimes u$ and also for Y_q. ∎

The algebra of operators on $k\langle V\rangle$ generated by H_q, X_q, Y_q commutes with the operations of $S_m(\gamma_q)_L$ on each $V^{\otimes m}$, but, as observed by Montgomery and Smith, to introduce the correct comultiplication we must enlarge it. First we need

Lemma 9.2. *Let V be any vector space $\mu : k\langle V\rangle \otimes k\langle V\rangle \to k\langle V\rangle$ be the multiplication in the free (= tensor) algebra $k\langle V\rangle$, $\mathfrak{C} \subset \operatorname{End} k\langle V\rangle$ be a subspace of the linear endomorphisms, and $\Delta : \mathfrak{C} \to \mathfrak{C}\otimes\mathfrak{C}$ be a linear map such that $\mu\Delta f(a\otimes b) = f(ab)$ for all $f\in\mathfrak{C}$, $a, b \in k\langle V\rangle$. Then Δ is coassociative, i.e. $\mathfrak{C}$ is a coalgebra.*

Proof. Let $\mu_3 : k\langle V\rangle \otimes k\langle V\rangle \otimes k\langle V\rangle \to k\langle V\rangle$ be the (iterated) multiplication map, write $\Delta f = \sum f_{(1)} \otimes f_{(2)}$ and $(\Delta\otimes 1)\Delta f = \sum f_{(11)} \otimes f_{(12)} \otimes f_{(2)}$, and similarly for $(1\otimes\Delta)\Delta f$. Then $\mu_3(\Delta\otimes 1)\Delta f(a\otimes b\otimes c) = \sum f_{(11)}a \cdot f_{(12)}b \cdot f_{(2)}c$. But $f_{(1)} \in \mathfrak{C}$, so this is $\sum f_{(1)}(ab)f_{(2)}(c) = f(abc)$. But $\mu_3(1\otimes\Delta)\Delta f(a\otimes b\otimes c)$ is the same, so $((\Delta\otimes 1)\Delta - (1\otimes\Delta)\Delta)f(a\otimes b\otimes c)$ is in the kernel of μ_3, which is 0, so $(\Delta\otimes 1)\Delta f = (1\otimes\Delta)\Delta f$. ■

Since H_q is a semi derivation for σ^2 and $H_q u = qu$ we have $H_q u^{\otimes r} = q(1+q^2+\ldots+q^{2r-2})u^{\otimes r}$.

The *q-binomial* coefficients are defined by

$$\binom{r}{i}_q = \frac{(1-q^r)(1-q^{r-1})\cdots(1-q^{r-i+1})}{(1-q^i)(1-q^{i-1})\cdots(1-q)}.$$

for $r \neq 0$, and $\binom{0}{i}_q = 0$. We write also $(r)_q$ for $\binom{r}{1}_q = (1-q^r)/(1-q)$. Note that $(-r)_q = -q^{-1}\cdot(r)_{q^{-1}}$. With this, $H_q u^{\otimes r} = q\cdot(r)_{q^2}u^{\otimes r}$ and $H_q v^{\otimes r} = -q^{-1}\cdot(r)_{q^{-2}}v^{\otimes r} = q\cdot(-r)_{q^2}v^{\otimes r}$. It follows that $H_q u\otimes v = 0$ and by repeated application of these rules, that if $\alpha \in V^{\otimes m}(i,j)$ then $H_q\alpha = q\cdot(i-j)_{q^2}\alpha$.

We can deduce the following commutation relations by showing that both sides are semiderivations for the same automorphism, and that they agree on u and v.

$$qX_qY_q - q^{-1}Y_qX_q = H_q \quad \text{(definition)}$$
$$q^{-2}H_qX_q - q^2X_qH_q = (q+q^{-1})X_q$$
$$q^2H_qY_q - q^{-2}Y_qH_q = -(q+q^{-1})Y_q$$

We call the algebra generated by X_q and Y_q, subject to these relations the *Woronowicz quantization* $U_q^W\mathfrak{sl}(2)$ since it is isomorphic to one introduced by him ([35], p. 150) as part of a non-commutative calculus. It is this algebra which will be used in section 11 to deform $\mathbf{M}(n)$. As noted in [26], it is this quantization whose representation theory is most like that of $U\mathfrak{sl}(2)$.

As mentioned, the Woronowicz quantization is not a bialgebra. However, it does have a "natural" enlargement to a bialgebra. Although we shall not need this in subsequent sections, we pause here to consider it. Generally speaking, we expect an element of a bialgebra which acts as a σ-derivation on a module algebra to be a σ-primitive [26]. Thus, denoting by σ, as before, the automorphism sending $\alpha \in V^{\otimes m}(i,j)$ to $q^{i-j}\alpha$ we should like to define ΔX_q to be $X_q\otimes 1+\sigma\otimes X_q$, ΔY_q to be $Y_q\otimes 1+\sigma\otimes Y_q$, and ΔH_q to be $H_q\otimes 1+\sigma^2\otimes H_q$. Unfortunately, the automorphism σ is not in the algebra generated by X_q and Y_q. On the other hand, $H_q\alpha = q(i-j)_{q^2}\alpha$ and $(q^{-1}-q)q\cdot(r)_{q^2} = 1-q^{2r}$, so $(1-(q-q^{-1})H_q)^{\frac{1}{2}}$ formally behaves like σ. In fact viewed as a power series in $(q^{-1}-q)H_q$ it is a well-defined operator on the sum of all $V^{\otimes m}(i,j)$ for which $|\, 1-q^{2(i-j)} \,| < 1$. We therefore introduce a new element K_q with $K_q^2 = 1-(q^{-1}-q)H_q$, i.e., K_q represents σ. For $\alpha \in V^{\otimes m}(i,j)$, we have $X_q\alpha \in V^{\otimes m}(i+1, j-1)$ so

$$\sigma X_q\alpha = q^{i-j+2}X_q\alpha, \quad \text{i.e. } \sigma X_q\alpha = q^2X_q\sigma\alpha \text{ so}$$

$$\sigma X_q = q^2 X_q \sigma, \quad \text{and similarly}$$
$$\sigma Y_q = q^{-2} Y_q \sigma.$$

We may now define the quantized universal enveloping (bi)algebra $U_q\mathfrak{sl}(2)$: As an algebra it is $k\langle X_q, Y_q, K_q\rangle$ modulo the commutation rules

$$K_q X_q = q^2 X_q K_q, \qquad K_q Y_q = q^{-2} Y_q K_q \quad \text{and}$$
$$q X_q Y_q - q^{-1} Y_q X_q = (q^{-1} - q)^{-1}(1 - K_q^2)\ (\text{ i.e.} = H_q).$$

Since σ^2 is already represented by the element $1 - (q^{-1} - q)H_q$ we have $\Delta H_q = H_q \otimes 1 + [1 - (q^{-1} - q)H_q] \otimes H_q$ from which one finds that $1 - (q^{-1} - q)H_q$ is group-like. Define the comultiplication in $U_q\mathfrak{sl}(2)$ by

$$\Delta X_q = X_q \otimes 1 + K_q \otimes X_q, \quad \Delta Y_q = Y_q \otimes 1 + K_q \otimes Y_q, \quad \Delta K_q = K_q \otimes K_q.$$

Notice that $U_q\mathfrak{sl}(2)$ is still a subalgebra of $\mathrm{End}\, k\langle V\rangle$ and $q^{-1} - q = 2\tan t$, so $K_q = [1 - (q^{-1} - q)H_q]^{\frac{1}{2}} = [1 - 2\tan t \cdot H_q]^{\frac{1}{2}}$ can be written as a formal power series in t with coefficients in $U\mathfrak{sl}(2)$. If we do this and view t as a variable, then we can view $U_q\mathfrak{sl}(2)$, which is now defined over $k[\![t]\!]$, as constructed on the same underlying vector space as $U\mathfrak{sl}(2)[\![t]\!]$. (Note, however, that we have not exhibited it as a deformation.)

For $n > 2$ define $X_{iq}, Y_{iq}, H_{iq}, K_{iq}$ acting as before on u_i, u_{i+1} exclusively, treating all other u_j as constants. (If $\alpha \in V^{\otimes m}(m_1, \ldots, m_n)$ then $\sigma_i \alpha = q^{m_i - m_{i+1}}\alpha$ and K_i represents σ_i.) All H_{iq} commute and H_{iq} commutes with X_{iq} for $| i - j | > 1$. We have further

$$q H_{jq} X_{j+1,q} - q^{-1} X_{j+1,q} H_{jq} = -X_{j+1,q}$$
$$q^{-1} H_{jq} Y_{j+1,q} - q Y_{j+1,q} H_{jq} = Y_{j+1,q}$$

which may be checked by showing that both sides are semiderivations for the same σ and agree on the generators. Similarly,

$$X_{iq} Y_{i+1,q} = q Y_{i+1,q} X_{iq}, \quad \text{and}$$
$$Y_{iq} X_{i+1,q} = q^{-1} X_{i+1,q} Y_{iq}.$$

The algebra generated by $\{X_{iq}, Y_{iq}\}$ with these relations is the Woronowicz quantization $U_q^W\mathfrak{sl}(n)$. The larger bialgebra, $U_q\mathfrak{sl}(n)$ generated by $\{X_{iq}, Y_{iq}, K_{iq}\}$ constructed here is essentially that introduced by Drinfel'd [7] and Jimbo [22]. To see this return to the case $n = 2$ and define τ to be $\sigma^{\frac{1}{2}}$, so if $\alpha \in V^{\otimes m}(i,j)$, then $\tau\alpha = q^{(1/2)(i-j)}$. Clearly τ commutes with $(12)_t$, hence so does $X'_q = \tau^{-1} X_q$. For this we have $X'_q(\alpha \otimes \beta) = X'_q\alpha \otimes \tau^{-1}\beta + \tau\alpha \otimes X'_q\beta$, so

$$\Delta X'_q = X'_q \otimes K^{-\frac{1}{2}} + K^{\frac{1}{2}} \otimes X'_q.$$

Defining Y'_q similarly, we obtain the Drinfel'd-Jimbo quantization.

10. Conditions for quantization

With our previous notation we have

$$\Delta x_{ij} = \Delta(u_i \otimes e_j) = \Delta(u_i \otimes u_j^*) = \sum_k (u_i \otimes u_k) \otimes (u_k^* \otimes u_j^*).$$

Denote the basis element $u_{i_1} \otimes u_{i_2} \otimes \cdots \otimes u_{i_m}$ of $V^{\otimes m}$ by u_I. On $(M(n)^*)^{\otimes m} = V^{\otimes m} \otimes (V^*)^{\otimes m}$ we then have

$$\Delta(u_I \otimes u_J^*) = \sum_K (u_I \otimes u_K) \otimes (u_K^* \otimes u_J^*).$$

Now suppose that W is any vector space with basis $w_1, \ldots, w_p$, that $w_1^*, \ldots, w_p^*$ is the dual basis, and that $T \in GL(W)$. Then $(T \otimes T^{-*} \sum w_i \otimes w_i^* = \sum w_i \otimes w_i^*$. For notice that $\sum w_i \otimes w_i^*$ is the identity map of W; it does not depend on the choice of basis, and that the dual basis to $\{Tw_i\}$ is $\{(T^{-*})w_i^*\}$.

Theorem 10.1. *If $T_L \in GL(V^{\otimes m}), T_R = (T_L^{-*})$, and*

$$T = T_L \otimes T_R : (M(n)^*)^{\otimes m} \to (M(n)^*)^{\otimes m},$$

then

$$\Delta T = (T \otimes T)\Delta,$$

i.e., T is a coalgebra morphism.

Proof. Applying the left side to $u_I \otimes u_J^*$ gives

$$\Delta(T_L u_I \otimes T_R u_J^*) = \sum_K (T_L u_I) \otimes u_K \otimes u_K^* \otimes (T_R u_J^*)$$

$$= T_L u_I \otimes (\sum_K u_K \otimes u_K^*) \otimes T_R u_J^* = T_L u_I \otimes (\sum_K T_L u_K \otimes T_R u_K^*) \otimes T_R u_J^*,$$

with the right side. ∎

If $(-|-)$ is any non-degenerate symmetric bilinear form on V, then every element of S_m acts orthogonally on $V^{\otimes m}$, and the form is non-degenerate on $\mathrm{sym}(V^{\otimes m})$. If $W \subset V^{\otimes m}$ then we set $W^\perp = \{\alpha \in V^{\otimes m} | (\alpha|W) = 0\}$. Choose $\gamma \in \bigwedge^2 M(n)$, thereby defining $S_m(\gamma)_L$ for every m, let $\mathrm{sk}_q(k\langle V\rangle)$ be the ideal of $k\langle V\rangle$ generated by all $u \otimes u' - (12)_t u \otimes u' \in V^{\otimes 2}$, and set $\mathrm{sk}_q(V^{\otimes m}) = \mathrm{sk}_q(k\langle V\rangle) \cap V^{\otimes m}$. We call the elements of $\mathrm{sk}_q(k\langle V\rangle)$ *quantum skew.* Clearly $\mathrm{sk}_q(k\langle V\rangle)$ is homogeneous and is the direct sum of all $\mathrm{sk}_q(V^{\otimes m})$.

Lemma 10.2. *For all m we have* $\dim(\mathit{sym}_q(V^{\otimes m})) \le \dim(\mathit{sym}(V^{\otimes m}))$ *and* $\dim(\mathit{sk}_q(V^{\otimes m})) \ge \dim(\mathit{sk}(V^{\otimes m}))$.

Proof. For the first part, observe that we can always choose a basis for $\mathrm{sym}_q(V^{\otimes m})$, the elements of which are power series in t, such that the constant terms, which are in $\mathrm{sym}(V^{\otimes m})$ are linearly independent. For the second, if we have independent elements of $\mathrm{sk}(V^{\otimes m})$, then replacing (12) in each by $(12)_t$ gives independent elements of $\mathrm{sk}_q(V^{\otimes m})$.

Definition 10.3. A $\gamma \in \bigwedge^2 M(n)$ will be called *quantizing* if

1. there is a non-degenerate symmetric bilinear form $(-|-)$ with respect to which it is skew, and
2. for every m there is a direct sum decomposition $V^{\otimes m} = \bigoplus M_i$ of $V^{\otimes m}$ into simple S_m modules and a mapping $T_L^m = T_L^m(t) \in Gl(V^{\otimes m})$ which is analytic in t with $T_L^m(0) = $ id such that (for all small or generic t), $(T_L^m)^{-1}M_i$ is a $S_m(\gamma)_L$ module (necessarily simple) and $(T_L^m)^{-1}M_i \cong (T_L^m)^{-1}M_j$ if and only if $M_i \cong M_j$.

If γ is quantizing, then *a fortiori* $\dim(\mathrm{sym}_q(V^{\otimes m})) = \dim(\mathrm{sym}(V^{\otimes m}))$, since the trivial module appears as often in the $S_m(\gamma)_L$-decomposition of $V^{\otimes m}$ as in the S_m-decomposition. The isomorphism classes of simple S_m-submodules of $V^{\otimes m}$ correspond to partitions $\mathfrak{p} = (m_1, \dots, m_n)$ of m into n (nonnegative) parts some of which may be zero. A partition $\mathfrak{p}$ is a *nonincreasing* if $m_1 \geq \cdots \geq m_n \geq 0$. If the order of the parts is important then we will call $\mathfrak{p}$ an *ordered* partition. Let $M_\mathfrak{p}$ denote any simple submodule of $V^{\otimes m}$ in the class $\mathfrak{p}$, set $qM_\mathfrak{p} = (T_L^m)^{-1}M_\mathfrak{p}$, and let $V_\mathfrak{p}^{\otimes m}$ denote the sum of all S_m submodules of $V^{\otimes m}$ isomorphic to $M_\mathfrak{p}$. Then $V^{\otimes m} = \bigoplus V_\mathfrak{p}^{\otimes m}$; this is the "primary decomposition" of $V^{\otimes m}$. It is unique, and the direct summands $V_\mathfrak{p}^{\otimes m}$ are mutually orthogonal with respect to the form $(-|-)$. (It makes no difference what the form is, as long as it is symmetric and non-degenerate, in which case its restriction to $V_\mathfrak{p}$ is also necessarily non-degenerate.) Letting $qV_\mathfrak{p}^{\otimes m}$ be the direct sum of all $S_m(\gamma)_L$ submodules of $V^{\otimes m}$ isomorphic to $qM_\mathfrak{p}$ we have $qV_\mathfrak{p}^m = (T_L^m)^{-1}V_\mathfrak{p}^m$ and $V^{\otimes m} = \bigoplus qV_\mathfrak{p}^{\otimes m}$, the summands again being mutually orthogonal. If γ is skew relative to the form, then $\mathrm{sk}_q(V^{\otimes m}) \subset (\mathrm{sym}_q(V^{\otimes m}))^\perp$. For any element of $\mathrm{sk}_q(V^{\otimes m})$ is a sum of elements of the form $\alpha = \alpha_1(u \otimes u' - (12)_t u \otimes u')\alpha_2$ with $\alpha_1 \in V^{\otimes i-1}$ and $\alpha_2 \in V^{\otimes m-i-1}$ for some i. If $\beta \in \mathrm{sym}_q(V^{\otimes m})$ then

$$-(\alpha|\beta) = ((i, i+1)_t\alpha|\beta) = (\alpha|(i, i+1)_t\beta) = (\alpha|\beta)$$

so $(\alpha|\beta) = 0$. Using the preceding lemma, we therefore have

Lemma 10.4. *If γ is quantizing then $V^{\otimes m} = sk_q(V^{\otimes m}) \bigoplus sym_q(V^{\otimes m})$ for all m and $k\langle V\rangle = sk_q(k\langle V\rangle) \bigoplus sym_q(k\langle V\rangle)$, the sums being orthogonal. Also, T_L^m carries $sym_q(k\langle V\rangle)$ into $sym(k\langle V\rangle)$ and $sk_q(k\langle V\rangle)$ into $sk(k\langle V\rangle)$.* ∎

Denote by $T_L : k\langle V\rangle \to k\langle V\rangle$ the map assembled from all the T_L^m, and introduce a new product, $\circledast_L$, on $k\langle V\rangle$ by $\alpha \circledast_L \beta = T_L(T_L^{-1}\alpha \otimes T_L^{-1}\beta)$. Then T_L is an isomorphism $(k\langle V\rangle, \otimes) \to (k\langle V\rangle, \circledast_L)$. Since $\mathrm{sk}_q(V)$ was by construction an ideal of $(k\langle V\rangle, \otimes)$, it follows that $\mathrm{sk}(k\langle V\rangle)$ is an ideal of $(k\langle V\rangle, \circledast_L)$. As a vector space, $(k\langle V\rangle, \circledast_L)/\mathrm{sk}(k\langle V\rangle)$ is still just $\mathrm{sym}(k\langle V\rangle)$, which we have identified with SV or $k[V]$, but it now has a new multiplication which we will denote by $*_L$. We denote by $(S_qV, \times_L)$ the algebra $(k\langle V\rangle, \otimes)/\mathrm{sk}_q(k\langle V\rangle)$, where S_qV may be identified, as a vector space, with $\mathrm{sym}_q(k\langle V\rangle)$ and $\times_L$ is the multiplication induced by $\otimes$. We then have

Theorem 10.5. *There is a commutative diagram*

$$\begin{array}{ccccccccc}
0 & \longrightarrow & sk_q(k\langle V\rangle) & \longrightarrow & (k\langle V\rangle, \otimes) & \longrightarrow & (S_qV, \times_L) & \longrightarrow & 0 \\
 & & \downarrow{\scriptstyle T_L} & & \downarrow{\scriptstyle T_L} & & \downarrow{\scriptstyle \overline{T}_L} & & \\
0 & \longrightarrow & sk(k\langle V\rangle) & \longrightarrow & (k\langle V\rangle, \circledast_L) & \xrightarrow{\mu} & (k[V], *_L) & \longrightarrow & 0
\end{array}$$

where $\overline{T}_L$ is the algebra isomorphism induced by T_L. ■

The algebra $(k\langle V\rangle, *_L)$ is a deformation of $k[V]$: It clearly has the same underlying vector space, the multiplication $*_L$ is analytic in t (since $\circledast_L$ is), and it is the original commutative multiplication at $t = 0$. Of course, $(k\langle V\rangle, \circledast_L)$ is a deformation of $(k\langle V\rangle, \otimes)$, but it is a trivial deformation. (The map T_L exhibits its triviality.) The deformation is non-trivial, however, on the quotient algebra. To pass from this to a preferred deformation of the bialgebra $\mathbf{M}(n)$ we need some elementary lemmas. If M is a module over an arbitrary group G then so is M^*, where $g \in G$ operates as $(g^*)^{-1} = (g^{-1})^*$, which, as earlier, we write simply as g^{-*}.

Lemma 10.6. *If M is a finite-dimensional module over a group G, then $M \otimes M^*$ contains the trivial module.*

Proof. Choose any basis $\alpha_1, \ldots, \alpha_r$ of M. Then $\sum \alpha_i \otimes \alpha_i^*$ (which does not depend on the choice of basis) is invariant under G. ■

When $k = \mathbb{R}$, G is finite, and M is finite dimensional, there is always an invariant non-degenerate symmetric bilinear form on M, so $M \cong M^*$ and $M \otimes M$ contains the trivial module. For $G = S_m$ we have more precisely

Lemma 10.7. *Suppose that M and M' are simple modules over S_m. Then $M \otimes M'$ contains the trivial module if and only if $M \cong M'$, in which case it contains it exactly once.*

Proof. Let χ and χ' be the respective characters of M and M'. Then the character of $M \otimes M'$ is $\chi\chi'$ so the number of times that the trivial module appears in $M \otimes M'$ is

$$\frac{1}{m!} \sum_{\sigma \in S_m} \chi(\sigma)\chi'(\sigma).$$

The simple representations of S_m are rational, hence in particular real, so this is identical with

$$\frac{1}{m!} \sum_{\sigma} \chi(\sigma)\chi'(\sigma^{-1}),$$

which is 0 or 1 according as $\chi \neq \chi'$ or $\chi = \chi'$. ■

Suppose that γ is quantizing. As usual, let $T_R^m : V^{*\otimes m} \to V^{*\otimes m}$ be $(T_L^m)^{-*}$, define

$$T^m : V^{\otimes m} \otimes V^{*\otimes m} \longrightarrow V^{\otimes m} \otimes V^{*\otimes m}$$

to be $T_L^m \otimes T_R^m$. Let $T_R : k\langle V^*\rangle \to k\langle V^*\rangle$ be the map assembled from all the T_R^m, and let

$$T : k\langle M(n)^*\rangle \to k\langle M(n)^*\rangle$$

be the map assembled from all the T^m. (Recall that $M(n)^* = k\langle V\rangle \otimes k\langle V^*\rangle$). Notice that $V \otimes V^*$ has a natural non-degenerate symmetric bilinear form, hence so do all the $V^{\otimes m} \otimes V^{*\otimes m}$, and all $(i, i+1)_t = (i, i+1)_t^L \otimes ((i, i+1)_t^L)^{-*}$ are orthogonal, as is $T^m = T_L^m \otimes (T_L^m)^{-*}$. Each T^m is a coalgebra isomorphism by the first theorem of this section, hence so is T. Defining $\circledast_R$ on $k\langle V^*\rangle$ by $\alpha^* \circledast_R \beta^* = T_R(T_R^{-1}\alpha^* \circledast T_{R^{-1}}\beta^*)$, we have a new multiplication $\circledast = \circledast_L \otimes \circledast_R$ on $k\langle M(n)^*\rangle\rangle$ and a bialgebra morphism

$$T : (k\langle M(n)^*\rangle, \otimes) \to T(k\langle M(n)^*\rangle, \circledast).$$

Define the symmetric and skew (both ordinary and quantum) elements precisely as before. Then we shall see that $T^m(\mathrm{sym}_q(V^{\otimes m} \otimes V^{*\otimes m})) = \mathrm{sym}(V^{\otimes m} \otimes V^{*\otimes m})$. First, we must count the elements on the right. Let $M_\mathfrak{p}$ again be a module corresponding to the partition $\mathfrak{p}$ and let $n_\mathfrak{p}$ be the multiplicity of $M_\mathfrak{p}$ in $V^{\otimes m}$. The number of times that $M_\mathfrak{p} \otimes M_\mathfrak{p}^*(\cong M_\mathfrak{p} \otimes M_\mathfrak{p})$ appears in $V^{\otimes m} \otimes V^{*\otimes m}$ is then $n_\mathfrak{p}^2$, so by the preceding lemma,

$$\dim(\mathrm{sym}(V^{\otimes m} \otimes V^{*\otimes m})) = \sum_\mathfrak{p} n_\mathfrak{p}^2.$$

Since γ is quantizing, the right side is also the number of times that a $S_m(\gamma)$ module of the form $qM_\mathfrak{p} \otimes qM_\mathfrak{p}^*$ appears in $V^{\otimes m} \otimes V^{*\otimes m}$. The number of quantum symmetric elements is not more than the number of ordinary symmetric ones, so we have

$$\dim(\mathrm{sym}_q(V^{\otimes m} \otimes V^{*\otimes m})) = \dim(\mathrm{sym}(V^{\otimes m} \otimes V^{*\otimes m})).$$

Lemma 10.4 implies that $\mathrm{sym}(V^{\otimes m} \otimes V^{*\otimes m}) = \bigoplus_\mathfrak{p} \mathrm{sym}(V_\mathfrak{p}^{\otimes m} \otimes V_\mathfrak{p}^{*\otimes m})$. If $v_1, \ldots, v_r$ is a basis for $M_\mathfrak{p}$, then the unique (up to scalar multiple) symmetric element in $M_\mathfrak{p} \otimes M_\mathfrak{p}^*$ is $\sum v_i \otimes v_i^*$, where $v_1^*, \cdots, v_r^*$ is the dual basis of $M_\mathfrak{p}^*$. The $n_\mathfrak{p}^2$ such elements in $V_\mathfrak{p}^{\otimes m} \otimes V_\mathfrak{p}^{*\otimes m}$ are the images of the similarly generated quantum symmetric elements in $qV_\mathfrak{p}^{\otimes m} \otimes qV_\mathfrak{p}^{*\otimes m}$. From the equalities of dimensions we see that

$$T^m(\mathrm{sym}_q(V^{\otimes m} \otimes V^{*\otimes m})) = \mathrm{sym}(V^{\otimes m} \otimes V^{*\otimes m}).$$

Let $\mathrm{sk}_q(k\langle M(n)^*\rangle) = \mathrm{sk}_q((k\langle V\rangle)\otimes(k\langle V^*\rangle))$ be the ideal spanned by all $w - (12)_t w$ with $w \in V^{\otimes 2} \otimes V^{*\otimes 2}$. This is the orthogonal complement of $\mathrm{sym}_q(\langle M(n)^*\rangle) = \mathrm{sym}_q(\langle\langle V \otimes V^*\rangle\rangle)$ since $T = T_L \otimes T_R$ preserves the form induced by $(-|-)$. It is easy to check that $\mathrm{sk}(\langle M(n)^*\rangle)$ is a coideal, so we have

Theorem 10.8. *There is a commutative diagram of isomorphisms*

$$\begin{array}{ccccccccc}
0 & \longrightarrow & sk_q(\langle M(n)^*\rangle) & \longrightarrow & (k\langle M(n)^*\rangle, \otimes) & \longrightarrow & (S_q M(n)^*, \times) & \longrightarrow & 0 \\
 & & \downarrow T & & \downarrow T & & \downarrow \overline{T} & & \\
0 & \longrightarrow & sk(\langle M(n)^*\rangle) & \longrightarrow & (k\langle M(n)^*\rangle, \circledast) & \stackrel{\mu}{\longrightarrow} & (k[X], *) & \longrightarrow & 0
\end{array}$$

(Here X is the matrix of $x_{ij} = u_i \otimes e_j = u_i \otimes u_j^* \in V \otimes V^*$, i.e., $k[X] = k[V \otimes V^*] = k[M(n)^*]$). ■

Corollary 10.9. *The bialgebra $(k[X], *)$ is a preferred deformation of $\mathbf{M}(n) = k[X]$ (with the usual commutative multiplication).* ■

With this preferred deformation of $\mathbf{M}(n)$, Theorem 10.1 says that $(k[V], *_L)$ is a corresponding preferred deformation of the left algebra comodule $k[V]$. That is, if $\alpha, \alpha' \in$

$k[V]$ and $\Delta_L\alpha = \sum w_{(1)} \otimes \alpha_{(2)}, \Delta_L\alpha' = \sum w'_{(1)} \otimes \alpha'_{(2)}$ with $w_{(1)}, w'_{(1)} \in k[X]$, then $\Delta_L(\alpha *_L \alpha') = \sum(w_{(1)} * w'_{(1)}) \otimes (\alpha_{(2)} *_L \alpha'_{(2)})$, and similarly for Δ_R and $(k[V], *_R)$.

We have therefore shown how, starting with a quantizing $\gamma \in \bigwedge^2 M(n)$, to construct matching preferred deformations of $\mathbf{M}(n)$ and of $k[V]$. Note that if $w, w' \in M(n)^*$ (i.e., if they are degree 1 elements of $\mathbf{M}(n)$ then

$$w \circledast w' = (\exp(t \cdot \delta_c^{A\otimes A}\gamma))(w \otimes w'),$$

and $w * w' = \mu(w \circledast w')$. This justifies the formula given in theorem 11 of [18], which, although it does not give an associative multiplication on all of $\mathbf{M}(n)$, is correct on generators and therefore gives the correct "commutation relations".

While it is convenient to identify $k[X]$ with the symmetric elements of $k\langle M(n)^*\rangle$, that identification, which works best in characteristic 0, is not essential to the theory. If the infinitesimal γ is nilpotent, then $S_m(\gamma)_L$ will be defined for all characteristics greater than the index of nilpotence. The γ giving the Dipper-Donkin group (cf. [6]) has $\gamma^2 = 0$ and for it $S_m(\gamma)_L$ is defined over the integers. To apply these techniques to study quantum groups over a ring R we need representation theory of S_m over R and corresponding information for $S_m(\gamma)_L$.

Note finally that $\delta_c^{A\otimes A}\gamma$ depends in an essential way on the coalgebra we are considering. For example $e_{12} \wedge e_{21}$ is quantizing, as we shall prove, as an element of $\bigwedge^2 M(2)$ but is obstructed as an element of $\bigwedge^2 M(3)$ and its coboundary is different.

11. The "standard" quantization of M(n)

The "standard" $\gamma_q = \sum_{i<j} e_{ij} \wedge e_{ji}$ is skew with respect to the standard bilinear form on V. To show that it is quantizing, we use $U_q^W\mathfrak{sl}(n)$ to show that each $V^{\otimes m}$ decomposes under $S_m(\gamma_q)_L$ exactly as it does under S_m. Denote the standard basis of $V(= (k^n)^*)$ as usual by $u_1(= e_1^*), \dots, u_n(= e_n^*)$ and set $u_i\partial/\partial u_{i+1} = X_i, u_{i+1}\partial/\partial u_i = Y_i, u_i\partial/\partial u_i - u_{i+1}\partial/\partial u_{i+1} = H_i, i = 1, \dots, n-1$. These are derivations of $k\langle V\rangle$ carrying each $V^{\otimes m}$ into itself. We have already defined their quantized counterparts (which are semiderivations). Let $\mathfrak{p} = (m_1, \dots, m_n)$ be an ordered partition of m and let $V^{\otimes m}(m_1, \dots, m_n) = V^{\otimes m}(\mathfrak{p}) = V(\mathfrak{p})$ be the k-subspace of $V^{\otimes m}$ spanned by all words of total length m in $u_1, \dots, u_n$ which contain u_i a total of m_i times. Set $X_i\mathfrak{p} = (m_1, \dots, m_i + 1, m_{i+1} - 1, \dots, m_n)$, and $Y_i\mathfrak{p} = (m_1, \dots, m_i - 1, m_{i+1} + 1, \dots, m_n)$ when these are meaningful; if either is not, set it equal to 0. Then $X_i : V^{\otimes m}(\mathfrak{p}) \to V^{\otimes m}(X_i\mathfrak{p})$ and $Y_i : V^{\otimes m}(\mathfrak{p}) \to V^{\otimes m}(Y_i\mathfrak{p})$. Each $V^{\otimes m}(\mathfrak{p})$ consists of eigenvectors for every H_i. Extend the natural partial order on nonincreasing partitions to all $\mathfrak{p}$ by setting $(m_1, \dots m_n) \geq (m'_1, \dots, m'_n)$ if $m_1 + \dots + m_i \geq m'_1 + \dots + m'_i$ for $i = 1, \dots, n$. Then $X_i\mathfrak{p} \geq \mathfrak{p} \geq Y_i\mathfrak{p}$. If $\mathfrak{p} = (m_1, \dots, m_n)$, set $\|\mathfrak{p}\|^2 = m_1^2 + \dots + m_n^2$. Then $Y_i : V^{\otimes m}(\mathfrak{p}) \to V^{\otimes m}(Y_i\mathfrak{p})$ is a monomorphism if $\|Y_i\mathfrak{p}\|^2 \leq \|\mathfrak{p}\|^2$ and an epimorphism if $\|Y_i\mathfrak{p}\|^2 \geq \|\mathfrak{p}\|^2$. If M is an S_m submodule of any $V^{\otimes m}(\mathfrak{p})$ then in this section $M^\perp$ will denote its orthogonal complement *inside* $V^{\otimes m}(\mathfrak{p})$ relative to the restriction of the standard form on $V^{\otimes m}$. It is again an S_m-submodule.

As remarked before, to every nonincreasing partition $\mathfrak{p}$ of m into n parts there is associated an isomorphism class of simple S_m modules. The module $V^{\otimes m}(\mathfrak{p})$ contains a single member of this class, for which we now reserve the notation $M_\mathfrak{p}$. This $M_\mathfrak{p}$ is just

$(\sum_i Y_i V^{\otimes m}(X_i \mathfrak{p}))^{\perp}$. Note that because $\mathfrak{p}$ is nonincreasing, if $X_i\mathfrak{p}$ is meaningful then $Y_i|V^{\otimes m}(X_i\mathfrak{p}) \to V^{\otimes m}(\mathfrak{p})$ is a monomorphism and $X_iY_i|V^{\otimes m}(X_i\mathfrak{p}) \to V^{\otimes m}(X_i\mathfrak{p})$ is an isomorphism. It follows that

$$M_{\mathfrak{p}} = \bigcap_i \ker(X_i|V^{\otimes m}(\mathfrak{p})).$$

To get an explicit decomposition of every $V^{\otimes m}(\mathfrak{p}')$ (and hence of all of $V^{\otimes m}$) as a direct sum of simple modules, observe that a module isomorphic to $M_{\mathfrak{p}}$ appears in $V^{\otimes m}(\mathfrak{p}')$ only if $\mathfrak{p} \geq \mathfrak{p}'$. (Here $\mathfrak{p}'$ need not be nonincreasing, and $M_{\mathfrak{p}}$ need not appear at all.) We can choose a set $\mathcal{Z} = \mathcal{Z}(\mathfrak{p}, \mathfrak{p}')$ of words in $X_1, \ldots, X_{n-1}$ such that $Z\mathfrak{p} = \mathfrak{p}'$ for all $Z \in \mathcal{Z}$ and such that the $M_{\mathfrak{p}}$-primary component of $V^{\otimes m}(\mathfrak{p}')$ (i.e., the sum of all submodules isomorphic to $M_{\mathfrak{p}}$) is $\bigoplus_{Z\in\mathcal{Z}} ZM_{\mathfrak{p}}$. Since $V^{\otimes m}(\mathfrak{p}')$ is the direct sum of its primary components, doing this with fixed $\mathfrak{p}'$ and all nonincreasing $\mathfrak{p}$ gives a decomposition. The foregoing can be "quantized" if q is generic, i.e., transcendental over $\mathbb{Q}$. (In fact, it is sufficient that it not be a root of unity.) If Z is a word in $X_1, \ldots, X_{n-1}$, let Z_q denote the word obtained by replacing every X_i by X_{iq}. Set $qM_{\mathfrak{p}} = (\sum_i Y_{iq} V^{\otimes m}(X_i\mathfrak{p}))^{\perp}$. *A priori* this may be 0. Clearly $\dim(qM_{\mathfrak{p}}) \leq \dim(M_{\mathfrak{p}})$ since $\dim(\sum_i Y_{iq}V^{\otimes m}(X_i\mathfrak{p})) \geq \dim(\sum_i Y_iV^{\otimes m}(X_i\mathfrak{p}))$. As $q \to 1$ all Z_q tend to the corresponding Z and $qM_{\mathfrak{p}}$ therefore tends to a submodule of $M_{\mathfrak{p}}$. Since $M_{\mathfrak{p}}$ is simple, it follows that if $qM_{\mathfrak{p}} \neq 0$ then it is simple (else there would be too many quantum symmetric elements) and $\dim(qM_{\mathfrak{p}}) = \dim(M_{\mathfrak{p}})$. Therefore, to show that $qM_{\mathfrak{p}}$ is simple and has the same dimension as $M_{\mathfrak{p}}$, it is sufficient to exhibit a single $\beta \neq 0$ in $V^{\otimes m}(\mathfrak{p})$ with $X_{iq}\beta = 0$ for all i. We do this by induction on m; it is trivial for $m = 1$. If $\mathfrak{p} = (m_1, \ldots, m_n)$, choose the least j such that $\mathfrak{p}_0 = (m_1, \ldots, m_j - 1, \ldots, m_n)$ is still nonincreasing. Then $\mathfrak{p}_0$ has the form $(r, \ldots, r, m_j, \ldots, m_n)$ where $r > m_j$. By hypothesis there is an $\alpha \in V^{\otimes m-1}(\mathfrak{p}_0)$ with $X_{iq}\alpha = 0$, all i. It follows by direct calculation that if $r < j$ then $X_{rq}Y_{rq}Y_{r+1,q}\cdots Y_{j-1,q}\alpha = Y_{r+1,q}\cdots Y_{j-1,q}\alpha$ but $X_{sq}Y_{rq}Y_{r+1,q}\cdots Y_{j-1,q}\alpha = 0$ for $s > r$. Therefore, we may take

$$\beta = (u_j - q^{-1}u_{j-1}Y_{j-1,q} + q^{-2}u_{j-2}Y_{j-2,q}Y_{j-1,q} - q^{-3}u_{j-3}Y_{j-3,q}Y_{j-2,q}Y_{j-1,q} + \ldots)\alpha.$$

With this we therefore have

Theorem 11.1. *The "standard" $\gamma_q = \sum_{i<j} e_{ij} \wedge e_{ji}$ is quantizing.* ■

To understand the procedure it may be useful to contrast $\gamma = e_{12} \wedge e_{21}$ viewed as an infinitesimal of $\mathbf{M}(2)$, where it gives the classical deformation to $\mathbf{M}_q(2)$ with the same γ as an infinitesimal of $\mathbf{M}(3)$, where it is obstructed. Let $\dim V = 3$ with basis $u_1 = u, u_2 = v, u_3 = w$. If α is any word in u, v, w, let $\operatorname{inv}\alpha$ (the "inversions" of α) denote the number of times that u appears before v in α, without regard to any intervening w' s; e.g. $\operatorname{inv}(u\otimes u\otimes v\otimes w\otimes v) = 4$. There is a unique quantum symmetric element in each $V^{\otimes m}(m_1, m_2, m_3)$, namely $\sum q^{\operatorname{inv}\alpha}\alpha$ where α runs through all words in u, v, w of total degree m_1 in u, m_2 in v, and m_3 in w. Each $V^{\otimes 3}(\mathfrak{p})$ (of which there are 10) thus contains one copy of the trivial $S_3(\gamma)_L$ module. In $V^{\otimes 3}(2,0,1)$, $V^{\otimes 3}(0,2,1)$, $V^{\otimes 3}(1,0,2)$ and $V^{\otimes 3}(0,1,2,)$ the orthogonal complement of this trivial module is actually a simple two-dimensional module $M = M(2,1,0)$ over S_3 on which $S_3(\gamma)_L$ operates by the epimorphism $S_3(\gamma)_L \to S_3$. That is, M has a basis α, β with

$$(12)_t\alpha = \alpha + \beta, \quad (12)_t\beta = -\beta \quad (23)_t\alpha = -\alpha, \quad (23)_t\beta = \alpha + \beta.$$

In $V^{\otimes 3}(2,1,0)$ and $V^{\otimes 3}(1,2,0)$ however, the complementary module qM has basis $\widehat{\alpha}, \widehat{\beta}$ with action

$$(12)_t\alpha = \alpha + (\cos t)\beta, \quad (12)_t\beta = -\beta, \quad (23)_t\alpha = -\alpha, \quad (23)_t\beta = (\cos t)\alpha + \beta.$$

(Note that the actions agree modulo t^2.) Finally, $V^{\otimes 3}(1,1,1)$ has dimension 6 and contains the trivial module spanned by an element z_0, two copies, M_1 and M_2, of M, and a one-dimensional module spanned by an element z_{sgn} on which $(12)_t$ and $(23)_t$ act as -1; these are given by the following table of coefficients:

	$u\otimes v\otimes w$	$v\otimes u\otimes w$	$u\otimes w\otimes v$	$v\otimes w\otimes u$	$w\otimes u\otimes v$	$w\otimes v\otimes u$
z_0 :	q	1	q	1	q	1
α_1 :	q	1	$-q$	-1	0	0
β_1 :	0	0	q	1	$-q$	-1
α_2 :	-1	q	-1	q	0	0
β_2 :	0	0	1	$-q$	1	$-q$
z_{sgn} :	1	$-q$	-1	q	1	$-q$

Here it is not possible to quantize $\mathbf{SL}(3)$ by $\gamma = e_{12}\wedge e_{21}$. The quantizing conditions 1. clearly continues to hold but 2. fails, and $\delta_c^{A\otimes A}\gamma$ now is obstructed even as an *algebra* deformation. It is easy to check that $qM\otimes M^*$ does not contain the trivial module. Over S_3, all eight two-dimensional modules which appear are isomorphic to $M(2,1,0)$, so for the nonincreasing partitions $\mathfrak{p}$ of 3 into 3 parts we have the following module counts in $V^{\otimes 3}$:

$$n(3,0,0) = 10, \; n(2,1,0) = 8, \; n(1,1,1) = 1.$$

The total number of (linearly independent) symmetric elements of degree 3 in $k[X]$ is therefore $10^2 + 8^2 + 1^2 = 165 = 9\cdot 10\cdot 11/3!$. Over $S_m(\gamma)_L$, however, there are two classes of two-dimensional simple submodules of $V^{\otimes 3}$ having 6 and 2 members, respectively, so the total number of quantum symmetric elements is only $10^2 + 6^2 + 2^2 + 1^2 = 141$.

12. The standard quantum Yang-Baxter matrix

Returning to the standard quantization of $\mathbf{M}(n)$ by $\gamma_q = \sum_{i<j} e_{ij}\wedge e_{ji}$, if we are concerned only with the commutation relations amongst the generators, then these are given by $Q^2(X_1X_2) = (X_2X_1)Q^2$. It is easy to check, as was shown in [18], that any other matrix R giving the same commutation relations must have the form $R = QCQ$ where C is invertible and commutes with the generic symmetric element of $M(n)\otimes M(n)$. This C must have the form $\exp(t\beta)$ for some symmetric $\beta\in M(n)\otimes M(n)$, so

$$R = \exp(t\gamma_q)\exp(t\beta)\exp(t\gamma_q).$$

The C which produces the standard quantum Yang-Baxter R for any n was given in [18]; for $n = 2$ it is

$$C = \begin{pmatrix} \sec t - \tan t & 0 & 0 & 0 \\ 0 & \sec t & -\tan t & 0 \\ 0 & -\tan t & \sec t & 0 \\ 0 & 0 & 0 & \sec t - \tan t \end{pmatrix}$$

Now let

$$\kappa = \frac{1}{2}(e_{11} - e_{22}) \otimes (e_{11} - e_{22}) + e_{12} \otimes e_{21} + e_{21} \otimes e_{12}$$

be the Casimir element of $\mathfrak{sl}(2) \otimes \mathfrak{sl}(2)$. Recalling that $q = \sec t - \tan t$ one can check that

$$C = \exp((\ln q)(\frac{1}{2} I \otimes I + \kappa)).$$

That is, the C given in [18] is $q^{1/2} \exp(\kappa \ln q)$. Since the scalar factor $q^{1/2}$ is immaterial, the standard quantum Yang-Baxter matrix for $\mathbf{M}_q(2)$ is given, up to scalar multiple, by

$$R = (\exp(t\gamma_q)) \cdot (\exp(\kappa \ln q)) \cdot (\exp(t\gamma_q)) = Qq^{\kappa}Q$$

where q^{κ} symbolically denotes $\exp(\kappa \ln q)$. This formula, however, is meaningful for all semisimple (hence also for all reductive) Lie algebras $\mathfrak{g}$ since γ_q and κ are intrinsically defined.

13. The Preferred quantum linear space

The formula for the preferred deformation $*$ on all of $k[X]$ has so far been obtained only for $n = 2$, but that for $*_L$ on the quantum linear space $k[V]$ is straightforward. Recall that the deformation is described by the diagram of Theorem 10.5. There is some arbitrariness in the choice of T_L; we may take it to be orthogonal on $\mathrm{sym}_q(V)$. To compute $a * b$ in $k[V]$, let $\widehat{a}$ and $\widehat{b}$ be the ordinary symmetric elements in $k\langle V\rangle$ with $\mu\widehat{a} = a, \mu\widehat{b} = b$. Then $T_L^{-1}\widehat{a}, T_L^{-1}\widehat{b}$ are quantum symmetric. If $\alpha \in k\langle V\rangle$, let $\mathrm{sym}_q\alpha$ denote its projection on the quantum symmetric elements. Then

$$a * b = \mu(T_L \,\mathrm{sym}_q(T_L^{-1}\widehat{a} \otimes T_L^{-1}\widehat{b})).$$

To compute the product, first define the q-factorial $m!_q$ (which is *not* identical with $(m!)_q$ to be

$$m!_q = \frac{(1-q^m)(1-q^{m-1})\cdots(1-q)}{(1-q)^m},$$

and define the q-multinomial coefficients to be

$$(m : m_1, \ldots, m_n)_q = \binom{m}{m_1, \ldots, m_n}_q = \frac{m!_q}{m_1!_q \cdots m_n!_q}.$$

For both the monomial $a = u_1^{m_1}u_2^{m_2}\cdots u_n^{m_n} \in k[V]$ and the associated partition $\mathfrak{p} = (m_1, m_2, \ldots, m_n)$ we write $\langle a\rangle_q = \langle\mathfrak{p}\rangle_q = (m : m_1, \ldots, m_n)_q$. Without the subscript, this will mean the usual multinomial coefficient. It is easy to exhibit explicitly the quantum symmetric elements of $k[V]$: If $\mathfrak{p} = (m_1, \ldots, m_n)$ and α is a word in $u_1, \ldots, u_n$ lying in $V^{\otimes m}(\mathfrak{p})$, let $\mathrm{inv}\,\alpha$ denote the total number of times any u_i precedes a u_j with $j > i$. Then $\sum q^{\mathrm{inv}\alpha}\alpha$, where the sum runs over all words $\alpha \in V^{\otimes m}(\mathfrak{p})$ is the unique (up to scalar multiple) quantum symmetric element in $V^{\otimes m}(\mathfrak{p})$. Its norm is $\sum_\alpha q^{2\mathrm{inv}\alpha} = \langle\mathfrak{p}\rangle_{q^2}$. Therefore, denoting by $e_\mathfrak{p}$ and $e_{\mathfrak{p},q}$ the unique symmetric, respectively quantum symmetric elements of norm 1 in $V^{\otimes m}(\mathfrak{p})$, we have

$$e_\mathfrak{p} = \langle\mathfrak{p}\rangle^{-1/2}\sum\alpha \quad \text{and} \quad e_{\mathfrak{p}q} = \langle\mathfrak{p}\rangle_{q^2}^{-1/2}\sum q^{\mathrm{inv}\alpha}\alpha.$$

(The sum again runs through all words $\alpha \in V(\mathfrak{p})$.) If $a \in k[V]$ is a monomial $u_1^{m_1} \cdots u_n^{m_n}$, which we may write simply as $u^{\mathfrak{p}}$ where $\mathfrak{p} = (m_1, \dots, m_n)$, then $\widehat{a} = \langle \mathfrak{p} \rangle^{-1/2} e_{\mathfrak{p}} = \langle \mathfrak{p} \rangle^{-1} \sum \alpha$ is the symmetric element of $V(\mathfrak{p})$ with $\mu\widehat{a} = a$. If $b = u^{\mathfrak{p}'}$ where $\mathfrak{p}' = (m_1', \dots, m_n')$, then to compute $a * b$ we need only compute $\mathrm{sym}_q(e_{\mathfrak{p}q} \otimes e_{\mathfrak{p}'q})$. To this end set $\mathrm{inv}(a:b) = \mathrm{inv}(\mathfrak{p}:\mathfrak{p}') = m_1(m_2' + \dots + m_n') + m_2(m_3' + \dots + m_n') + \dots + m_{n-1}m_n'$. This is the number of times that a factor u_i of $\alpha \in V(\mathfrak{p})$ precedes a factor u_j of $\beta \in V(\mathfrak{p}')$ with $j > i$ in the product $\alpha \otimes \beta$. It follows that for such α and β we have $\mathrm{inv}\alpha \otimes \beta = \mathrm{inv}\alpha + \mathrm{inv}\beta + \mathrm{inv}(a:b)$. Now $\mathrm{sym}_q e_{\mathfrak{p}q} \otimes e_{\mathfrak{p}'q} = c e_{(\mathfrak{p}+\mathfrak{p}')q}$ for some scalar c, where $\mathfrak{p} + \mathfrak{p}' = (m_1 + m_1', \dots, m_n + m_n')$; this c is just $(e_{\mathfrak{p}q} \otimes e_{\mathfrak{p}'q} | e_{(\mathfrak{p}+\mathfrak{p}')q})$. If α is a word in $V(\mathfrak{p})$ and β one in $V(\mathfrak{p}')$, then $\alpha \otimes \beta$ appears in $e_{\mathfrak{p}q} \otimes e_{\mathfrak{p}'q}$ with coefficient

$$\langle \mathfrak{p} \rangle_{q^2}^{-1/2} \langle \mathfrak{p}' \rangle_{q^2}^{-1/2} q^{\mathrm{inv}\alpha + \mathrm{inv}\beta},$$

and in $e_{(\mathfrak{p}+\mathfrak{p}')q}$ with coefficient $\langle \mathfrak{p} + \mathfrak{p}' \rangle_{q^2}^{-1/2} q^{\mathrm{inv}\alpha\beta} = \langle \mathfrak{p} + \mathfrak{p}' \rangle_{q^2}^{-1/2} q^{\mathrm{inv}\alpha + \mathrm{inv}\beta + \mathrm{inv}(\mathfrak{p}:\mathfrak{p}')}$. Therefore,

$$c = \langle \mathfrak{p} \rangle_{q^2}^{-1/2} \langle \mathfrak{p}' \rangle_{q^2}^{-1/2} \langle \mathfrak{p} + \mathfrak{p}' \rangle_{q^2}^{-1/2} q^{\mathrm{inv}(\mathfrak{p}:\mathfrak{p}')} \sum_{\alpha} q^{2\mathrm{inv}\alpha} \sum_{\beta} q^{2\,\mathrm{inv}\,\beta}.$$

But, as noted above, $\sum_\alpha q^{2\,\mathrm{inv}\alpha}$ is just $\langle \mathfrak{p} \rangle_{q^2}$, so

$$c = \langle \mathfrak{p} \rangle_{q^2}^{1/2} \langle \mathfrak{p}' \rangle_{q^2}^{1/2} \langle \mathfrak{p} + \mathfrak{p}' \rangle_{q^2}^{-1/2} q^{\mathrm{inv}(\mathfrak{p}:\mathfrak{p}')}.$$

Since $T_L e_{\mathfrak{p}q} = e_{\mathfrak{p}}, T_L e_{\mathfrak{p}'q} = e_{\mathfrak{p}'}$, we have finally the following:

Theorem 13.1. *The preferred deformation for quantum linear space is given by the products*

$$a * b = \left(\frac{\langle a \rangle_{q^2} \langle b \rangle_{q^2} \langle ab \rangle}{\langle a \rangle \langle b \rangle \langle ab \rangle_{q^2}} \right)^{1/2} q^{\mathrm{inv}(a:b)} ab. \ \blacksquare$$

As presented here, this formula arises as a direct application of a general theory. However, as mentioned in the forward, the second author discovered this formula by other means first ([19]). Analysis of the formula (and the means by which it was first derived) led to the general theory. In view of its importance, the first and third authors call it *Giaquinto's Formula.* This formula depends on our choosing T_L so that its restriction to $\mathrm{sym}_q(k\langle V \rangle)$ is orthogonal, so that it carries every $e_{\mathfrak{p}q}$ to $e_{\mathfrak{p}}$. The only possible change is to have $T_L e_{\mathfrak{p}q} = c_{\mathfrak{p}} e_{\mathfrak{p}}$ where $c_{\mathfrak{p}}$ is a scalar depending on t with value 1 at $t = 0$. If $a = u^{\mathfrak{p}}$ then we can write c_a for $c_{\mathfrak{p}}$, and for the new multiplication we have

$$a *' b = c_{ab} c_a^{-1} c_b^{-1} a * b.$$

In Giaquinto's formula, if we define the "normalized" monomial $\bar{a} = (\langle a \rangle / \langle a \rangle_{q^2})^{1/2} a$, then the multiplication formula becomes

$$\bar{a} * \bar{b} = q^{\mathrm{inv}(a:b)} \overline{ab}$$

The change of T_L preserves this form, but with a different normalization. Giaquinto's formula, which was derived under the hypothesis that q was generic, must hold formally for any q for which it is defined, i.e., for all q for which n_{q^2} is defined for all n. But $n_{q^2} = 0$ implies that $q^{2n} = 1$. The formula therefore holds unless q is a root of unity. The multiplication $\circledast$ on $k\langle X\rangle$ was defined precisely as on $k\langle V\rangle$, namely, $\alpha \circledast \beta = T(T^{-1}\alpha \otimes T^{-1}\beta)$, and we had the diagram of Theorem diag2. As with $k[V]$, to compute $*$ explicitly, all we need to know are the images under T of the quantum symmetric elements. But for that we now need to know explicitly the *full* decomposition of each $V^{\otimes m}$ under $S_m(\gamma_q)$, not just the symmetric part. The effect of any arbitrary choices made in the construction of T_L disappear in T since $T = T_L \otimes T_L^{-*}$ is automatically orthogonal. The preferred deformation $*$ of $\mathbf{M}(n)$ is therefore canonical. Now recall that $k[V]$ is a left comodule algebra over $\mathbf{M}(n)$ (since V was a left module over $M(n)$.) The multiplication $*$ is defined on the original vector space SV of $k[V]$; the comultiplication is unaffected. Therefore, we have not only a preferred deformed multiplication $*$ on $\mathbf{M}(n)$ (although the explicit formula for $n = 2$ will appear elsewhere), but $(k[V], *)$ is also a left comodule algebra over $(\mathbf{M}(n), *)$ with the original Δ. In this sense we have a preferred deformation of the comodule algebra $k[V]$. All of the foregoing applies, of course, to $k[V^*]$, which is a right comodule algebra over $\mathbf{M}(n)$ and $(\mathbf{M}(n), *)$ is determined by these two quantizations. This is similar to Manin's construction of quantum groups in [M] where he builds $\mathbf{M}_q(n)$ from associated quantizations of V and its dual V^*.

14. Diagonal deformations

Let A be any algebra of characteristic zero and $\mathfrak{h}$ be a linear space of mutually commuting derivations of A. If $\omega \in \wedge^2\mathfrak{h}$ and $\mu : A \otimes A \to A$ is the multiplication, then

$$\mu_t = \mu(\exp(t\omega)) : A \otimes A \to A$$

is a deformation of A, [12]. (This is true for an $\omega \in \mathfrak{h}\otimes\mathfrak{h}$, but all such are cohomologous to their skew parts, so we only need consider $\bigwedge^2 \mathfrak{h}$.) For a bialgebra $A = \mathcal{O}(G)$ with G reductive and $\omega \in \bigwedge^2 \mathfrak{g}$ one gets a preferred deformation by taking $\mu_t = \mu(\exp(t \cdot \delta_c^{A\otimes A}\omega))$. We call these *diagonal* or *Moyal* deformations. Now suppose that $(\mathbf{M}(n), *)$ is the canonical preferred deformation just constructed using the infinitesimal $\gamma_q = \sum_{i<j} e_{ij} \wedge e_{ji}$, and $\mathfrak{h}$ is the Cartan subalgebra of $\mathfrak{sl}(n)$ relative to which all of our constructions have tacitly taken place. That is, $\mathfrak{h}$ is spanned by all diagonal matrices of trace 0. If $h \in \mathfrak{h}$ then $[h, \gamma_q] = 0$. It is easy to see from this (and also is immediately evident from Giaquinto's formula) that every element of $\mathfrak{h}$ remains a derivation with respect to the deformed multiplication. Moreover, *since $*$ is a preferred deformation*, the coalgebra structure and hence $\delta_c^{A\otimes A}$ have not changed. Therefore we have

Theorem 14.1. *If $\omega \in \bigwedge^2 \mathfrak{h}$ then*

$$*_s = \exp(s \cdot \delta_c^{A\otimes A}\omega) \circ *$$

is a preferred deformation of $\mathbf{M}(n)$. ■

We have used s as the deformation parameter here since $*$ is itself already a function of t. The space of infinitesimals of the family of canonical preferred deformations just constructed consists of all $\lambda\gamma_q + \omega$ with $\lambda \in k$. One can show that these deformations are inequivalent (and, in particular, that γ_q is not a jump deformation, unlike γ_∞, cf. next section). The dimension of the family is therefore $1+\dim(\bigwedge^2 \mathfrak{h})$. The infinitesimals $\gamma_q + \omega$ all lie in $\bigwedge^2 \mathfrak{sl}(n)$, so the deformations just described may be restricted to $\mathbf{SL}(n)$. (Under the preferred deformation, the quantum determinant is identical, as an element of the underlying vector space, with the ordinary determinant.) Since the procedure for constructing diagonal deformations is available for all reductive G, the foregoing actually shows the following:

Let the rank of G be r. Then $\mathcal{O}(G)$ has a $1 + \binom{r}{2}$ dimensional family of deformations to quantum groups.

Let $\mathfrak{h}_+$ be the space of all diagonal matrices of $M(n)$. The infinitesimals $\lambda\gamma_q + \omega$ with $\omega \in \bigwedge^2 \mathfrak{h}_+$ give a family of deformations of dimension $1 + \binom{n}{2}$. This contains the Dipper-Donkin deformation, whose infinitesimal

$$\gamma_D = \gamma_q + \sum_{i<j} e_{ii} \wedge e_{jj} = \sum_{i<j} (e_{ij} \wedge e_{ji} + e_{ii} \wedge e_{jj})$$

is distinguished by the fact that $\gamma_D^2 = 0$. Therefore,

$$Q^{-2} = \exp(2t\gamma_D) = 1 - t\sum_{i<j} (e_{ij} \otimes e_{ji} - e_{ji} \otimes e_{ij} + e_{ii} \otimes e_{jj} - e_{jj} \otimes e_{ii})$$

is defined over the integers, and, so, in every characteristic. Therefore, so are $(12)_t^L = Q^{-2}(12)$ and $S_m(\gamma_D)_L$.

It is an easy matter to adapt Giaquinto's formula to all infinitesimals of the form $\gamma_q + \omega$. (We need only consider the case $\lambda = 1$.) Suppose first that $\omega = e_{ii} \wedge e_{jj}$ for some $i < j$, and that $a \in V^{\otimes m}(\mathfrak{p})$ and $b \in V^{\otimes r}(\mathfrak{p}')$ with $\mathfrak{p} = (m_1, \dots, m_n)$ and $\mathfrak{p}' = (r_1, \dots, r_n)$. If we restrict attention to $V^{\otimes m}$ then e_{ii} operates as $u_i\partial/\partial u_i$, so $\omega(a \otimes b) = \frac{1}{2}(m_i r_j - m_j r_i) a \otimes b$ and

$$\exp(s\delta_c^{A\otimes A}(e_{ii} \wedge e_{jj}))a \otimes b = \exp\left(\frac{s(m_i r_j - m_j r_i)}{2}\right) a \otimes b.$$

Writing $a * b = \nu(a,b) q^{\mathrm{inv}(a:b)} ab$, where $\nu(a,b)$ is the "norm factor", we have $\nu(ha, hb) = \nu(a,b)$ and $\mathrm{inv}(ha : hb) = \mathrm{inv}(a : b)$. Therefore we have

$$a *_\omega b = \nu(a,b) q^{\mathrm{inv}(a:b)} \mu(\exp(s(\delta_c^{A\otimes A}\omega))(a \otimes b)).$$

If $\omega = \sum c_{ij} e_{ii} \wedge e_{jj}$ then we have finally

$$a *_\omega b = \nu(a,b) q^{\mathrm{inv}(a:b)} [\exp(\frac{1}{2} \sum c_{ij}(m_i r_j - m_j r_i))] ab.$$

(Here $\omega \in \wedge^2 \mathfrak{h}$ if and only if $\sum_i c_{ij} = 0 = \sum_j c_{ij}$.)

15. At the boundary

The infinitesimal

$$\gamma_\infty = 2\sum_{j=2}^{n-1} e_{1j} \wedge e_{jn} + (e_{11} - e_{nn}) \wedge e_{1n}$$

is skew with respect to the "antidiagonal" form on V given by $\langle u_i, u_j\rangle = \delta_{i,n-j}$. Since γ_∞ is classical Yang-Baxter and $\gamma_\infty^3 = 0$, Q is quantum Yang-Baxter and $S_m(\gamma_\infty)_L$ collapses, as we have seen, to the ordinary symmetric group S_m. *A fortiori* the decomposition of $V^{\otimes m}$ under $S_m(\gamma)_L$ is the same as under S_m, so γ_∞ is quantizing. We denote the preferred deformation which it defines by $(\mathbf{M}(n), *_\infty)$. In particular, $(\mathbf{M}(2), *_\infty)$ is the preferred form of the "Jordan" quantization (cf. [5]) of $\mathbf{M}(2)$. Let

$$\mathfrak{h}_\infty = \{h \in \mathfrak{h} | [h, \gamma_\infty] = 0\} = \{\sum \lambda_i e_{ii} | \lambda_1 = \lambda_n\}$$

It is easy to check that all $h \in \mathfrak{h}_\infty$ remain derivations of the deformed algebra $(\mathbf{M}(n), *_\infty)$. As in the previous section, every infinitesimal of the form $\gamma_\infty + \omega_\infty$ with $\omega_\infty \in \bigwedge^2 \mathfrak{h}_\infty$ therefore induces a preferred deformation of $\mathbf{M}(n)$. However, the preferred deformation with infinitesimal γ_∞ is a jump deformation because γ_∞, being nilpotent, is conjugate to any non-zero multiple of itself. The bialgebra $(\mathbf{M}(n), *_\infty)$ does not depend on the value of the parameter t for sufficiently small $t \neq 0$. As remarked in section 7, the infinitesimal γ_∞ is not a coboundary at $t = 0$. On the other hand, by the general theory of jump deformations outlined in section 5 and developed in [13], it becomes a coboundary for all small $t \neq 0$, the necessary and sufficient condition for which, by the theory of [13] is that it be the obstruction to lifting a derivation of the original algebra to a derivation of the deformed one. When $n = 2$, for example, $\gamma_\infty = (e_{11} - e_{22}) \wedge e_{12}$, and the primary obstruction to lifting e_{12} is $[e_{12}, \gamma_\infty] = -2\gamma_\infty$. The number of parameters of the foregoing deformations at the boundary of the deformation space of $\mathbf{M}(n)$ in thus only $\dim \bigwedge^2 \mathfrak{h}_\infty$. We conjecture that we have now in fact found all deformations of $\mathbf{SL}(n)$ and $\mathbf{M}(n)$ up to the equivalence defined earlier, these being given by γ_q or γ_∞ followed by suitable diagonal deformations. The analogous statement should hold for all $\mathcal{O}(G)$ with G reductive. As remarked earlier, to each γ_∞ one can associate a universal deformation formula (udf) as follows: Write $\beta' = (e_{11} - e_{nn}) \otimes e_{1n}$, $\beta'' = 2\sum_{1<j<n} e_{1j} \otimes e_{jn}$, and $\beta = \beta' + \beta''$. Then γ_∞ is the skew part of β and therefore is cohomologous to β. Set $e_{11} - e_{nn} = H_+$ and write $H_+^{[m]} = H_+(H_+ + 1)\cdots(H_+ + m - 1), m > 0$. Then $H_+^{[m]} \otimes e_{1n}$ and β'' are commuting elements of $\bigotimes U\mathfrak{sl}(n)$. Set

$$\overline{\mu}_m = H_+^{[m]} \otimes e_{1n}^m + \binom{m}{1}(H_+^{[m-1]} \otimes e_{1n}^{m-1})\beta'' + \binom{m}{2}(H_+^{[m-2]} \otimes e_{1n}^{m-2})(\beta'')^2 + \ldots + (\beta'')^m.$$

Then $\overline{\mu}_1 = \beta$ and direct computation shows that

$$\bar{\mu}_t = 1 \otimes 1 + t\bar{\mu}_1 + \frac{t^2}{2!}\bar{\mu}_2 + \ldots, \qquad \text{where } \bar{\mu} = 1 \otimes 1.$$

is a udf based on $\mathfrak{sl}(n)$. It is equivalent to one with $\bar{\mu}_1 = \gamma_\infty$, but we do not have a closed form. There is also a udf based on every $\gamma_\infty + \omega$ with $\omega \in \bigwedge^2 \mathfrak{h}_\infty$, but again we do not have a closed form. On the other hand, it is clear how to extend these ideas to all simple Lie algebras.

While the explicit form of the preferred deformation of the associated linear space is not known, its commutation relations are

$$u_1u_n - u_nu_1 = u_1^2 \quad \text{and} \quad u_ju_n - u_nu_j = 2u_1u_j \quad \text{for } j = 2, \ldots n-1$$

where $\{u_i\}$ is the basis for V. (All other pairs of basis elements commute.) For $n = 2$ this is the algebra $k\langle u_1, u_2\rangle$ with relation $[u_1, u_2] = u_1^2$ which is a (noncommutative) *regular* algebra in the sense of Artin and Schelter (cf. [1]). The only other noncommutative regular algebra of dimension 2 is the quantum plane. We conjecture that the quantum linear spaces associated with γ_∞ are all regular. (The standard quantum linear spaces associated with γ_q are all regular).

16. Afterword

As we mentioned at the outset, this is a preliminary paper – it raises more questions than it answers. Algebraic deformation theory supplies a series of obstructions to an infinitesimal. They were not encountered when we showed that γ_q and γ_∞ were quantizing. What role do they play? In [18] we sought a formula for the preferred deformed multiplication in $(\mathbf{M}(n), *)$. We still have not found it for $n > 2$, even though - now that we know $\mathbf{M}_q(n)$ is a deformation - our cohomology theory for bialgebras shows trivially that such a formula exists. (We have found, but not presented, explicit formulas for the preferred deformation of all of $\mathbf{M}(2)$; they involve a special basis for $k[a, b, c, d]$.) We have shown by a trivial example that the commutation relations defined by a quantum Yang-Baxter matrix R may produce a bialgebra with a smaller Hilbert series from that of the original polynomial ring, and which therefore can not be a deformation in the sense of this paper. What are the true conditions that R yield a deformation? Does every quantum Yang-Baxter matrix have an "obstruction", and if so, is there only one? What implications does our present theory have for the broader theory of deformations of polynomial rings (without comultiplication)? Do the families of deformations we have exhibited really exhaust all those of $\mathbf{SL}(n)$ and $\mathbf{M}(n)$, and by analogy, of $\mathcal{O}(G)$ for all reductive G? What is the preferred deformation of $U\mathfrak{sl}(n)$ giving the Drinfel'd-Jimbo $U_q\mathfrak{sl}(n)$? What happens over rings of finite characteristic or over finite fields.? (Giaquinto's formula is meaningful over p-adic rings for $p \neq 2$.) Finally, what is the physical meaning (if any) of the symmetry breaking associated with a universal deformation formula? It is curious (but perhaps fundamental) that our theory simultaneously presents "quantum groups" as a manifestation of quantum symmetry but also as a breaking of ordinary symmetry. The existence of universal deformation formulas suggest that symmetry under the action of a continuous group is generally unstable. Is this truly a feature of the physical universe?

References

[1] M.Artin and W.Schelter , *Graded Algebras of Global Dimension 3*, Adv. Math. **66** (1987), 171 .
[2] K.Brown , *Buildings*, Springer Verlag , New York , 1990 .
[3] V.Coll , *Universal Deformation Formulae*, Ph.D. thesis, University of Pennsylvania (1990).
[4] V.Coll, M.Gerstenhaber and A.Giaquinto , *An Explicit Deformation Formula With Noncommuting Derivations*, Israel Math. Conf. Proc., vol. 1, Weizmann Science Press , 1989, pp. 396–403 .

[5] E.E.Deminov, Yu.I.Manin, E.E.Mukhin, and D.V.Zhdanovich , *Non-Standard Quantum Deformations of* GL(n) *and Constant Solutions of the Yang–Baxter Equation*, Kyoto University preprint (May 1990).

[6] R.Dipper and S.Donkin , *Quantum GL(n)*, preprint (1990).

[7] V.G.Drinfel'd , *Quantum Groups*, Proc. ICM-86, Berkeley **1** (1987), 798–820 .

[8] L.D.Faddeev, N.Y.Reshetikhin and L.A.Takhtajan , *Quantization of Lie Groups and Lie Algebras*, Leningrad Math. J. **1** (1990), 193–225 .

[9] A.Froelicher and A.Nijenhuis , *A Theorem on Stability of Complex Structures*, Proc. Nat. Acad. Sci. **43** (1957), 239–241 .

[10] M.Gerstenhaber , *The Cohomology Structure of an Associative Ring*, Ann. Math. **78** (1963), 267–288 .

[11] M.Gerstenhaber , *On the Deformation of Rings and Algebras*, Ann. Math. **79** (1964), 59–103 .

[12] M.Gerstenhaber , *On the Deformation of Rings and Algebras III*, Ann. Math. **88** (1968), 1–34 .

[13] M.Gerstenhaber , *On the Deformation of Rings and Algebras IV*, Ann. Math. **99** (1974), 257 .

[14] M.Gerstenhaber and S.D.Schack , *Relative Hochschild Cohomology, Rigid Algebras, and the Bockstein*, J. Pure and Appl. Alg. **43** (1986), 53–74 .

[15] M.Gerstenhaber and S.D.Schack , *A Hodge-type Decomposition for Commutative Algebra Cohomology*, J. Pure and Appl. Alg. **48** (1987), 229–247 .

[16] M.Gerstenhaber and S.D.Schack , *The Cohomology of Presheaves of Algebras I: Presheaves over a Partially Ordered Set*, Trans. Amer. Math. Soc. **310** (1988), 135–165 .

[17] M.Gerstenhaber and S.D.Schack , *Algebraic Cohomology and Deformation Theory*, Deformation Theory of Algebras and Structures and Applications, M. Hazewinkel and M. Gerstenhaber, eds., Kluwer , Dordrecht , 1988 , pp. 11–264 .

[18] M.Gerstenhaber and S.D.Schack , *Bialgebra Cohomology, Deformations and Quantum Groups*, Proc. Nat. Acad. Sci. **87** (1990), 478–481 .

[19] A.Giaquinto , *Deformation Methods in Quantum Groups*, Ph. D. Thesis (1991).

[20] G.Hochschild, B.Kostant, and A.Rosenberg , *Differential Forms on Regular Affine Algebras*, Trans. Amer. Math. Soc. **102** (1962), 383–408 .

[21] D.Knuth , *The Art of Computer Programming, V. 3*, Addison Wesley , 1973 .

[22] M.Jimbo , *A q-difference analogue of* $U(\mathfrak{g})$, Lett. Math. Phys. **10** (1985), 63–69 .

[23] A.Lichnerowitz , *Quantum Mechanics and Deformations of Geometrical Dynamics*, Quantum Theory, Groups, Fields, and Particles, A.O.Barut, ed., Reidel , Dordrecht .

[24] G.Lusztig , *Quantum Groups at Roots of 1*, Geometriae Dedicata **35** (1990), 89–113 .

[25] Yu.Manin , *Quantum Groups and Non-Commutative Geometry*, Les Publ. du CRM **1561** (1988), Universite de Montreal .

[26] S.Montgomery and S.P.Smith , *Skew Derivations and* $U_q(\mathfrak{sl}(2))$, Israel J. Math. **72** (1990), 158 .

[27] J.E.Moyal , *Quantum Mechanics as a Statistical Theory*, Proc. Cambridge Phil. Soc. **45** (1949), 99–124 .

[28] A.Nijenhuis and R.Richardson , *Cohomology and Deformations of Graded Lie Algebras*, Bull. Amer. Math. Soc. **72** (1966), 1–29 .

[29] R.Richardson , *On the Rigidity of Semi-direct Products of Lie Algebras*, Pac. J. Math. **22** (1967), 339–344 .

[30] W.Singer , *Extension Theory for Connected Hopf Algebras*, J. Alg. **21** (1972), 1–16 .

[31] L.Sudberry , *Consistent Multiparametric Deformations of* GL(n), preprint (1990).

[32] L.A.Takhtajan , *Introduction to Quantum Groups*, LNP **370** (1990), 3–28 .

[33] O.Villamayor and D.Zelinsky , *Galois Theory for Rings with Finitely Many Idempotents*, Nagoya Math. J. **27** (1966), 721–731 .

[34] W.Waterhouse , *Introduction to Affine Group Schemes*, Springer Verlag , New York , 1979 .

[35] S.L.Woronowicz , *Twisted* $SU(2)$*-group: An Example of a Non-commutative Differential Calculus*, Publ. RIMS Kyoto Univ. **23** (1987), 117–181 .

Department of Mathematics, David Rittenhouse Laboratory, University of Pennsylvania, 209 South 33rd Street, Philadelphia PA 19104-6395, USA

YANG–BAXTER EQUATION AND DEFORMATION OF ASSOCIATIVE AND LIE ALGEBRAS

D.GUREVICH AND V.RUBTSOV

Moscow Branch of Center "Sophus Lie"

ABSTRACT. We construct some cocycles (Hochschild and cyclic ones) connected with a classical R-matrix on associative and Lie algebras and "quantize" them. We treat "S-traces" (S is a solution of Yang-Baxter equation) on deformed algebras as a result of the quantization. The generalization of this construction is discussed.

Introduction

The great attention to the Yang-Baxter equations (YBE) both classical and quantum was firstly stimulated by the inverse scattering problem method. Then it turned out that many of other mathematical objects and physical models are connected with the YBE.

The most famous objects associated with an R-matrix (under the "R-matrix" we understand as usual a solution of the YBE) are "quantum groups". These objects arise in particular from quantization of some Poisson brackets on "usual" Lie groups (another types of quantum groups are given in [G3]). These brackets were introduced by E.Sklyanin and studied more carefully by V.Drinfeld. They have form

$$\{f,g\} = r^{ij}(\partial_i f \partial_j g - \partial'_i f \partial'_j g), \ \ f,g \in C^\infty(G)$$

where G is a Lie group, ∂_i (∂'_i) is right- (left-) invariant vector field on G corresponding to the element $X_i \in \mathcal{G}$, $\mathcal{G}$ is the Lie algebra of Lie group G and $R = r^{ij} X_i \otimes X_j \in \Lambda^2(\mathcal{G})$ is a solution of modified (M) YBE i.e. the element

$$[\![R,R]\!] = [R^{12}, R^{13}] + [R^{12}, R^{23}] + [R^{13}, R^{23}]$$

is $\mathcal{G}$-invariant. Here $R^{12} = R \otimes id$ and so on. We want to stress that it is naturally to treat the element $[\![R,R]\!]$ as Schouten tensor for R.

If $[\![R,R]\!] = 0$ (this equation is called classic (C) YBE) one can define another Poisson bracket on G as follows

$$\{f,g\} = r^{ij} ad^*_{X_i} f \ ad^*_{X_j} g, \quad f,g \in C^\infty(G).$$

So-called "monoidal" group is obtained by quantization of this bracket (see [G3]). One can define Poisson bracket on $C^\infty(\mathcal{G}^*)$ in a similar way and quantize it. As a result we get an associative (non-commutative) algebra $A_\hbar$ with a deformed multiplication of the following type $f \circ_\hbar g = f \circ g + \frac{\hbar}{2}\{f,g\} + \dots$ where $\circ$ is commutative multiplication in the initial algebra $A = A_0$. It means that the multiplication $\circ_\hbar$ is the result of a deformation of the multiplication $\circ$ "in direction" of the bracket $\{,\}$.

In this paper we review an analogous construction for non-commutative algebra A. Let $\mathcal{G}$ be a subalgebra in the Lie algebra $\mathcal{D}er(A)$ of derivations of an associative algebra A. We deform a multiplication $\circ$ "in direction" of some Hochschild two-cocycle associated with an R-matrix R. Our deformation takes an "intermediate" position between A.Lichnerovicz's $*$-deformation and deformation of associative structure in sense of M.Gerstenhaber.

If there exists a trace on A, i.e. a functional $\mathrm{tr} : A \to k$ ($k = \mathbf{R}$ or $\mathbf{C}$) such that $\mathrm{tr}(a \circ b) = \mathrm{tr}(b \circ a)$ or an involution (a conjugation) $\mathrm{inv} : a \to a^*$ these operators (if they are $\mathcal{G}$-invariant) can be deformed too.

We obtain an associative algebra A endowed with a"quantum R-matrix" S: $A^{\otimes^2} \to A^{\otimes^2}$, an S-trace tr_S: $A \to k$ and S-conjugation conj_S: $A \to A$ as the result of this deformation. In a similar way one can deform a Lie algebra structure (see below).

Note in conclusion that it is very natural to consider on algebras A and $\mathcal{G}$ S-cyclic (co)homology and S-Chevalley (co)homology as a generalization of the "classical" ones. Definitions and investigation of "S-(co)homology" will be done elsewhere.

We wish to thank the Euler International Mathematical Institute for kind invitation to take part in the Workshop on Quantum groups. This paper is an enlarged version of our Workshop talks.

1. Deformation of Lie algebras

Let $\mathcal{G}$ be a fixed Lie algebra over the field k, $R \in \mathcal{G}^{\otimes^2}$ and ρ be the adjoint representation of $\mathcal{G}$, i.e. $\rho(X)(Y) = [X,Y]$. Consider the operator $\mathcal{R} = (\rho \otimes \rho)R$ acting from $\mathcal{G}^{\otimes^2}$ into itself. Fix a basis $\{X_i\}$ in $\mathcal{G}$. Then $R = r^{ij} X_i \otimes X_j$ and $\mathcal{R}(X \otimes Y) = r^{ij}[X_i, X] \otimes [X_j, Y]$.

For a linear space V denote by $\Lambda_+(V)$ ($\Lambda_-(V)$) the symmetric (exterior) algebra of V and by $\Lambda^i_\pm(V)$ the homogeneous component of $\Lambda_\pm(V)$ of degree i.

Put also $\mathcal{R}_\pm = (\mathrm{id} \pm \sigma)\mathcal{R}/2$ where σ is the permutation ($\sigma(X \otimes Y) = Y \otimes X$). If $R \in \Lambda^2_-(\mathcal{G})$ an operator $\mathcal{R}_+ : \Lambda^2_-(\mathcal{G}) \to \Lambda^2_+(\mathcal{G})$ is well defined. Further on we will suppose that $R \in \Lambda^2_-(\mathcal{G})$.

Proposition 1. *The operator $\mathcal{R}_+ : \Lambda^2_-(\mathcal{G}) \to \Lambda^2_+(\mathcal{G})$ satisfies the following equation*

$$\mathcal{R}_+([X,Y] \otimes Z) - [X, \mathcal{R}_+(Y \otimes Z)] + \curvearrowright = 0$$

where $\curvearrowright$ means summing of all cyclic permutations and therefore $\mathcal{R}_+$ is a $\Lambda^2_+(\mathcal{G})$-valued two-cocycle on $\mathcal{G}$ ($\mathcal{G}$ acts on $\Lambda^2_+(\mathcal{G})$ in adjoint way).

Hence we have interpreted $\mathcal{R}$ as an element of $H^2(\mathcal{G}, \Lambda^2_+(\mathcal{G}))$ where $H^2(\mathcal{G}, M)$ means cohomology group of the algebra $\mathcal{G}$ with coefficients in $\mathcal{G}$-module M.

Let $\langle , \rangle : \mathcal{G}^{\otimes^2} \to k$ be a symmetric invariant pairing on $\mathcal{G}$ for instance $\langle X, Y \rangle = \mathrm{tr}\, \rho(X)\rho(Y)$ where ρ is the adjoint representation.

Consider the map

$$\Lambda^2_-(\mathcal{G}) \overset{\mathcal{R}_+}{\to} \Lambda^2_+(\mathcal{G}) \overset{\langle,\rangle}{\to} k.$$

Denote $c_R(X,Y) = \langle , \rangle \mathcal{R}_+(X \otimes Y)$. Thus c_R is k-valued form on $\mathcal{G}$.

Proposition 2 [G2]. *$c_R(X,Y)$ is k-valued cocycle on $\mathcal{G}$ i.e.*

$$c_R([X,Y],Z) + \curvearrowright = 0 \quad \forall X,Y,Z \in \mathcal{G}.$$

Moreover, this cocycle is the coboundary

$$c_R(X,Y) = \langle M, [X,Y] \rangle, \ M = \frac{1}{2} r^{ij} [X_i, X_j].$$

In a general case (see remark below) $c_R(X,Y)$ is not a coboundary.

Suppose now that R satisfies the CYBE $[\![R,R]\!] = 0$ i.e. R is a classical R-matrix and deform the Lie algebra structure and the invariant pairing $\langle , \rangle : \mathcal{G}^{\otimes^2} \to k$ "in direction": defined by this R-matrix.

Consider a series

$$F_\hbar(X,Y) = 1 \otimes 1 + \frac{\hbar}{2} R + \sum_{|\alpha|+|\beta|\geq 2} p_{\alpha,\beta}(\hbar) X^\alpha \otimes Y^\beta \in U(\mathcal{G})^{\otimes^2}[[\hbar]]$$

where $X^\alpha = X_1^{\alpha_1} \ldots X_n^{\alpha_n}, p_{\alpha,\beta}(\hbar) = \mathcal{O}(\hbar^2)$, satisfying to the conditions $F_\hbar(0,X) = F_\hbar(X,0) = 1$, $F_\hbar(X, Y+Z) F_\hbar(Y,Z) = F_\hbar(X+Y, Z) F_\hbar(X,Y)$.

This series was constructed in [D]. Let us introduce a new family of operators

$$\mathcal{F}_\hbar(X \otimes Y) = X \otimes Y + \frac{\hbar}{2} \mathcal{R}(X \otimes Y) + \sum_{\alpha\beta} p_{\alpha,\beta}(\hbar) ad^\alpha X \otimes ad^\beta Y \tag{1}$$

$(ad^\alpha X = ad_{X_1}^{\alpha_1} \ldots ad_{X_n}^{\alpha_n} X), \quad S = S_\hbar = \mathcal{F}_\hbar^{-1} \sigma \mathcal{F}_\hbar, \quad [,]_S = [,]\mathcal{F}, \quad \langle , \rangle_S = \langle , \rangle \mathcal{F}.$

The operators $S_\hbar$, $[,]_S$ and $\langle , \rangle_S$ act from $\mathcal{G}^{\otimes^2}$ into $\mathcal{G}^{\otimes^2}[[\hbar]]$, $\mathcal{G}[[\hbar]]$ and $k[[\hbar]]$ correspondingly but we omit the dependence on $\hbar$ in notations. Obviously $S^2 = id$.

Proposition 3 [G2], [GRZ]. *The operator S satisfies to the QYBE*

$$S^{12} S^{23} S^{12} = S^{23} S^{12} S^{23}$$

The space $\mathcal{G}$ equipped with the operators S and $[,]_S$ is an S-Lie algebra, i.e. the relations

1) $[,]_S = -[,]_S S$;
2) $[,]_S([,]_S \otimes id)(id + S^{12}S^{23} + S^{23}S^{12}) = 0$;
3) $S([,]_S \otimes id) = (id \otimes [,]_S) S^{12} S^{23}$

hold. The pairing $\langle , \rangle_S$ satisfies the equalities

1) $\langle , \rangle_S = \langle , \rangle_S S$;
2) $\langle , \rangle_S \otimes id = (id \otimes \langle , \rangle_S) S^{12} S^{23}$;
3) $\langle , \rangle_S([,] \otimes id)(id + S^{23}) = 0$.

We shall call symmetry any unitary ($S^2 = id$) solution S of the QYBE.

The linear space V equipped with the S-Lie algebra structure we denote by $\mathcal{G}_S$. The later condition means that the pairing $\langle , \rangle$ is $\mathcal{G}_S$-invariant.

S-Lie algebras (or generalized Lie algebras) were introduced by one of the authors (see [G1]).

There exists S-Lie algebras which can not be obtained by means of deformation of a usual Lie algebra (see Sec.3). The examples of such algebras are given in [G3]. In this paper a notion of a representation of S-Lie algebra is defined too. Recall it.

Let $\mathcal{G}_S$ be a S-Lie algebra and V be a linear space. We say that a map $\rho : \mathcal{G}_S \to \mathrm{End}(V)$ is a representation of S-Lie algebra $\mathcal{G}_S$ if the symmetry $S : \mathcal{G}^{\otimes^2} \to \mathcal{G}^{\otimes^2}$ can be extended upon the symmetry $S : (\mathcal{G}_S \oplus V)^{\otimes^2} \to (\mathcal{G}_S \oplus V)^{\otimes^2}$ in such a way that an operator $X \otimes v \to \rho(X)v$ is S-invariant and the condition

$$\rho([,]_S) = \mu(\rho \otimes \rho)(id - S)$$

holds (here μ is the multiplication in $\mathrm{End}(V)$).

If there exist two representations of S-Lie algebra $\mathcal{G}_S$ $\rho_i : \mathcal{G}_S \to \mathrm{End}(V_i)$ one can define their tensor product $(\rho_1 \otimes \rho_2) : \mathcal{G}_S \to \mathrm{End}(V_1 \otimes V_2)$ as follows

$$(\rho_1 \otimes \rho_2)\ \mathcal{G}_S \to \mathrm{End}(V_1 \otimes V_2),$$
$$(\rho_1 \otimes \rho_2) X\,(v_1 \otimes v_2) = \rho_1(X)v_1 \otimes v_2 + \widetilde{v}_1 \otimes \rho_2(\widetilde{X})v_2$$

where $\widetilde{v} \otimes \widetilde{X} = S(X \otimes v)$, $v_i \in V_i$, $X \in \mathcal{G}_S$.

If S-Lie algebra $\mathcal{G}_S$ is obtained by a deformation of an usual Lie algebra $\mathcal{G}$ (as above) one can deform the tensor category of the representations of $\mathcal{G}$ with the help of the intertwining operator $F_\hbar$. We put $\rho_S(X)v = ev(\rho \otimes id)\mathcal{F}_\hbar(X \otimes v)$. Here $\mathcal{F}_\hbar : \mathcal{G} \otimes V \to \mathcal{G} \otimes V$ is defined in similar way to (1) (we need only to put $\rho^\alpha = \rho^{\alpha_1}(X_1) \ldots \rho^{\alpha_n}(X_n)$ instead of ad^α_X).

In this way it is possible to define an operator $\mathcal{F}_\hbar : V_1 \otimes V_2 \to V_1 \otimes V_2$ for every two $\mathcal{G}$-modules V_1 and V_2. Introducing a new operator $S = \mathcal{F}_\hbar^{-1} \circ \sigma \circ \mathcal{F}_\hbar$ where $\sigma : V_1 \otimes V_2 \to V_2 \otimes V_1$ is usual permutation we deform a tensor category of representations of initial algebra $\mathcal{G}$ into one of $\mathcal{G}_S$.

Remark. We can generalize the constructions above supposing R to be an element of $\Lambda^2_-(\mathcal{J})$, $\mathcal{J} \subseteq \mathrm{Der}(\mathcal{G})$ if $\mathrm{Der}(\mathcal{G}) \neq \mathcal{G}$ where $\mathrm{Der}(\mathcal{G})$ is the algebra of derivations of $\mathcal{G}$. In this case one can deform the initial objects if they are $\mathcal{J}$-invariant.

2. Deformation of associative algebras

Now consider a deformation of associative structures generated by a classical R-matrix.

Let A be an associative algebra with a multiplication $\mu(a \otimes b) = ab$, $\mu : A^{\otimes^2} \to A$ and the unit 1. Let $\mathcal{G} \subseteq Der(A)$ be a subalgebra of derivations of algebra A and $R \in \mathcal{G}^{\otimes^2}$. Define operator $\mathcal{R} : A^{\otimes^2} \to A^{\otimes^2}$ as follows. If $R = r^{ij} X_i \otimes X_j$, where $\{X_i\}$ is a basis in $\mathcal{G}$ we put

$$\mathcal{R}(a \otimes b) = r^{ij} X_i a \otimes X_j b.$$

Define $c_R(a, b) = \mu\mathcal{R}(a \otimes b)$. It is well known that c_R is an A-valued Hochschild two-cocycle in A (this fact can be verified directly) i.e.

$$a \circ c_R(b, c) - c_R(a \circ b, c) + c_R(a, b \circ c) - c_R(a, b) \circ c = 0.$$

Suppose there exists a symmetric pairing on A, i.e. an operator $\langle,\rangle:A^{\otimes^2} \to k$ such that $\langle a,b\rangle=\langle b,a\rangle$. In a particular case when A is endowed with a trace $tr:A \to k$ we take $\langle a,b\rangle = tr(a \circ b)$.

Define a k-valued bilinear form $\bar{c}_R:A^{\otimes^2} \to k$ as follows $\bar{c}_R(a,b) = \langle,\rangle\mathcal{R}(a\otimes b)$ or $\bar{c}_R(a,b) = tr\, c_R(a,b)$ if the trace exists.

Assume that $R \in \Lambda^2_-(\mathcal{G})$. Then $\bar{c}_R(a,b) = -\bar{c}_R(b,a)$.

Proposition 4. *$\bar{c}_R$ is a cyclic one-cocycle on A, i.e.*

$$\bar{c}_R(a\circ b,c) - \bar{c}_R(a,b\circ c) + \bar{c}_R(c\circ a,b) = 0$$

(or in the cyclic form $\bar{c}_R(a\circ b,c)+ \curvearrowright = 0$).

If all derivations X_i are inner: $X_i(a) = a_i \circ a - a \circ a_i$ then this cocycle is the coboundary $\bar{c}_R(a,b) = \langle M, a\circ b - b\circ a\rangle$, $M = r^{ij} a_i \circ a_j$.

Now suppose the pairing $\langle,\rangle$ be $\mathcal{G}$-invariant i.e. satisfies the equality $\langle Xa,b\rangle + \langle a,Xb\rangle = 0$ for all $X\in\mathcal{G}$ and quantize the cocycles c_R and $\bar{c}_R$.

Let $F_\hbar$ be as above. Define an operator $\mathcal{F}_\hbar : A^{\otimes^2} \to A^{\otimes^2}$ in the similar way and put $S = S_\hbar = \mathcal{F}_\hbar^{-1}\circ\sigma\circ\mathcal{F}_\hbar$, $\mu_S = \mu\mathcal{F}_\hbar$, $\langle,\rangle_S = \langle,\rangle\mathcal{F}_\hbar$. It is obvious that $\mu_S(f\otimes 1) = \mu_S(1\otimes f) = f$.

Proposition 5. *$S_\hbar : A^{\otimes^2} \to A^{\otimes^2}$ is a unitary solution of the QYBE i.e. symmetry, $\mu_S : A^{\otimes^2} \to A$ is an associative multiplication, $\langle,\rangle_S : A^{\otimes^2} \to k$ is an S-symmetric pairing, i.e. $\langle,\rangle_S = \langle,\rangle S$. The multiplication μ_S and the pairing $\langle,\rangle_S$ are S-invariant, i.e. the following diagrams*

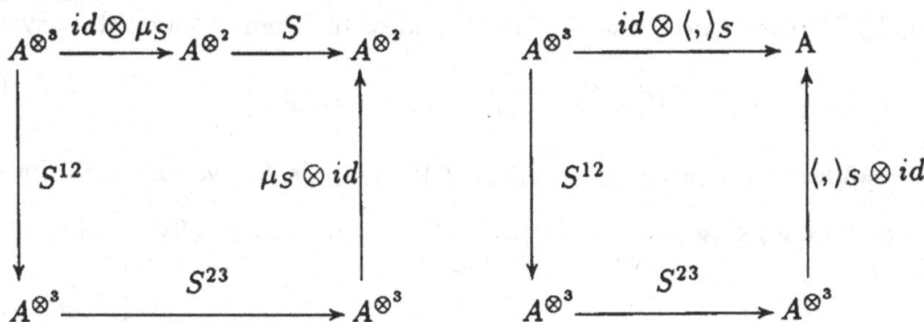

are commutative.

If the initial pairing has the form $\langle a,b\rangle = tr\, a\circ b$ then the new pairing equals to $\langle a,b\rangle_S = tr\,\mu\mathcal{F}_\hbar(a\otimes b)$.

Suppose that $H^2(A,A) = 0$ and therefore all deformations of the algebra A are trivial i.e. there exists an isomorphism $\alpha : A_S \to A$ such that $\mu(\alpha\otimes\alpha) = \alpha\mu_S$. Here we denote by A_S the algebra A equipped with the symmetry S. Consider the new symmetry $S' = (\alpha\otimes\alpha)S(\alpha^{-1}\otimes\alpha^{-1})$ and the new pairing $\langle,\rangle_{S'} = \langle,\rangle_S(\alpha^{-1}\otimes\alpha^{-1})$. We obtain the associative algebra A with the initial multiplication $\mu : a\otimes b \to a\circ b$ but equipped with the symmetry S' and the pairing $\langle,\rangle_{S'}$. This pairing is S'-symmetric, S'-invariant and the initial multiplication μ is S'-invariant too.

Consider now a deformation of an involution.

Suppose that there exists a morphism inv: $A \to A$ (we use also a notation $a^* = inv\, a$) such that $inv^2 = id$. Let $inv_{S'} = \alpha\, inv\, \alpha^{-1}$.

Proposition 6. *The equality*

$$S'(inv_{S'} \otimes (id)) = ((id) \otimes inv_{S'})S'$$

hold. If the relations $(a \circ b)^* = b^* \circ a^*$ *and* $tr\, a^* = tr\, a$ *are valid then the relations*

$$inv_{S'}(a \circ b) = \mu_{S'}(inv_{S'} \otimes inv_{S'})S' \; ;$$
$$tr_{S'} inv_{S'} = tr_{S'}$$

are valid too.

A conjugation is a particular case of an involution. In the next section we consider S-conjugation in a general case including the symmetries which can not be obtained by means of deformation.

3. Generalization, example

In previous section we endowed the space End(V) with a symmetry S, S-trace tr_S, S-involution inv_S (if there exists a "good" initial one inv) and S-conjugation $conj_S$ as a particular case of the deformed classical objects: permutation σ, trace tr, and involution inv (and analogous for Lie algebras). We are going to examine a more general situation without having supposed that S is a result of a deformation.

Consider a linear space V equipped with a "closed" symmetry $S : V^{\otimes^2} \to V^{\otimes^2}$. "Closed" means that S can be extended upon a symmetry $S : (V \oplus V^*)^{\otimes^2} \to (V \oplus V^*)^{\otimes^2}$ in a such way that the pairing operator $\langle , \rangle : V \oplus V^* \to k$ is S-invariant. Here V^* is called "right" dual to V (one can define the "left" dual too). Then there exists a symmetry

$$S : (V \oplus \mathrm{End}(V))^{\otimes^2} \to (V \oplus \mathrm{End}(V))^{\otimes^2}$$

such that evaluation operator $ev : V \otimes \mathrm{End}(V) \to V$ is S-invariant i.e. the diagram

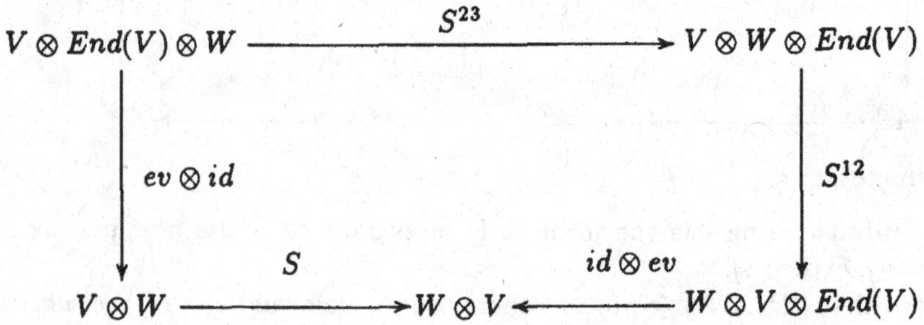

$$\begin{array}{ccccc}
V \otimes End(V) \otimes W & & \xrightarrow{\quad S^{23} \quad} & & V \otimes W \otimes End(V) \\
\Big\downarrow ev \otimes id & & & & \Big\downarrow S^{12} \\
V \otimes W & \xrightarrow{\quad S \quad} & W \otimes V & \xleftarrow{\quad id \otimes ev \quad} & W \otimes V \otimes End(V)
\end{array}$$

is commutative, where $W = V$ or End(V) or any element of tensor rigid category generated by the space V (see [L]). Here End(V) is treated as a space of the "right" endomorphisms.

This symmetry is constructed with the help of the endomorphism $\mathrm{End}(V) \cong V^* \otimes V$. The multiplication operator $\mu : \mathrm{End}(V)^{\otimes^2} \to \mathrm{End}(V)$ is S-invariant by construction.

There exists also S-trace on End(V), i.e. morphism $tr_S : \mathrm{End}(V) \to k$ such that this morphism is S-invariant ($S(id \otimes tr_S) = (tr_S \otimes id)S$) and the corresponding pairing $\langle a, b \rangle_S = tr_S\, a \circ b$ is S-symmetric ($\langle , \rangle_S = \langle , \rangle_S S$).

Furthermore the diagrams

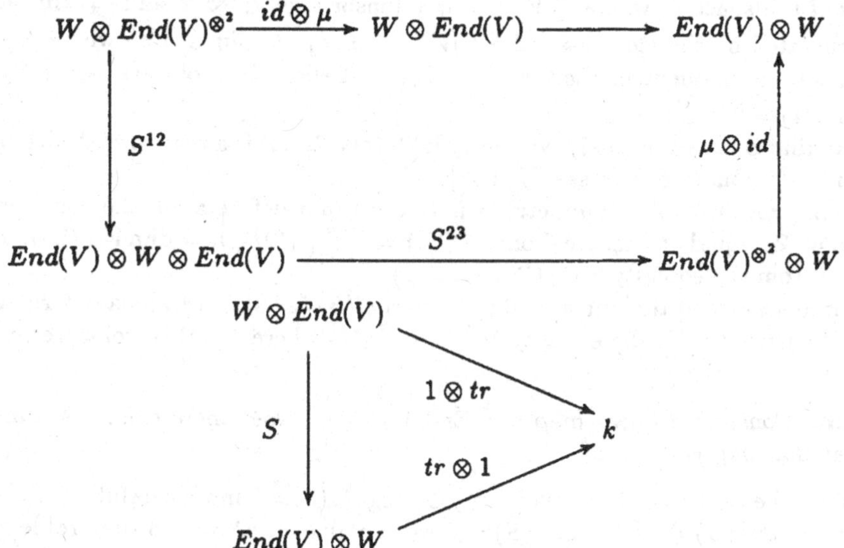

are commutative for all W.

If there exists a "good" pairing on the space V one can construct a conjugation on the space End(V) as follows.

Let $\langle,\rangle_V : V^{\otimes^2} \to k$ be a nondegenerate S-symmetric and S-invariant pairing i.e. the relations

1) $\langle,\rangle_V = \langle,\rangle_V S$;
2) $\langle,\rangle_V \otimes id = (id \otimes \langle,\rangle_V) S^{12} S^{23}$

hold.

Firstly we define the operator conj_S on the space $V^{\otimes^2}$ by $\mathrm{conj}_S = S$, then we define the operator conj_S on the space End(V) using the isomorphism $V^{\otimes^2} \simeq \mathrm{End}(V)$. This isomorphism is given by $a \otimes b \to A_{a\otimes b}$ where $A_{a\otimes b} v = a\langle b, v\rangle$.

In [GRZ] it is proved that the relations

1) $S(\mathrm{conj}_S \otimes id) = (id \otimes \mathrm{conj}_S) S$;
2) $\mathrm{conj}_S\, \mu = \mu(\mathrm{conj}_S \otimes \mathrm{conj}_S) S$;
3) $tr_S\, \mathrm{conj}_S = tr_S$

hold.

Remark. One can also define a conjugation if a pairing $\langle,\rangle_S$ is S-antisymmetric.

It is easy to see that the operator $[,]_S : \mathrm{End}(V)^{\otimes^2} \to \mathrm{End}(V)^{\otimes^2}$ defined as follows $[a, b]_S = a \circ b - \mu S(a \otimes b)$ endows the space End(V) with an S-Lie algebra structure. We denote by $gl(S)$ this S-Lie algebra. One can show that the linear space $\mathcal{J} \subset gl(S)$ of S-traceless elements ($a \in \mathcal{J} \leftrightarrow tr_S a = 0$) is S-Lie subalgebra i.e. $\mathcal{J}$ is S-invariant and $[a, b]_S \in \mathcal{J}$ if $a, b \in \mathcal{J}$. We denote by $sl(S)$ this S-Lie algebra.

Not every symmetry S (and S-Lie algebra) can be obtained by means of deformation of classical objects. The following question is very natural: how to describe the deformation? There is no complete answer. But there are some invariants of the deformation. They are so called Poincaré series of the space V.

Consider the "symmetric" $\Lambda_+(V) = T(V)/I_-$ and the "exterior" $\Lambda_-(V) = T(V)/I_+$ algebras of the space V where $T(V)$ is a free tensor algebra of V and $I_\pm$ are ideals in $T(V)$ generated by the elements $(id \pm S)V^{\otimes^2}$. Let $r_\pm^{(i)} = \dim \Lambda_\pm^i(V)$ where $\Lambda_\pm^i(V)$ is homogeneous component of the algebra $\Lambda_\pm(V)$. Define the Poincaré series $\mathcal{P}_\pm(t)$ as follows $\mathcal{P}_\pm(t) = \sum r_\pm^{(i)} t^i$.

We say that S is a even (odd) symmetry if $\mathcal{P}_-(t)$ ($\mathcal{P}_+(t)$) is a polynomial with highest coefficient 1 (for more details see [L], [G3]).

One can prove that if a symmetry S is a result of a deformation the corresponding polynomial $\mathcal{P}_-$ equals to classical one: $\mathcal{P}_-^{cl}(t) = \sum_{i=0}^n \binom{n}{i} t^i$, $n = \dim V$ ($\mathcal{P}_+(t)$ can be computed from the equality $\mathcal{P}_+(t)\mathcal{P}_-(-t) = 1$).

The necessary condition for an S-Lie algebra $gl(S)$ (or $sl(S)$) to be a deformation of classical Lie structure is the equality $\mathcal{P}_-(t) = \mathcal{P}_-^{cl}(t)$ where $\mathcal{P}_-(t)$ is Poincaré series for $\Lambda_-(V)$.

Problem. *Consider a linear map* $\alpha : \mathrm{End}(V) \to k$. *Does there exist a symmetry S such that this map is S-trace?*

Consider the case of an even symmetry S, $\deg \mathcal{P}_-(t) = 2$ more carefully (in this case we say that $rank(S) = 2$). If $rank(S) = 2$ we can give an answer to the problem using the results of [G3].

Let us fix a basis $\{h_i^j\} \in \mathrm{End}(V)$ such that $h_i^j(e_p) = e_i\delta_p^j$. Denote $tr_S h_i^j = M_i^j$ and consider a matrix $M = \|M_i^j\|$.

Proposition 7. *If S is an even symmetry and $rank(S) = 2$ then the matrix M satisfies the conditions*

1) *$tr.M = 2$ (here tr is usual trace);*
2) *If Jordan form of the matrix M contains a cell $K_{\lambda,m}$ with an eigenvalue λ dimension m, then it contains a cell $K_{\lambda^{-1},m}$ with the same multiplicity.*

Conversely if M satisfies the conditions above there exists a symmetry $S : V^{\otimes^2} \to V^{\otimes^2}$ such that a map $h_i^j \to M_i^j$ is the corresponding S-trace on the space $\mathrm{End}(V)$.

In this case the S-Lie bracket on $\mathrm{End}(V)$ is given by

$$[h_r^l, h_q^n] = 2\delta_r^n h_q^l - M_q^l \delta_r^n h_i^i - M_r^l h_q^n - M_q^n h_r^l + M_r^l M_q^n h_i^i$$

where h_i^j means the operator such that $h_i^j(e_p) = e_i\delta_p^j$ and $M = \|M_i^j\|$ is the operator satisfying the conditions above (see [G3]).

It is very surprisingly that the operator $[,]_S$ depends only on the operator M (in example below the bracket depends on a parameter $\beta = -(p+q)$ though the symmetry S depends on two parameters p and q).

When $\dim V = 2$, the operator M has the following form in some basis: $\begin{pmatrix} 1 & \beta \\ 0 & 1 \end{pmatrix}$.

The corresponding symmetry is

$$S = \begin{pmatrix} 1 & p & -p & pq \\ 0 & 0 & 1 & -q \\ 0 & 1 & 0 & q \\ 0 & 0 & 0 & 1 \end{pmatrix}$$

where $\beta = -(p+q)$ and the multiplication is given by the following multiplication table

	h_1^1	h_1^2	h_2^1	h_2^2
h_1^1	0	$-h_1^2+\beta h_2^2$	h_2^1	0
h_1^2	$h_1^2-\beta h_1^1$	$-2\beta h_1^2+\beta^2(h_1^1+h_2^2)$	$h_2^2-h_1^1-\beta h_2^1$	$-h_1^2+\beta h_1^1$
h_2^1	$-h_2^1$	$h_1^1-h_2^2-\beta h_2^1$	0	h_2^1
h_2^2	0	$h_1^2-\beta h_2^2$	$-h_2^1$	0

For instance $[h_1^2, h_2^1] = h_2^2 - h_1^1 - \beta h_2^1$.

Note that $tr_S h_1^1 = tr_S h_2^2 = 1$, $tr_S h_2^1 = 0$, $tr_S h_1^2 = \beta$.

Consider now S-Lie algebra $sl(S)$. Fix in $sl(S)$ the following basis $H = h_1^1 - h_1^2$, $X = h_2^1$, $Y = h_1^2 - \beta h_1^1$ (it is easy to see that $tr_S H = tr_S X = tr_S Y = 0$).

The multiplication table for H, X, Y is following

	H	X	Y
H	0	$2X$	$-2Y-2\beta H$
X	$-2X$	0	H
Y	$2Y$	$-H-2\beta X$	$-2\beta Y$

if $\beta = 0$ we obtain the classical Lie algebra sl_2. Construct now S-involution inv_S: $\mathrm{End}(V) \to \mathrm{End}(V)$ as follows. There exists a decomposition of $sl(S)$-modules $\mathrm{End}(V) = sl(S) \oplus k \cdot \mathbb{1}$, where $\mathbb{1} = h_1^1 + h_2^2$. Put $inv_S = id$ on $k \cdot \mathbb{1}$ and $inv_S = -id$ on $sl(S)$. In the base $\{h_i^j\}$ this S-involution has form

$$h_1^1 \to h_2^2, \quad h_1^2 \to -h_1^2 + \beta \circ \mathbb{1}, \quad h_2^1 \to -h_2^1, \quad h_2^2 \to h_1^1.$$

If $p = q$ this involution arises from a conjugation. To demonstrate it one has to introduce an S-antisymmetric pairing on V as follows $\langle e_i, e_j \rangle = u_{ij}$ (see the remark above). Let us underline that the S-Lie algebra $sl(S)$ can be obtained by quantization of the following solution of the CYBE $R = a(H \otimes X - X \otimes H)$ for some a.

References

[D] Drinfeld V., *On constant, quasiclassical solutions of the Yang-Baxter quantum equation*, Sov. Math., Dokl. **28** (1983), 667–671.

[G1] Gurevich D., *The Yang-Baxter equation and generalization of formal Lie theory*, Sov. Math., Dokl. **33** (1986), 758–762.

[G2] Gurevich D., *Equation de Yang-Baxter et quantification des cocycles*, Compt. Rend. Ac. Sci. Paris, Ser.1 **310** (1990), 845–848.

[G3] Gurevich D., *Algebraic aspects of the quantum Yang-Baxter equation*, Algebra Anal **2** no. 4 (1990), 119–148. (in Russian)

[GRZ] Gurevich D., Rubtsov V., Zobin N., *Quantization of Poisson pairs: the R-matrix approach*, Geometry and Physics (to appear).

[L] Lyubashenko V., *Vectorsymmetries*, Rep. Dep. Math., Univ. of Stockholm **19** (1987), 1–77.

QUANTUM G–SPACES AND HEISENBERG ALGEBRA

L.I.KOROGODSKY AND L.L.VAKSMAN

Department of Mathematics, Rostov State University

ABSTRACT. In this paper we construct an isomorphism between the quantum Heisenberg algebra and a quantum function algebra. Some applications to the representation theory of quantum groups $SU(n,1)$ and $SU(n+1)$ are given.

1. The development of the quantum inverse scatterring method (QISM) gave rise in the 1980's to the basic notions of the quantum group theory given in the papers of V.G. Drinfeld and M. Jimbo [1,2].

It became evident that various results of the real semisimple Lie group theory can be generalized to the quantum case [21] (we can mention, for example, harmonic analysis on homogeneous G-spaces,the geometric realization of the discrete series representations, the character theory). The most interesting results are those which display the differences between quantum and classical groups. Let us describe some of these differences.

In the paper by N.Yu.Reshetikhin, L.A.Takhtajan and L.D.Faddeev [3] a quantum function algebra of Hermitian space has been introduced. It can be defined with the generators $z_0, z_1, ..., z_{n+1}, z_0^*, z_1^*, ..., z_{n+1}^*$ and the fundamental relations (2) (see below). It has turned out [22] that the subalgebra of this function algebra generated by $z_0, z_1, \ldots, z_n, z_0^*, z_1^*, \ldots, z_n^*$ and $\varkappa = z_{n+1}z_{n+1}^*$ is isomorphic to the quantum Heisenberg algebra.

We shall show that this isomorphism can be used to solve some problems in the representation theory of the quantum group $SU(n,1)$. It is also used in discussion of difficulties related to the notion of tensor product of representations in the case of the quantum group $SU(1,1)$.

2. Let X be a smooth manifold with the action of a connected real Lie group G. For each element of the Lie algebra $\mathfrak{g}$ there is a corresponding vector field on X, and for each element of the universal enveloping algebra $U\mathfrak{g}$ there is a corresponding differential operator acting on $C^\infty(X)$.

A quantum function algebra of X in example considered below is an algebra $\mathfrak{F}$ over $\mathbb{C}$ endowed with a left $U_h\mathfrak{g}$ -module structure(where $U_h\mathfrak{g}$ stands for quantum universal enveloping algebra). The multiplication $\mathfrak{F}\otimes\mathfrak{F} \to \mathfrak{F}$ is a morphism, so $\mathfrak{F}$ is a $U_h\mathfrak{g}$ -module algebra (see [16] for the definitions). To quantize a real manifold X we need a conjugate linear involution in $\mathfrak{F}$ compatible with the involution in $U_h\mathfrak{g}$ in the following sense:

$$(\xi f)^* = (S(\xi))^* f^*, \qquad (\xi \in U_h\mathfrak{g}, f \in \mathfrak{F}), \tag{1}$$

where S is the antipode (recall that $U_h\mathfrak{g}$ is a Hopf $*$-algebra), i.e. it is endowed with a conjugate linear involution such that

$$(\xi\eta)^* = \eta^*\xi^*, \qquad \Delta(\xi^*) = \Delta(\xi)^{*\otimes *}, \qquad \varepsilon(\xi) = \overline{\varepsilon(\xi)},$$

where Δ is the comultiplication, ε is the counit [11,21].

In the present paper we introduce examples of quantum G-spaces with $G = SU(n+1)$ or $G = SU(n,1)$. But some of the results can be obtained in the more general case of $G = SU(m,n)$ [22].

We use h (or $q = e^{-h}$) as the notation for the deformation parameters. Let $h > 0$.

Recall the definition of $U_h sl(n+1)$ [1,2] and its real forms $U_h su(n+1)$ and $U_h su(n,1)$. $U_h su(n+1)$ is generated by $E_i, F_i, K_i^{\pm}$ $(i = 1,2,...,n)$ with the fundamental relations

$$K_i^+K_i^- = 1, \qquad [K_i^{\pm}K_j^{\pm}] = [K_i^{\pm}K_j^{\mp}] = 0,$$

$$K_i^{\pm}E_j = e^{\pm a_{ij}h/4}E_jK_i^{\pm}, \quad K_i^{\pm}F_j = e^{\mp a_{ij}h/4}F_jK_i^{\pm},$$

$$E_iF_j - F_jE_i = \delta_{ij}(K_i^2 - K_i^{-2})/(2\sinh h/2), \quad [E_i,E_j] = [F_i,F_j] = 0(|i-j| \neq 1),$$

$$E_i^2E_{i\pm1} - 2\cosh h/2E_iE_{i\pm1}E_i + E_{i\pm1}E_i^2 = 0, \; F_i^2F_{i\pm1} - 2\cosh h/2F_iF_{i\pm1}F_i + F_{i\pm1}F_i^2 = 0,$$

where $a_{ii} = 2, a_{ii\pm1} = -1$ and $a_{ij} = 0(|i-j| > 1)$. The Hopf algebra structure is given by[1]

$$\Delta(E_i) = E_i \otimes 1 + K_i^{-2} \otimes E_i, \; \Delta(F_i) = F_i \otimes K_i^2 + 1 \otimes F_i, \; \Delta(K_i^{\pm}) = K_i^{\pm} \otimes K_i^{\pm},$$

$$S(E_i) = -K_i^2E_i, \; S(F_i) = -F_iK_i^{-2}, \; S(K_i^{\pm}) = K_i^{\mp}, \; \varepsilon(E_i) = 0, \; \varepsilon(F_i) = 0, \; \varepsilon(K_i^{\pm}) = 1.$$

The involution given by $E_i^* = K_i^{-2}F_i, F_i^* = E_iK_i^{+2}, (K_i^{\pm})^* = K_i^{\pm}, (i = 2,3,...,n)$, $E_1^* = \pm K_1^{-2}F_1, F_1^* = \pm E_1K_1^{+2}, (K_1^{\pm})^* = K_1^{\pm}$ defines the Hopf $*$-algebra $U_h su(n+1)$ (in the case of the upper sign) or $U_h su(n,1)$ (in the case of lower sign).

3. We give a construction of quantum G-spaces which are not necessarily realized as quotient space of the form G/H. The first examples of this kind were obtained by P.Podleś [5] (the so-called quantum Podleś 2-spheres).

A family of quantum 3-spheres was introduced in [7]. These quantum spaces can be obtained via our construction in the simplest case of $G = SU(2)$.

This construction can be illustrated by the following geometric picture.

Suppose $h = 0$ and $G = SU(n+1)$. Let us consider the $SU(n+2)$ space $\mathbb{C}^{n+2}$ on which the group acts as on the carrier space of the first fundamental representation. By embedding $SU(n+1)$ into $SU(n+2)$ we equip $\mathbb{C}^{n+2}$ with the $SU(n+1)$-space structure.

The complex hyperplanes $z_{n+1} = const$ are $SU(n+1)$-subspaces (where $z_0, z_1, ..., z_n, z_{n+1}$ are coordinates defined by (3)), the sphere and the hyperboloid (or the point and the cone) defined by

$$|z_0|^2 + |z_1|^2 + \ldots |z_n|^2 \pm |z_{n+1}|^2 = \sigma, \text{ where } \sigma \geq 0$$

[1] Recall that the tensor product of the representations π_1, π_2 of a Hopf algebra is the representation defined by $a \mapsto \pi_1 \otimes \pi_2\Delta(a)$.

are homogeneous $SU(n+2)$ and $SU(n+1,1)$-spaces respectively. Their sections by the hyperplanes $z_{n+1} = const$ are easily seen to be homogeneous $SU(n+1)$-spaces, namely $(2n+1)$-spheres.

The quantum analogue of this construction provides us with a family of quantum S^{2n+1}. We can also construct a family of quantum $\mathbb{CP}^n$ which are quantum analogues of the base of the Hopf fibration $S^{2n+1} \to \mathbb{CP}^n$. When $G = SU(n,1)$ we get a family of quantum $(2n+1)$-and $(2n)$-hyperboloids.

4. The function algebra $\mathbb{C}[\mathbb{C}^N]_{q,1/q}$ of the quantum G-space $\mathbb{C}^N$ (where $G = SU(N)$ or $G = SU(N-1,1)$) has been introduced in [3] as a $*$-algebra generated by the matrix elements $z_i = t_{0i} (i = 0,1,...,N-1)$ of the first fundamental representation (see [3]). It can be described by the fundamental relations

$$
\begin{gathered}
z_i z_j = e^{-h/2} z_j z_i, \quad z_i^* z_j^* = e^{h/2} z_j^* z_i^* \ (i<j), \\
z_i z_j^* = e^{-h/2} z_j^* z_i \ (i \neq j), \\
z_i z_i^* - z_i^* z_i = (e^h - 1) \sum_{k>i} z_k z_k^* \ (i = 1,2,...,N-1) \\
z_0 z_0^* - z_0^* z_0 = \pm (e^h - 1) \sum_{k>0} z_k z_k^*.
\end{gathered} \tag{2}
$$

(Here the upper (the lower) sign corresponds to the case of $G = SU(N)$, $(G = SU(N-1,1))$.) The $U_h\mathfrak{g}$-module structure is determined by

$$
\begin{gathered}
E_i : z_j \mapsto \delta_{ij} e^{h/4} z_{j-1}, \qquad F_i : z_j \mapsto \delta_{i-1,j} e^{-h/4} z_{j+1}, \\
K_i^{\pm} : z_{i-1} \mapsto e^{\pm h/4} z_{i-1}, \quad z_i \mapsto e^{\mp h/4} z_i, \quad z_j \mapsto z_j \ (j \notin \{i-1, i\})
\end{gathered} \tag{3}
$$

and by (1).

Suppose $N = n+2$. Let us consider $\mathbb{C}[\mathbb{C}^{n+2}]_{q,1/q}$ as an $U_h sl(n+1)$-module algebra, with this structure being defined by the embedding $U_h sl(n+1)$ into $U_h sl(n+2)$ such that $E_i \mapsto E_i, F_i \mapsto F_i, K_i^{\pm} \mapsto K_i^{\pm}$ $(i = 1,2,...,n)$.

The subalgebra M_q^{2n+3} generated by $z_i, z_i^* (i = 0,1,...,n)$ and $\varkappa = z_{n+1} z_{n+1}^*$ is an $U_h sl(n+1)$-module algebra. It can be described by the following fundamental relations (between the generators $z_0, z_1, ..., z_n, z_0^*, z_1^*, ..., z_n^*, \varkappa, \sigma$):

$$
\begin{gathered}
z_i z_j = e^{-h/2} z_j z_i, \quad z_i^* z_j^* = e^{h/2} z_j^* z_i^* \ (i<j), \\
z_i z_j^* = e^{-h/2} z_j z_i^* \ (i \neq j), \\
z_0 z_0^* - z_0^* z_0 = \pm (e^h - 1) (\sum_{k>0} z_k z_k^* + \varkappa), \\
z_i z_i^* - z_i^* z_i = (e^h - 1) (\sum_{k>i} z_k z_k^* + \varkappa) \ (i = 1,2,...,N-1), \\
z_i \varkappa = e^{-h} \varkappa z_i, \quad z_i^* \varkappa = e^h \varkappa z_i^*, \quad \varkappa^* = \varkappa, \\
\pm z_0 z_0^* + \sum_{k>0} z_k z_k^* = e^h \sigma + \varkappa, \quad \sigma^* = \sigma, \\
z_i \sigma = \sigma z_i, \quad z_i^* \sigma = \sigma z_i^* \ (i = 0,1,...,n), \quad \varkappa \sigma = \sigma \varkappa.
\end{gathered} \tag{4}
$$

The elements $\varkappa$ and σ generate the subalgebra of $U_h sl(n+1)$-invariants: $\{f \in M_q^{2n+3} \mid \xi f = \varepsilon(\xi)f, \forall \xi \in U_h sl(n+1)\}$. They can be regarded as parameters of a family of quantum $(2n+1)$ -spheres or $(2n+1)$-hyperboloids respectively (though $\varkappa$ does not belong to the center of M_q^{2n+3}), and M_q^{2n+3} can be regarded as a function algebra on the total space of the family.

Note that when $\varkappa = 0$ and $\sigma = 1$ we get the "ordinary" quantum S^{2n+1} introduced in [20] as a quotient space.

After the work had been finished, we learned at the conference in EIMI that M.Nuomi and K.Mimachi also obtained the family of quantum S^{2n+1} in a quite different way.

The automorphisms of M_q^{2n+3}, such that $z_j \mapsto e^{i\phi}z_j$, $z_j^* \mapsto e^{-i\phi}z_j^*$, $\varkappa \mapsto \varkappa$ are also $U_h\mathfrak{g}$-module morphisms (where $\mathfrak{g} = su(n+1)$ or $\mathfrak{g} = su(n,1)$), so the subalgebra of invariants is an $U_h\mathfrak{g}$-module algebra(note that $\varkappa$ belongs to its center).

Assuming that $\varkappa$ and σ are real, we get an $U_h\mathfrak{g}$-module algebra denoted $\mathbb{C}[X_{\varkappa,\sigma}]_q$ which can be regarded as a function algebra of a quantum G-space $X_{\varkappa,\sigma}$. When $G = SU(n+1)$, these are quantum $\mathbb{CP}^n$, when $G = SU(n,1)$, these are quantum $(2n)$ -hyperboloids. When $G = SU(1,1)$, we can put also $\varkappa = \overline{\sigma} \notin \mathbb{R}$, then we get a quantum one-sheet 2-hyperboloid.

It is easy to see that $X_{\alpha\varkappa,\alpha\sigma} \simeq X_{\varkappa,\sigma} (\alpha \in \mathbb{R} \setminus 0)$ and, when $n = 1$, $X_{\varkappa,\sigma} = X_{\sigma,\varkappa}$. In the general case $X_{\varkappa,\sigma}$ are not equivalent unlike the classical case.

When $G = SU(n+1)$, the quantum $SU(n+1)$-space $X_{\varkappa,\sigma}$ is not empty (i.e. there is a $*$-representation of $\mathbb{C}[X_{\varkappa,\sigma}]_q$) in the following cases only:

$$1) \ \frac{\varkappa}{\sigma} = q^{(n+1)l+n}, (n+1)l \in \mathbb{Z}/\{-1,-2,...,-n\}; \ 2) \ \varkappa\sigma \le 0.$$

5. The subalgebra of K_i-invariant functions $\{f \in \mathbb{C}[X_{\varkappa,\sigma}]_q \mid K_i^{\pm} f = f, i = 1,2,...,n\}$ is generated by the elements

$$x_i = e^{-h/2}(\sum_{k \ge i} z_k z_k^* + \varkappa) \qquad (i = 1,2,...,n).$$

In [22] we have introduced an $U_h\mathfrak{g}$-module algebra $Fun(X_{\sigma,\varkappa})_q$ obtained by adding functions $f(x_1,...,x_n)$ to $\mathbb{C}[X_{\varkappa,\sigma}]_q$.

Let $\mathfrak{F}$ be a function algebra of a quantum G-space X. Let us call invariant integral a linear functional $\mathfrak{F} \to \mathbb{C}$ such that $\int (\xi f) d\nu = \varepsilon(\xi) \int f d\nu$ for every $\xi \in U_h\mathfrak{g}, f \in \mathfrak{F}$. It is easy to see that the coimage of invariant integral is embedded into the space of K_i-invariant functions $f(x_1,...,x_n)$.

Invariant integrals on the quantum spaces $X_{\varkappa,\sigma}$ in the general case will be obtained later on. Now we consider the case of $G = SU(1,1)$. The following formula defines an invariant integral on $X_{\varkappa,\sigma}$: $\int f(x_1) d\nu = 2 \sinh \frac{h}{2} \sum_{t \in \mathfrak{M}} t f(t)$, where $\mathfrak{M}$ is a union of geometric progressions with the ratio $q = e^{-h}$. In the different cases $\mathfrak{M}$ is as follows $(\sigma > 0)$:

1) $\mathfrak{M}_\beta = \{-q^{k+\beta}\}_{k \in \mathbb{Z}} \cup \{\varkappa q^{k+1/2}\}_{k \in \mathbb{Z}_+}$, $\beta \in [0,1)$, the case of $\varkappa = 0$ included,
2) $\mathfrak{M}_\beta = \{\sigma q^{-(k+1/2)}\}_{k \in \mathbb{Z}_+}$,
3) Let $\sigma \ge \varkappa > 0$, $\log_q(\frac{\varkappa}{\sigma}) < 1$, $\mathfrak{M}_{\beta',\beta''} = \{-q^{k+\beta'}\}_{k \in \mathbb{Z}} \cup \{q^{k+\beta''}\}_{k \in \mathbb{Z}}$, where $\beta', \beta'' \in [0,1)$, and β'' satisfies the condition that there is no $k \in \mathbb{Z}$ such that $\varkappa < q^{k+\beta''+1/2} < \sigma$,

4) $\varkappa = \overline{\sigma} \notin \mathbf{R}$, $\mathfrak{M}_{\beta',\beta''} = \{-q^{k+\beta'}\}_{k\in\mathbf{Z}} \cup \{q^{k+\beta''}\}_{k\in\mathbf{Z}}, \beta', \beta'' \in [0,1)$,
5) $\varkappa = \sigma = 0$, $\mathfrak{M}_{\beta',\beta''} = \{-q^{k+\beta'}\}_{k\in\mathbf{Z}} \cup \{q^{k+\beta''}\}_{k\in\mathbf{Z}}$, $\beta', \beta'' \in [0,1)$.

Harmonic analysis in the cases 1) and 2) has been studied in [21,22] , in [21] the Plancherel formula for the quantum group $SU(1,1)$ has been obtained (a part of these results has been obtained also in [12]).

In the case 1) there is a quantum effect that invariant integral on the sheet of quantum hyperboloid is not unique. The parameter β appears in the Plancherel formula under the sign of theta-function.

Indeed, it is a general fact that harmonic analysis on quantum $SU(1,1)$-spaces in complete analogy with the classical case is reduced to spectral analysis for the second order Casimir operator. Asymptotic behavior of its eigenfunctions determines the spectral measure. And it was G.D.Birkhoff who had first noticed long ago [13] that asymptotic behavior of solutions of q-difference equations may be expressed in terms of theta-functions.

Another quantum effect is appearance of discrete spectrum of the Casimir operator. It is determined by zeroes of theta-functions and depends on β. This corresponds to the fact that in this case decomposition of the quasi-regular representation[2] includes the so-called "strange" series representations (which have no classical analogue) [11,12,21,22] as direct summands.

In the case 2) the second order Casimir operator is bounded. Its spectrum is continuous and "strange" series representations are not contained in the quasi-regular representation.

6. Now let as consider another subject. We introduce the quantum Heisenberg algebra as an extension of the quantum function algebra $\mathbf{C}[\mathbf{C}^{n+1}]_{q,1/q}$.

Let $V \simeq \mathbf{C}^{n+1}$ be the carrier space of the first fundamental representation of $U_h sl(n+1)$, V^* be the dual $U_h sl(n+1)$-module[3], $\{z_0, ..., z_n\}$ and $\{\hat{z}_0, ..., \hat{z}_n\}$ be the dual bases of V and V^* respectively (see (3)).

Let us consider the tensor algebra $T(V \oplus V^*)$. $\mathbf{C}[\mathbf{C}^{n+1}]_{q,1/q}$ can be described as the quotient $U_h sl(n+1)$-module algebra $T(V \oplus V^*)/J(L)$ where ideal $J(L)$ is generated by the $U_h sl(n+1)$-submodule $L \subset (V \oplus V^*)^{\otimes 2}$(see (2)). The element

$$\sum_{k=0}^{n} z_k \otimes \hat{z}_k - \sum_{k=0}^{n} e^{kh} \hat{z}_k \otimes z_k$$

generates the trivial submodule of L, when $n > 1$(of $L \cap (V \otimes V^* \oplus V^* \otimes V\ )$, when $n = 1$).

Let L' be a complementary submodule. We define $\mathcal{H}_q(n+1) = T(V \oplus V^*)/J(L')$. $\mathcal{H}_q(n+1)$ is a quantum analogue of the Heisenberg algebra. It can be described by the

[2] We call quasi-regular representation on a quantum G-space X the representation of $U_h\mathfrak{g}$ on a quantum algebra $\mathfrak{F}$.

[3] Recall that $\xi_{|V^*} = (S(\xi)|_V)^*$.

following fundamental relations:

$$z_i z_j = e^{-h/2} z_j z_i, \quad \hat{z}_i \hat{z}_j = e^{h/2} \hat{z}_j \hat{z}_i \ (i < j),$$
$$z_i \hat{z}_j = e^{-h/2} \hat{z}_j z_i \ (i \neq j),$$
$$z_j \hat{z}_j - \hat{z}_j z_j = \tilde{\varkappa} + (e^h - 1) \sum_{k>j} z_k \hat{z}_k, \quad z_j \tilde{\varkappa} = e^{-h} \tilde{\varkappa} z_j, \hat{z}_j \tilde{\varkappa} = e^h \tilde{\varkappa} z_j, \tag{5}$$
$$\text{where} \quad \tilde{\varkappa} = \frac{e^{(n+1)h} - 1}{e^h - 1} \left(\sum_{k=0}^{n} z_k \hat{z}_k - \sum_{k=0}^{n} e^{kh} \hat{z}_k z_k \right).$$

The quantum analogue of the Heisenberg algebra introduced in [9] is isomorphic to the algebra generated by $\varkappa^{-1/2} z_i, \hat{z}_i \varkappa^{-1/2}$ $(i = 0, 1, ..., n)$[4]. Let us note, however , that now the quantum Heisenberg algebra turned out to be an $U_h sl(n+1)$-module algebra.

The automorphism of $\mathcal{H}_q(n+1)$ such that $z_i \mapsto \alpha z_i, \hat{z}_i \mapsto \alpha \hat{z}_i, \tilde{\varkappa} \mapsto \alpha^2 \tilde{\varkappa} (\alpha \in \mathbf{R} \setminus \{0\})$ are $U_h sl(n+1)$-module morphisms. Let us extend $\mathcal{H}_q(n+1)$ by adding $x_i^{\pm 1/4}$ $(i = 0, 1, ..., n+1)$ where

$$x_i = e^{-h/2} \left(\sum_{k \geq i} z_k \hat{z}_k + \frac{\tilde{\varkappa}}{e^h - 1} \right)$$

and consider the subalgebra of invariants. It is generated by $\psi_i, \hat{\psi}_i, \omega_i^{\pm 1}$ $(i = 0, 1, ..., n)$ such that $\omega_i^{\pm 1} = (e^{-h} x_i x_{i+1}^{-1})^{\pm 1/4}$,

$$\psi_i = (e^h - 1)^{-1/2} (x_i x_{i+1})^{-1/4} z_i, \hat{\psi}_i = (e^h - 1)^{-1/2} \hat{z}_i (x_i x_{i+1})^{-1/4}$$

and isomorphic to the quantum Weyl algebra $\mathcal{A}_{q'}^-(n+1)$ (where $q' = e^{h/4}$) introduced in [8].

Let us note, however, that now the quantum Weyl algebra turned out to be an $U_h sl(n+1)$-module algebra. Note also that the involutions of $\mathcal{H}_q(n+1)$ and $\mathcal{A}_{q'}^-(n+1)$ respectively such that $z_i^* = \hat{z}_i, z_0^* = \pm z_0$, $\tilde{\varkappa}^* = \tilde{\varkappa}$, $\psi_i^* = \hat{\psi}_i, \psi_0^* = \pm \hat{\psi}_0, (i = 1, 2, ..., n)$, $(\omega_j^{\pm 1})^* = \omega_j^{\pm 1}$ $(j = 0, 1, ..., n)$ are compatible with (1) where $\mathfrak{g} = su(n+1)$ or $\mathfrak{g} = su(n, 1)$ according to the upper or the lower sign in the above formulae respectively.

7. The quantum universal enveloping algebra $U_h \mathfrak{g}$ can be equipped with an $U_h \mathfrak{g}$-module algebra structure by means of the well-known quantum adjoint representation:

$$(ad_q \xi) \eta = \sum_k \xi_k^{(1)} \eta S(\xi_k^{(2)})$$

where $\Delta(\xi) = \sum_k \xi_k^{(1)} \otimes \xi_k^{(2)}$. In the classical case $(h = 0)$ the representation of $U_h su(n+1)$ in the Weyl algebra is equivalent to the tensor product $T \otimes T^*$ where T stands for a $*$-representation of $U su(n+1)$ namely the Shale-Weil representation. (It is due to a homomorphism from $U su(n+1)$ into the Weyl algebra.)

[4] For $n = 1$ see also [10].

In the quantum case the situation is similar. The homomorphism $\Gamma : U_h su(n+1, \mathbb{C}) \to A^-_{q'}(n+1)$ such that

$$\Gamma : E_i K_i^+ \mapsto \psi_{i-1}\hat{\psi}_i;\ K_i^- F_i \mapsto \psi_i \hat{\psi}_{i-1};\ K_i^{\pm} \mapsto (\omega_{i-1}^{-1}\omega_i)^{\pm 1}$$

was first introduced in [8]. As shown in [22], Γ is an $U_h sl(n+1)$-module algebra morphism.

In [8] the morphism Γ was used to construct the oscillator representation of $U_h sl(n+1)$ in the form of $T = \pi \circ \varkappa$ where π is an irreducible $*$-representation of $A^-_{q'}(n+1)$ It is easy to show [22] that the representation of $U_h su(n+1)$ on $A^-_{q'}(n+1)$ is equivalent to $T \otimes T^*$.

In complete analogy with the classical case the carrier space of T was equipped in [22] with an $Usl(n+1)$ module algebra structure so that it is nothing but the function algebra of the quantum $SL(n+1, \mathbb{C})$-space $\mathbb{C}^{n+1}$, namely the subalgebra of $\mathbb{C}[\mathbb{C}^{n+1}]_{q,1/q}$ generated by $z_i (i = 0, ..., n)$.

Thus the quantum Weyl algebra may be regarded as the quantum analogue of the algebra of holomorphic(antiholomorphic) differential operators with polynomial coefficients on the $SL(n+1, \mathbb{C})$-space $\mathbb{C}^{n+1}$.

8. Now it is time to notice that the quantum function algebra M_q^{2n+3} and the quantum Heisenberg algebra $\mathcal{H}_q(n+1)$ are isomorphic as $U_h\mathfrak{g}$ -module algebras(compare (4) with (5)). The isomorphism is given by $z_i \mapsto z_i, z_i^* \mapsto z_i^*, \tilde{\varkappa} \mapsto (e^h - 1)\varkappa$.

This fact enables us to obtain a series of the following results. For example [22], there is the general observation that some of the quotient algebras of the $U_h\mathfrak{g}$ -module algebra turn out to be isomorphic to the quantum function algebras introduced above (see e.g. (6)).

9. Now we can easily obtain invariant integrals on the quantum spaces $X_{\varkappa,\sigma}$ [22]. We use the $U_h sl(n+1)$ module algebra morphism $\Gamma_{\varkappa,\sigma} : U_h sl(n+1) \to Fun(X_{\varkappa,\sigma})_q$ induced by Γ when $\varkappa\sigma \neq 0$.

Suppose, for example, $G = SU(n+1)$[5]. Let π be an irreducible $*$-representation of $Fun(X_{\varkappa\sigma})$

$$\int f d\nu = tr\pi(f\Gamma_{\varkappa,\sigma}(e^{\rho h/2}))$$

where ρ stands for the half of sum of all positive roots for $sl(n+1)$. We have

$$\Gamma_{\varkappa,\sigma}(e^{\rho h/2}) = (\varkappa\sigma)^{-n/2} x_1 x_2 \cdots x_n,$$

so the formula

$$\int f d\nu_\pi = tr\pi(f \cdot x_1 x_2 \cdots x_n)$$

defines an invariant integral on $X_{\varkappa,\sigma}$ (which, in general, may be zero).

[5] In [22] it has been done also for the analogous quantum $SU(m,n)$-spaces, namely the quantum analogous of the minimal-dimensional coadjoint orbits.

Invariant integral on the quantum $SU(n+1)$-space $X_{\varkappa,\sigma}$ is unique up to a constant and is given by

$$\int f(x_1,...,x_n)d\nu = \int_{\sqrt{q}\sigma}^{\sqrt{q}\varkappa} d_q x_n \cdot \int_{\sqrt{q}\sigma}^{x_n} d_q x_{n-1} \cdots \int_{\sqrt{q}\sigma}^{x_2} d_q x_1 f(x_1,...,x_n)^6 \qquad (6)$$

(recall that the coimage of invariant integral is embedded into the space of K_i-invariant functions). The same formula defines a $\mathbb{C}[\varkappa,\sigma]$-valued invariant integral of the family of quantum S^{2n+1}.

The invariant integrals on the quantum Podleś 2-spheres ($\mathbb{CP}^1$), on the family of quantum 3-spheres, and on the "ordinary" quantum S^{2n+1} (when $\varkappa = 0$ and $\sigma = 1$) were first obtained in [6],[7],[20] respectively (in another way).

In complete analogy we can obtain invariant integrals on the quantum $SU(n,1)$-spaces $X_{\varkappa\sigma}$. There is no uniqueness in this case, for instance, for invariant integrals on the quantum $SU(1,1)$-spaces.

10. It is well known that in classical case the following problems are closely related: harmonic analysis on groups or on homogeneous spaces and decomposition of tensor product of two irreducible unitary group representations.

For instance, there exists a formula expressing asymptotic behavior of Clebsch-Gordan coefficients for $SU(2)$ in terms of the Jacobi polynomials, i.e. spherical functions on $SU(2)$ [14].

In the quantum case the analogy becomes much closer. Suppose $G = SU(n+1)$.

Let us consider a discrete family of quantum $\mathbb{CP}^n$, namely the family of quantum $SU(n+1)$-spaces $X_{\varkappa,\sigma}$ determined by the condition $\varkappa/\sigma = q^{(n+1)l+n}, (n+1)l \in \mathbb{Z} \setminus \{-1,-2,...,-n\}$. In [22] the following results have been obtained.

In this case $Fun(X_{\varkappa,\sigma})$ has a unique irreducible $*$-representation $\pi_{\varkappa,\sigma}$. It is finite-dimensional, so $Fun(X_{\varkappa,\sigma})_q$ is finite-dimensional, too.

It is easy to see that $\Gamma_{\varkappa,\sigma}$ is a $*$-morphism, when $\varkappa\sigma > 0$. Therefore, $\pi_{\varkappa,\sigma} \circ \Gamma_{\varkappa,\sigma}$ is an irreducible $*$-representation of $U_h su(n+1)$, namely $T_{0,...,0,(n+1)l}$ (when $l \geq 0$) or $T_{-(n+1)(l+1),0,...,0}$ (when $l \leq -1$), where $T_{p_1,...,p_n}$ stands for the irreducible $*$-representation of $U_h su(n+1)$ with the highest weight $(p_1,...,p_n) \in \mathbb{Z}_+^n$.

It follows that $Fun(X_{\varkappa,\sigma})_q \simeq U_h su(n+1)/KerT$ (where $T = \pi_{\varkappa,\sigma} \circ \Gamma_{\varkappa,\sigma}$) as $U_h su(n+1)$-module algebras. We see that the quasi-regular representation of $U_h su(n+1)$ on $Fun(X_{\varkappa,\sigma})_q$ is equivalent to $T \otimes T^*$ where $T = T_{0,...,0,(n+1)l}$ (when $l \geq 0$) or $T = T_{-(n+1)(l+1),0,...,0}$ (when $l \leq -1$).

This fact enables us to identify the corresponding quantum Clebsch-Gordan coefficients (QCGC) regarded as orthogonal polynomials in discrete variable with spherical functions on $X_{\varkappa,\sigma}$. Indeed, the QCGC may be obtained as a result of harmonic analysis on the $U_h su(n+1)$-module algebra $U_h su(n+1)/KerT$.

Therefore we can obtained the QCGC by using explicit formulae for spherical functions on $X_{\varkappa,\sigma}$ [6,7] (compare with [17-19]).

In complete analogy we can deal with the tensor products $T_{0,...,0,p'} \otimes T_{p'',0,...,0}$ [22] (recall that $T^*_{p_1,...,p_n} \simeq T_{p_n,...,p_1}$). The corresponding QCGC regarded as orthogonal

[6] $\int_a^b f(t)d_q t = (1-q)\{\sum_{k=0}^{\infty} bq^k f(bq^k) - \sum_{k=0}^{\infty} aq^k f(aq^k)\}$ is the Jackson's q-integral.

polynomials in discrete variable may be identified with spherical functions on the family of quantum spheres S^{2n+1}.

We see that the asymptotic equality becomes exact in the quantum case.

11. We have considered above the geometric realization of irreducible $*$-representations of $U_h su(n+1)$ in the form of $\pi_{\varkappa,\sigma} \circ \Gamma_{\varkappa,\sigma}$ where $\pi_{\varkappa,\sigma}$ is an irreducible $*$-representation of $Fun(X_{\varkappa,\sigma})_q$.

Similar results may be obtained also for irreducible $*$-representations of $U_h su(n+1)$ [22]. They have been described in [11,22] and parametrized as follows:

(1) $T_{l,\varepsilon}$ where $l = -\frac{1}{2} + i\rho, \rho \in [0, \frac{\pi}{h}]$ and $\varepsilon \in [0,1) \setminus \{\frac{1}{2}\}$ or $\rho \in (0, \frac{\pi}{h}]$ and $\varepsilon = \frac{1}{2}$ (the principal series),
(2) $T^+_{l,\varepsilon} : l - \varepsilon \in \mathbf{Z}, l \le -\frac{1}{2}, \varepsilon \in [0,1)$, (the representations with the highest weight)
(3) $T^-_{l,\varepsilon} : l + \varepsilon \in \mathbf{Z}, l \le -\frac{1}{2}, \varepsilon \in [0,1)$, (the representations with the lowest weight)
(4) $T_{l,\varepsilon} : l = \alpha + i\frac{\pi}{h}, \alpha < -\frac{1}{2}, \varepsilon \in [0,1)$, (the strange series)
(5) $T_{l,\varepsilon} : l \in (-1, -\frac{1}{2}), \varepsilon \in [0,1) \setminus [l+1, -l]$ (the supplemental series)
(6) The trivial representation

(explicit formulae can be found in [11,22]).

The special case is determined by the condition $\varepsilon \in \{0, \frac{1}{2}\}$ These representations are quantum analogues of unitary representations of the universal covering group.

Note that each of geometric progressions in $\mathfrak{M}_\beta$ corresponds to an irreducible $*$-representation of $Fun(X_{\varkappa\sigma})_q$ and $\Gamma_{\varkappa\sigma} : U_h su(1,1) \to Fun(X_{\varkappa\sigma})_q$ is $*$-morphism, when $\varkappa\sigma > 0$.

The geometric realization of $T_{l\varepsilon}$ and $T^\pm_{l\varepsilon}$ may be illustrated as follows:

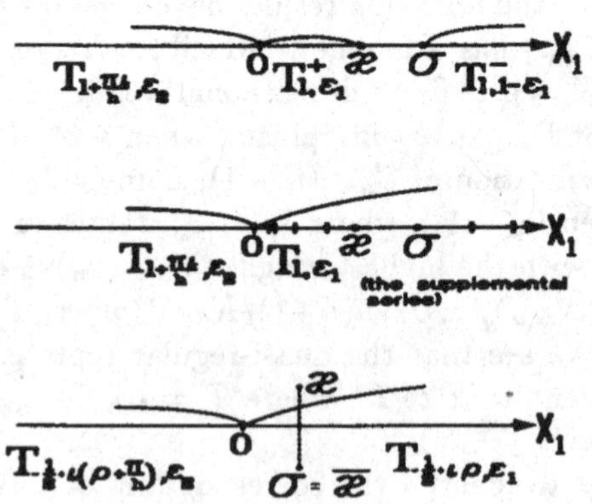

Fig. 1

(recall that $T_{l+\frac{2\pi i}{h},\varepsilon} \simeq T_{l,\varepsilon} \simeq T_{-l-1,\varepsilon}$). We have $T_{l,\varepsilon} \simeq \pi^\varepsilon_{\varkappa,\sigma} \circ \Gamma^{(\pm)}_{\varkappa,\sigma}$ where $\frac{\varkappa}{\sigma} = q^{-2l-1}$, ε corresponds to a choice of geometric progression, $\pi^\varepsilon_{\varkappa,\sigma}$ is the corresponding irreducible $*$-representation, $\Gamma^+_{\varkappa,\sigma} = \Gamma_{\varkappa,\sigma}$, $\Gamma^-_{\varkappa,\sigma} = \Gamma_{\varkappa,\sigma} \circ g$ where g is an automorphism of $U_h su(1,1)$ such that $g(E_1) = E_1, g(F_1) = -F_1,\ g(K_1^{\pm 2}) = -K_1^{\mp 2}$, $\Gamma^{(\pm)}_{\varkappa,\sigma}$ appears in the case when the geometric progression is embedded into $\mathbb{R}_\pm$.

This geometric realization enables us to realize in [22] the irreducible $*$-representation of $U_h su(1,1)$ with the highest or the lowest weight as representations on the corresponding Hilbert spaces of functions holomorphic in the unit disc in $\mathbb{C}$. This is the quantum analogue of the construction called by F.A. Berezin "quantization in the Lobachevsky plane" [4]. In complete analogy we have realized in [22] some of the strange series representations as representations on the Hilbert spaces of functions holomorphic in a ring in $\mathbb{C}$. Note that the reproducing kernel in this case is expressed in terms of Ramanujan's Ψ-function. After the work had been finished, we were informed about paper [23] where analogous results were obtained.

12. In the final section we shall say a few words on some still unfinished results. We saw that the quasi-regular representation on the quantum $SU(n+1)$-space $X_{\varkappa,\sigma}$ in the case $\varkappa\sigma > 0$ is equivalent to the tensor product $T \otimes T^*$ of an irreducible $*$-representation T of $U_h su(n+1)$. It sets one thinking to use the obtained geometric realization of irreducible $*$-representation of $U_h su(1,1)$ in order to connect tensor product of two ones with harmonic analysis on the quantum 2-hyperboloids. There is, however, an interesting quantum effect in this case. First of all, there exists a problem to define tensor product of representations, because $U_h su(1,1)$ is represented in them by unbounded operators. Without going into details, we say that it is possible to give a meaning to such expressions as

$$T^+_{l,\varepsilon_1} \otimes T^-_{l,1-\varepsilon_1} \oplus T_{l+i\frac{\pi}{h},\varepsilon_2} \otimes T_{l+i\frac{\pi}{h},1-\varepsilon_2},\ T^-_{l,1-\varepsilon_2} \otimes T^+_{l,1-\varepsilon_1},$$

$$T_{l,\varepsilon_1} \otimes T_{l,1-\varepsilon_1} \oplus T_{l+i\frac{\pi}{h},\varepsilon_2} \otimes T_{l+i\frac{\pi}{h},1-\varepsilon_2}\ (l \in (-1,-\frac{1}{2}), \varepsilon \in [0,1) \setminus [l-1,l])^7$$

$$T_{-\frac{1}{2}+i\rho,\varepsilon_1} \otimes T_{-\frac{1}{2}+i\rho,1-\varepsilon_1} \oplus T_{-\frac{1}{2}+i(\rho+\frac{\pi}{h}),\varepsilon_2} \otimes T_{-\frac{1}{2}+i(\rho+\frac{\pi}{h}),1-\varepsilon_2},$$

which have to be understood as indivisible symbols. Further investigation shows out that these representations are equivalent to the quasi-regular representations on $X_{\varkappa\sigma}$ in the cases 1), 2), 3), 4) (when $\varkappa\sigma > 0$). See Fig.1. We have explicit formulae for the unitary operators which are, in a sense, intertwiners.

Let us consider, for example, the first case. We have

$$T^+_{l,\varepsilon_1} \otimes T^-_{l,1-\varepsilon_1} \oplus T_{l+i\frac{\pi}{h},\varepsilon_2} \otimes T_{l+i\frac{\pi}{h},1-\varepsilon_2} \simeq$$
$$\int_0^{i\frac{\pi}{h}} T_{-\frac{1}{2}+i\rho,0}\, d\rho \oplus \sum_{k\in\mathbb{Z},-\varepsilon_2+k+l<-\frac{1}{2}} T_{k+l+i\frac{\pi}{h}-\varepsilon_2,0}. \tag{7}$$

It is possible to define the corresponding QCGC as "matrix elements" of the unitary operator which connects two canonical "bases" of the carrier space of the representation[8]

It turns out that relations between the QCGC in this case differs in principle from the relations between the classical CGC. We have for example,

$$\sum_k \begin{bmatrix} (l,\varepsilon_1,+) & (l,1-\varepsilon_1,-) & l' \\ k & m-k & m \end{bmatrix}_{q\varepsilon_2} \cdot \begin{bmatrix} (l,\varepsilon_1,+) & (l,1-\varepsilon_1,-) & l' \\ k & n-k & n \end{bmatrix}_{q\varepsilon_2} +$$
$$+\sum_k \begin{bmatrix} (l+i\frac{\pi}{h},\varepsilon_2) & (l+i\frac{\pi}{h},1-\varepsilon_2) & l' \\ k & m-k & m \end{bmatrix}_q \cdot \begin{bmatrix} (l+i\frac{\pi}{h},\varepsilon_2) & (l+i\frac{\pi}{h},1-\varepsilon_2) & l' \\ k & n-k & n \end{bmatrix}_q = \delta_{mn} \tag{8}$$

---

[7] Recall that $T^*_{l,\varepsilon} \simeq T_{l,1-\varepsilon}, (T^{\pm}_{l,\varepsilon})^* \simeq T^{\mp}_{l,1-\varepsilon}$.

[8] Note that the QCGC $\begin{bmatrix} (l,\varepsilon_1,+) & (l,1-\varepsilon_1,-) & l' \\ k & m-k & m \end{bmatrix}_{q\varepsilon_2}$ depends on ε_2. Indeed, even the discrete spectrum of the second order Casimir operator (and, therefore, l') depends on ε_2 (see (7)).

where none of the sums equals zero.

Let, for instance, $f(x_1)$ be a K_i-invariant eigenfunction of the second order Casimir operator such that the corresponding eigenvalue is $\sinh(l'h/2)\sinh((l'+1)h/2)/\sinh^2(h/2)$. Then the values of $f(x_1)$ in the points of the geometric progression $\{\varkappa q^{k+1/2}\}_{k\in\mathbf{Z}_+}$ (recall that $\frac{\varkappa}{\sigma}=q^{-2l-1}$) are up to constant the QCGC which appear in the first sum in (8), while the values of $f(x_1)$ in the points of the geometric progression $\{-q^{k+\beta}\}_{k\in\mathbf{Z}}$ (where $\beta=1-\varepsilon_2$) are up to constant the QCGC which appears in the second sum in (8).

References

[1] Drinfeld V.G., *Quantum groups*, Proc. ICM-86 (Berkeley) **1** (1987), 798–820.

[2] Jimbo N., *Quantum R-matrix related to the generalized Toda system: an algebraic approach*, Lect. Notes in Phys. **246** (1985), 335–360, Springer.

[3] Reshetikhin N.Y., Takhtajan L.A., Faddeev L.D., *Quantization of Lie groups and Lie algebras*, Algebra Anal. **1** (1989), 178–206. (in Russian)

[4] Berezin F.A., *A general concept of quantization*, Commun. Math. Phys. **40** (1975), 153–174.

[5] Podleś P., *Quantum spheres*, Lett. Math. Phys. **14** (1987), 193–202.

[6] Mimachi K., Nuomi M., *Quantum 2-spheres and big q-Jacobi polynomials*, Commun. Math. Phys. **128** (1990), 521–531.

[7] Mimachi K., Nuomi M., *Spherical functions on a family of quantum 3-spheres*, Preprint (1990).

[8] Hayashi T., *Q-analogues of Clifford and Weyl algebras. Spinor and oscillator representation of quantum enveloping algebras*, Commun. Math. Phys. **127** (1990), 129–144.

[9] Pusz W., Woronowicz S.L., *Twisted second quantization*, Rep.Math.Phys. **27** no. 2 (1989), 231–257.

[10] Biedenharn L.C., *The quantum group $SU_q(2)$ and a q-analogue of the boson operator*, J.Phys.A.:-Math.Gen. **22** (1989), L873–878.

[11] Masuda T.,Mimachi K., Nakagami Y., Nuomi M., Saburi Y., Ueno K., *Unitary representations of the quantum group $SU_q(1,1)$*, I,II.-Lett.Math.Phys. **19** (1990), 187–204.

[12] Ueno K., *Spectral analysis for the Casimir operator on the quantum group $SU_q(1,1)$*, Proc. Japan Acad.,ser. A. **66** (1990), 42–44.

[13] Birkhoff G.D., *The generalized Riemann problem for linear differential and q-difference equations*, Proc. Amer. Acad. Arts and Sci. **49** (1913), 521–568.

[14] Vilenkin N.Ya., *Special functions and group representation theory.*, Transl. of Math. Monographs Amer.Math.Soc. **22** (1968).

[15] Titchmarsch E.C., *Eigenfunction expansions associated with second-order differential equitions.- Vol.1*, Oxford University Press, 1946.

[16] Majid Sh., *Quasi-triangular Hopf algebras and Yang-Baxter equations*, Int.J.Mod.Phys.A **5** (1990), 1-91.

[17] Koelink H.T., Koornwinder T.H., *The Clebsch-Gordan coefficients for the quantum group $SU_q(2)$ and q-Hahn polynomials*, Proc.Kon.Ned.Akad.van Wetensch **A 92** no. 4 (1989), 443–456.

[18] Kachurik I.I., Klimyk A.U., *On Clebsch-Gordan coefficients of quantum algebra $U_q(SU_2)$*, Preprint of Inst. for Theor. Phys. ITP-89-51E (1989), Kiev.

[19] Vaksman L.L., *Q-analogues of Clebsch-Gordan coefficients and a function algebra of the quantum group $SU(2)$*, Soviet Math. Dokl. **306** (1989), 269–271. (in Russian)

[20] Vaksman L.L., Soibelman Ya.S., *On algebras of functions on quantum group $SU(N)$ and odd dimensional quantum spheres.*, Algebra Anal. **5** (1990), 101–120. (in Russian)

[21] Vaksman L.L., Korogodsky L.I., *Harmonic analysis on quantum hyperboloids*, Preprint of Inst. for Theor. Phys. ITP-90-27P (1990). (in Russian)

[22] Korogodsky L.I., *Quantum projective spaces,spheres and hyperboloids*, Preprint of Inst. for Theor. Phys. ITP-90-27P (1991), Kiev.

[23] Jurco B., *On coherent states for the simplest quantum groups*, Lett.Math.Phys. **21** (1991), 51–58.

Department of Mathematics, Rostov State University, Rostov-na-Donu, USSR

REAL AND IMAGINARY FORMS OF QUANTUM GROUPS

VLADIMIR LYUBASHENKO

Kiev Polytechnical Institute, Kiev

ABSTRACT. Existing definition of a real form of quantum group as a $*$-Hopf algebra is not quite satisfactory from the categorical point of view. In this paper another definition is proposed, which essentially coincides with the previous one for $q \in \mathbb{R}$ and yields new examples for $|q| = 1$. The last case is important because of applications of quantum groups to conformal field theory.

1. Categories

Let us use the following acronyms:

- caTegory = tensor category = category $(\mathfrak{C}, \otimes)$ with a functor $\otimes : \mathfrak{C} \times \mathfrak{C} \to \mathfrak{C}$;
- cATegory = associative tensor category $(\mathfrak{C}, \otimes, a)$= monoidal category: associativity constraint $a : X \otimes (Y \otimes Z) \to (X \otimes Y) \otimes Z$ satisfies the pentagon identity;
- CATegory = commutative associative tensor category $(\mathfrak{C}, \otimes, a, c)$ = braided monoidal category =quasitensor category; commutativity constraint c: $X \otimes Y \to Y \otimes X$ satisfies two hexagon identities.

All the cATegories are supposed to have a unit object I with isomorphism m: $I \otimes I \to I$. We'll consider *rigid* cATegories $\mathfrak{C}$,which means that for every object $X \in \mathfrak{C}$ there exist objects $X^\vee$, ${}^\vee X \in \mathfrak{C}$ and morphism of evaluation and coevaluation (cf. [M])

$$ev: \ X \otimes X^\vee \to I, \quad ev: \ {}^\vee X \otimes X \to I$$
$$coev: \ I \to X^\vee \otimes X, \quad coev: \ I \to X \otimes^\vee X$$

They must satisfy the following identities (cf. [M])

$$\left(X \simeq X \otimes I \overset{1\otimes coev}{\longrightarrow} X \otimes (X^\vee \otimes X) \overset{a}{\to} (X \otimes X^\vee) \otimes X \overset{ev\otimes 1}{\to} I \otimes X \simeq X\right) = \mathrm{id}_X$$

$$\left(X^\vee \simeq I \otimes X^\vee \overset{coev\otimes 1}{\longrightarrow} (X^\vee \otimes X) \otimes X^\vee \overset{a^{-1}}{\to}\right.$$
$$\left.\overset{a^{-1}}{\to} X^\vee \otimes (X \otimes X^\vee) \overset{1\otimes ev}{\to} X^\vee \otimes I \simeq X^\vee\right) = \mathrm{id}_{X^\vee}$$

$$\left(X \simeq I \otimes X \overset{coev\otimes 1}{\longrightarrow} (X \otimes^\vee X) \otimes X \overset{a^{-1}}{\to} X \otimes ({}^\vee X \otimes X) \overset{1\otimes ev}{\to} X \otimes I \simeq X\right) = \mathrm{id}_X$$

$$\left({}^\vee X \simeq^\vee X \otimes I \overset{1\otimes coev}{\longrightarrow} {}^\vee X \otimes (X \otimes^\vee X) \overset{a}{\to}\right.$$
$$\left.\overset{a}{\to} ({}^\vee X \otimes X) \otimes^\vee X \overset{ev\otimes 1}{\to} I \otimes^\vee X \simeq^\vee X\right) = \mathrm{id}_{{}^\vee X}$$

Functors

$$\cdot^{\vee} : \mathfrak{C} \to \mathfrak{C}^{op}, \quad X \mapsto X^{\vee}, \quad f \mapsto f^{t}$$
$$^{\vee}\cdot : \mathfrak{C}^{op} \to \mathfrak{C} \quad X \mapsto {}^{\vee}X, \quad f \mapsto {}^{t}f$$

are quasi-inverse equivalence. Moreover, we can suppose that these functors are inverse to each other without loss of generality. We denote $X^{(n\vee)} = X^{\vee\cdots\vee}$ (n times) and $X^{(-n\vee)} = {}^{\vee\cdots\vee}X$ (n times) for $n \geq 0$.

We need also the notion of coboundary cATegory (the name is borrowed from [D]).

Definition. A coboundary cATegory $(\mathfrak{C}, \otimes, a, \xi)$ is a cATegory equipped with a functorial isomorphism ξ: $X \otimes Y \to Y \otimes X$ such that $\xi^2 = 1$ and the diagram

$$\begin{array}{ccccc} (Z \otimes Y) \otimes X & \xrightarrow{\xi \otimes 1} & (Y \otimes Z) \otimes X & \xrightarrow{\xi} & X \otimes (Y \otimes Z) \\ {\scriptstyle a^{-1}}\downarrow & & & & \downarrow{\scriptstyle a} \\ Z \otimes (Y \otimes X) & \xrightarrow{1 \otimes \xi} & Z \otimes (X \otimes Y) & \xrightarrow{\xi} & (X \otimes Y) \otimes Z \end{array} \tag{1.1}$$

is commutative for all $X, Y, Z \in \mathfrak{C}$

2.Tangles

We need the CATegory $\mathcal{PT}an$ of "framed" tangles [Lyu] which can be represented by planar tangles with elementary pieces shown at the left hand side of (2.1). The difference from the definitions of [R,T] is that a part of thread (between two critical points) carries not only color X, but also a number $m \in \mathbf{Z}$, which is interpreted as a number of $\cdot^{\vee}$'s. The allowed changes of the number m can be seen from (2.1). The category $\mathcal{PT}an$ is the universal rigid CATegory. It means that any mapping colors of tangles $\to Ob\mathfrak{C}$, where $\mathfrak{C}$ is a rigid CATegory, extends to the functor

$$\mathcal{PT}an \to \mathfrak{C} \tag{2.1}$$

$$(X, m) \quad \mapsto \quad X^{(m\vee)}$$

$$(X, m) \cup (X, m+1) \quad \mapsto \quad X^{(m\vee)} \otimes X^{(m+1\vee)} \xrightarrow{ev} I$$

$$(X, m+1) \cap (X, m) \quad \mapsto \quad I \xrightarrow{coev} X^{(m+1\vee)} \otimes X^{(m\vee)}$$

$(X, m) \quad (Y, n)$

$$\mapsto c_{X^{(m\vee)}, Y^{(n\vee)}} \quad : \quad X^{(m\vee)} \otimes Y^{(n\vee)} \longrightarrow Y^{(n\vee)} \otimes X^{(m\vee)}$$

$(X, m) \quad (Y, n)$

$$\mapsto \left(c_{Y^{(n\vee)}, X^{(m\vee)}}\right)^{-1} \quad : \quad X^{(m\vee)} \otimes Y^{(n\vee)} \to Y^{(n\vee)} \otimes X^{(m\vee)}$$

Pictures are read downwards.
In particular, for every object $X \in \mathfrak{C}$ there are functorial isomorphisms

$$u^2_{\pm 1} \quad {}_X : X \to X^{\vee\vee}, \quad u^{-2}_{\pm 1} \quad {}_X : X \to {}^{\vee\vee}X$$

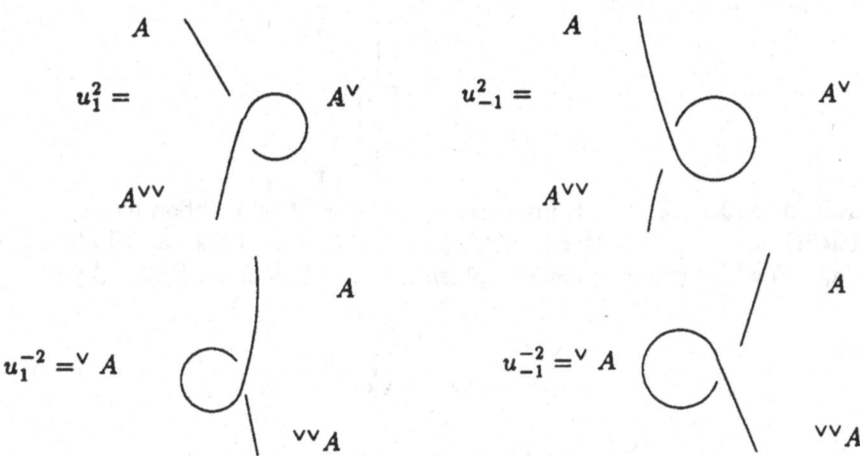

These drawings must be interpreted as follows:
$u_1^2 : X \simeq X \otimes I \overset{1\otimes coev}{\longrightarrow} X \otimes (X^{\vee\vee} \otimes X^\vee) \overset{a}{\to} (X \otimes X^{\vee\vee}) \otimes X^\vee \overset{c\otimes 1}{\to} (X^{\vee\vee} \otimes X) \otimes X^\vee \overset{a^{-1}}{\to}$ $X^{\vee\vee} \otimes (X \otimes X^\vee) \overset{1\otimes ev}{\to} X^{\vee\vee} \otimes I \simeq X^{\vee\vee}$ etc.
Let's denote the product of these commuting isomorphisms by

$$u^a_b = u^{a_1}_{b_1} \circ \cdots \circ u^{a_n}_{b_n} \quad : \quad X \to X^{(a\vee)}.$$

$a = \sum a_i$, $b = \sum b_i$, $a_i = \pm 2$, $b_i = \pm 1$. We get, in particular, automorphisms $u^0_{2X} = u^0_2 = u^2_1 \circ u_1^{-2} \quad : \quad X \to X$. Then

$$\left(u^0_{2X}\right)^t = u^0_{2(X^\vee)}$$

and the diagram

$$\begin{array}{ccc} X \otimes Y & = & X \otimes Y \\ {\scriptstyle u^0_2 \otimes u^0_2} \downarrow & & \downarrow {\scriptstyle u^0_2} \\ X \otimes Y & \overset{c^4}{\longrightarrow} & X \otimes Y \end{array}$$

commutes.

3. Ribbon Tangles

Definition A *ribbon CATegory* $(\mathfrak{C}, \otimes, a, c, \nu)$ is a rigid CATegory $\mathfrak{C}$ equipped with functorial isomorphism $\nu \equiv u^0_1 \in \operatorname{Aut} \mathfrak{C}$, $\nu_X \quad : X \to X$ such that [Lyu]

$$\begin{array}{ccc} X \otimes Y & = & X \otimes Y \\ {\scriptstyle \nu \otimes \nu} \downarrow & & \downarrow {\scriptstyle \nu} \\ X \otimes Y & \overset{c^2}{\longrightarrow} & X \otimes Y \end{array} \qquad (3.1)$$

$$\nu^2 \equiv (u_1^0)^2 = u_2^0, \tag{3.2}$$

$$(\nu_X)^t = \nu_{X^\vee}. \tag{3.3}$$

For instance, the CATegory $\mathcal{RT}an$ of *ribbon tangles* is obtained from $\mathcal{PT}an$ by adding the isomorphism

$$\nu_x = \nu \quad \begin{array}{c} (X,m) \\ \boxed{\ } \\ (X,m) \end{array}$$

constrained to (3.1)-(3.3). ν is interpreted as the twist of a ribbon by 2π.

Definition. A *π-ribbon CATegory* $(\mathfrak{C}, \otimes, a, c, \zeta)$ is a ribbon CATegory $(\mathfrak{C}, \otimes, a, c, \nu)$ with functorial automorphism $\zeta \ : \ X \to X$ such that [Lyu]

$$\zeta_I = 1_I,$$

$$\zeta_{X^\vee} = \zeta_X^t,$$

$$\zeta^2 = \nu$$

We can also write $\zeta = u_{1/2}^0$. The isomorphism ζ can be interpreted as a twist of a ribbon by π.

Proposition. *Let $\mathfrak{C}$ be a π-ribbon CATegory. Define a functorial isomorphism ξ : $X \otimes Y \longrightarrow Y \otimes X$ by*

$$\xi \ : \ X \otimes Y \xrightarrow{\zeta^{-1}} X \otimes Y \xrightarrow{c} Y \otimes X \xrightarrow{\zeta \otimes \zeta} Y \otimes X \tag{3.4}$$

Then $(\mathfrak{C}, \otimes, a, \xi)$ is a coboundary cATegory.

4. Quantum groups

For CATegory $\mathfrak{C} = U_q - mod$ of modules over quasitriangular algebra (U_q, R) [D] we have $c = P \circ R$, where P is a standard permutation. For quantum groups $U_q(\mathfrak{g})$ [D,J] the R-matrix is computed by Kirillov and Reshetikhin [KR], Levendorskii and Soibelman [LS]. Khoroshkin and Tolstoy [TKh] computed the R-matrix also for superalgebras. Let T_i denote the Lusztig automorphism of $U_q(\mathfrak{g})$ [Lu]. Choose a reduced decomposition of maximal length element $w_0 = s_{i_1} \dots s_{i_k}$ of the Weyl group of $\mathfrak{g}$. Let $t_0 = \sum I_n \otimes I^n \in \mathfrak{h} \otimes \mathfrak{h}$ be the tensor, dual to the Killing form restricted to Cartan subalgebra $\mathfrak{h} \subset \mathfrak{g}$. Then the R-matrix is given by such formulas:

$$R = q^{t_0} \cdot X_{i_1}^{-1} \cdot (T_{i_1} \otimes T_{i_1})(X_{i_2}^{-1}) \cdot \dots \cdot (T_{i_1} \dots T_{i_{k-1}} \otimes T_{i_1} \dots T_{i_{k-1}})(X_{i_k}^{-1}) \tag{4.1}$$

$$R = q^{t_0} \cdot (T_{i_1}^{-1} \dots T_{i_{k-1}}^{-1} \otimes T_{i_1}^{-1} \dots T_{i_{k-1}}^{-1})(X_{i_k}^{-1}) \cdot \dots \cdot (T_{i_1}^{-1} \otimes T_{i_1}^{-1})(X_{i_2}^{-1}) \cdot X_{i_1}^{-1} \tag{4.2}$$

$$R = \Psi_{i_1}^{21} \cdot (T_{i_1} \otimes T_{i_1})(\Psi_{i_2}^{21}) \cdot \dots \cdot (T_{i_1} \dots T_{i_{k-1}} \otimes T_{i_1} \dots T_{i_{k-1}})(\Psi_{i_k}^{21}) \cdot q^{t_0} \tag{4.3}$$

$$R = (T_{i_1}^{-1} \dots T_{i_{k-1}}^{-1} \otimes T_{i_1}^{-1} \dots T_{i_{k-1}}^{-1})(\Psi_{i_k}^{21}) \cdot \dots \cdot (T_{i_1}^{-1} \otimes T_{i_1}^{-1})(\Psi_{i_2}^{21}) \cdot \Psi_{i_1}^{21} \cdot q^{t_0} \tag{4.4}$$

where

$$X_i^{-1} = \exp_{q_i^{-2}}((q_i - q_i^{-1})k_i e_i \otimes f_i k_i^{-1}) \tag{4.5}$$

$$\Psi_i^{21} = \exp_{q_i^{-2}}((q_i - q_i^{-1})k_i^{-1} e_i \otimes f_i k_i) \tag{4.6}$$

with generic

$$q = e^{h/2}, \quad q_i = q^{(\alpha_i,\alpha_i)/2}$$

It follows from [KR] that the isomorphism $u_0^2 = u_{-1}^2 \circ u_1^0 : X \to X^{\vee\vee}$ is determined by the equation

$$\langle q^{-2H_\rho}\mathrm{x},\mathrm{x}^\vee\rangle = \langle \mathrm{x}^\vee, u_0^2(\mathrm{x})\rangle,$$

where $\mathrm{x} \in X$, $\mathrm{x}^\vee \in X^\vee$, $H_\rho = \check{\rho} \in \mathfrak{h}$ is a half-sum of positive roots. The automorphism $\nu : X \to X$ is given by the central element N

$$N = u_1^0 = q^{2H_\rho} R^{(2)} \gamma^2(R^{(1)}) = \gamma^2(R^{(1)}) R^{(2)} q^{-2H_\rho}$$

($\gamma : U_q(\mathfrak{g}) \to U_q(\mathfrak{g})$ is the antipode). The value of ν on an irreducible module $L(\Lambda)$ with the highest weight Λ is [KR]

$$u_1^0|_{L(\Lambda)} = q^{(\Lambda+2\rho,\Lambda)} \tag{4.7}$$

Finally, there exists a root $\zeta = \nu^{1/2}$ (or $Z = N^{1/2}$)(in the sense of formal power series). This allows to introduce a coboundary constraint $\xi : X \otimes Y \longrightarrow Y \otimes X$ as $\xi = P \circ \Xi$, $\Xi \in U_q(\mathfrak{g})^{\otimes 2}$,

$$\Xi = Z \otimes Z \circ R \circ \Delta Z^{-1}.$$

Thus, $U_q(\mathfrak{g})$ – mod is a π-ribbon CATegory

5. Quantum reflections

Definition. *Tensor functor* for ribbon (corr. π-ribbon) CATegories

$$(F,\psi) : (\mathfrak{C},\otimes,a,c,\nu) \to (\mathfrak{C}',\otimes',a',c',\nu')$$

(corr. ζ,ζ' instead of ν,ν') is a tensor functor

$$F \;:\; \mathfrak{C} \to \mathfrak{C}', \quad \psi \;:\; FX \otimes' FY \to F(X \otimes Y)$$

coherent with commutativity, associativity and unity [DM, Sa], such that for all $X \in \mathfrak{C}$

$$F(\nu_X) = \nu'_{FX} : FX \to FX$$
$$(\text{corr. } F(\zeta_X) = \zeta'_{FX} : FX \to FX)$$

Suppose we have an algebra automorphism $s : U_q \to U_q$ of quasitriangular algebra (U_q, R). We can make a self-equivalence $\varsigma : \mathfrak{C} \to \mathfrak{C}$ of the category $\mathfrak{C} = U_q$-mod: for $M \in \mathfrak{C}$ let $\varsigma(M)$ denote the same vector space M with new action of U_q, $u \otimes m \mapsto u \curvearrowright m = s(u).m$ and $\varsigma(f) = f$. In other words, $\varsigma = s$-mod. The question is: which element $\Psi \in U_q \otimes U_q$ can we take to obtain a tensor functor (ς, ψ)? Here $\psi(x \otimes y) = \Psi^{(1)}.x \otimes \Psi^{(2)}.y$. Axioms lead to the following equations:

$$\begin{gathered}
(s \otimes s)\Delta u = \Psi^{-1}\Delta(su)\Psi \\
(1 \otimes \Delta)\Psi \; 1 \otimes \Psi = (\Delta \otimes 1)\Psi \; \Psi \otimes 1 \\
\Psi^{21}(s \otimes s)R = R\Psi \\
\varepsilon(s(u)) = \varepsilon(u) \\
(\varepsilon \otimes \varepsilon)\Psi = 1 \\
s(N) = N \quad (\text{or } s(Z) = Z\ )
\end{gathered} \tag{5.1}$$

Example. Let $U_q = U_q(\mathfrak{g})$. Denote (ς, ψ) by (s, Ψ). Then

$$\varsigma_i = (T_i, \Psi_i) \quad \text{and} \quad \varsigma_i^{-1} = (T_i^{-1}, X_i)$$

are tensor functors, where

$$\begin{gathered}
X_i = \exp_{q_i^2}((q_i^{-1} - q_i)k_i e_i \otimes f_i k_i^{-1}) \\
\Psi_i = \exp_{q_i^{-2}}((q_i - q_i^{-1})f_i k_i \otimes k_i^{-1} e_i)
\end{gathered}$$

are the same as in (4.5), (4.6). ς_i is called *quantum reflection* (cf.[So]).

Two tensor functors can be composed:

$$(\varsigma', \psi') \circ (\varsigma, \psi) \equiv (s', \Psi') \circ (s, \Psi) = (s \circ s', \Psi(s \otimes s)\Psi').$$

The formulas (4.3) and (4.1) can be interpreted as follows:

$$(s_0, \Psi_0) \circ (T_{i_k}, \Psi_{i_k}) \circ \cdots \circ (T_{i_1}, \Psi_{i_1}) = (\theta_1, R^{21}), \tag{4.3'}$$

$$(s_0^{-1}, \Psi_0^{-1}) \circ (T_{i_k}^{-1}, X_{i_k}) \circ \cdots \circ (T_{i_1}^{-1}, X_{i_1}) = (\theta_2, R^{-1}) \tag{4.1'}$$

where

$$\begin{gathered}
s_0(u) = q^{t_1/2} u q^{-t_1/2}, \quad t_1 = \sum I_k I^k, \\
\Psi_0 = q^{t_0}, \\
\theta_1 = T_{i_1} \ldots T_{i_k} s_0, \quad \theta_2 = T_{i_1}^{-1} \ldots T_{i_k}^{-1} s_0^{-1}.
\end{gathered}$$

Remark. If $(\varsigma, \psi) = (s, \Psi)$ is a tensor self-equivalence of U_q-mod, then any functor isomorphic to (ς, ψ) is also a tensor self-equivalence. Thus, for any invertible $\Lambda \in U_q$, $\varepsilon(\Lambda) = 1$, a pair $(\varsigma', \psi') = (s', \Psi')$ with

$$\begin{gathered}
s'(u) = \Lambda \; s(u) \; \Lambda^{-1} \\
\Psi' = \; \Delta\Lambda \; \Psi \; \Lambda^{-1} \otimes \Lambda^{-1}
\end{gathered}$$

is also a tensor functor.

6. Descent

We wish to introduce $\mathbb{R}$–linear tensor categories using the idea of descent [DM, Sa]. Assume that $\mathfrak{C} = (\mathfrak{C}, \otimes, a, c, \nu, \zeta)$ is a $\mathbb{C}$–linear abelian ribbon (or π-ribbon) CATegory. Then there is a CATegory $\mathfrak{C}^- = (\mathfrak{C}, \otimes, a, c^{-1}, \nu^{-1}, \zeta^{-1})$ with

$$(c^{-1})_{XY} = (c_{XY})^{-1} \ : \ X \otimes Y \to Y \otimes X.$$

Definition. A *descent* is the following data (cf. [DM,Sa])
a) $\mathbb{C}$–linear ribbon (corr. π-ribbon) CATegory $\mathfrak{C}$;
b) antilinear tensor functor

$$(\overline{\cdot}, \phi) \ : \ \mathfrak{C} \to \mathfrak{C}$$

$$(\text{corr. } (\overline{\cdot}, \phi) \ : \ \mathfrak{C} \to \mathfrak{C}^-)$$

c) isomorphism of tensor functors

$$(Id,\ id) \xrightarrow{\mathfrak{m}} (\overline{\cdot}, \phi)^2,$$

namely, $\mathfrak{m}_X \ : \ X \to \overline{\overline{X}}$, such that

$$\mathfrak{m}_{\overline{X}} = \overline{\mathfrak{m}_X}.$$

More scrupulously: we need
an exact antilinear functor $\overline{} : \mathfrak{C} \to \mathfrak{C}$, $f \mapsto \overline{f}$,
a functorial isomorphism $\phi = \phi_{XY} : \overline{X} \otimes \overline{Y} \to \overline{X \otimes Y} \in \mathfrak{C}$,
a functorial isomorphism $\mathfrak{m} = \mathfrak{m}_X : X \to \overline{\overline{X}} \in \mathfrak{C}$,
an isomorphism $b : I \to \overline{I} \in \mathfrak{C}$, such that

$$\begin{array}{ccccc}
\overline{X} \otimes (\overline{Y} \otimes \overline{Z}) & \xrightarrow{1 \otimes \phi} & \overline{X} \otimes \overline{(Y \otimes Z)} & \xrightarrow{\phi} & \overline{X \otimes (Y \otimes Z)} \\
{\scriptstyle a}\downarrow & & & & \downarrow{\scriptstyle \overline{a}} \\
(\overline{X} \otimes \overline{Y}) \otimes \overline{Z} & \xrightarrow{\phi \otimes 1} & \overline{(X \otimes Y)} \otimes \overline{Z} & \xrightarrow{\phi} & \overline{(X \otimes Y) \otimes Z}
\end{array}$$

$$\begin{array}{ccc}
\overline{X} \otimes \overline{Y} & \xrightarrow{\phi} & \overline{X \otimes Y} \\
{\scriptstyle (c^{\pm})_{XY}}\downarrow & & \downarrow{\scriptstyle \overline{c_{XY}}} \\
\overline{Y} \otimes \overline{X} & \xrightarrow{\phi} & \overline{Y \otimes X}
\end{array}$$

$$\begin{array}{ccccc}
I \otimes I & \xrightarrow{b \otimes b} & \overline{I} \otimes \overline{I} & \xrightarrow{\phi_{II}} & \overline{I \otimes I} \\
{\scriptstyle m}\downarrow & & & & \downarrow{\scriptstyle \overline{m}} \\
I & & \longrightarrow & & \overline{I}
\end{array}$$

$$\begin{array}{ccc}
X \otimes Y & \xrightarrow{\mu_{X \otimes Y}} & X \otimes Y \\
{\scriptstyle \mu_X \otimes \mu_Y}\downarrow & & \uparrow{\scriptstyle \overline{\phi_{XY}}} \\
\overline{\overline{X}} \otimes \overline{\overline{Y}} & \xrightarrow{\phi_{\overline{X}\,\overline{Y}}} & \overline{\overline{X} \otimes \overline{Y}}
\end{array}$$

are commutative diagrams and

$$\overline{\nu} = \nu \quad (\text{corr. } \overline{\zeta} = \zeta^{-1})$$

$$\mathfrak{m}_{\overline{X}} = \overline{\mathfrak{m}_X} \;:\; \overline{X} \to \overline{\overline{X}}.$$

Suppose we have a descent, denoted briefly as $(\overline{}, \phi, \mathfrak{m}) : \mathfrak{C} \to \mathfrak{C}^{\pm}$.

Definition. Define an $\mathbb{R}$–linear category $\mathcal{D}$

$$\operatorname{Ob}\mathcal{D} = \{(X, \alpha_X : X \to \overline{X}) | \overline{\alpha} \circ \alpha = \mathfrak{m}_X \;:\; X \to \overline{\overline{X}}\}$$

$$\operatorname{Mor}\mathcal{D} = \{f \;:\; X \to Y \quad \left| \begin{array}{ccc} X & \xrightarrow{f} & Y \\ {\scriptstyle \alpha_X}\downarrow & & \downarrow{\scriptstyle \alpha_Y} \\ \overline{X} & \xrightarrow{\overline{f}} & \overline{Y} \end{array} \right\}$$

$\mathcal{D}$ is called *a real form* of $\mathfrak{C}$ if it is constructed from descent $(\overline{}, \phi, \mathfrak{m}) : \mathfrak{C} \to \mathfrak{C}$. $\mathcal{D}$ is called *an imaginary form* of $\mathfrak{C}$ if it comes from $(\overline{}, \phi, \mathfrak{m}) : \mathfrak{C} \to \mathfrak{C}^{-}$.

Theorem. *A real form is an abelian $\mathbb{R}$–linear ribbon CATegory. An imaginary form is an abelian $\mathbb{R}$–linear coboundary cATegory.*

Proof. For real case we have a commutative diagram

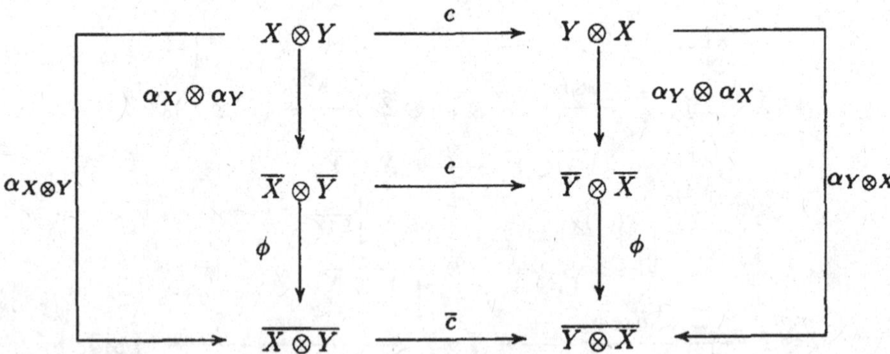

which proves that $c : X \otimes Y \to Y \otimes X \in \mathcal{D}$. For imaginary case this diagram must not commute, but will do if c is substituted by ξ from (3.4).

7. Real and imaginary forms of quasitriangular algebras

Let (U_q, R) be a $\mathbb{C}$–linear quasitriangular Hopf algebra. Let $\sigma : U_q \to U_q$ be an antilinear algebra automorphism. Then we construct a functor $\overline{} : \mathfrak{C} \to \mathfrak{C} \equiv U_q$-mod, mapping a module V to a conjugate complex space $\overline{V}$ with new action of U_q : $u \otimes \overline{v} \mapsto \overline{\sigma(u)v}$. The question is: for which elements do $\Phi \in U_q \otimes U_q$, $\mathfrak{M} \in U_q$ mappings

$$\phi(\overline{x} \otimes \overline{y}) = \overline{\Phi^{(1)}x \otimes \Phi^{(2)}y}, \quad \mathfrak{m}(x) = \overline{\overline{\mathfrak{M}x}}$$

define a descent? They must form a solution of the following equations:

$$\left.\begin{array}{c}(\sigma\otimes\sigma)\Delta u = \Phi^{-1}\ \Delta\sigma(u)\ \Phi \\ (1\otimes\Delta)\Phi\ 1\otimes\Phi = (\Delta\otimes 1)\Phi\ \Phi\otimes 1 \\ (\varepsilon\otimes\varepsilon)\Phi = 1 \\ \varepsilon(\sigma(u)) = \overline{\varepsilon(u)}\end{array}\right| \tag{7.1}$$

$$\left.\begin{array}{c}\Phi^{21}\ (\sigma\otimes\sigma)R = R\ \Phi \\ \sigma(N) = N\end{array}\right| \begin{array}{c}\text{real}\\ \text{form}\end{array}$$

or

$$\left.\begin{array}{c}R^{21}\ \Phi^{21}\ (\sigma\otimes\sigma)R = \Phi \\ \sigma(Z) = Z^{-1}\end{array}\right| \begin{array}{c}\text{imaginary}\\ \text{form}\end{array}$$

$$\left.\begin{array}{c}\sigma^2(u)\ = \mathfrak{M}\ u\ \mathfrak{M}^{-1} \\ \Delta\mathfrak{M} = \Phi\ (\sigma\otimes\sigma)\Phi\ \mathfrak{M}\otimes\mathfrak{M} \\ \sigma(\mathfrak{M})\ =\ \mathfrak{M}\end{array}\right| \tag{7.2}$$

Don't forget that isomorphic triples $(\overline{}, \phi, \mathrm{m})$ make a descent simultaneously. Hence, if $(\sigma, \Phi, \mathrm{m})$ is a solution of these equations, then for any invertible $\Lambda \in U_q$, $\varepsilon(\Lambda) = 1$, the triple $(\sigma', \Phi', \mathrm{m}')$ with

$$\begin{gathered}\sigma'(u) = \Lambda\ \sigma(u)\Lambda^{-1} \\ \Phi'(u) = \Delta\Lambda\ \Phi\ \Lambda^{-1}\otimes\Lambda^{-1} \\ \mathfrak{M}' = \Lambda\ \sigma(\Lambda)\ \mathfrak{M}\end{gathered} \tag{7.3}$$

is also a solution. Assuming the existence of such Λ that $\sigma(\Lambda) = \Lambda$, $\Lambda^2 = \mathfrak{M}^{-1}$, we should obtain $\mathfrak{M}' = 1$. We really do this in imaginary case and make weaker assumption in real case: $\mathfrak{M}'$ belongs to center of U_q. Thus (7.2) simplifies to

$$\sigma^2 = \mathrm{id}$$

$$\left.\begin{array}{c}\Delta\mathfrak{M} = \Phi\ (\sigma\otimes\sigma)\Phi\ \mathfrak{M}\otimes\mathfrak{M} \\ \sigma(\mathfrak{M}) = \mathfrak{M} \in \text{ Center } U_q\end{array}\right| \begin{array}{c}\text{real}\\ \text{form}\end{array}$$

$$(\sigma\otimes\sigma)\Phi = \Phi^{-1} \quad \text{-imaginary form}$$

Example 1 (of real form). Let U_q be a $*$-Hopf algebra. The equation

$$* = \gamma\sigma$$

determines an antilinear involutive mapping $\sigma : U_q \to U_q$ which is automorphism of algebra, antiautomorphism of coalgebra U_q. Assume that

$$(\sigma\otimes\sigma)R = R^{21} \tag{7.4}$$

Then $\sigma(u_2^0) = u_2^0$, $N = \sqrt{(u_2^0)}$. Formally, we can conclude that

$$\sigma(N) = N$$

Equations of this Section are satisfied by (equivalent) solutions

$$(\sigma, R^{-1}, N^{-1})$$
$$(\sigma, R^{21}, N)$$
$$(\sigma, \Xi^{21}, 1)$$

The equivalence (7.3) between them is given by $\Lambda = Z^{\pm 1}$, $N^{\pm 1}$, $Z = N^{1/2}$.
Example 2 (of imaginary form). Let $\sigma : U_q \to U_q$ be an antilinear involutive automorphism of algebra and coalgebra. Assume that

$$(\sigma \otimes \sigma) R^{21} = R^{-1}. \tag{7.5}$$

Then $\sigma(u_2^0) = (u_2^0)^{-1}$, $Z = (u_2^0)^{1/4}$. Formally, we get

$$\sigma(Z) = Z^{-1}.$$

Equations of this Section are satisfied by

$$(\sigma, 1, 1).$$

8. Classification

Let $\mathfrak{g}$ be a complex simple Lie algebra, $U_q = U_q(\mathfrak{g})$. We want to classify all real and imaginary forms of $\mathfrak{C} = U_q(\mathfrak{g})$-mod up to a conjugation by tensor self-equivalence of $\mathfrak{C}$. We look for antilinear automorphism of algebra σ: $U_q(\mathfrak{g}) \to U_q(\mathfrak{g})$ which can be given by formal power series in h, $\Phi = 1 + h + \ldots$. In particular, specialization $h := 0$ induces antilinear Hopf automorphism $\sigma_0 : U_q(\mathfrak{g}) \to U_q(\mathfrak{g})$.

First of all, for any simple highest weight module $L(\Lambda) \in \mathfrak{C}$ we should have $\overline{L(\Lambda)} \simeq L(\Lambda)^\vee$. Looking at values of ν in (4.7) we conclude that $h \in \mathbf{R}$ for real forms, and $h \in i\mathbf{R}$ for imaginary forms. Next classification steps imitate Cartan classification in the case of classical Lie algebras (see e.g.[GG]). In particular, the quantum reflections of Sec. 5 are used.

Theorem. *To any real form of a complex Lie algebra $\mathfrak{g}$ correspond exactly one real and one imaginary form of quantum universal enveloping algebra $U_q(\mathfrak{g})$. Any real form can be realized by $*$-Hopf algebra structure as in example 7.1. Any imaginary form can be realized as in example 7.2.*

The list of real forms (corresponding to tables in [GG]) as in example 7.1, $h \in \mathbf{R}$:
1) Compact form

$$\sigma_c(H_i) = -H_i, \quad \sigma_c(e_i) = -q_i f_i, \quad \sigma_c(f_i) = -q_i^{-1} e_i$$

$$H_i^* = H_i \quad e_i^* = f_i \quad f_i^* = e_i$$

2) Inner type.

$$\sigma = \eta_i \circ \sigma_c, \quad \eta_i = \exp(ad\, \check{H}_i/2)$$

η_i is a **C**–linear automorphism, i is such that $m_i = 1$ or 2 (table I Chapter 8 [GG])

$$\eta_i(H_j) = H_j, \quad \eta_i(e_i) = (-1)^{\delta_{ij}} e_j, \quad \eta_i(f_j) = (-1)^{\delta_{ij}} f_j$$

$$H_j^* = H_j, \quad e_j^* = (-1)^{\delta_{ij}} f_j, \quad f_j^* = (-1)^{\delta_{ij}} e_j.$$

3) Dynkin's diagram automorphism type.

$$\sigma = \rho \circ \sigma_c,$$

ρ is **C**-linear automorphism given by Dynkin's diagram automorphism of degree 2.

$$\rho(H_j) = H_{\rho(j)}, \quad \rho(e_j) = e_{\rho(j)}, \quad \rho(f_j) = f_{\rho(j)},$$

$$H_j^* = H_{\rho(j)}, \quad e_j^* = f_{\rho(j)}, \quad f_j^* = e_{\rho(j)}.$$

4) External type.

$$\sigma = \rho \eta_i \sigma_c$$

(i can be found in table II Chapter 8 [GG])

$$H_j^* = H_{\rho(j)}, \quad e_j^* = (-1)^{\delta_{ij}} f_{\rho(j)}, \quad f_j^* = (-1)^{\delta_{ij}} e_{\rho(j)}.$$

The list of imaginary forms (corresponding to tables in [GG]) as in example 7.2, $h \in i\mathbb{R}$:
1) Pseudo-compact form

$$\sigma_{pc}(H_i) = -H_i, \quad \sigma_{pc}(e_i) = -f_i, \quad \sigma_{pc}(f_i) = -e_i.$$

2) Inner type

$$\sigma = \eta_i \sigma_{pc} = \sigma_{pc} \eta_i$$

(i from table I Chapter 8 [GG]).
3) Dynkin's diagram automorphism type

$$\sigma = \rho \sigma_{pc} = \sigma_{pc} \rho.$$

4) External type

$$\sigma = \rho \eta_i \sigma_{pc}$$

(i from table II Chapter 8 [GG]).

Remark. There are examples of imaginary structures realized by $*$-Hopf algebras (with some Φ). Not all imaginary structures can be realized in that way. Indeed, for $|q| = 1$ where q is not a root of unity there are two imaginary structures on $U_q(\mathfrak{sl}(2))$-mod and there is only one structure of $*$-Hopf algebra on $U_q(\mathfrak{sl}(2))$.

Acknowledgements. I thank Professor S.Majid for explaining some basic results in the theory of rigid categories.

References

[DM] Deligne P., Milne J.S., *Tannakian Categories*, Lect.Notes Math. **900** (1982), 101–228, Springer, Berlin a.o..

[D] Drinfeld, V.G., *Quantum groups*, Proc. of the ICM-86 1 (1987), 798–820, Berkeley.

[GG] Goto, M., Grosshans, F.D., *Semisimple Lie algebras*, Lect.Notes Pure Appl.Math. **38** (1978), Marsel Dekker, N.Y. and Basel.

[J] Jimbo, M., *A q-Difference Analogue of $U_q(\mathfrak{g})$ and the Yang-Baxter Equation*, Lett.Math.Phys. **10** (1985), 63–69.

[KhT] Khoroshkin, S.M., Tolstoy, V.N., *Universal R-matrix for quantum supergroups* (to appear).

[KR] Kirillov, A.N., Reshetikhin, N.Yu., *q-Weyl Group and a Multiplicative Formula for Universal R-matrices*, Commun. Math. Phys. **134** (1991), 421–431.

[LS] Levendorskii, S., Soibelman, Ya., *Some applications of the quantum Weyl group*, Preprint (1989).

[Lu] Lusztig, G., *Finite dimensional Hopf algebras arising from quantum groups*, Preprint.

[Lyu] Lyubashenko, V.V., *Tensor categories and RCFT. I.Hopf algebras in rigid categories*, Preprint no. ITP-90-30E (1990), Kiev; *Tensor categories and RCFT. II. Modular transformations*, Preprint no. ITP-90-59E (1990), Kiev.

[M] Majid S., *Duals and Doubles of Monoidal Categories*, Preprint no. DAMTP/89-41 (1989).

[R] Reshetikhin, N.Yu., *Quasitriangular Hopf algebras, solutions of the Yang-Baxter equation and invariants of links*, Algebra Anal **1** no. 2 (1989), 169–194. (in Russian)

[Sa] Saavedra, R.N., *Catégories Tannakiennes*, Lect.Notes Math. **265** (1972), 420, Springer, Heidelberg.

[So] Soibelman, Ya.S., *Quantum Weyl group and some of its applications*, Supl.Rend.Circ.Mat.Palermo (1990), (to appear).

[T] Turaev, V.G., *Operator invariants of links and R-matrices*, Izv.Akad.Nauk SSSR, Ser.Mat. **53** no. 5 (1989), 1073–1107. (in Russian)

Kiev Polytechnical Institute, KPI-2606, prospect Pobedy,37, Kiev 252056, USSR

RANK OF QUANTUM GROUPS AND BRAIDED GROUPS IN DUAL FORM

SHAHN MAJID

Department of Applied Mathematics
& Theoretical Physics
University of Cambridge

ABSTRACT. We give a dual formulation of recent work on the representation theory of general quantum groups. These form a rigid quasitensor category $\mathcal{C}$ to which is associated a braided group $\mathrm{Aut}(\mathcal{C})$ of braided-commutative "co-ordinate functions" analogous to the ring of functions on a group or supergroup. Every dual quasitriangular Hopf algebra A gives rise to such a braided group $\underline{A}$. We give the example of the braided group $BSL(2)$ in detail. We also give the rank of quantum groups in dual form and explain its connection with the partition function of simple quantum systems.

1. Introduction

Many authors have studied the representation theory of quantum groups. For a general quasitriangular Hopf algebra $H, \mathcal{R}$ (or quantum group in the strict sense, i.e. with quasitriangular structure $\mathcal{R}$) the finite-dimensional modules ${}_H\mathcal{M}^{f.d.}$ form a rigid quasitensor category or "braided monoidal category with conjugates". See [3, Sec. 7.5.] for an early treatment. If A is a dual quasitriangular Hopf algebra (analogous to the ring of functions on a group, but generally non-commutative), the finite-dimensional comodules ${}^A\mathcal{M}^{f.d.}$ again form a rigid quasitensor category. The relevant definitions are reviewed in the Preliminaries.

In this note we describe two general constructions on the rigid quasitensor categories $\mathcal{C}$. Because the constructions are very general, they are very natural or "intrinsic" for quantum groups and their generalizations. We compute them in detail in the case $\mathcal{C} = {}^A\mathcal{M}^{f.d.}$ associated to a dual quasitriangular Hopf algebra A.

The main new construction, given in Section 2, associates to any cocomplete rigid quasitensor categories $\mathcal{C}$ a braided group $\mathrm{Aut}(\mathcal{C})$ analogous to the ring of functions on a group. Such a theorem and the resulting braided groups were given in the non-dual case in [1] [2] (analogously to the group algebra or universal enveloping algebra). In this note we now present the results in the case analogous to functions on groups and supergroups. As such, braided groups are like supergroups except that the ± 1 statistics for supergroups is replaced by braid statistics. Every dual quasitriangular Hopf algebra

SERC Research Fellow and Drapers Fellow of Pembroke College, Cambridge

A gives such a braided group $\underline{A} = \mathrm{Aut}({}^A\mathcal{M}^{f.d.})$. We spell out the formulae in this case in Section 3, including the example of the braided group $BSL(2)$.

Finally, in any rigid quasitensor categories there is an intrinsic notion of rank(X) for any object X. In the example where $\mathcal{C} = {}_H\mathcal{M}^{f.d.}$, the rank (X) basically recovers the familiar quantum dimension of X but is defined more generally. The rank or "quantum order", $|H|$, of H is the category-theoretic dimension of the left-regular or "Fock space" representation of H [4][3, Sec. 7.6]. In some joint work with Ya. S. Soibelman the rank of quantum groups $|U_q(g)|$ [5][6] was found as

$$|U_q(g)| = \frac{\sum_{\Lambda\in P} q^{-(\Lambda,\Lambda)}}{\prod_{\alpha>0}(1-q^{-2(\alpha,\rho)})}. \tag{1}$$

Here the sum is over all weights Λ and the product is over all positive roots α, $\rho = \frac{1}{2}\sum_{\alpha>0}\alpha$ and the inner product is determined from the Killing form. The numerator here is the partition function for a quantum particle confined to a certain bounded domain, namely the alcove of g [6]. Apart from applications to Chern-Simons theory and number theory[6], this is satisfying because it does mean that $|H|$ is a good analog of the order $|G|$ of a finite group: it counts the "points" in the underlying quantum group by counting quantum states in the sense of a partition function. In Section 4 we give the relevant dual formula for $|A|$ and details for the ranks of the quantum groups $A(R)$ of [7] and their quotients. This motivates the problem of finding dual quasitriangular structures and ranks for the quantum groups associated to vertex models.

Acknowledgements. It is a pleasure to thank the director and organizers; L.D. Faddeev, P.P. Kulish and others for an excellent and stimulating programme at the Euler Institute. I also want to thank all participants for discussions. During the conference I learned that V. Lyubashenko has also obtained a Hopf algebra from a braided category in very interesting recent preprint [25]. This overlaps with [1] and its dual form (part of Corollary 2.2 below).

Preliminaries. A category $\mathcal{C}$ is *monoidal* if there is a functor $\otimes : \mathcal{C}\times\mathcal{C}\to\mathcal{C}$ and functorial isomorphisms $\Phi_{X,Y,Z} : X\otimes(Y\otimes Z)\to(X\otimes Y)\otimes Z$ for all objects X,Y,Z, and a unit object $\underline{1}$ with functorial isomorphisms $l_X : X\to\underline{1}\otimes X, r_X : X\to X\otimes\underline{1}$ for all objects X. The Φ should obey a well-known pentagon coherence identity while the l and r obey triangle identities of compatibility with Φ. A monoidal category $\mathcal{C}$ is *rigid* if for each object X, there is an object X^* and functorial morphisms $ev_X : X^*\otimes X\to\underline{1}, \pi_X : \underline{1}\to X\otimes X^*$ such that

$$X\cong\underline{1}\otimes X\xrightarrow{\pi\otimes\mathrm{id}}(X\otimes X^*)\otimes X\cong X\otimes X\otimes(X^*\otimes X)\xrightarrow{\pi\otimes\mathrm{id}}X\otimes\underline{1}\cong X \tag{2}$$

$$X^*\cong X^*\otimes\underline{1}\xrightarrow{\mathrm{id}\otimes\pi}X^*\otimes(X\otimes X^*)\cong X\otimes(X^*\otimes X)\otimes X^*\xrightarrow{ev\otimes\mathrm{id}}\underline{1}\otimes X^*\cong X^* \tag{3}$$

compose to id_X and id_{X^*} respectively. A monoidal category is *quasitensor* (or "braided monoidal" [9]) if there is a collection of functorial isomorphisms $\Psi_{X,Y} : X\otimes Y)\to Y\otimes X$ obeying two hexagon coherence identities with Φ and triangle coherence identities with l and r. See [3, Sec.7] where they arose in the context of quantum groups independently of [9]. Other authors also realised the connection with quantum groups

[8]. If Ψ^2 =id we have an ordinary *tensor* category as in [10]. Finally, a monoidal functor between monoidal categories is one that respects $\otimes$ by functorial isomorphisms $c_{X,Y}: F(X)\otimes F(Y)\cong F(X\otimes Y)$, as well as Φ, l, r.

Let k denote a field. Most of what we say also works(with care to use projective modules) over a commutative ring. ${}_k\mathcal{M}$ denotes the tensor category of modules over k. Here $\Psi = \Psi^{Vec}$ is the obvious twist map. ${}_k Super\mathcal{M}$ is similar but with a new symmetry $\Psi_{V,W}: V\otimes W \to W\otimes V$ defined by $\Psi(v\otimes w) = (-1)^{|v||w|} w\otimes v$ on elements v, w homogeneous of degree $|v|, |w|$. The finite-dimensional versions $\mathcal{C} = {}_k\mathcal{M}^{f.d.}$, $\mathcal{C} = {}_k \text{Super}\mathcal{M}^{f.d.}$ are examples of rigid tensor categories. Finally, the category of finite-dimensional H-modules of quasitriangular Hopf algebra H [11] form a rigid quasitensor categories ${}_H\mathcal{M}^{f.d.}$ as explained in detail in [3, Sec. 7].

We now describe the less familiar dual setting. A dual quasitriangular Hopf algebra $A, \mathcal{R}$ is a Hopf algebra $A, \mathcal{R} \in (A\otimes A)^*$ obeying some obvious axioms [12,Sec. 4] obtained by dualizing Drinfeld's. Explicitly,

$$\mathcal{R}(ab\otimes c) = \sum \mathcal{R}(a\otimes c_{(1)})\mathcal{R}(b\otimes c_{(2)}), \mathcal{R}(a\otimes bc) = \sum \mathcal{R}(a_{(1)}\otimes c)\mathcal{R}(a_{(2)}\otimes b) \quad (4)$$

$$\sum b_{(1)}a_{(1)}\mathcal{R}(a_{(2)}\otimes b_{(2)}) = \sum \mathcal{R}(a_{(1)}\otimes b_{(1)})a_{(2)}b_{(2)}, \forall\, a,b,c\in A. \quad (5)$$

A right A-comodule V is a vector space V and a linear structure map $\beta_V : V \to V\otimes A$ required to obey $(\beta_V\otimes \mathrm{id})\circ\beta_V = (\mathrm{id}\otimes\Delta)\beta_V$ and $(\mathrm{id}\otimes\epsilon)\circ\beta_V$ =id. We often write such comodules in the standard notation $\beta_V(v) = \sum v^{\bar 1}\otimes v^{\bar 2}$. The category of right A-comodules is denoted ${}^A\mathcal{M}$ and is a quasitensor one. The finite-dimensional comodules $\mathcal{C} = {}^A\mathcal{M}^{f.d.}$ form a rigid quasitensor category. Explicitly, the tensor product of comodules is determined by the algebra structure. Here

$$\beta_{V\otimes W} = (\mathrm{id}\otimes\cdot)\circ\Psi^{Vec}_{A,W}\circ(\beta_V\otimes\beta_W). \quad (6)$$

Here $\Psi^{Vec}_{A,W}$ is the symmetry or usual twist map in ${}_k\mathcal{M}$. The quasisymmetry is defined by

$$\Psi_{V,W} = (\Psi^{Vec}_{V,W}\otimes\mathcal{R})\circ\Psi^{Vec}_{A,W}\circ(\beta_V\otimes\beta_W). \quad (7)$$

The duals require the existence of an antipode: β_{V^*} is defined by $\beta_{V^*}(f) = (f\otimes S)\circ\beta_V$. Here the left hand side in $V^*\otimes A$ is viewed as a map $V\to A$.

This completes the description of various examples of rigid tensor or quasitensor categories $\mathcal{C}$. Algebras can of course be defined in any monoidal category: an algebra in $\mathcal{C}$ is an object A of $\mathcal{C}$ with morphisms $A\otimes A\to \underline{1}, \eta : \underline{1}\to A$ obeying axioms of associativity and unity (η plays the role of unit). Algebra maps are assumed unital. Note that the quasisymmetry of $\mathcal{C}$ is relevant only if we want a tensor product algebra structure $A_1\otimes A_2$. Namely,

$$(A_1\otimes A_2)\otimes(A_1\otimes A_2)\xrightarrow{\Psi_{A_2,A_1}}(A_1\otimes A_1)\otimes(A_2\otimes A_2)\xrightarrow{\cdot\otimes\cdot}A_1\otimes A_2. \quad (8)$$

Likewise, coalgebras make sense in any monoidal category: a coalgebra in $\mathcal{C}$ is an object C of $\mathcal{C}$ with morphisms $\Delta : C\otimes C\to C$, $\epsilon : C\to\underline{1}$ obeying axioms of coassociativity and counity (as familiar for quantum groups). We write the coproduct formally as $\Delta c = \sum c_{(1)}\otimes c_{(2)}$ in the usual way. Coalgebra maps are assumed counital. Again, $\mathcal{C}$

needs to have a quasisymmetry if we want to have a tensor product coalgebra $C_1 \otimes C_2$ defined like (8) with arrows reversed.

Hopf algebra make sense in any tensor or quasitensor category. The tensor case is well known. For the quasitensor case see [1][13]: A Hopf algebra in the category $\mathcal{C}$ is an object H of $\mathcal{C}$ which is both an algebra and a coalgebra in $\mathcal{C}$ for which these are compatible in the sense (omitting Φ for brevity) of commutativity of

$$\begin{array}{ccccc} H\otimes H & \longrightarrow & H & \xrightarrow{\Delta} & H\otimes H \\ \downarrow{\scriptstyle \Delta\otimes\Delta} & & & & \uparrow{\scriptstyle \cdot\otimes\cdot} \\ H\otimes H\otimes H\otimes H & & \xrightarrow{\mathrm{id}\otimes\Psi_{H,H}\otimes\mathrm{id}} & & H\otimes H\otimes H\otimes H \end{array} \tag{9}$$

The notion of antipode is as usual. The idea of working over a general rigid tensor category $\mathcal{C}$ is nothing new, e.g. [22]. What is new idea of braided quantum groups is to generalize further to $\mathcal{C}$ rigid quasitensor.

For any coalgebra or Hopf algebra in a category $\mathcal{C}$, we define a comodule in the category $\mathcal{C}$ as an object V in $\mathcal{C}$ and morphism $\beta_V : V \to V \otimes C$ obeying the same axioms $(\beta_V \otimes \mathrm{id}) \circ \beta_V = (\mathrm{id} \otimes \Delta)\beta_V$ and $(\mathrm{id} \otimes \epsilon) \circ \beta_V = \mathrm{id}$ (as morphisms and suppressing Φ). Instead of Ψ^{Vec} in (6) we use now the quasisymmetry in $\mathcal{C}$. Likewise a dual quasitriangular structure $\mathcal{R}$ on a Hopf algebra in $\mathcal{C}$ is a morphism $A \otimes A \to \underline{1}$ obeying axioms similar to the unbraided case (4)-(5). Twist maps implicit there must be replaced by Ψ, Ψ^{-1}. In the present paper we only need the case of trivial quantum group structure, see below. All categories are assumed equivalent to small categories.

2. Braided groups

In this section we want to prove the theorem that every rigid quasitensor categories has a quantum group of automorphisms Aut($\mathcal{C}$). By braided group we mean a Hopf algebra in a quasitensor categories that is braided-commutative in the sense cf.[2]

$$(\mathrm{id} \otimes \cdot) \circ \beta_V = (\mathrm{id} \otimes \cdot) \circ Q_{V,A} \circ \Psi_{A,A} \circ \beta_V \tag{10}$$

on $V \otimes A$ for all right comodules β_V in the category with some class (which to be specified). Here $Q_{X,Y} = \Psi_{Y,X} \circ \Psi_{X,Y}$. This definition (10) is motivated by considering dual quasitriangular Hopf algebras in quasitensor categories as explained above, with trivial $\mathcal{R}$. It is the most novel aspect of the paper. These results are the relevant dual versions of the results applied in [2] and proven in [1, Sec 4.].

Theorem 2.1. *Let $\mathcal{C}$ be a monoidal category and $\mathcal{V}$ a rigid quasitensor categories cocomplete over $\mathcal{C}$. Let $F : \mathcal{C} \to \mathcal{V}$ be a monoidal functor (one that respects the tensor product up to isomorphism, and associativity). Then there exists a $\mathcal{V}$-Hopf algebra, $A = Aut(\mathcal{C}, F, \mathcal{V})$ such that F factorizes monoidally as $\mathcal{C} \to {}^A\mathcal{V} \to \mathcal{V}$. Here ${}^A\mathcal{V}$ is the category of A-comodules in $\mathcal{V}$ and the second factor is the forgetful functor. A it is universal with this property. That is, if A' is any other such $\mathcal{V}$-Hopf algebra then there exists a $\mathcal{V}$-Hopf algebra map $A \to A'$ inducing a functor ${}^A\mathcal{V} \to {}^{A'}\mathcal{V}$ compatible with* *he given functors $\mathcal{C} \to {}^{A}\mathcal{V}, \mathcal{C} \to {}^{A'}\mathcal{V}$. If $\mathcal{C}$ is suitably rigid then A has an antipode. If ' is quasitensor then A is dual quasitriangular.*

Proof. The full details of the proof will be given elsewhere [14]. Here we give only a ketch. The key step is to consider the functor $\tilde{F} : \mathcal{V} \to Ens \ : \ V \mapsto \mathrm{Nat}(F, F_V)$. Here Nat denotes natural transformation and $F_V \ : \mathcal{C} \to \mathcal{V}$ is defined by $F_V(X) = F(X) \otimes V$ We first show that $\tilde{F}$ is representable, i.e. there exist an object A in $\mathcal{V}$ and functorial somorphisms

$$\theta_V \ : \mathrm{Hom}(A, V) \cong \mathrm{Nat}(F, F_V). \tag{11}$$

To do this, note that in $\mathcal{V}$ any objects $U^* \otimes W$ are like "matrices" (dual internal hom) in he sense $\theta_V \ : \mathrm{Hom}(U^* \otimes W, V) \cong \mathrm{Hom}(W, U \otimes V)$ functorially in V. (This follows easily using π, ev obeying (2)(3) in $\mathcal{V}$). Hence we need to take $A = \coprod_X F(X)^* \otimes F(X)/ \sim$ where we must quotient by an equivalence relation $\sim$ to ensure that the collection Hom$(F(X), F(X) \otimes V)$ corresponding to Hom(A, V) is natural. If $\mathcal{V}$ is concrete then the desired equivalence relation is therefore just

$$F(\phi)^* y^* \otimes x \sim y^* \otimes F(\phi) x$$

for all $\phi \ : X \to Y, x \in F(X), y^* \in F(Y)^*$. This is just the dual of the "coherent matrix-valued functions on $\mathcal{C}$" in [1, Secs 2,4]. There is a fancy way to say this for abstract categories: Let $A = \varinjlim A_{X,Y,\phi}$ be the colimit over $X, Y, \phi : X \to Y$ of $A_{X,Y,\phi}$ obtained as coequalizer of the diagram

$$\coprod_{X',Y'} F(Y')^* \otimes F(X') \overset{f_{X,Y,\phi}}{\underset{g_{X,Y,\phi}}{\rightrightarrows}} \coprod_{X''} F(X'')^* \otimes F(X'') \to \ A_{X,Y,\phi} \to 0.$$

Here $f_{X,Y,\phi} = F(\phi)^* \otimes \mathrm{id}$ and $g_{X,Y,\phi} = \mathrm{id} \otimes F(\phi)$ if $X' = X, Y' = Y$ and are otherwise zero. This is very similar to the well-known case when $\mathcal{V} = Vec$[15,Sec. I.1.3.2.1] or [16].

Once we have obtained an object A obeying (11) the rest follows in a standard way by repeated use of (11) without recourse to details of A. This is an important advantage of the present approach. Thus, let $\beta_X = \theta_A(\mathrm{id}_A)_X \ : F(X) \to F(X) \otimes A$ correspond in (11) to the identity $A \to A$. From this, θ_V is recovered as $\theta_V(f)_X = (\mathrm{id} \otimes f) \circ \beta_X$ for all $f \in \mathrm{Hom}(A, V)$. Let $\beta^2 \in \mathrm{Nat}(F, F_{A \otimes A}$ be defined as $\beta^2_X = (\beta_X \otimes \mathrm{id}) \circ \beta_X$ Then $\Delta = \theta^{-1}_{A \otimes A}(\beta^2)$ defines a coproduct on A in the usual way. Likewise the functor $\tilde{F}^2$ corresponding to $F^2 : \mathcal{C} \times \mathcal{C} \to \mathcal{V} : (X, Y) \mapsto F(X) \otimes F(Y)$ is representable by $A \otimes A$. To prove this, we define morphisms $\theta^2 : \mathrm{Hom}(A \otimes A, V) \cong \mathrm{Nat}(F^2, F^2_V)$ by

$$\theta^2_V(f)_{X,Y} = (\mathrm{id} \otimes \mathrm{id} \otimes f) \circ \Psi^{\mathcal{V}}_{A, F(Y)} \circ (\beta_X \otimes \beta_Y), \ \forall f \in \mathrm{Hom}(A \otimes A, V). \tag{12}$$

Then the element of $\mathrm{Nat}(F^2, F^2_A)$ defined by $(c^{-1}_{X,Y} \otimes \mathrm{id})\beta_{X \otimes Y} \circ c_{X,Y}$ has inverse image under θ^2_A a morphism $\cdot : A \otimes A \to A$. This defines the product. Similarly for the other Hopf algebra structures and dual quasitriangular structure when $\mathcal{C}$ is quasitensor. The functor $\mathcal{C} \to^A \mathcal{V}$ is $X \mapsto (F(X), \beta_X)$. Finally, suppose A' is another such Hopf algebra in $\mathcal{V}$ and we are given a functor $\beta' : \mathcal{C} \to {}^{A'}\mathcal{V}$ as in the theorem. This means a collection maps $\beta'_X : F(X) \to F(X) \otimes A'$. View this as an element of $\mathrm{Nat}(F, F_{A'})$. By (11) this corresponds via $\theta_{A'}$ to a map $A \to A'$. This gives the universal property. This is similar to the familiar case over Vec [17]. The novel ingredient is that we must be careful to put in the braiding $\Psi^{\mathcal{V}}$ correctly and see that it doesn't get "tanglet up". The proof is already spelled out in detail in the module framework (assuming representability) in [1,Sec.4].

Corollary 2.2. *Let $\mathcal{C}$ be a cocompleted rigid quasitensor category. Then $\mathrm{Aut}(\mathcal{C}) = \mathrm{Aut}(\mathcal{C}, id, \overline{\mathcal{C}})$ is a braided group, i.e. braided-commutative in the sense of (10). We call it the automorphism braided group of $\mathcal{C}$. $A = \mathrm{Aut}(\mathcal{C})$ is the universal object with the property that the forgetful functor ${}_A\overline{\mathcal{C}} \to \overline{\mathcal{C}}$ has a section (i.e. a monoidal functor $\mathcal{C} \to {}_A\overline{\mathcal{C}}$ composing with the forgetful functor to the identity functor).*

Proof. Since $\mathcal{C}$ is rigid quasitensor we can use $\mathcal{V} = \bar{\mathcal{C}}$ and the identity functor $\mathcal{C} \to \bar{\mathcal{C}}$ in Theorem 2.1. In general in Theorem 2.1 the quasitriangular structure is the ratio of the quasisymmetry in $\mathcal{C}$ and that in $\mathcal{V}$ [1,Sec. 4]. This means [2] that in the present case the quasitriangular structure $\mathcal{R} : A \otimes A \to \underline{1}$ is trivial. By this we mean that $\theta^2_{\underline{1}}(\mathcal{R})$ is the identity in $\mathrm{Nat}(F^2, F^2)$. In particular, this implies (10).

3. Examples of braided groups

The preceding corollary solves (in principle) a problem posed by algebraic quantum field theory in low dimension [18]. In four dimensions every such theory leads to a tensor (i.e. symmetric monoidal) category and using this [19] were able identify $\mathcal{C}$ as the category of representations of a compact group of internal symmetries. In dimensions two or three the category $\mathcal{C}$ is instead a quasitensor one so that it cannot be the representations of any group. There have been many conjectures as to what should be the correct object replacing the compact group. Corollary 2.2 says that in principle the right object is the braided group $\mathrm{Aut}(\mathcal{C})$ [2].

We can also use Corollary 2.2 to associate to any quantum group a braided group analog. We introduced this process of *transmutation* in [2, Sec.3]. We now give its dual formulation.

Theorem 3.1 [14]. *Let A be a dual quasitriangular Hopf algebra in the usual sense. Then there is a braided group $\underline{A}$ in the category ${}^A\mathcal{M}$ describes as follows in terms of A. As a linear space and coalgebra, $\underline{A}$ it coincides with A. The algebra structure is transmuted to*

$$a\underline{\cdot}b = \sum a_{(2)}b_{(3)}\mathcal{R}(a_{(3)} \otimes Sb_{(1)})\mathcal{R}(a_{(1)} \otimes b_{(2)}). \tag{13}$$

$\underline{A}$ is an object in ${}^A\mathcal{M}$ by the adjoint right coaction

$$\beta_{\underline{A}}(a) = \sum a_{(2)} \otimes (Sa_{(1)})a_{(3)} = \sum a^{(\bar{1})} \otimes a^{(\bar{2})}. \tag{14}$$

As such, there is an action of the braid group on tensor powers of $\underline{A}$ defined by $\Psi_{\underline{A},\underline{A}}$ in [7]. This makes $\underline{A}$ into a braided Hopf algebra. It is braided-commutative in the sense

$$b\underline{\cdot}a = \sum a^{(\bar{1})}\underline{\cdot}b^{(\bar{1})}_{(2)}\mathcal{R}(b^{(\bar{2})}_{(2)} \otimes a^{(\bar{2})}_{(2)})\mathcal{Q}(b_{(1)} \otimes a^{(\bar{2})}_{(1)}). \tag{15}$$

Proof. In the case that A is finite-dimensional, this follows from Corollary 2.2 applied to $C = {}^A\mathcal{M}^{f.d.}$. The resulting formulae (13)-(15) then hold even when A is infinite dimensional [14].

This is an important application because it transmutes a non-commutative object A in the ordinary category of vector spaces into commutative object, albeit in a braided category. This is a shift in view-point from non-commutative geometry in the sense of [21] to the philosophy of supergeometry and its extensions. That is, the braided group $\underline{A}$ is braided-commutative , i.e. a classical and not quantum object. Thus a braided group is like a supergroup but in an even more non-commutative category (since the quasisymmetry no-longer has square one). Because of the many successes of the super philosophy, it does seem worthwhile to transmute quantum groups in this way.

The results so far have been somewhat abstract. Therefore we now describe them in the concrete setting of matrix braided group. Thus, let $\{R^i{}_j{}^k{}_l\} \in M_n \otimes M_n$ be a regular invertible solution of the quantum Yang-Baxter equations (QYBE), i.e. $R_{12}R_{13}R_{23} = R_{23}R_{13}R_{12}, R$ invertible and R^{t_2} invertible. Here t_2 denotes transpose in the second factor. To every such R we will associate "braided matrices" $B(R)$. We first define the matrices $\Psi, \mathfrak{D}$ in $M_{n^2} \otimes M_{n^2}$ by

$$\Psi^{(k_0,k_1)}{}_{(l_0,l_1)}{}^{(i_0,i_1)}{}_{(j_0,j_1)} = R^{i_0}{}_a{}^d{}_{l_0} R^{-1a}{}_{j_0}{}^{l_1}{}_b R^{j_1}{}_c{}^b{}_{k_1} \tilde{R}^c{}_{i_1}{}^{k_0}{}_d \tag{16}$$

$$\mathfrak{D}^{(m_0,m_1)}{}_{(l_0,l_1)}{}^{(i_0,i_1)}{}_{(n_0,n_1)} = \delta^{i_1}_{n_1} R^{l_1}{}_a{}^e{}_b R^b{}_c{}^a{}_{m_1} \tilde{R}^c{}_{n_0}{}^{m_0}{}_d R^{-1d}{}_{l_0}{}^{i_0}{}_e \tag{17}$$

Here $\tilde{R} = ((R^{t_2})^{-1})^{t_2}$ and i_0, i_1, etc. run over $1, \ldots, n$. We write $I = (i_0, i_1), J = (j_0, j_1)$ as multiindices running from (1,1),(1,2),...,(n,n). We now define $B(R)$ as the associative algebra generated by 1 and n^2 indeterminates $\{u^i{}_j\}$ modulo the relations

$$u^I u^K = u^L u^J \Psi^K{}_M{}^N{}_J \mathfrak{D}^M{}_L{}^I{}_N \tag{18}$$

where $u^I = u^{i_0}{}_{i_1}$ etc.

Theorem 3.2 [20]. *The matrix $\{\Psi^I{}_J{}^K{}_L\} \in M_{n^2} \otimes M_{n^2}$ obeys the QYBE, i.e. defines an action of the braid group on the $\{u^I\}$ (extending to an action on all of B(R)) by*

$$\Psi(u^I \otimes u^K) = u^L \otimes u^J \Psi^K{}_L{}^I{}_J. \tag{19}$$

With this action, the matrix coproduct and counit

$$\Delta u^i{}_j = u^i{}_k \otimes u^k{}_j, \ \epsilon(u^i{}_j) = \delta^i{}_j \tag{20}$$

extend to B(R) as a braided bialgebra ("braided matrices"). B(R) has a quotient which is a braided group (i.e. has an antipode, at least formally).

Proof. This is a version of Theorem 3.1 applied to the braided groups $A(R)$ of [7] and their quotients as studied for general R in [24][3, Sec. 3]. The dual quasitriangular structure was obtained in [3, Sec. 3] as $\mathcal{R}(u^i{}_j = u^i{}_k \otimes u^k{}_l) = R^i{}_j{}^k{}_l$. The antipode in Theorem 3.1 disappears from the final formulae. (15) reduces to (18) while (14) in (7) gives $\Psi_{\underline{A},\underline{A}}$ as (16). We then adopt (18) as a definition and check that everything is consistent. Note that Ψ extends to the whole of $B(R)$ by $\Psi(fg\otimes h)=(\mathrm{id}\otimes\cdot)\circ\Psi((f\otimes g)\otimes h)$. Here Ψ on higher tensor products of $\underline{A}$ is determined by consistency under the braided group, i.e. $\Psi_{X\otimes Y,Z}=\Psi_{X,Z}\circ\Psi_{Y,Z}$ etc. for any tensor products X,Y,Z of $\underline{A}$. The extension of $\Delta : B(R)\to B(R)\otimes B(R)$ means as an algebra map where $B(R)\otimes B(R)$ has a braided-tensor product algebra structure (8). That Δ respects this braided tensor product algebra structure that is respects the underlying braided manifold. After quotienting suitably, it is not hard (at lest formally) to define an antipode (obeying the usual axioms) on the generators. It extends from the generators by the rule $S(fg)=\cdot\Psi(Sf\otimes Sg)$ for all f,g. The algebra structure (18) of $B(R)$ is very natural: it just says that $B(R)$ is indeed braided-commutative in the sense of (10) or (15). Indeed, for the triangular case where the braid group action in actually a symmetric group action (e.g. super vector spaces), i.e. if $R_{21}R_{12}=1\otimes 1$, then $\mathfrak{D}^M{}_L{}^I{}_N=\delta^M{}_N\delta^I{}_N$ in which case the braided-commutative relations (18) are just

$$\cdot\Psi(f\otimes g)=fg \tag{21}$$

The extra $\mathfrak{D}$ factor thus adjusts for the failure of $\Psi^2=id$ i.e. the degree of quasitriangularity. It is new ingredient coming out of the abstract category theory in Sec.2 above. In the sense then, B(R) really is braided-commutative, i.e. a classical (not quantized) object in a braided or quasitensor category!

Example 3.3 BRAIDED SL(2). For the standard R matrix for SL(2) we obtain by the above a braided group which we denote $BSL(2)$. It is the algebra generated by 1 and four indeterminates $u=\begin{pmatrix}a&b\\c&d\end{pmatrix}$ with the algebra structure

$$ba=q^2ab,\ ca=q^{-2}ac,\ da=ad,\ db=bd+(1-q^{-2})ab \tag{22}$$

$$cd=dc+(1-q^{-2})ca,\ bc=cb+(1-q^{-2})(ad-a^2) \tag{23}$$

$$ad-q^2cb=1. \tag{24}$$

Here (22)-(23) are just the braided-commutative relations (18) and with the coproduct (20) they define the *braided-matrices*, $BM(2)$. The left hand side of the additional relation (24) is the q-determinant for $BSL(2)$. It is central and grouplike in $BM(2)$. $BSL(2)$ has an antipode given by

$$Sa=q^2d+(1-q^2)a,\ Sb=-q^2b,\ Sd=a,\ Sc=-q^2c. \tag{25}$$

The braid group action on generators is

$$\Psi(a\otimes a)=a\otimes a+(1-q^2)b\otimes c,\Psi(a\otimes b)=b\otimes a,\ \Psi(a\otimes c)=c\otimes a+(1-q^2)(d-a)\otimes c$$

$$\Psi(a\otimes d)=d\otimes a+(1-q^2)b\otimes c,\ \Psi(b\otimes a)=a\otimes b+(1-q^2)b\otimes(d-a),\ \Psi(b\otimes b)=q^2b\otimes b$$

$$\Psi(b\otimes c)=q^{-2}c\otimes b+(1+q^2)(1-q^{-2})b\otimes c-(1-q^{-2})(d-a)\otimes(d-a)$$

$$\Psi(b\otimes d)=d\otimes b+(1-q^{-2})b\otimes(d-a),\ \Psi(c\otimes a)=a\otimes c,\ \Psi(c\otimes b)=q^{-2}b\otimes c$$

$$\Psi(c\otimes c)=q^2c\otimes c,\Psi(c\otimes d)=d\otimes c,\ \Psi(d\otimes a)=a\otimes d+(1-q^{-2})b\otimes c$$

$\Psi(d\otimes b) = b\otimes d, \Psi(d\otimes c) = c\otimes d+(1-q^{-2})(d-a)\otimes c,\ \Psi(d\otimes d) = d\otimes d-q^{-2}(1-q^{-2})b\otimes c.$

This makes $BSL(2)$ a (braided) Hopf algebra (c.f. coordinate functions on a classical group or supergroup). Here the matrix form $\Delta\ : BSL(2) \to BSL(2)\otimes BSL(2)$ respects the algebra structure (i.e. the underlying braided-manifold) provided $BSL(2) \to BSL(2)\otimes BSL(2)$ has the algebra structure (8). Thus we have explicitly transmuted the quantum group $SL_q(2)$ into a braided group. When this is done it takes the form of a braided-commutative space obtained by quotienting the braided-commutative 4-plane (18) by the determinant (24).

It is very important to realize that the braided group does not exist in isolation-it is just one object in a quasitensor categories viewed as a generalization of the category of superspaces. Such notions as non-commutative differential calculus now become as similar as possible to the classical or super case, using Ψ whenever a transposition is needed. In the present case, for example, the objects in the category are the representations of $U_q(su(2))$ (or $SU_q(2)$ comodules). The role of the decomposition of super vector spaces into bosonic and fermionic parts is now played by the spectral decomposition under the action. All the maps of $BSL(2)$ above, i.e. $\cdot, \Delta, \Psi$ are morphisms in the category, i.e. respect this decomposition. We do not have space to discuss these and other topics here: they are presented elsewhere[20][14]. Finally, note that we have applied here only the case of the identity functor in Theorem 2.1: other monoidal functors would give rise to general braided quantum groups, generalizing the notion of super-quantum groups[23]. For example, if A_1 and A_2 are dual quasitriangular Hopf algebras and $A_1 \to A_2$ is a Hopf algebra map, this induces a braided quantum group $B(A_1, A_2)$ in the category ${}^{A_2}\mathcal{M}$.

4. Rank of dual quantum groups

To complete the dual of the present paper we now give the formula for the rank of dual quasitriangular Hopf algebra and compute a class of examples. The category-theoretic rank that we use is defined in any rigid quasitensor categories in the same way as in the tensor case[15][10], as the composite

$$\mathrm{rank}(X):\ \underline{1} \xrightarrow{\pi} X\otimes X^* \xrightarrow{\Psi_{X,X^*}} X^*\otimes X \xrightarrow{ev} \underline{1}. \tag{26}$$

We now apply it in the case $\mathcal{C} = {}^{A}\mathcal{M}^{f.d.}$ for any dual quasitriangular Hopf algebra A. In this case $\underline{1} = k$ so that $rank(X)$ is in k (e.g. a complex number). This gives the category-theoretic dimension of any finite dimensional A-comodule. In this setting we put (6)-(7) into (26) to obtain

$$\mathrm{rank}(V) = \mathrm{Tr}(\mathrm{id}\otimes \underline{u})\circ \beta_V,\ \text{ where } \underline{u} = \sum \mathcal{R}(a_{(2)}\otimes Sa_{(1)}). \tag{27}$$

Here (V, β_V) is any finite-dimensional right comodule and the trace is that of a map from V to V. Now, A is an A-comodule by the right regular comodule structure $\beta_A = \Delta : A \to A\otimes A$. So if A is finite dimensional we define

$$|A| = \mathrm{rank}(A) = \mathrm{Tr}(\mathrm{id}\otimes \underline{u})\circ \Delta. \tag{28}$$

We now compute $|\check{A}|$. Note that we really need the dual quasitriangular structure and antipode but the latter does not show up in the final answer provided $\tilde{R}=((r^{t_2})^{-1})^{t_2}$ exists. As a result, the formula can be modified to give a natural definition also $|A(R)|$ where $A(R)$ is the bialgebra of quantum matrices of [7]. To describe the result we suppose that $\check{A}(R)$ and $A(R)$ have a basis of the form $u^{i_1}{}_{j_1}\dots u^{i_n}{}_{j_n}$ for the collection $(I,J)\in\mathcal{L}$. Here $I=(i_1\cdots,i_n)$ etc. and $\mathcal{L}$ is some multi-indexing set depending on the quantum group in question. We let $\{E^{j_1,\dots,j_n}_{i_1,\dots,i_n}\}$ denote a dual basis. Then in (28) we have

$$|\check{A}|=\sum_{(I,J)\in\mathcal{L}}E^{j_1,\cdots,j_n}_{i_1,\cdots,i_n}((u^{i_1}{}_{j_1}\cdots u^{i_n}{}_{j_n})_{(1)})\mathcal{R}((u^{i_1}{}_{j_1}\cdots u^{i_n}{}_{j_n})_{(3)}\otimes S(u^{i_1}{}_{j_1}\cdots u^{i_n}{}_{j_n})_{(2)}).$$

Using the standard matrix coproduct and (4) this evaluates as

$$|\check{A}|=\sum_{(I,J)\in\mathcal{L}}\mathcal{E}^{I\,J}_{K}Z_{\tilde{R}}\begin{pmatrix} & K & \\ L & \Box & J\\ & L & \end{pmatrix} \tag{29}$$

where $\mathcal{E}^{i_1,\cdots,i_n,j_1,\cdots,j_n}_{k_1,\cdots,k_n}=E^{j_1,\cdots,j_n}_{i_1,\cdots,i_n}(u^{i_1}{}_{k_1}\cdots u^{i_n}{}_{k_n})$ (no summation) and where

$$Z_{\tilde{R}}\begin{pmatrix} & K & \\ I & \Box & J\\ & L & \end{pmatrix}$$

$$=\tilde{T}^{i_n}{}_{j_n}{}^{K}{}_{M_1}T^{i_{n-1}}{}_{j_{n-1}}{}^{M_1}{}_{M_2}\cdots T^{i_1}{}_{j_1}{}^{M_{n-1}}{}_{L};\ \tilde{T}^{i}{}_{j}{}^{K}{}_{L}=\tilde{R}^{i}{}_{m_1}{}^{k_1}{}_{l_1}\tilde{R}^{m_1}{}_{m_2}{}^{k_2}{}_{l_2}\cdots\tilde{R}^{m_{n-1}}{}_{j}{}^{k_n}{}_{l_n}.$$

Here $\tilde{T}$ is the single-row transfer matrix and $Z_{\tilde{R}}$ the partition function for a vertex model with Boltzmann weight $\tilde{R}$ and boundary conditions as shown. This is similar to [3, Sec. 5.2] where we also obtained partition functions in this way. The main difference is that because of the antipode in (27) we have weight $\tilde{R}$ rather than R. The boundary coefficients $\mathcal{E}$ depend on the indexing set $\mathcal{L}$. For example, it might be that the indexing set is all of the (I,J). In this ideal case, $|\check{A}(R)|=Z_{\tilde{R}}\begin{pmatrix} & K & \\ L & \Box & J\\ & L & \end{pmatrix}$. The general case nevertheless has the same flavour: $|\check{A}(R)|$ (and $|A(R)|$) have the character of a partition function. That is, they count quantum states in some sense.

The significance of these computations is that they provide a strategy or approach for understanding the origin of number theory in exactly solvable lattice models. The strategy is to try to identify the partition functions of more realistic physical systems (e.g. the six-vertex model) as an intrinsic quantum order or rank of some algebraic structure, $|A|=Z$. The number-theoretic properties of Z would then arise from the combinatorics of the algebraic structure. Because the rank should be an analog of counting the points in a group, such combinatorics would show up in $|A|$. Certainly this is confirmed in the computation (1)[6], where this strategy was proposed. As a first step it is therefore natural to ask if the familiar bialgebras arising in such vertex models (generated by $u^{i}{}_{j}(\lambda)$ where λ is a spectral parameter) are to some extent dual quasitriangular, and if so, what is their rank.

References

[1] S.Majid, *Reconstruction Theorems and rational conformal field theories*, to appear Int.J.Mod. Phys.A. (1989).

[2] S.Majid, *Braided group and algebraic quantum field theories*, Preprint (1990); *Some physical applications of category theory*, Lec. Notes Phys., Springer (to appear).

[3] S.Majid, *Quasitriangular Hopf algebras and Yang-Baxter Equations*, Int. J. Modern Physics A **5(1)** (1990), 1-91.

[4] S.Majid, *Representation theoretic rank and double Hopf algebras*, Comm. Algebra **18** (1990); 3705–3712.

[5] S.Majid and Ya.S. Soibelman, *Rank of quantized universal enveloping algebras and modular functions*, to appear Comm. Math. Phys. (1990).

[6] S.Majid and Ya.S. Soibelman, *Chern-Simons theory, modular functions and quantum mechanics in an alcove*, to appear Int. J. Math. Phys. (1990).

[7] L.D.Faddeev, N.Yu. Reshetikhin, L.A. Takhtajan, *Quantization of Lie groups and Lie algebras*, Algebra Anal. **1(1)** (1989), 178–206.

[8] N.Yu.Reshetikhin, V.G.Turaev, *Ribbon graphs and their invariants derived from quantum groups*, Comm. Math. Phys. **127** (1990), 1–26.

[9] A.Joyal and R.Street, *Braided monoidal categories*, Macquarie Reports 86008 (1986).

[10] P.Deligne and J.S.Milne, *Tannakian categories*, Lec. Notes in Math. 900 Springer (1982).

[11] V.G.Drinfeld, *Quantum groups*, Proc. ICM-86(Berkeley) **1** (1987), 798–820.

[12] S.Majid, *Quantum groups and quantum probability*, to appear in Quantum probability and Applications, Trento, Kluwer (1989).

[13] S.Majid, *Representations, duals and quantum doubles of monoidal categories*, to appear in Suppl. Rend. Circ. Mat. Palermo (1989).

[14] S.Majid, *Braided groups*, Preprint (1990).

[15] N.Saavedra Rivano, *Catégories Tannakiennes*, Lec. Notes in Math. 265 Springer (1972).

[16] D.Yetter, *Quantum groups and representations of monoidal categories*, Math Proc. Cam. Phil. Soc. **108** (1990), 261–290.

[17] K.-H.Ulbrich, *On Hopf algebras and rigid monoidal categories*, to appear Isr. J. Math. (1989).

[18] K.Fredenhagen,K.-H.Rehren,B.Schroer, *On Hopf algebras and rigid monoidal categories*, Comm. Math.Phys. **125**, 201.
R.Longo, Comm. Math.Phys. **126** (1989), 217.

[19] S.Doplicher and J.E.Roberts, *A new duality theorem for compact groups*, Inv. Math.Phys. **985** (1989), 157–218.

[20] S.Majid, *Examples of braided groups and braided matrices*, Preprint (1990).

[21] A.Connes, *Non-commutative differential geometry*, IHES **62** (1986).

[22] N.Ya.Manin, *Noncommutative geometry and quantum groups*, Monthreal Notes (1988).
A.Rozenberg, *Hopf algebras and Lie algebras in categories with multiplication*, preprint (1978). (in Russian)
D.Gurevich, Stockholm Math. Rep. **24** (1986), 33-122 **25** (1988), 1-16.

[23] P.P. Kulish, *Quantum superalgebra* $osp(2|1)$, preprint RIMS-615 (1988).
M.Chaichan and P.P.Kulish, Phys.Lett.B **234** (1990), 72–80.

[24] S.Majid, *More examples of double cross product Hopf algebras*, Isr.J.Math. **71(3)** (1990).

[25] V.V.Lyubashenko, *Tensor categories and RCFT,I,II*, preprints (1990).

DEPARTMENT OF APPLIED MATHEMATICS & THEORETICAL PHYSICS UNIVERSITY OF CAMBRIDGE, CAMBRIDGE, CB3, 9EW, U.K.

YANGIANS OF THE "STRANGE" LIE SUPERALGEBRAS

M.L. NAZAROV

Department of Mathematics, Moscow State University

ABSTRACT. Consider the matrix Lie superalgebra $A = \mathfrak{gl}(n|n, \mathbb{C})$ with the standard generators $e_{ij}; i, j \in \mathbb{Z}_{2n}$. Define an automorphism π of A by $\pi(e_{ij}) = e_{i+n,j+n}$. The automorphisms π and $-\pi \circ t$, where t denotes the matrix supertransposition, are involutory. Dissimilarly to the even case,

$$\mathfrak{g} = \{x(\lambda) \in A[\lambda] : x(\lambda)^{\varphi} = x(-\lambda)\}; \ \varphi = \pi, -\pi \circ t$$

acquires a natural superalgebra structure. A quantization of the co-Poisson Hopf superalgebra $U(\mathfrak{g})$ is constructed. It gives rise to new solutions of the Yang-Baxter equation.

The classical case

Let $m, n \in \mathbb{N}$ be fixed. Consider the general linear Lie superalgebra $\mathfrak{gl}\,(m|n, \mathbb{C}) = A$. Let e_{ij}, $1 \leq i, j \leq m+n$ be standard generators of A:

$$[e_{ij}, e_{kl}] = \delta_{kj} e_{il} - \delta_{il} e_{kj} (-1)^{p(e_{ij}) p(e_{kl})};$$

the $\mathbb{Z}_2$-gradation on A is defined by $e_{ij} \mapsto \bar{i} + \bar{j}$, where $\bar{i} = 0$ or 1 depending on whether $1 \leq i \leq m$ or $m+1 \leq i \leq m+n$.

If $m \neq n$ the group of outer automorphisms Out$A$ is generated [1] by the element

$$-t : \ e_{ij} \mapsto -e_{ji}(-1)^{\bar{i}(\bar{j}+1)}.$$

If $m = n$ the group Out$A$ enlarges: there exists an extra automorphism

$$\pi : \ e_{ij} \mapsto e_{i+n,j+n},$$

where i, j are considered modulo $2n$. The elements π and $-t$ generate the group Out$A$ if $m = n > 2$.

From now on I shall assume $m = n$. The automorphisms π and $-\pi \circ t$ of A are involutory; let φ be one of these two automorphisms. The "strange" Lie superalgebras are defined to be the fixed point superalgebras of A with respect to φ:

$$\mathfrak{q}(n, \mathbb{C}) = A^{\pi}, \qquad \mathfrak{U}(n, \mathbb{C}) = A^{-\pi \circ t}.$$

Consider the polynomial current Lie superalgebra $A[\lambda]$. The *Casimir element* of $(A^{\varphi})^{\otimes 2}$ vanishes:

$$\sum_{i,j} (e_{ij} + \varphi(e_{ij})) \otimes (e_{ji} + \varphi(e_{ji}))(-1)^{\bar{j}} = 0.$$

Thus one cannot define a Lie superbialgebra structure on $A^{\varphi}[\lambda]$ as it was done in [2] for $\mathfrak{B}[\lambda]$, where $\mathfrak{B}$ is a simple finite-dimensional Lie algebra over $\mathbb{C}$. I propose the following construction instead.

Let c be the Casimir element of $A^{\otimes 2}$:

$$c = \sum_{i,j} e_{ij} \otimes e_{ji}(-1)^{\bar{j}}.$$

Introduce the following operators in $A^{\otimes 2}$: ${}^1\varphi = \varphi \otimes \mathrm{id}$, ${}^2\varphi = \mathrm{id} \otimes \varphi$;

$$\sigma : e_{ij} \otimes e_{kl} \mapsto e_{kl} \otimes e_{ij}(-1)^{p(e_{ij})p(e_{kl})};$$

being the permutation. Then

$$\sigma(c) = c, \qquad \sigma \circ {}^1\varphi(c) = {}^2\varphi(c) = -{}^1\varphi(c). \tag{1}$$

Define the *classical r-matrix* $r : \mathbb{C} \times \mathbb{C} \mapsto A^{\otimes 2}$ by (cf. [3,4])

$$r(\lambda_1, \lambda_2) = \frac{c}{\lambda_1 - \lambda_2} + \frac{\varphi(c)}{\lambda_1 + \lambda_2}. \tag{2}$$

Due to (1) it is *antisymmetric* and obeys the *classical Yang-Baxter identity* [5]:

$$\sigma(r(\lambda_1, \lambda_2)) = -r(\lambda_2, \lambda_1), \tag{3}$$

$$[{}^{12}r(\lambda_1, \lambda_2), {}^{13}r(\lambda_1, \lambda_3)] + [{}^{12}r(\lambda_1, \lambda_2), {}^{23}r(\lambda_2, \lambda_3)] + [{}^{13}r(\lambda_1, \lambda_3), {}^{23}r(\lambda_2, \lambda_3)] = 0. \tag{4}$$

Moreover,

$${}^1\varphi(r(\lambda_1, \lambda_2)) = r(-\lambda_1, \lambda_2), \ {}^2\varphi(r(\lambda_1, \lambda_2)) = r(\lambda_1, -\lambda_2). \tag{5}$$

Let $A = A^{\varphi} \oplus A^{-\varphi}$ be the special decomposition of A with respect to φ. Consider the twisted polynomial current Lie superalgebra

$$\mathfrak{g} = \{x(\lambda) \in A[\lambda] : \ \varphi(x(\lambda)) = x(-\lambda)\} = \{\sum_{s \geq 0} x_s \lambda^s : x_s \in A^{(-1)^s \varphi}\}.$$

Identify $A[\lambda]^{\otimes 2}$ with $A^{\otimes 2}[\lambda_1, \lambda_2]$,then

$$\mathfrak{g}^{\otimes 2} = \{y(\lambda_1, \lambda_2) \in A^{\otimes 2}[\lambda_1, \lambda_2] : \ {}^1\varphi(y(\lambda_1, \lambda_2)) = y(-\lambda_1, \lambda_2), \\ {}^2\varphi(y(\lambda_1, \lambda_2)) = y(\lambda_1, -\lambda_2).$$

For each $x(\lambda) \in \mathfrak{g}$ put

$$\delta(x(\lambda)) = [x(\lambda_1) \otimes 1 + 1 \otimes x(\lambda_2), \ r(\lambda_1, \lambda_2)]. \tag{6}$$

Since the element $c \in A^{\otimes 2}$ is A-invariant the r.h.s. of (6) has no pole at $\lambda_1 = \lambda_2$, nor has it a pole at $\lambda_1 = -\lambda_2$:

$$[x(\lambda_1) \otimes 1 + 1 \otimes x(\lambda_2), \frac{{}^2\varphi(c)}{\lambda_1 + \lambda_2}] = {}^2\varphi([x(\lambda_1) \otimes 1 + 1 \otimes x(-\lambda_2), \frac{c}{\lambda_1 + \lambda_2}]).$$

Moreover, $\delta(\mathfrak{g}) \subset \Lambda^2 \mathfrak{g}$ due to (3,5). Thus (4) implies

PROPOSITION. *The map $\delta : \mathfrak{g} \to \Lambda^2 \mathfrak{g}$ defines a Lie superbialgebra structure on $\mathfrak{g}$.*

In the next section I provide a quantization of the Lie superbialgebra $\mathfrak{g}$ in the sense of [2]. Now I would like to comment on the even case. If A is a finite-dimensional Lie algebra over $\mathbb{C}$ and c, φ are its Casimir element and an involutory automorphism, respectively then the r.h.s. of (2) is never antisymmetric (see [4]). Thus there is no natural Lie bialgebra structure on the twisted polynomial current Lie algebra $\mathfrak{g}$. For $A = sl(n, \mathbb{C})$ and $\varphi = -t$, the appropriate additional structure on $\mathfrak{g}$ and its quantization are constructed in [11].

The quantum case

Let E_{ij}, $1 \le i,j \le 2n$ be the canonical generators of the matrix superalgebra $\mathrm{Mat}(n|n,\mathbb{C})$: $(E_{ij})_{kl} = \delta_{ik}\delta_{jl}$. Let $E \in \mathrm{Mat}(n|n,\mathbb{C})$ be the unit matrix and $P \in \mathrm{Mat}(n|n,\mathbb{C})^{\otimes 2}$ be the permutation matrix: $P_{ik,jl} = \delta_{il}\delta_{jk}(-1)^{\bar{i}\cdot\bar{k}}$. Introduce the standard representation

$$\rho : A \to \mathrm{Mat}(n|n,\mathbb{C}) : e_{ij} \mapsto E_{ij}.$$

Let r be the classical r-matrix (2). For any $Q \in \mathrm{Mat}(n|n,\mathbb{C})^{\otimes 2}$ define

$$Q^{\natural} \in \mathrm{Mat}(n|n,\mathbb{C})^{\otimes 2} \;:\; Q^{\natural}_{ik,jl} = Q_{jl,ik}.$$

Introduce the *quantum R-matrix* $R \;:\; \mathbb{C}\times\mathbb{C} \to \mathrm{Mat}(n|n,\mathbb{C})^{\otimes 2}$,

$$R(u_1,u_2) = (E^{\otimes 2} - \rho^{\otimes 2} r(u_1,u_2)))^{\natural}. \tag{7}$$

It obeys the *quantum Yang-Baxter identity* [5]

$${}^{12}R(u_1,u_2)^{13}R(u_1,u_3)^{23}R(u_2,u_3) = {}^{23}R(u_2,u_3)^{13}R(u_1,u_3)^{12}R(u_1,u_2) \tag{8}$$

and the conditions of *symmetry* and *unitarity*:

$$P\;R(u_1,u_2)\;P \;= R(-u_2,-u_1); \tag{9}$$

$$R(u_1,u_2)R(-u_1,-u_2) = \begin{cases} 1-(u_1-u_2)^{-2}-(u_1+u_2)^{-2} & ,\varphi = \pi; \\ 1-(u_1-u_2)^{-2} & ,\varphi = -\pi\circ t. \end{cases} \tag{10}$$

The identities (8) for $R(u_1,u_2)$ and $R^{\natural}(u_1,u_2)$ are equivalent. If $\varphi = \pi$ then these R-matrices coincide, but it is not so for $\varphi = -\pi\circ t$.

To display the R-matrix (7) I shall introduce a $\mathbf{Z}_2$-graded analogue of the Brauer algebra [10]. At first I recall the original definition.

Let $N \in \mathbb{N}$ be fixed. Consider the algebra $Br(2n)$ over $\mathbb{C}$ with the linear basis $\{e(\gamma)\}$, where γ runs through all the patterns like this

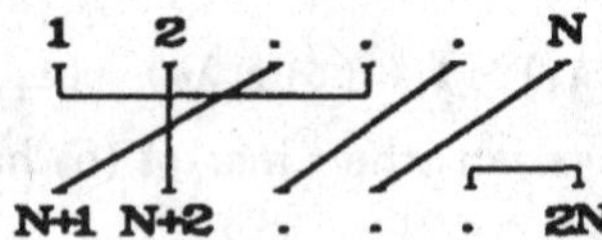

there are $2N$ "vertices" on the plane linked by N "threads" so that:

- each vertex is incident to exactly one thread;
- each thread is incident to exactly two vertices.

These patterns are subject to the following equivalence relations. One can move threads through each other and remove loops:

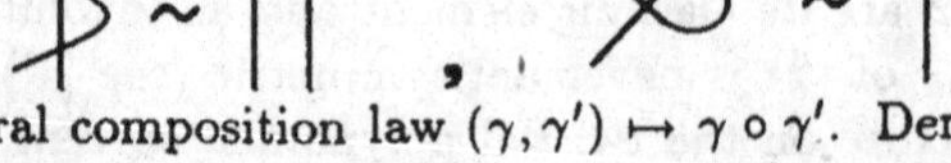

$$\sim \qquad , \qquad \sim \tag{11}$$

There is a natural composition law $(\gamma,\gamma') \mapsto \gamma\circ\gamma'$. Denote by $\gamma \sqcup \gamma'$ the pattern obtained by linking up the vertices $N+1,\ldots,2N$ of γ with the vertices $1,\ldots,N$ of γ'

respectively. Erasing all the cycles from $\gamma \sqcup \gamma'$, we get $\gamma \circ \gamma'$. For instance,

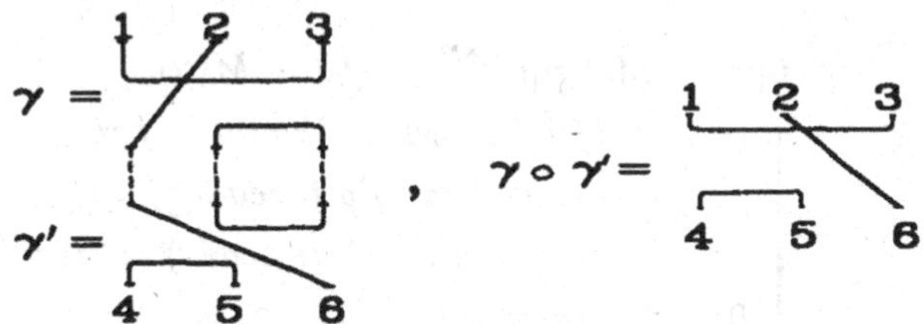

Denote by $c(\gamma, \gamma')$ the quantity of the cycles erased. Then

$$e(\gamma) \cdot e(\gamma') = (2n)^{c(\gamma,\gamma')} e(\gamma \circ \gamma').$$

We have a natural representation

$$Br(2n) \to \mathrm{Mat}(2n, \mathbb{C})^{\otimes N} : e(\gamma) \mapsto M(\gamma),$$

$$M(\gamma)_{i_1 \ldots i_N, i_{N+1} \ldots i_{2N}} = \begin{cases} 1 & \textit{if (the vertices } k \textit{ and } l \textit{ of } \gamma \\ & \textit{are linked by a thread)} \Rightarrow (i_k = i_l); \\ 0 & \textit{otherwise.} \end{cases}$$

To introduce the version of the algebra $Br(2n)$ related to the Lie superalgebras $\mathfrak{q}(n, \mathbb{C})$ and $\mathfrak{U}(n, \mathbb{C})$ I amend the above construction as follows:

- each thread of γ is allowed to carry several "beads" of two kinds: "black" and "white":

- instead of the equivalence relations (11) one should use the following ones:

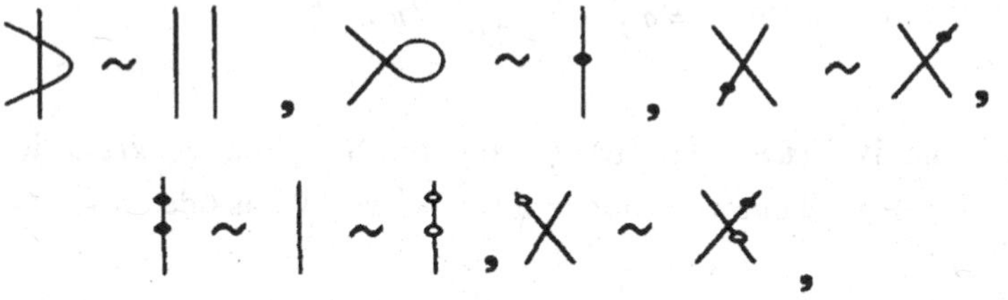

(12)

- if patterns γ, γ' differ by a transposition of two diverse adjacent beads then $e(\gamma) + e(\gamma') = 0$;
- defining the product $e(\gamma) \cdot e(\gamma')$ we erase from $\gamma \sqcup \gamma'$ only the cycles without beads; if a cycle of $\gamma \sqcup \gamma'$ carries exactly one or two beads of the different kinds then $e(\gamma) \cdot e(\gamma') = 0$.

Denote the algebra so obtained by $Br(n|n)$; the $\mathbb{Z}_2$-gradation on $Br(n|n)$ is defined by $e(\gamma) \mapsto$ (*the number of white beads in* γ) mod 2.

The above alternations result in a natural representation

$$Br(n|n) \mapsto \mathrm{Mat}(n|n)^{\otimes N} : \ e(\gamma) \mapsto M(\gamma),$$

$$M(\gamma)_{i_1 \dots i_N, i_{N+1} \dots i_{2N}} = \begin{cases} (-1)^d & \textit{if (the vertices } k \textit{ and } l \textit{ of } \gamma \\ & \textit{are linked by a thread} \\ & \textit{carrying } m \textit{ white beads)} \Rightarrow (i_k = i_l + m \cdot n); \\ 0 & \textit{otherwise.} \end{cases}$$

Due to (12) we can assume that any thread carries no more than one white bead. To determine the number d, for each point g of a thread of γ put $\tilde{g} = \tilde{i}_k$ if there is no white beads between g and the vertex k of the thread. Then

$$d = \sum_{\substack{\text{there is} \\ \text{a black} \\ \text{bead at } g}} \tilde{g} + \sum_{\substack{\text{two threads cross} \\ \text{at } g \text{ and } g' \\ \text{respectively}}} \tilde{g}\tilde{g}'.$$

Now I can display the R-matrix (7):

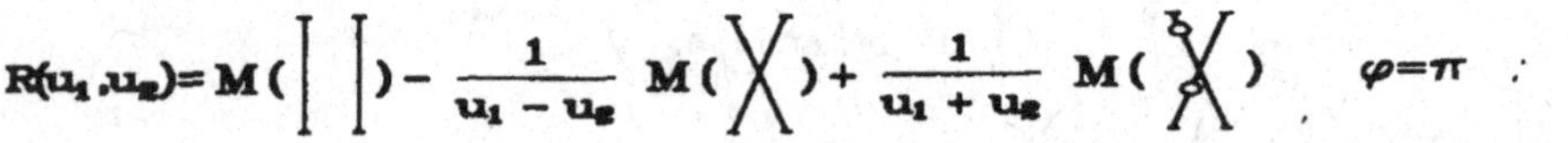

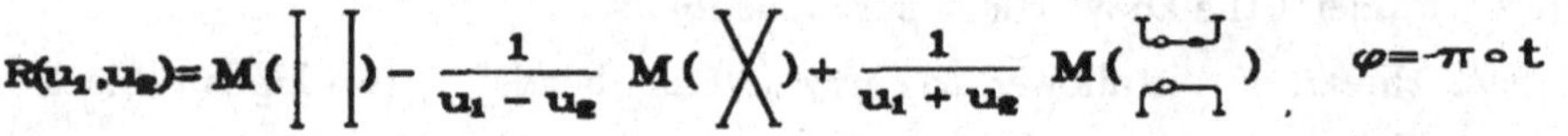

$$R(u_1,u_2) = M(|\,|) - \frac{1}{u_1 - u_2} M(X) + \frac{1}{u_1 + u_2} M(X \text{ with bead}), \qquad \varphi = \pi \ ;$$

$$R(u_1,u_2) = M(|\,|) - \frac{1}{u_1 - u_2} M(X) + \frac{1}{u_1 + u_2} M(\text{cup/cap}), \qquad \varphi = -\pi \circ t \ .$$

Using the relations (12) one can easily verify the equalities (8) to (10).

Define an associative superalgebra $Y(A, \varphi) = G$ over $\mathbb{C}$ with generators $T_{ij}^{(s)}$, where $1 \le i, j \le 2n$ and $s \ge 1$, as follows. Put

$$T_{ij}(u) = \delta_{ij} + \sum_{s \ge 0} T_{ij}^{(s+1)} u^{-s-1}.$$

and consider the matrix $T(u) = [T_{ij}(u)] \in \mathrm{Mat}(n|n, G)$. The $\mathbb{Z}_2$-gradation on G is defined by $T_{ij}^{(s)} \mapsto \tilde{i} + \tilde{j}$. Thus the matrix $T(u)$ is even. Consider also the matrices $T^{\pi}(u)$ and $T^{\pi \circ t}(u)$:

$$T_{ij}^{\pi}(u) = T_{i+n, j+n}(u), \qquad T_{ij}^{\pi \circ t}(u) = T_{i+n, j+n}(u)(-1)^{(\tilde{i}+1)\tilde{j}}.$$

Then (cf. [6,7]) the defining relations of the algebra G are

$$R(u_1, u_2)\, {}^1T(u_1)\, {}^2T(u_2) = {}^2T(u_2)\, {}^1T(u_1) R(u_1, u_2); \tag{13}$$

$$\begin{cases} T^{\pi}(u) = T(-u) & , \quad \varphi = \pi \ ; \\ T^{\pi \circ t}(u) = T^{-1}(-u) & , \quad \varphi = -\pi \circ t \ . \end{cases} \tag{14}$$

A Hopf superalgebra structure on G is defined in the standard way; the following formulae determine co-multiplication Δ, co-unit ε and antipode S on G:

$$\Delta(T(u)) = T(u)\dot{\otimes}T(u),$$
$$\varepsilon(T(u)) = E,$$
$$S(T(u)) = T^{-1}(u),$$

where $\dot{\otimes}$ means the matrix multiplication combined with tensor product of their elements.

The antipode S is an anti-automorphism of the superalgebra G, hence S^2 is an automorphism. Put

$$C(u) = \sum_k T_{ki}(u)(T^{-1})_{ik}(u)(-1)^{\bar{i}+\bar{k}}. \tag{15}$$

PROPOSITION. *(cf. [7,8]). The r.h.s. of (15) does not depend on i and belongs to the center of G.*

To prove this proposition consider the matrix $K \in \mathrm{Mat}(n|,\mathbb{C})^{\otimes 2}$ with $K_{ik,jl} = \delta_{ik} \cdot \delta_{jl}(-1)^{\bar{j}}$,

$$\mathbf{K} = \mathbf{M}(\quad)$$

Note that $K^2 = 0$ and $P^{t_2} = K$, there $Q^{t_2}_{ik,jl} = Q_{il,jk} \cdot (-1)^{(\bar{k}+1)\bar{l}}$ for any $Q \in \mathrm{Mat}(n|,\mathbb{C})^{\otimes 2}$. We obtain from (13) the relation

$$R^{t_2}(u_1,u_2)^{-1} \cdot {}^1T(u_1) \cdot {}^2T^{-1}(u_2)^t = {}^2T^{-1}(u_2)^t \cdot {}^1T(u_1) \cdot R^{t_2}(u_1,u_2)^{-1}. \tag{16}$$

The residue of $R^{t_2}(u_1,u_2)^{-1}$ at $u_1 = u_2$ is equal to K, and we come to the relation

$$K \cdot {}^1T(u) \cdot {}^2T^{-1}(u)^t = {}^2T^{-1}(u)^t \cdot {}^1T(u) \cdot K.$$

Taking the $(11, ii)$ - entry of this matrix relation we prove that the r.h.s. of (15) does not depend on i. Moreover, it implies that

$$K \cdot {}^1T(u) \cdot {}^2T^{-1}(u)^t = C(u) \cdot K. \tag{17}$$

Actually $C(u) = 1$ if $\varphi = -\pi \cdot t$. Indeed, due to (14) we have

$$\sum_k T_{ki}(u) \cdot (T^{-1})_{ik}(u) \cdot (-1)^{\bar{i}+\bar{k}}$$
$$= \sum_k (T^{-1})^{\pi o t}_{ki}(-u) \cdot T^{\pi o t}_{ik}(-u)^{\bar{i}+\bar{k}}$$
$$= \sum_k (T^{-1})_{i+n,k+n}(-u) \cdot T_{k+n,i+n}(-u) = 1.$$

Assume now that $\varphi = \pi$. It follows from (13),(16) and (17) that

$$\begin{aligned} &{}^{23}K \cdot {}^{13}R^{t_2}(u_1,u_2)^{-1} \cdot {}^{12}R(u_1,u_2) \cdot {}^{1}T(u_1) \cdot {}^{2}T(u_2) \cdot {}^{3}T^{-1}(u_2)^{t} \\ &= {}^{23}K \cdot {}^{2}T(u_2) \cdot {}^{3}T^{-1}(u_2)^{t} \cdot {}^{1}T(u_1) \cdot {}^{13}R^{t_2}(u_1,u_2)^{-1} \cdot {}^{12}R(u_1,u_2) \qquad (18) \\ &= C(u_2) \cdot {}^{1}T(u_1)^{23} \cdot K \cdot {}^{13}R^{t_2}(u_1,u_2)^{-1} \cdot {}^{12}R(u_1,u_2). \end{aligned}$$

Using the equality

$${}^{23}K \cdot {}^{13}R^{t_2}(u_1,u_2)^{-1} \cdot {}^{12}R(u_1,u_2) = (1 - (u_1 - u_2)^{-2} - (u_1 + u_2)^{-2}) \cdot {}^{23}K$$

we get from (18) the relation $T(u_1)C(u_2) = C(u_2)T(u_1)$, so $C(u)$ belongs to the center of G. Using the monomorphism

$$G \to U(\mathfrak{g}(n)) : \; T_{ij}^{(1)} \mapsto (e_{ij} + e_{i+n,j+n}) \cdot (-1)^{\bar{i}}; \; T_{ij}^{(s)} \mapsto 0, \; s \geq 2$$

one can show that $C(u) \neq 1$ for $\varphi = \pi$.

If $\varphi = \pi$ then $C(u) = C(-u)$: due to (14) we have

$$\begin{aligned} C(u) &= \sum_k T_{ki}(u) \cdot (T^{-1})_{ik}(u) \cdot (-1)^{\bar{i}+\bar{k}} \\ &= \sum_k T^{\pi}_{\bar{k}\bar{i}}(-u) \cdot ({}^{-1})T^{\pi}_{\bar{i}\bar{k}}(-u) \cdot (-1)^{\bar{i}+\bar{k}} \\ &= \sum_k T_{k+n,i+n}(-u) \cdot (T^{-1})_{i+n,k+n}(-u) \cdot (-1)^{\bar{i}+\bar{k}} = C(-u), \end{aligned}$$

Since $C(u)$ does not depend on i. Thus $C(u) = 1 + C^{(1)}u^{-2} + C^{(3)}u^{-4} + \cdots$ for some $C^{(1)}, C^{(3)}, \cdots \in Y(A,\pi)$.

CONJECTURE. *The elements $C^{(1)}, C^{(3)}, \ldots$ generate the center of the superalgebra $Y(A,\pi)$.*

PROPOSITION. $S^2(T(u)) = C(u)^{-1}T(u)$.

Indeed, the definition of the antipode S implies that

$$\sum_k S^2(T_{ki}(u)) \cdot (T^{-1})_{jk}(u) \cdot (-1)^{(\bar{i}+\bar{k})(\bar{j}+\bar{k})} = \delta_{ij},$$

while the relation (17) means that

$$\sum_k T_{ki}(u) \cdot (T^{-1})_{jk}(u) \cdot (-1)^{(\bar{i}+\bar{j}+1)\bar{k}} = C(u) \cdot \delta_{ij}(-1)^{\bar{j}}.$$

Comparing of the last two relations proves the proposition.

COROLLARY. $\Delta(C(u)) = C(u) \otimes C(u)$.

To conclude this section I shall explain why the Hopf superalgebra G is a quantization of the Lie superbialgebra $\mathfrak{g}$.

Let h be an even formal variable. Denote by $G(h)$ the superalgebra of $G \otimes \mathbb{C}[[h]]$ generated by $T_{ij}^{(s+1)}h^s$; $1 \leq i,j \leq 2n$; $s \geq 0$. The factor-algebra $G(h)/hG(h)$ and the universal enveloping algebra $U(\mathfrak{g})$ enjoy a co-Poisson Hopf superalgebra structure [2].

PROPOSITION. *The map*

$$T_{ij}^{(s+1)}h^s \mapsto (e_{ij} + (-1)^s\varphi(e_{ij}))\lambda^s \cdot (-1)^{\bar{j}}$$

extends to an isomorphism $G(h)/hG(h) \to U(\mathfrak{g})$ *of co-Poisson Hopf superalgebras.*

As in [2], uniqueness of the quantization is achieved by a minor reduction of the algebras $\mathfrak{g}$ and G, but here I cannot go into details.

Acknowledgements

I am grateful to I.V. Cherednik and G.I. Olshansky for fruitful discussions. I am also grateful to V.G. Drinfeld, P.P. Kulish and E.K. Sklyanin for their kind interest in this work.

References

[1] Leites D.A., Serganova V.V. and Feigin B.L., *Kac-Moody superalgebras*, Grouptheoretical Methods in Physics, vol. 1, Nauka, Moscow, 1983, pp. 274–278. (in Russian)

[2] Drinfeld V.G., *Hopf algebras and the quantum Yang-Baxter equation*, Soviet Math. Dokl. **32** (1985), 254–258.

[3] Reyman A.G. and Semenov-Tian-Shansky M.A., *Lie algebras and the Lax equations with spectral parameter on the elliptic curve*, Zap. Nauchn. Sem. LOMI **150** (1986), 104–118. (in Russian)

[4] Avan J. and Talon M., *Rational and trigonometric constant non-antisymmetric R-matrices*, Phys. Lett. **B241** (1990), 77.

[5] Kulish P.P. and Sklyanin E.K., *On the solutions of the Yang-Baxter equation*, Zap. Nauchn. Sem. LOMI **95** (1980), 129–160. (in Russian)

[6] Kulish P.P. and Reshetikhin N.Yu., *Universal R-matrix of the quantum superalgebra osp(2|1)*, Lett. Math. Phys. **18** (1989), 143–149.

[7] Reshetikhin N.Yu. and Semenov-Tian-Shansky M.A., *Central extensions of the quantum current groups*, Lett. Math. Phys. **19** (1990), 133–142.

[8] Nazarov M.L., *Quantum Berezinian and the classical Capelli identity*, Lett.Math.Phys. **21** (1991), 123–131.

[9] Leites D.A. and Serganova V.V., *Solutions of the classical Yang-Baxter equation for simple Lie superalgebras*, Teor.Math.Phys. **58** (1984), 26. (in Russian)

[10] Brauer R., *On algebras which are connected with semisimple continuous groups*, Ann.Math. **38** (1937), 857–872.

[11] Olshansky G.I., *Twisted Yangians and infinite-dimensional classical Lie algebras*, article in this volume.

DEPARTMENT OF MATHEMATICS, MOSCOW STATE UNIVERSITY, 119899, MOSCOW, USSR

ASKEY–WILSON POLYNOMIALS AS SPHERICAL FUNCTIONS ON $SU_q(2)$

MASATOSHI NOUMI[1] AND KATSUHISA MIMACHI[2]

[1]Department of Mathematics, College of Arts and Sciences University of Tokyo
[2]Department of Mathematics, Nagoya University

Quantum groups give a good framework for q-orthogonal polynomials. We emphasize, in this short note, that "semisimple" twisted primitive elements of quantized universal enveloping algebra $U_q(s\ell(2))$ play an interesting role in the theory of quantum G-spaces and their spherical functions.

1. Semisimple twisted primitive elements in $U_q(s\ell(2;\mathbf{C}))$

We start with a remark on "semisimple" twisted primitive elements in the quantized universal enveloping algebra $U_q(s\ell(2;\mathbf{C}))$. Here we take the following conventions: $U_q(s\ell(2;\mathbf{C}))$ is the $\mathbf{C}$-algebra generated by the elements $X_\pm$, $q^{\pm H/2}$ with relations

$$q^{H/2}X_\pm q^{-H/2} = q^{\pm 1}X_\pm, \quad X_+X_- - X_-X_+ = \frac{q^H - q^{-H}}{q - q^{-1}}.$$

The Hopf algebra structure on $U_q(s\ell(2;\mathbf{C}))$ will be fixed so that

$$\Delta(q^{H/2}) = q^{H/2}\otimes q^{H/2}, \quad \Delta(X_\pm) = X_\pm \otimes q^{H/2} + q^{-H/2}\otimes X_\pm.$$

We say that an element D of $U_q(s\ell(2;\mathbf{C}))$ is a *twisted primitive elements* of type (m,n) if D satisfies

$$\Delta(D) = D\otimes q^{mH/2} + q^{nH/2}\otimes D, \quad \varepsilon(D) = 0,$$

where $m,n \in \mathbf{Z}$. For the moment, let us denote by $\Theta(m,n)$ the vector space of all twisted primitive elements of type (m,n). If q is not a root of unity, all such twisted primitive elements of $U_q(s\ell(2;\mathbf{C}))$ are completely classified as follows:

a) If $m = n$, then $\Theta(m,n) = 0$.
b) If $m - n \neq 0, 2$, then $\Theta(m,n) = \mathbf{C}(q^{mH/2} - q^{nH/2})$.
c) $\Theta(1,-1) = \mathbf{C}X_+ \oplus \mathbf{C}(q^{H/2} - q^{-H/2}) \oplus \mathbf{C}X_-$.
d) If $m - n = 2$, then $\Theta(m,n) = q^{(m-1)H/2}\Theta(1,-1)$.

We say that an element of $U_q(s\ell(2;\mathbf{C}))$ is *semisimple* if it is diagonalizable on every finite dimensional $U_q(s\ell(2;\mathbf{C}))$-module. In the case of the complex Lie algebra $s\ell(2;\mathbf{C})$, any semisimple element must be a constant multiple of $\mathrm{Ad}(g)H$ for some $g \in SL(2;\mathbf{C})$. As is recognized by Koornwinder [K3], the algebra $U_q(s\ell(2;\mathbf{C}))$ has some semisimple twisted primitive elements besides those in the form $c(q^{mH/2} - q^{nH/2})$ ($m, n \in \mathbf{Z}$, $c \in \mathbf{C}$). Here we construct an analogue of the adjoint orbit $\{\mathrm{Ad}(g)H \; ; \; g \in SL(2;\mathbf{C})\}$ in $U_q(s\ell(2;\mathbf{C}))$.

For any element $g = \begin{pmatrix} \alpha & \beta \\ \gamma & \delta \end{pmatrix}$ in $GL(2;\mathbf{C})$, we define the twisted primitive element $\theta(g)$ of type $(1,-1)$ by

$$\theta(g) = -\alpha\beta q^{-1/2} X_+ + (\alpha\delta + \beta\gamma)\frac{q^{H/2} - q^{-H/2}}{q - q^{-1}} + \gamma\delta q^{1/2} X_-.$$

This element $\theta(g)$ corresponds to $\mathrm{Ad}(g)H/2$ of the classical case.

Theorem 1. *Assume that q is not a root of unity. Let $g = \begin{pmatrix} \alpha & \beta \\ \gamma & \delta \end{pmatrix}$ be an element in $GL(2;\mathbf{C})$ such that $\alpha\delta - q^{2k}\beta\gamma \neq 0$ for all $k \in \mathbf{Z}$. Then the twisted primitive element $D = q^{H/2}\theta(g)$ of type $(2,0)$ is semisimple. On each spin j representation V_j ($j \in \frac{1}{2}\mathbf{N}$), The element $D = q^{H/2}\theta(g)$ is diagonalizable with distinct eigenvalues $\lambda_m(g)$ ($m = j, j-1, \ldots, -j$), where*

$$\lambda_m(g) = \frac{q^m - q^{-m}}{q - q^{-1}} \cdot (\alpha\delta q^m - \beta\gamma q^{-m}).$$

This fact can be proved by constructing the eigenvectors of $q^{H/2}\theta(g)$ in the algebra of functions $A(\mathbf{C}_q^2) = \mathbf{C}[z, w | qzw = wz]$ on the quantum plane. The algebra $A(\mathbf{C}_q^2)$ has a natural structure of left $U_q(s\ell(2;\mathbf{C}))$-module such that

$$q^{H/2} \cdot (z, w) = (zq^{1/2}, wq^{-1/2}),\ X_+ \cdot (z, w) = (0, z),\ X_- \cdot (z, w) = (w, 0).$$

Then one can show that, for any $r, s \in \mathbf{N}$, the q-shifted product

$$(z\alpha + w\gamma)(z\alpha + w\gamma q^{-1}) \ldots (z\alpha + w\gamma q^{-r+1})(z\beta^{-r} + w\delta)(z\beta^{-r+1} + w\delta) \ldots (z\beta^{-r+s-1} + w\delta)$$

is an eigenvector of $q^{H/2}\theta(g)$ with eigenvalue $\lambda_m(g)$ for $m = (r-s)/2$.

We remark that one can also construct such a family of semisimple twisted primitive elements of type $(0,-2)$. However, the situation becomes much complicated in the case of twisted primitive elements of type $(1,-1)$.

2. Podles' quantum 2-spheres

An interesting feature of quantum groups is the existence of "non-standard homogeneous spaces", namely "homogeneous spaces" that are not directly written as quotient spaces G/K. Podles' quantum 2-spheres ([P]) were the first example of such non-standard homogeneous spaces.

Let $A(G)$ be a Hopf $*$-algebra and $A(X)$ a $*$-algebra. Then we say that X is a quantum right G-space over the quantum group G if $A(X)$ has a right $A(G)$-comodule structure such that the coaction $R_G{:}A(X) \to A(X) \otimes A(G)$ is a homomorphism of $*$-algebra. We denote by

$$A(X/G) = \{\varphi \in A(X); \quad R_G(\varphi) = \varphi \otimes 1\}$$

the $*$-subalgebra of right G-invariants, considering $A(X/G)$ as the algebra of functions on the quotient space X/G. We will say that X is homogeneous if $A(X/G) = \mathbf{C}$. We do not repeat similar definitions for quantum left G-spaces. In the framework of quantized universal enveloping algebras, algebra of functions on quantum G-spaces give rise to QGS-algebras in the sense of Korogodsky [Ko].

Hereafter we assume that q is a real number with $|q| \neq 0, 1$. Following the notations of [NM1], we denote by $S_q^2(c,d)$ $(c, d \in \mathbf{R})$ Podles' quantum 2-spheres. The algebra of functions $A(S_q^2(c,d))$ is the $\mathbf{C}$-algebra generated by the three elements ξ, z, η with relations

$$q^2\xi z = z\xi, \quad q^2 z\eta = \eta z, \quad -q\xi\eta = (c-z)(d+z), \quad -q\eta\xi = (c-q^2z)(d+q^2z).$$

We take the $*$-operation on $A(S_q^2(c,d))$ such that $\xi^* = -q^{-1}\eta$, $z^* = z$. Podles' quantum 2-spheres $S_q^2(c,d)$ $(c, d \in \mathbf{R})$ are a family of homogeneous quantum left $SU_q(2)$-spaces in the above sense (see [P, NM1]).

We denote by K diagonal subgroup of the quantum group $SU_q(2)$. The algebra of functions of K is the $*$-algebra of Laurent polynomials $\mathbf{C}[t, t^{-1}]$ with $t^* = t^{-1}$. The embedding of K into $SU_q(2)$ is specified by the "restriction mapping" $A(SU_q(2)) \to A(K)$ such that $t_{11} \mapsto t$, $t_{12} \mapsto 0$, $t_{21} \mapsto 0$, $t_{22} \mapsto t^{-1}$, where t_{ij} $(1 \leq i, j \leq 2)$ are the canonical coordinates for $SU_q(2)$. So long as one takes the viewpoint of quotient Hopf algebras, the diagonal is the only place in $SU_q(2)$ where a group isomorphic to $U(1)$ can be embedded. As for the diagonal subgroup K, the quotient space $SU_q(2)/K$ gives rise to the Podles' quantum 2-sphere $S_q^2(1,0)$ by setting $\xi = t_{11}t_{12}$, $z = -q^{-1}t_{12}t_{21}$, $\eta = t_{21}t_{22}$. However, one can also show that Podles' quantum 2-spheres $S_q^2(c,d)$ for general (c,d) cannot be realized as quotient spaces of $SU_q(2)$ by quantum subgroups in the above sense.

Recall that the algebra of functions $A(SU_q(2))$ has a natural structure of two-sided module over $U_q(s\ell(2;\mathbf{C}))$. We will denote by $U_q(su(2))$ the Hopf $*$-algebra $U_q(s\ell(2;\mathbf{C}))$ endowed with the $*$-operation such that $(q^{H/2})^* = q^{H/2}$ and $X_+^* = X_-$. Returning to discussion of section 1, we consider the twisted primitive element $q^{H/2}\theta(g)$ $(g \in GL(2;\mathbf{C}))$. For the compatibility with the $*$-operation, we assume that the element g satisfies the condition $\overline{\alpha} = \delta$, $\overline{\gamma} = -\beta$. This condition is equivalent to saying that $g \in SU(2) \times \mathbf{R}_{>0}$. In this setting, the twisted primitive element $q^{H/2}\theta(g)$ can be regarded as an analogue of the infinitesimal generator of the subgroup $K(g) := gKg^{-1}$ in $SU(2)$, where $K = U(1)$ is the diagonal subgroup of $SU(2)$. We now define a quantum analogue of the quotient space $SU(2)/K(g)$ by

$$A_q(SU(2)/K(g)) = \{\varphi \in A(SU_q(2)); \quad q^{H/2}\theta(g) \cdot \varphi = 0\}.$$

This is a $*$-subalgebra of $A(SU_q(2))$ and a left $A(SU_q(2))$-subcomodule of $A(SU_q(2))$. Accordingly, it can be regarded as the algebra of functions on a quantum left $SU_q(2)$-space.

oposition 2. *The left quantum $SU_q(2)$-space represented by $A_q(SU(2)/K(g))$ is entified with Podles' quantum sphere $S_q^2(c,d)$ with $c = |\alpha|^2$, $d = |\gamma|^2$. An explicit entification between $A_q(SU(2)/K(g))$ and $A(S_q^2(c,d))$ is given by*

$$\begin{aligned} \xi &= t_{11}^2\alpha\beta q^{-1} + t_{11}t_{12}(\alpha\delta+\beta\gamma) + t_{12}^2\gamma\delta \\ z &= -t_{11}t_{21}\alpha\beta q^{-1} - q^{-1}t_{12}t_{21}(\alpha\delta+\beta\gamma) - t_{12}t_{22}\gamma\delta \\ \eta &= t_{21}^2\alpha\beta q^{-1} + t_{21}t_{22}(\alpha\delta+\beta\gamma) + t_{22}^2\gamma\delta. \end{aligned}$$

This shows that the Podles' quantum 2-spheres $S_q^2(c,d)$ with $c > 0$, $d > 0$ are uotient spaces" of $SU_q(2)$ by semisimple twisted primitive elements in $U_q(su(2))$.

3. Askey-Wilson polynomials as spherical functions

We now consider a quantum analogue of the two-sided action of a pair of subgroups $(g_1), K(g_2))$ on $SU(2)$. Hereafter we fix two elements $g_i = \begin{pmatrix} \alpha_i & \beta_i \\ \gamma_i & \delta_i \end{pmatrix}$ $(i = 1,2)$ in $J(2) \times \mathbf{R}_{>0}$ and assume that $\alpha_i \neq 0$ and $\gamma_i \neq 0$ $(i = 1,2)$. By the action of the two misimple twisted primitive elements

$$D_1 = \theta(g_1)q^{H/2} \quad \text{and} \quad D_2 = q^{H/2}\theta(g_2),$$

om the right side and the left side respectively, the algebra of functions $A(SU_q(2))$ is composed into the subspaces of two-sided relative invariants. For any couple (m,n) half integers, we set

$$\mathcal{A}_{mn} = \{\varphi \in A(SU_q(2));\ (D_2 - \lambda_n(g_2))\cdot\varphi = \varphi\cdot(D_1 - \lambda_m(g_1)) = 0\},$$

that $A(SU_q(2)) = \bigoplus_{m,n} \mathcal{A}_{m,n}$. Then the subalgebra

$$\mathcal{A}_{00} = \{\varphi \in A(SU_q(2));\quad \theta(g_2)\cdot\varphi = \varphi\cdot\theta(g_1) = 0\}$$

n be regarded as the algebra of functions on a quantization of the double coset space $(g_1)\backslash SU(2)/K(g_2)$.

roposition 3. *(1) The subalgebra $\mathcal{A}_{00}$ of $A(SU_q(2))$ is a $*$-algebra isomorphic to the olynomial ring in one indeterminate. A generator $x(g_1,g_2)$ for this algebra is given by*

$$2q^{-1}|\alpha_1\gamma_1\alpha_2\gamma_2|x(g_1,g_2) =$$
$$= (-\gamma_1\delta_1, \alpha_1\delta_1+\beta_1\gamma_1, -\alpha_1\beta_1 q^{-1}) \begin{pmatrix} t_{11}^2 & t_{11}t_{12} & t_{12}^2 \\ t_{11}t_{21} & q^{-1}t_{12}t_{21} & t_{12}t_{22} \\ t_{21}^2 & t_{21}t_{22} & t_{22}^2 \end{pmatrix} \begin{pmatrix} \alpha_2\beta_2 q^{-1} \\ \alpha_2\delta_2+\beta_2\gamma_2 \\ \gamma_2\delta_2 \end{pmatrix}$$

?) If a couple (m,n) of half integers satisfies the condition $m - n \in \mathbf{Z}$, the subspace $_{mn}$ is a free left $\mathcal{A}_{00}$-module of rank 1. Otherwise $\mathcal{A}_{mn} = 0$.

For each $j \in \frac{1}{2}\mathbf{N}$ and $m,n \in \frac{1}{2}\mathbf{Z}$, we say that an element ψ in $A(SU_q(2))$ is a *(doubly ssociated) spherical function* of type $(j;m,n)$ with respect to (g_1,g_2) if it satisfies

$$(C - \rho_j)\cdot\psi = 0 \tag{1}$$

$$(D_2 - \lambda_n(g_2)) \cdot \psi = 0, \ \psi \cdot (D_1 - \lambda_m(g_1)) = 0. \tag{2}$$

Here

$$C = X_- X_+ + \frac{1}{(q-q^{-1})^2}(q^{H/2} - q^{-H/2})(qq^{H/2} - q^{-1}q^{-H/2})$$

is the Casimir element of $U_q(su(2))$ and $\rho_j = (q^j - q^{-j})(q^{j+1} - q^{-j-1})/(q-q^{-1})^2$. Note that equation (1) requires that ψ should be a matrix element of the spin j representation of $SU_q(2)$. Spherical functions of type $(j;0,0)$ are the *zonal* spherical functions studied by Koornwinder [K3,K4].

One can show that, for each triple $(j;m,n)$ with $j \in \frac{1}{2}\mathbf{N}$, $m,n \in \frac{1}{2}\mathbf{Z}$, $m-n \in \mathbf{Z}$, there is a nonzero spherical function ψ^j_{mn} of type $(j;m,n)$, unique up to constant multiples if $j \geq \min\{|m|,|n|\}$. Setting $e_{mn} = \psi^{\min(|m|,|n|)}_{mn}$, one can express ψ^j_{mn} in the form ψ^j_{mn} $= p^j_{mn}(\mathrm{x})e_{mn}$. Here $p^j_{mn}(\mathrm{x})$ is a polynomial in the variable $\mathrm{x}=\mathrm{x}(g_1,g_2)$ of Proposition 3, which we call the *zonal part* of ψ^j_{mn}.

The zonal parts of our spherical functions are expressed by the *Askey-Wilson polynomials*

$$p_k(\mathrm{x};a,b,c,d|q) = a^{-k}(ab,ac,ad;q)_k \ {}_4\phi_3\left(\begin{matrix} q^{-k}, \ abcdq^{k-1}, \ az, \ az^{-1} \\ ab, \ ac, \ ad \end{matrix} ;q,q\right),$$

where $\mathrm{x}= (z+z^{-1})/2$ (see [AW])

Theorem 4. *For each $(j;m,n)$ with $j \in \frac{1}{2}\mathrm{N}$, $m,n \in \frac{1}{2}\mathbf{Z}$, $m-n \in \mathbf{Z}$, the zonal part of the doubly associated spherical function ψ^j_{mn} is a constant multiple of the Askey-Wilson polynomial*

$$p_k(\mathrm{x}(g_1,g_2); \frac{s}{t}q^{2|m-n|+1}, \frac{t}{s}q, -stq^{2|m+n|+1}, -\frac{1}{st}q \,|\, q^2),$$

where $k = j - \max\{|m|,|n|\}$ and the parameters s,t are determined as follows:
Case (1) $m+n \geq 0$, $m \geq n$: $(s,t) = (|\alpha_1/\gamma_1|, |\alpha_2/\gamma_2|)$,
Case (2) $m+n \geq 0$, $m \leq n$: $(s,t) = (|\alpha_2/\gamma_2|, |\alpha_1/\gamma_1|)$,
Case (3) $m+n \leq 0$, $m \geq n$: $(s,t) = (|\gamma_2/\alpha_2|, |\gamma_1/\alpha_1|)$,
Case (4) $m+n \leq 0$, $m \leq n$: $(s,t) = (|\gamma_1/\alpha_1|, |\gamma_2/\alpha_2|)$.

In our realization, the equation $(C-\rho_j)\cdot\psi^j_{mn} = 0$ for the Casimir element C corresponds to the q-difference equation for the Askey-Wilson polynomial. The orthogonality of matrix elements implies the orthogonality relations for the Askey-Wilson polynomials. From the fact that ψ^j_{mn} are matrix elements of the spin j representation, we can also derive an addition formula for the zonal spherical functions $p_j(\mathrm{x}; \frac{s}{t}q, \frac{t}{s}q, -stq, -\frac{1}{st}q \,|\, q^2)$ (see [NM5]).

We also remark that the elements ψ^j_{m0} ($j \in \mathrm{N}$, $m = j, j-1, \ldots, -j$) give the associated spherical functions on Podles' quantum 2-sphere $S^2_q(|\alpha_2|^2, |\gamma_2|^2)$ with respect to the right action of the twisted primitive element $D_1 = \theta(g_1)q^{H/2}$ under the identification of Proposition 2. If $|\alpha_i| = |\gamma_i|$ $(i = 1,2)$, the zonal parts of these associated spherical functions are Rogers' q-ultraspherical polynomials (see also [NM4]).

References

[AW] Askey R. and Wilson J., *Some basic hypergeometric orthogonal polynomials that generalize Jacobi polynomials*, Mem. Am. Math. Soc. **54** no. 319 (1985).

[D] Drinfeld V.G., *Quantum groups*, Proc. IMC-86,1 (1987), 798–820, Berkely.

[J] Jimbo M., *A q-difference analogue of $U(\mathcal{G})$ and the Yang-Baxter equation*, Lett. Math. Phys. **10** (1985), 63–69.

[K1] Koornwinder T.H., *Representations of the twisted $SU(2)$ quantum group and some q-hypergeometric orthogonal polynomials*, Proc. K. Ned. Akad. Wet., Ser.A **92** (1989), 97–117.

[K2] Koornwinder T.H., *Continuous q-Legendre polynomials as spherical matrix elements of irreducible representations of the quantum $SU(2)$ group*, CWI Quaterly **2** (1989), 171–173.

[K3] Koornwinder T.H., *Orthogonal polynomials in connections with quantum groups, in Orthogonal Polynomials, Theory and Practice ed. P.Nevai*, NATO ASI Series (1990), 257–292, Kluwer Academic Publishers.

[K4] Koornwinder T.H., *Askey-Wilson polynomials as zonal spherical functions on the quantum group $SU_q(2)$*, Preprint (1990).

[Ko] Korogodsky L.I., *Quantum projective spaces, spheres and hyperboloids*, Preprint (1990).

[M] Masuda T.,Mimachi K., Nakagami Y., Noumi M. and Ueno K., *Representations of the quantum group $SU_q(2)$ and the little q-Jacobi polynomials*, J. Funct. Anal. (to appear).

[NM1] Noumi M. and Mimachi K., *Quantum 2-spheres and big q-Jacobi polynomials*, Commun. Math. Phys. **128** (1990), 521–531.

[NM2] Noumi M. and Mimachi K., *Big q-Jacobi polynomials, q-Hahn polynomials and a family of quantum 3-spheres*, Lett. Math. Phys. **19** (1990), 299–305.

[NM3] Noumi M. and Mimachi K., *Spherical functions on a family of quantum 3-spheres*, Preprint (1990).

[NM4] Noumi M. and Mimachi K., *Rogers' q-ultraspherical polynomials on a quantum 2-sphere*, Preprint (1990).

[NM5] Noumi M. and Mimachi K., *Askey-Wilson polynomials and the quantum group $SU_q(2)$*, Proc. Japan Acad. Ser.A **66** (1990), 146–149.

[P] Podles P., *Quantum spheres*, Lett. Math. Phys. **14** (1987), 193–202.

[VS] Vaksman L.L. and Soibelman Ya.S., *Algebra of functions on the quantum $SU(2)$ group*, Funkz. Anal. Pril. **22** (1988), 1036–1040. (in Russian)

[W] Woronowicz S.L., *Twisted $SU(2)$ group. An example of non-commutative differential calculus*, Publ. RIMS **23** (1987), 117–181, Kyoto Univ..

(1) Department of Mathematics, College of Arts and Sciences University of Tokyo, Komaba, Meguro-Ku, Tokyo 153, Japan

(2) Department of Mathematics, Nagoya University, Furo-Cho, Chikusa-Ku, Nagoya 464-01, Japan

TWISTED YANGIANS AND INFINITE–DIMENSIONAL CLASSICAL LIE ALGEBRAS

G.I.OLSHANSKIĬ

Institute of Geography, Moscow

ABSTRACT. The Yangians are quantized enveloping algebras of polynomial current Lie algebras and twisted Yangians should be their analogs for twisted polynomial current Lie algebras. We define and study certain examples of twisted Yangians and describe their relationship to a problem which arises in representation theory of infinite-dimensional classical groups.

0. Introduction

The Yangians form a remarkable family of quantized enveloping algebras, see e.g. Drinfeld [7]. Let $\mathfrak{a}$ be a simple finite-dimensional complex Lie algebra and $\mathfrak{a}[x]$ denote the polynomial current Lie algebra $\mathfrak{a}\otimes\mathbb{C}[x]$. Then the Yangian $Y(\mathfrak{a})$ is a certain deformation of $\mathcal{U}(\mathfrak{a}[x])$ in the class of Hopf algebras.

The aim of this paper is to study certain *twisted* versions of Yangians which emerge in the context of the representation theory of infinite-dimensional groups and Lie algebras.

The term "twisted" means that we shall consider deformations of the enveloping algebras of the form $\mathcal{U}(\mathfrak{a}[x]^\sigma)$ where $\mathfrak{a}[x]^\sigma$, the twisted polynomial current Lie algebra, is an involutive subalgebra of $\mathfrak{a}[x]$. In fact we shall deal only with the case when $\mathfrak{a}=\mathfrak{sl}(N)$ and the involution σ corresponds to either $\mathfrak{o}(N)\subset\mathfrak{sl}(N)$ or to $\mathfrak{sp}(N)\subset\mathfrak{sl}(N)$.

Now I shall explain how this subject turns out to be related to representation theory. Let us suppose that we are studying representations of an infinite-dimensional group G or of its Lie algebra $\mathfrak{g}$. Then we are often need to investigate some operators in the space of a representation (e.g. analogs of Laplace operators) which are not contained in the image of $\mathcal{U}(\mathfrak{g})$ but are, in a certain sense "affiliated" to it. For example, in representation theory of Kac–Moody Lie algebras a very important role is played by the quadratic Casimir operator which does not belong to the universal enveloping algebra. Such a situation arises because, for most interesting infinite-dimensional Lie algebras $\mathfrak{g}$, the enveloping algebra $\mathcal{U}(\mathfrak{g})$ turns out to be a bad object. So we would like to replace $\mathcal{U}(\mathfrak{g})$ by a certain "good extension" of it containing "affiliated" elements needed in representation theory.

For the Virasoro and Kac–Moody Lie algebras this problem was studied by Feigin and Fuks [8] and Kac [9]. For certain infinite-dimensional classical Lie algebras it was studied in author's papers [13,14,15]. There, in particular, a "good" extension A of

---

Completed while a guest of the CWI, Amsterdam .

$A(\mathfrak{gl}(\infty))$ was constructed and it turned out that this algebra A is closely related to the Yangians $Y(\mathfrak{sl}(N))$!

In the present paper this author's construction is generalized to the classical Lie algebras $\mathfrak{o}(2\infty)$, $\mathfrak{o}(2\infty+1)$ and $\mathfrak{sp}(2\infty)$ and it is shown that the corresponding extensions can be described by making use of certain "twisted Yangians" $Y^{\pm}(N)$ (here the upper sign refers to the orthogonal case and the lower sign to the symplectic case).

To conclude, I shall mention some properties of the twisted Yangians $Y^{\pm}(N)$:

(a) $Y^{\pm}(N)$ may be defined by generators and quadratic relations. Like in the case of the Yangian $Y(N) := Y(\mathfrak{gl}(N))$, these relations can be conveniently presented with the use of the R-matrix formalism. However, in our relations there appear both the Yang R-matrix $R(u-v)$ and the transposed matrix $R^t(-u-v)$, see (2.2) below. Similar (but non-identical) relations with two R-matrices have been investigated by Sklyanin [17], see also Cherednik [1], and Reshetikhin and Semenov-Tian-Shansky [16].

(b) $Y^{\pm}(N)$ is not a Hopf algebra but it may be realized as a subalgebra of $Y(\mathfrak{gl}(N))$ which is a left coideal. This means that the comultiplication in $Y(N)$ moves $Y^{\pm}(N)$ into $Y(\mathfrak{gl}(N)) \otimes Y^{\pm}(N)$.

(c) $Y^{\pm}(N)$ is a deformation of $\mathcal{U}(\mathfrak{gl}(N)[x]^{\sigma})$. Moreover one can define a notion of quantum determinant in $Y^{\pm}(N)$ which enables us to define a deformation of $\mathcal{U}(\mathfrak{sl}(N)[x]^{\sigma})$.

(d) Like the Yangians $Y(N)$, the twisted Yangians $Y^{\pm}(N)$ seem to be very natural and useful objects from the point of view of the classical representation theory (cf. Cherednik [2]).

Acknowledgement. During my work on this subject I had numerous stimulating discussions with I. V. Cherednik, V. G. Drinfeld and especially M. L. Nazarov. Some suggestions made by M. L. Nazarov played an important role in the achievement of my goal. A substantial part of my work was inspired by E. K. Sklyanin's paper [17]. I would like to express my deep gratitude to all these colleagues. I am very indebted to T. H. Koornwinder who organized my visit to CWI, Amsterdam and helped me very much in preparing the present variant of the paper. I would also like to thank the Department of Analysis, Algebra and Geometry of CWI and its head M. Hazewinkel for the hospitality.

1. Preliminaries: the Yangians $Y(N)$

I would like to stress that no result mentioned in this section is new nor due to the author. The main references are Drinfeld [4–7] and Kirillov and Reshetikhin [10].

In this section, we fix $N \in \{1, 2, \dots\}$.

Definition 1.1. The *Yangian* $Y(N) = Y(\mathfrak{gl}(N))$ is defined as the complex associative algebra with the unity 1, countably many generators $t_{ij}^{(1)}, t_{ij}^{(2)}, \dots$, where $1 \le i, j \le N$, and defining quadratic relations

$$[t_{ij}^{(M+1)}, t_{kl}^{(L)}] - [t_{ij}^{(M)}, t_{kl}^{(L+1)}] = t_{kj}^{(M)} t_{il}^{(L)} - t_{kj}^{(L)} t_{il}^{(M)} \tag{1.1}$$

where $M, L = 0, 1, 2, \dots$ and $t_{ij}^{(0)} := \delta_{ij} \cdot 1$.

The system (1.1) is equivalent to the following one:

$$[t_{ij}^{(M)}, t_{kl}^{(L)}] = \sum_{r=0}^{\min(M,L)-1} (t_{kj}^{(r)} t_{il}^{(M+L-1-r)} - t_{kj}^{(M+L-1-r)} t_{il}^{(r)}) \tag{1.2}$$

where $M, L = 1, 2, \dots, 1 \le i, j, k, l \le N$.

Consider the polynomial current Lie algebra $\mathfrak{gl}(N)[x] := \mathfrak{gl}(N) \otimes \mathbb{C}[x]$ and its universal enveloping algebra $\mathcal{U}(\mathfrak{gl}(N)[x])$.

Theorem 1.2. *The algebra $Y(N)$ is a (flat) deformation of the algebra $\mathcal{U}(\mathfrak{gl}(N)[x])$.*

Proof. Let $h \in \mathbb{C}^*$ be the deformation parameter. Consider in $Y(N)$ new generators which depend on h:

$$\tilde{t}_{ij}^{(M)} = h^{M-1} t_{ij}^{(M)}.$$

Then our relations take the following form:

$$\begin{aligned} [\tilde{t}_{ij}^{(M)}, \tilde{t}_{kl}^{(L)}] =& \delta_{kj} \tilde{t}_{il}^{(M+L-1)} - \delta_{il} \tilde{t}_{kj}^{(M+L-1)} \\ &+ h \sum_{r=1}^{\min(M,L)-1} (\tilde{t}_{kj}^{(r)} \tilde{t}_{il}^{(M+L-1-r)} - \tilde{t}_{kj}^{(M+L-1-r)} \tilde{t}_{il}^{(r)}) \end{aligned} \tag{1.3}$$

In the limit $h \to 0$ the last term disappears and we obtain precisely the commutation relations of the Lie algebra $\mathfrak{gl}(N)[x]$ in its basis $\{E_{ij} x^{M-1}\}$ where $\{E_{ij}\}$ denote the matrix units . It remains to verify the statement of flatness which means that our relations provide us with an algebra of the same "size" as $\mathcal{U}(\mathfrak{gl}(N)[x])$. In other words we have to prove a Poincaré-Birkhoff-Witt type theorem for $Y(N)$. This can be done in several ways one of which consists in using the results of the author's papers [14, 15].

Remark 1.3. There are at least two "natural" filtrations in the algebra $Y(N)$: they are defined by putting $\deg t_{ij}^{(M)} = M$ and $\deg t_{ij}^{(M)} = M-1$, respectively. The graded algebra associated to the first filtration is commutative and isomorphic to grad $\mathcal{U}(\mathfrak{gl}(N)[x])$ while the graded algebra associated to the second filtration is noncommutative and isomorphic to $\mathcal{U}(\mathfrak{gl}(N)[x])$.

Let us introduce the generating series for our generators:

$$t_{ij}(u) = \delta_{ij} + t_{ij}^{(1)} u^{-1} + t_{ij}^{(2)} u^{-2} + \cdots \in Y(N)[[u^{-1}]] \tag{1.4}$$

Then (1.1) can be rewritten as follows:

$$[t_{ij}(u), t_{kl}(v)] = \frac{1}{u-v}(t_{kj}(u) t_{il}(v) - t_{kj}(v) t_{il}(u)) \tag{1.5}$$

where $1 \le i, j, k, l \le N$.

Let $T(u) = [t_{ij}(u)]$ denote the $N \times N$ matrix with entries $t_{ij}(u)$, let P denote the permutation operator $\xi \otimes \eta \mapsto \eta \otimes \xi$ in $\mathbb{C}^N \otimes \mathbb{C}^N$, let $R(u) = 1 - Pu^{-1}$ be another operator in $\mathbb{C}^N \otimes \mathbb{C}^N$ (the Yang matrix), and let

$$T_1(u) = T(u) \otimes 1, \quad T_2(v) = 1 \otimes T(v) \tag{1.6}$$

be $N^2 \times N^2$ matrices with entries in $Y(N) \otimes \mathbb{C}[[u^{-1}]]$. Then the system (1.5) can be rewritten as a single equation for the T-matrix:

$$R(u-v)T_1(u)T_2(v) = T_2(v)T_1(u)R(u-v) \tag{1.7}$$

There exists a very convenient interpretation of (1.7). Let us suppose that the generators $t_{ij}^{(1)}, t_{ij}^{(2)}, \ldots$ operate in a vector space W. Then $T(u)$ may be regarded as an operator in $W \otimes \mathbb{C}^N$ depending on a (formal) parameter u and the both sides of (1.7) may be regarded as operators in $W \otimes \mathbb{C}^N \otimes \mathbb{C}^N$.

More generally, we adopt the following notation. For an arbitrary $m = 2, 3, \ldots$ we consider the tensor product space $W \otimes (\mathbb{C}^N)^{\otimes m}$ and we define the following operators in this space: $T_i(u_i), i = 1, \ldots, m$, where $u_1, \ldots, u_m$ are parameters, and $R_{ij}(u_i - u_j)$ where $1 \le i < j \le m$. By definition, $T_i(u_i)$ acts essentially as $T(u_i)$ in the tensor product of W and of the i-th copy of $\mathbb{C}^N$ while $R_{ij}(u_i - u_j)$ stands for the operator $R(u_i - u_j)$ acting in the tensor product of i-th and j-th copies of $\mathbb{C}^N$.

Now we have the following useful identity:

Lemma 1.4. *Set*

$$R(u_1, \ldots, u_m) := (R_{m-1,m})(R_{m-2,m}R_{m-2,m-1}) \cdots (R_{1m} \cdots R_{12}) \tag{1.8}$$

where $R_{ij} := R_{ij}(u_i - u_j)$. Then we have

$$R(u_1, \ldots, u_m)\, T_1(u_1) \cdots T_m(u_m) = T_m(u_m) \cdots T_1(u_1)\, R(u_1, \ldots, u_m). \tag{1.9}$$

Consider the natural action of the symmetric group S_m in $(\mathbb{C}^N)^{\otimes m}$ and denote by A_m the image of the antisymmetrizer $\frac{1}{m!} \sum \pm s \in \mathbb{C}[S_m]$ in $\mathrm{End}(\mathbb{C}^N)^{\otimes m}$.

Lemma 1.5. *If $u_i - u_{i+1} = 1$ for $i = 1, \ldots, m-1$ then we have*

$$R(u_1, \ldots, u_m) = m!\, A_m$$

Now take $m = N$ and remark that A_N is a one-dimensional projector in $\mathbb{C}[S_m]$. Then the two lemmas above imply the following result

Lemma 1.6. *The following identities hold:*

$$\begin{aligned} &A_N\, T_1(u) \cdots T_N(u-N+1) = \\ &= T_N(u-N+1) \cdots T_1(u)\, A_N \\ &= A_N\, T_1(u) \cdots T_N(u-N+1)\, A_N \\ &= A_N\, T_N(u-N+1) \cdots T_1(u)\, A_N \end{aligned} \tag{1.10}$$

Corollary 1.7. *There exists a formal series*

$$\det T(u) := 1 + d_1 u^{-1} + d_2 u^{-2} + \cdots \in Y(N)[[u^{-1}]] \tag{1.11}$$

such that (1.10) equals $\det T(u)\, A_N = A_N \det T(u)$.

Definition 1.8. det $T(u)$ is called the *quantum determinant* of the matrix $T(u)$.

Theorem 1.9. *det $T(u)$ lies in the center of $Y(N)$, i.e., all its coefficients are central elements.*

The idea of the proof is to use lemma 1.4 where we have to set $m = N+1, u_i = u-i+1$ for $i = 1, \dots, N$, and $u_{N+1} = v$.

Definition 1.10. The quotient algebra $Y(N)/(\det T(u) = 1)$ is called the *Yangian* of the Lie algebra $\mathfrak{sl}(N)$ and is denoted by $Y(\mathfrak{sl}(N))$ (here the relation det $T(u) = 1$ is an abridged form of the system of relations $d_1 = 0, d_2 = 0, \dots$).

Theorem 1.11 (compare to theorem 1.2). *$Y(\mathfrak{sl}(N))$ is a flat deformation of the algebra $\mathcal{U}(\mathfrak{sl}(N)[x])$.*

Theorem 1.12. *The center of $Y(\mathfrak{sl}(N))$ is trivial. Thus the coefficients of* det $T(u)$ *generate the center of $Y(N)$.*

Idea of proof: Using the second filtration of $Y(N)$ (see remark 1.3) we reduce our claim to the fact that the center of $\mathcal{U}(\mathfrak{sl}(N)[x])$ is trivial. (I learned this reasoning from V. G. Drinfeld.)

Theorem 1.13 (V. G. Drinfeld). *We can also realize $Y(\mathfrak{sl}(N))$ as a subalgebra in $Y(N)$. Moreover $Y(N)$ is isomorphic to the tensor product of its center and $Y(\mathfrak{sl}(N))$.*

Remark 1.14. For $Y(\mathfrak{sl}(N))$ there exists a *finite* system of generators (see Drinfeld [4, 5, 7]).

Theorem 1.15. *$Y(N)$ is a Hopf algebra with respect to the comultiplication Δ defined by*

$$\Delta(t_{ij}(u)) = \sum_{\alpha=1}^{N} t_{i\alpha}(u) \otimes t_{\alpha j}(u),$$

the antipode S defined by

$$S(T(u)) = T(u)^{-1}$$

and the counit $\varepsilon : Y(N) \to \mathbb{C}$ defined by

$$\varepsilon(T(u)) = 1$$

Moreover $Y(N)$ can be viewed as a deformation of $\mathcal{U}(\mathfrak{gl}(N)[x])$ in the class of Hopf algebras.

Theorem 1.16. *We have*

$$\Delta(\det T(u)) = \det T(u) \otimes \det T(u),$$

so $Y(\mathfrak{sl}(N))$ is also a Hopf algebra.

2. Twisted Yangians $Y^{\pm}(N)$

From now on we shall assume that the indices enumerating the basis vectors in $\mathbb{C}^N$ (as well as rows or columns of a $N\times N$ matrix) run through the set $\{-n,-n+1,\dots,n-1,n\}$ where $n=[N/2]$ and in case of even N the zero value is excluded. Let us endow $\mathbb{C}^N$ with either the symmetric form $<,>_+$ or the alternating form $<,>_-$ which are defined as follows:

$$< e_i, e_j >_+ = \delta_{i,-j} \quad \text{or} \quad < e_i, e_j >_- = \mathrm{sgn}(i)\,\delta_{i,-j}$$

where $\{e_j\}$ is the canonical basis of $\mathbb{C}^N$, $\mathrm{sgn}(i)=1$ for $i>0$ and $\mathrm{sgn}(i)=-1$ for $i<0$. In order to consider the both cases at the same time it will be convenient to use the symbol θ_{ij} where $\theta_{ij}\equiv 1$ in the symmetric case and $\theta_{ij}=\mathrm{sgn}(i)\cdot\mathrm{sgn}(j)$ in the alternating case. For a $N\times N$ matrix A we define its transpose A^t as follows: $(A^t)_{ij}=\theta_{ij}\,A_{-j,-i}$. Whenever a double sign $\pm$ or $\mp$ occurs it is supposed that the upper sign corresponds to the symmetric case and the lower sign to the alternating one.

Definition 2.1. Set $S(u)=T(u)T^t(-u)$ and let

$$s_{ij}(u)=\delta_{ij}+s_{ij}^{(1)}u^{-1}+s_{ij}^{(2)}u^{-2}+\cdots,\quad -n\le i,j\le n \tag{2.1}$$

be the entries of this matrix. The *twisted Yangian* $Y^{\pm}(N)$ is defined as the subalgebra of $Y(N)$ generated by all the elements $s_{ij}^{(1)},s_{ij}^{(2)},\dots,\ -n\le i,j\le n$.

For a $N^2\times N^2$ matrix (i.e. an operator in $\mathbb{C}^N\otimes\mathbb{C}^N$), let $(\cdot)^{t_1}$ and $(\cdot)^{t_2}$ denote the partial transpositions relative to the first or second copy of $\mathbb{C}^N$, respectively. Set $Q=P^{t_1}=P^{t_2}$ (this is a one-dimensional operator in $\mathbb{C}^N\otimes\mathbb{C}^N$) and let

$$R^t(u)=R^{t_1}(u)=R^{t_2}(u)=1-Qu^{-1}.$$

Theorem 2.2. *Define $S_1(u)$ and $S_2(u)$ by analogy with* (1.6). *Then we have*

$$R(u-v)S_1(u)R^t(-u-v)S_2(v)=S_2(v)R^t(-u-v)S_1(u)R(u-v) \tag{2.2}$$

$$S^t(-u)=S(u)\pm\frac{S(u)-S(-u)}{2u} \tag{2.3}$$

Moreover (2.2) *and* (2.3) *are precisely defining relations for the generators* (2.1) *of the algebra* $Y^{\pm}(N)$.

Note that (2.2) and (2.3) may be rewritten as follows:

$$\begin{aligned}[s_{ij}(u),\,s_{kl}(v)] &= \frac{1}{u-v}(s_{kj}(u)s_{il}(v)-s_{kj}(v)s_{il}(u)\\ &-\frac{1}{u+v}(\theta_{k,-j}s_{i,-k}(u)s_{-j,l}(v)-\theta_{i,-l}s_{k,-i}(v)s_{-l,j}(u))\\ &+\frac{1}{u^2-v^2}(\theta_{i,-j}s_{k,-i}(u)s_{-j,l}(v)-\theta_{i,-j}s_{k,-i}(v)s_{-j,l}(u))\end{aligned} \tag{2.4}$$

$$\theta_{ij}s_{-j,-i}(-u)=s_{ij}(u)\pm\frac{s_{ij}(u)-s_{ij}(-u)}{2u} \tag{2.5}$$

Let us introduce the involutive automorphism σ of the polynomial current Lie algebra $\mathfrak{gl}(N)[x]$:

$$(\sigma(f))(x) = -(f(x))^t, \quad f \in \mathfrak{gl}(N)[x]. \tag{2.6}$$

This involution determines a Lie subalgebra in $\mathfrak{gl}(N)[x]$ which will be denoted by $\mathfrak{gl}(N)[x]^\sigma$ and will be called the *twisted* polynomial current Lie algebra corresponding to $\mathfrak{o}(N) \subset \mathfrak{gl}(N)$ or to $\mathfrak{sp}(N) \subset \mathfrak{gl}(N)$. If

$$f = a_0 + a_1 x + \cdots a_k x^k$$

is an element of $\mathfrak{gl}(N)[x]^\sigma$ then the coefficients a_{2i} lie in the subalgebra $\mathfrak{o}(N) \subset \mathfrak{gl}(N)$ or $\mathfrak{sp}(N) \subset \mathfrak{gl}(N)$ while the coefficients a_{2i-1} lie in the orthogonal complement to such a subalgebra.

Theorem 2.3 (compare to theorem 1.2). *$Y^\pm(N)$ is a flat deformation of the universal enveloping algebra $\mathcal{U}(\mathfrak{gl}(N)[x]^\sigma)$.*

Definition 2.4. The algebras $Y^+(N)$ and $Y^-(N)$ will be called the *twisted Yangians* corresponding to the involutive subalgebras $\mathfrak{o}(N) \subset \mathfrak{gl}(N)$ and $\mathfrak{sp}(N) \subset \mathfrak{gl}(N)$, respectively.

Lemma 2.5 (compare to lemma 1.4). *Let us fix $m = 2, 3, \ldots$. As in the case of the Yangians $Y(N)$, introduce the operators*

$$S_1(u_1), \ldots, S_m(u_m)$$

and set

$$\tilde{R}_{ij} = R_{ij}(-u_i - u_j),\ 1 \le i, j \le m$$

Then we have

$$\begin{aligned} R(u_1, \ldots, u_m) S_1(u_1) \tilde{R}_{12} \cdots \tilde{R}_{1m} S_2(u_2) \tilde{R}_{23} \cdots \tilde{R}_{2m} S_3(u_3) \cdots \tilde{R}_{m-1,m} S_m(u_m) \\ = S_m(u_m) \tilde{R}_{m,m-1} \cdots \tilde{R}_{m1} S_{m-1}(u_{m-1}) \\ \cdot \tilde{R}_{m-1,m-2} \cdots \tilde{R}_{m-1,1} \cdots S_1(u_1) R(u_1, \ldots, u_m) \end{aligned} \tag{2.7}$$

Lemma 2.6. *An analog of lemma 1.6 holds where we have to insert $\tilde{R}$-matrices between S-matrices in the same way as in (2.7) :*

$$\begin{aligned} A_N S_1(u) \tilde{R}_{12} \cdots \tilde{R}_{1N} S_2(u-1) \tilde{R}_{23} \cdots \tilde{R}_{2N} S_3(u-2) \cdots \tilde{R}_{N-1,N} S_N(u-N+1) \\ = S_N(u-N+1) \tilde{R}_{N,N-1} \cdots \tilde{R}_{N1} S_{N-1}(u-N+2) \\ \cdot \tilde{R}_{N-1,N-2} \cdots \tilde{R}_{N-1,1} \cdots S_1(u) A_N \end{aligned} \tag{2.8}$$

Corollary 2.7 (compare to corollary 1.7). *There exists a formal series*

$$\text{ddet } S(u) := 1 + d_1 u^{-1} + d_2 u^{-2} + \cdots \in Y^\pm(N)[[u^{-1}]]$$

such that both sides of (2.8) are equal to ddet $S(u)\, A_N = A_N$ ddet $S(u)$.

Example 2.8. Let $N = 2$. Then we have

$$\text{ddet}\, S(u)A_2 = A_2 S_1(u)R^t(-2u+1)S_2(u-1) = S_2(u-1)R^t(-2u+1)S_1(u)A_2 \quad (2.9)$$

Definition 2.9. We will call ddet $S(u)$ the *quantum determinant* of the S-matrix.

I would like to stress that the construction of ddet has been inspired by the Sklyanin paper [17].

Theorem 2.10 (compare with theorem 1.11). *The quotient algebra* $[Y^{\pm}(N)] := Y^{\pm}(N)/\ \text{ddet}\, S(u) = 1$ *is a flat deformation of the algebra* $\mathcal{U}(\mathfrak{sl}(N)[x]^{\sigma})$.

Theorem 2.11 (compare with theorem 1.12). *The center of the algebra* $[Y^{\pm}(N)]$ *is trivial. Thus the coefficients of* ddet $S(u)$ *generate the whole center of the algebra* $Y^{\pm}(N)$.

Theorem 2.12 (compare with theorem 1.13). *We can realize* $[Y^{\pm}(N)]$ *as a subalgebra in* $Y^{\pm}(N)$. *Moreover* $Y^{\pm}(N)$ *is isomorphic to the tensor product of its center and the subalgebra* $[Y^{\pm}(N)]$.

Note that $T(u) \to T^t(-u)$ is an involutive automorphism of the algebra $Y^{\pm}(N)$ and denote it by σ.

Theorem 2.13. *We have*

$$\text{ddet}\ S(u) = \det T(u) \cdot \sigma(\det T(u)) = \det T(u) \cdot \det T(-u+N-1)$$

This result can be deduced directly from the definition of the algebra $Y^{\pm}(N)$ and we can use it as the starting point of another approach to "ddet".

It seems that there is no natural Hopf algebra structure in $Y^{\pm}(N)$. This agrees with the more general fact that the twisted polynomial current Lie algebras like $\mathfrak{gl}(N)[x]^{\sigma}$ seem to possess no natural Lie bialgebra structure in Drinfeld's sense. Instead of this the following result holds:

Theorem 2.14.

(i) $Y^{\pm}(N)$ *is a left coideal of the Hopf algebra* $Y(N)$, *i.e.,*

$$\Delta(Y^{\pm}(N)) \subset Y(N) \otimes Y^{\pm}(N)$$

(ii) *Similarly* $[Y^{\pm}(N)]$ *is a left coideal in* $Y(\mathfrak{sl}(N))$.

Remark 2.15. One can say that $Y^{\pm}(N)$ is the result of a quantization of the *pair* consisting of the Lie (bi)algebra $\mathfrak{gl}(N)[x]$ and its involutive Lie subalgebra $\mathfrak{gl}(N)[x]^{\sigma}$. It seems very probable that one can construct analogs of twisted Yangians corresponding to all the twisted polynomial current Lie algebras.

Remark 2.16. The fact that $Y^{\pm}(N)$ is a coideal in $Y(N)$ suggests the idea to study (say, finite-dimensional) representations of the algebra $Y^{\pm}(N)$ starting from representations of the algebra $Y(N)$.

Remark 2.17. It was recently shown by M.L.Nazarov [12] that certain "strange" twisted current polynomial Lie superalgebras possess a natural structure of a super Lie bialgebra and the corresponding twisted Yangians are Hopf superalgebras.

Theorem 2.18. *Let $\tilde{Y}^{\pm}(N)$ denote the algebra with the generators (2.1) subject to the relations (2.2) but not to (2.3). Then*

$$QS_1(-u)R(-2u)S_2^{-1}(u) = S_2^{-1}(u)R(-2u)S_1(-u)Q = \\ = Q\cdot(1\pm\frac{1}{2u})\cdot\delta(u), \tag{2.11}$$

$$\delta(u) = 1+\delta^{(1)}u^{-1}+\delta^{(2)}u^{-2}+\dots, \tag{2.12}$$

where $\delta^{(1)},\delta^{(2)},\dots$ belong to the center of $\tilde{Y}^{\pm}(N)$. Moreover in the definition of $Y^{\pm}(N)$ by the relations (2.2) and (2.3) it is possible to replace (2.3) by the single relation $\delta(u)=1$.

Note a certain similarity between this result and theorem 6 in Drinfeld [4].

Remark 2.19. There exists an involutive automorphism inv : $\tilde{Y}^{\pm}(N)\to\tilde{Y}^{\pm}(N)$ which transfers $S(u)$ into $S(-u-\frac{N}{2})^{-1}$. Let $z(u)$ denote the image of inv$(\delta(u))$ in $Y^{\pm}(N)$ $[[u^{-1}]]$. Note that $z(u)$ is nontrivial!

3. Highest weight modules and virtual Laplace operators

Let $\mathfrak{gl}(2\infty)$ (respectively, $\mathfrak{gl}(2\infty+1)$) denote the Lie algebra of all complex matrices $A=[A_{ij}]$ such that i,j run through $\mathbf{Z}\backslash\{0\}$ (respectively, $\mathbf{Z}$) and the number of non-zero A_{ij}'s is finite. In these Lie algebras there is a convenient basis $\{E_{ij}\}$ formed by the standard matrix units (recall that all the entries of E_{ij} are 0's except one 1 in the (i,j) position). Let us denote by $\mathfrak{g}$ any of the following Lie algebras:

$$\mathfrak{o}(2\infty) = \{A\in\mathfrak{gl}(2\infty) : A_{-j,-i} = -A_{ij}\}$$

$$\mathfrak{o}(2\infty+1) = \{A\in\mathfrak{gl}(2\infty+1) : A_{-j,-i} = -A_{ij}\}$$

$$\mathfrak{sp}(2\infty) = \{A\in\mathfrak{gl}(2\infty) : A_{-j,-i} = -\operatorname{sgn}(i)\operatorname{sgn}(j)A_{ij}\}$$

Thus $\mathfrak{g}$ is the algebra which leaves invariant $<,>_{\pm}$. The elements

$$F_{ij} := E_{ij} - \theta_{ij}E_{-j,-i}$$

form a basis of $\mathfrak{g}$.

Let $\mathfrak{h}$ denote the Cartan subalgebra of diagonal matrices in $\mathfrak{g}$. For a linear functional $x\in\mathfrak{h}^*$, set $x_i = x(F_{ii})$. Then we have $x_{-i}=-x_i$ so x is uniquely determined by the sequence $(\dots,x_{-2},x_{-1})$.

Let $\mathfrak{b}$ denote the Borel subalgebra of upper triangular matrices in $\mathfrak{g}$. For $\lambda\in\mathfrak{h}^*$, let M_λ denote the universal $\mathfrak{g}$-module with a cyclic vector v such that

$$Av = \sum_{i<0}\lambda_i\cdot A_{ii}\cdot v \quad \forall A\in\mathfrak{b}$$

(the Verma module) and let L_λ denote its unique nontrivial irreducible quotient module.

For $n=1,2,\dots$, set

$$\mathfrak{g}(n) = \oplus_{-n\le i,j\le n}\mathbf{C}F_{ij}\subset\mathfrak{g}$$

$$\mathfrak{b}_n = \{A \in \mathfrak{b} : A_{ij} = 0 \quad \text{when} \quad -n \leq i, j \leq n\} \subset \mathfrak{b}.$$

Then $\mathfrak{g}(n)$ is isomorphic to $\mathfrak{o}(2n)$, $\mathfrak{o}(2n+1)$ or $\mathfrak{sp}(2n)$ and $\mathfrak{g}(n) + \mathfrak{b}_n$ is a parabolic subalgebra of $\mathfrak{g}$.

For a $\mathfrak{g}$-module V and a weight $x \in \mathfrak{h}^*$, define an ascending chain of subspaces

$$V_n(x) = \{v \in V : Av = \sum_{i<0} x_i A_{ii} v \quad \forall A \in \mathfrak{b}_n\}.$$

Let $V_\infty(x) = \cup_{n>0} V_n(x)$ and note that $V_\infty(x)$ is a $\mathfrak{g}$-submodule of V because $V_n(x)$ is $\mathfrak{g}(n)$-stable for any n. Let us write $x \sim x'$ when $x_i = x'_i$ for $|i| \gg 0$. Then $x \sim x'$ implies $V_\infty(x) = V_\infty(x')$.

Definition 3.1. Fix a number $c \in \mathbf{C}$ and define $x^c \in \mathfrak{h}^*$ by setting $(x^c)_i = c \cdot \operatorname{sgn}(-i)$. Let us introduce the category $\Omega(c)$ of those $\mathfrak{g}$-modules V for which $V_\infty(x^c) = V$.

The category $\Omega(c)$ is an analog of the category $\mathcal{O}$ of Bernstein-Gelfand-Gelfand. It contains the modules M_λ, L_λ with $\lambda \sim x^c$.

Now suppose $\lambda \in \mathfrak{h}^*$ satisfies the following conditions:

(a) $\lambda \sim x^c$, where $c \in \{0, \frac{1}{2}, 1, \frac{3}{2}, 2, \dots\}$ for $\mathfrak{g} = \mathfrak{o}(2\infty)$ or $\mathfrak{g} = \mathfrak{o}(2\infty + 1)$, and $c \in \{0, 1, 2, \dots\}$ for $\mathfrak{g} = \mathfrak{sp}(2\infty)$.

(b) λ is a dominant weight i.e. the differences $\lambda_{-(i+1)} - \lambda_{-i}$ are nonnegative integers for $i = 1, 2, \dots$, and in addition $\lambda_{-2} \geq |\lambda_{-1}|$ for $\mathfrak{g} = \mathfrak{o}(2\infty)$.

Then L_λ may be represented as the inductive limit of irreducible finite dimensional $\mathfrak{g}(n)$-modules whose highest weights are obtained by restricting λ to $\mathfrak{h}(n) := \mathfrak{h} \cap g(n)$.

Note that we can associate to our infinite-dimensional Lie algebra $\mathfrak{g}$ a semi-infinite Dynkin diagram. Next, just as the finite-dimensional case, we can represent any $\lambda \in \mathfrak{h}^*$ as a collection of labels attached to nodes of this Dynkin diagram. In this representation, the conditions (a) and (b) express the fact that all the labels are nonnegative integers and the number of nonzero labels is finite.

Let $\rho \in \mathfrak{h}^*$ denote the half sum of positive roots of $\mathfrak{g}$ relative to $\mathfrak{b}$ (or which is the same, ρ is such that $\rho \mid \mathfrak{h}(n)$ is the half sum of positive roots of $\mathfrak{g}(n)$ for any n). Then we have for $i = 1, 2, \dots$

$$\rho_{-i} = -\rho_i = \begin{cases} i - 1, \mathfrak{g} = \mathfrak{o}(2\infty) \\ i - \dfrac{1}{2}, \mathfrak{g} = \mathfrak{o}(2\infty + 1) \\ i, \quad \mathfrak{g} = \mathfrak{sp}(2\infty) \end{cases}$$

Let $Z(n)$ denote the center of $\mathcal{U}(\mathfrak{g}(n))$. Using the Harish-Chandra isomorphism (see Dixmier [3]) we identify $Z(n)$ with the algebra of all the polynomial functions $f(\lambda_{-n}, \dots, \lambda_{-1})$ on $\mathfrak{h}(n)^*$ which satisfy the following conditions. Set $l_i = \lambda_i + \rho_i$ and consider f as a function of the variables $l_{-n}, \dots, l_{-1}$. Then f must be invariant relative to all the permutations of the variables and all the transformations of the type $l_{-i} \mapsto \pm l_{-i}$ where, in the case $\mathfrak{g}(n) = \mathfrak{o}(2n+1)$, the number of "$-$" has to be even. Note that the (unique) eigenvalue of a central element in a $\mathfrak{g}(n)$-module with a highest weight $(\lambda_{-n}, \dots, \lambda_{-1})$ equals the value at $(\lambda_{-n}, \dots, \lambda_{-1})$ of the corresponding polynomial. Note also that the canonical filtrations of $\mathcal{U}(\mathfrak{g}(n))$ and $\mathbf{C}[\lambda_{-n}, \dots, \lambda_{-1}]$ induce one and the same filtration on $Z(n)$.

Definition 3.2. Fix $c \in \mathbb{C}$ and define the morphisms $\pi_n : Z(n) \to Z(n-1)$ (which preserve the filtration) as follows:

$$(\pi_n f)(\lambda_{-(n-1)}, \ldots, \lambda_{-1}) = f(c, \lambda_{-(n-1)}, \ldots, \lambda_{-1}).$$

Using these morphisms we define the *algebra* $Z = Z(\mathfrak{g})$ *of virtual Laplace operators* as the projective limit of the filtered algebras $Z(n)$ as $n \to \infty$.

Note that $f(\lambda)$ is well defined for any $f \in Z$ and any $\lambda \in \mathfrak{h}^*$ such that $\lambda \sim x^c$.

Lemma 3.3. *The algebra Z is isomorphic to the algebra of polynomials in countably many indeterminates. As a system of algebraically independent generators of Z, one may take the elements $\Delta^{(2k)}$, $k = 1, 2, \ldots$, which are defined as follows*

$$\Delta^{(2k)}(\lambda) = \sum_{i=\pm1,\pm2,\ldots} (l_i^{2k} - (l_i^c)^{2k})$$

where $l = \lambda + \rho$, $l^c = x^c + \rho$.

Let $\mathrm{pr}_n : Z \to Z(n)$ denote the canonical projections.

Lemma 3.4. *For any $V \in \Omega(c)$, there exists a canonical action of Z in V which is characterized as follows: if $z \in Z$ and $v \in V$ then $zv = \mathrm{pr}_n(z) \cdot v$ as soon as n is large enough.*

Note that if $\lambda \sim x^c$ and $V = M_\lambda$ or $V = L_\lambda$ then $zv = z(\lambda)v$.

4. The main construction

We retain the notation of Secs. 2–3. Suppose $m \in \{0, 1, 2, \ldots\}$ for $\mathfrak{g} = \mathfrak{o}(2\infty)$, $\mathfrak{g} = \mathfrak{sp}(2\infty)$, and $m \in \{-1, 0, 1, 2, \ldots\}$ for $\mathfrak{g} = \mathfrak{o}(2\infty + 1)$. We shall assume $n \geq m$. Let us introduce the following notation:

(a) $\mathfrak{g}_m(n)$ denotes the subalgebra of $\mathfrak{g}(n)$ spanned by the F_{ij}'s subject to the condition $m < |i|, |j| \leq n$. It is isomorphic to $\mathfrak{o}(2(n-m))$ or $\mathfrak{sp}(2(n-m))$.

(b) $A_m(n)$ denotes the centralizer of $\mathfrak{g}_m(n)$ in $\mathcal{U}(\mathfrak{g}(n))$. Note that the center $Z(n)$ of $\mathcal{U}(\mathfrak{g}(n))$ coincides with $A_0(n)$ (resp., $A_{-1}(n)$) when $\mathfrak{g} \neq \mathfrak{o}(2\infty + 1)$ (resp., $\mathfrak{g} = \mathfrak{o}(2\infty + 1)$).

(c) $I(n)$ denotes the left ideal of $\mathcal{U}(\mathfrak{g}(n))$ generated by the elements

$$F_{in} + \delta_{in} c, \quad -n \leq i \leq n$$

and $J(n)$ denotes the right ideal generated by the "opposite" elements

$$F_{nj} + \delta_{jn} c, \quad -n \leq j \leq n.$$

(d) $\mathcal{U}(\mathfrak{g}(n))^0$ denotes the centralizer of F_{nn} in $\mathcal{U}(\mathfrak{g}(n))$. Note that $A_m(n) \subset \mathcal{U}(\mathfrak{g}(n))^0$, $m < n$.

Lemma 4.1.

(i) *There exists a homomorphism* $\pi_n : \mathcal{U}(\mathfrak{g}(n))^0 \to \mathcal{U}(\mathfrak{g}(n-1))$ *which is characterized as follows: for any* $a \in \mathcal{U}(\mathfrak{g}(n))^0$, *we have* $a - \pi_n(a) \in I(n)$.

(ii) π_n *preserves the filtration.*

(iii) *We have* $\pi_n(A_m(n)) \subset A_m(n-1)$ *for any* $m < n$.

Proof. By the Poincaré-Birkhoff-Witt theorem, we have

$$\mathcal{U}(\mathfrak{g}(n)) = (I(n) + J(n)) \oplus \mathcal{U}(\mathfrak{g}(n-1)).$$

Let π_n be the projection on the second component of this decomposition. It is easily verified that the intersection of $\mathcal{U}(\mathfrak{g}(n))^0$ with $I(n) + J(n)$, $I(n)$ or $J(n)$ is one and the same. Thus this intersection is a two-sided ideal in $\mathcal{U}(\mathfrak{g}(n))^0$ and so π_n yields a homomorphism of $\mathcal{U}(\mathfrak{g}(n))^0$ in $\mathcal{U}(\mathfrak{g}(n-1))$. This proves (i). Now (ii) is obvious and (iii) follows from the fact that the construction of π_n is equivariant relative to the adjoint action of the subgroup corresponding to the Lie subalgebra $\mathfrak{g}_m(n-1)$.

Note that this construction is similar to the construction of the Harish-Chandra homomorphism, see Dixmier [3]. Note also that the restriction of π_n to the center $Z(n)$, which is none other than $A_0(n)$ or $A_{-1}(n)$, coincides with the projection $\pi_n : Z(n) \to Z(n-1)$ defined in Sec. 3.

Definition 4.2. For any m, the algebra $A_m = A_m(\mathfrak{g})$ is defined as the projective limit as $n \to \infty$ of the chain of filtered algebras

$$A_m(m) \overset{\pi_{m+1}}{\leftarrow} A_m(m+1) \overset{\pi_{m+2}}{\leftarrow} A_m(m+2) \leftarrow \dots$$

By this definition, an element $a \in A_m$ is represented by a sequence $(a_n : n \geq m)$ such that $a_n \in A_m(n)$, $\pi_n(a_n) = a_{n-1}$ and $\sup \deg a_n < \infty$. The canonical projection $a \mapsto a_n$ defines a homomorphism of A_m in $A_m(n) \subset \mathcal{U}(\mathfrak{g}(n))$ which will be denoted by pr_n.

Definition 4.3. The algebra $A = A(\mathfrak{g})$ is defined as the inductive limit of the algebras A_m as $m \to \infty$ where, for any m, the canonical embedding $A_m \hookrightarrow A_{m+1}$ is determined by the obvious inclusions $A_m(n) \subset A_{m+1}(n)$.

Note that, for any $a \in A$, $\mathrm{pr}_n(a) \in \mathcal{U}(\mathfrak{g}(n))$ is well defined for all sufficiently large n. There is an obvious embedding $\mathcal{U}(\mathfrak{g}) \hookrightarrow A(\mathfrak{g})$: its image is formed by those elements $a \in A(\mathfrak{g})$ for which $\mathrm{pr}_n(a)$ stabilizes as $n \to \infty$.

Lemma 4.4. *Any* $\mathfrak{g}$*-module* $V \in \Omega(c)$ *can be endowed with a canonical structure of a* $A(\mathfrak{g})$*-module.*

Proof. The definitions of $A(\mathfrak{g})$ and of $\Omega(c)$ imply that, for any $a \in A(\mathfrak{g})$ and $v \in V$, the vector $v_n = \mathrm{pr}_n(a)v$ does not depend on n when n is large enough. Let us take the stable value of v_n as the definition of av. Using the fact that $I(n)\mathcal{U}(\mathfrak{g}(n-1)) = I(n)$ one easily verifies that this is indeed an action of $A(\mathfrak{g})$.

Note that the algebra $Z(\mathfrak{g})$ of virtual Laplace operators (see Sec. 3) is identified, in an obvious manner, with the subalgebra A_0 (if $\mathfrak{g} \neq \mathfrak{o}(2\infty+1)$) or A_{-1} (if $\mathfrak{g} = \mathfrak{o}(2\infty+1)$)

of the algebra $A(\mathfrak{g})$. Furthermore it is easily verified that $Z(\mathfrak{g})$ coincides with the center of $A(\mathfrak{g})$.

To state the main result I need some extra notation.

Suppose n runs through $\{1, 2, \dots\}$; $N = 2n$ for $\mathfrak{g} \neq \mathfrak{o}(2\infty+1)$ and $N = 2n+1$ for $\mathfrak{g} = \mathfrak{o}(2\infty+1)$; $F^{(n)}$ denotes the $N \times N$ matrix whose entries are F_{ij} where $-n \leq i, j \leq n$ (recall that the zero value of an index is allowed only for odd N). Now define $f \in Z(\mathfrak{g})[[u^{-1}]]$ as follows:

$$f(u)(\lambda) = \Pi_{j=\pm1,\pm2,\dots} \frac{u + l_j + \frac{1}{2}}{u + l_j^c + \frac{1}{2}} \cdot \begin{cases} \dfrac{u - c + \frac{1}{2}}{u + \frac{1}{2}} & \text{if } \mathfrak{g} = \mathfrak{o}(2\infty) \\ \dfrac{u - c}{u} & \text{if } \mathfrak{g} = \mathfrak{o}(2\infty + 1) \\ \dfrac{u - c - \frac{1}{2}}{u - \frac{1}{2}} & \text{if } \mathfrak{g} = \mathfrak{sp}(2\infty) \end{cases} \tag{4.1}$$

where $\lambda \in \mathfrak{h}^*$, $\lambda \sim x^c$, $l = \lambda + \rho$, $l^c = x^c + \rho$.

Further let us note that the defining relations of the algebras $Y(N)$ and $Y^{\pm}(N)$ do not depend on N, see (1.5) and (2.4), (2.5). So we can define the inductive limits $Y(\infty)$ and $Y^{\pm}(\infty)$ of these algebras as $n = [N/2] \to \infty$. (Note however that $Y^{\pm}(\infty)$ is not a subalgebra of $Y(\infty)$ because the embedding $Y^{\pm}(N) \to Y(N)$ depends on N !)

Theorem 4.5.

(i) *There exists an embedding $\phi : Y^{\pm}(\infty) \hookrightarrow A(\mathfrak{g})$ which is characterized as follows: Let i, j be fixed and n be sufficiently large (namely $n \geq \max(|\,i\,|, |\,j\,|)$). Then*

$$\mathrm{pr}_n(\phi(s_{ij}(u))) = \frac{u + c + \frac{N \mp 1}{2}}{u + \frac{N \mp 1}{2}} f(u) \left(1 - \frac{F^{(n)}}{u + \frac{N \mp 1}{2}}\right)^{-1}_{ij} \tag{4.2}$$

where, as usual, the upper (respectively, lower) sign is chosen for the orthogonal (respectively, symplectic) algebra.

(ii) *Let us identify $Y^{\pm}(\infty)$ with its image under ϕ. Then we have*

$$A(\mathfrak{g}) = Z(\mathfrak{g}) \otimes Y^{\pm}(\infty) \tag{4.3}$$

(iii) *Let $m = 1, 2, \dots$ and set $M = 2m$ when $\mathfrak{g} = \mathfrak{o}(2\infty)$ or $\mathfrak{g} = \mathfrak{sp}(2\infty)$ and $M = 2m + 1$ when $\mathfrak{g} = \mathfrak{o}(2\infty + 1)$. Then ϕ identifies $Y^{\pm}(M)$ with a subalgebra of $A_m(\mathfrak{g})$ and we have*

$$A_m(\mathfrak{g}) = Z(\mathfrak{g}) \otimes Y^{\pm}(M) \tag{4.4}$$

Remark 4.6. I would like to comment on the isomorphism (4.4) since it can be applied to certain "classical" problems of representation theory known as "missing label problems"

Let us recall that the algebra $A_m(\mathfrak{g})$ is the projective limit of the centralizers $A_m(n)$ as $n \to \infty$ (see the beginning of this section). The structure of the centralizers $A_m(n)$

s cumbersome but, as one sees from (4.4), we can describe the structure of their limit $A_m(\mathfrak{g})$.

On the other hand, it should be noted that the centralizers of the form $A_m(n)$ are very mportant objects because they control the decomposition of irreducible representations f Lie algebras $\mathfrak{g}(n)$ being restricted to subalgebras $\mathfrak{g}_m(n)$. For this reason we need to tudy irreducible representations of these centralizers.

Further we remark that any representation of a centralizer $A_m(n)$ can be viewed s a representation of the limit algebra $A_m(\mathfrak{g})$. But now we can obtain (at least, in rinciple), the representations of $A_m(\mathfrak{g})$ using the isomorphism (4.4) and the embedding $Y^{\pm}(N) \to Y(N)$.

There is a special case which seems to me especially interesting. Namely the algebra $Y^{-}(2)$ and its representations are closely related to the old unsolved problem of onstructing an analog of Gelfand-Tsetlin basis for irreducible finite-dimensional $\mathfrak{sp}(2n)$-nodules.

5. Outline of the proof of the main theorem

The main idea of the proof of theorem 4.5 is essentially the same as in the case of he Lie algebra $\mathfrak{g} = \mathfrak{gl}(\infty)$ which is discussed in the author's papers [14],[15] (theorems 5.3 and 2.1.15, respectively). However now its realization becomes substantially more lifficult.

Step 1. Rather long but elementary calculations show that the $N \times N$ matrix

$$1 + \frac{F^{(n)}}{u \pm \frac{1}{2}}$$

satisfies (2.4) and (2.5). (Moreover these relations are precisely equivalent to the commutation relations of the basis elements F_{ij} of the Lie algebra $\mathfrak{g}(n)$ combined with the symmetry relations $F_{-j,-i} = -\theta_{ij}F_{ij}$. Thus we found an R-matrix interpretation of the commutation relations in orthogonal and symplectic Lie algebras!)

Step 2. The result of step 1 provides us with a homomorphism of algebras $\tilde{Y}^{\pm}(N) \to \mathcal{U}(\mathfrak{g}(n))$. Next we combine it with the automorphism

$$\mathrm{inv}: \; S(u) \to S(-u - \frac{N}{2})^{-1}$$

of the algebra $\tilde{Y}^{\pm}(N)$ and obtain another homomorphism

$$s_{ij}(u) \to \left(1 - \frac{F^{(n)}}{u + \frac{N \mp 1}{2}}\right)^{-1}_{ij}, \quad -n \le i,j \le n$$

Finally we note that the right hand side can be multiplied by an arbitrary formal series $g_n(u)$ from $C[[u^{-1}]]$ whose constant term equals 1 (the precise form of this formal series will be specified in the next step). So we obtain a homomorphism $\tilde{Y}^{\pm}(N) \to \mathcal{U}(\mathfrak{g}(n))$ which will be denoted by ψ_n:

$$\psi_n(s_{ij}(u)) = g_n(u)\left(1 - \frac{F^{(n)}}{u + \frac{N \mp 1}{2}}\right)^{-1}_{ij}, \quad -n \le i,j \le n$$

Step 3 (compare to lemma 6.2 in [14] or lemma 2.1.6 in [15]). It can be shown that

$$\psi_n(s_{ij}(u)) - \psi_{n-1}(s_{ij}(u)) \in I(n)[[u^{-1}]], \quad n > \max(|i|, |j|)$$

provided

$$g_n(u) = \frac{u + c + \frac{N\mp 1}{2}}{u + \frac{N\mp 1}{2}}.$$

.Thus, for any m, the sequence $\{\psi_n : n \geq m\}$ determines a homomorphism

$$\chi_m : \tilde{Y}^{\pm}(M) \to A_m(\mathfrak{g})$$

where, as in the statement of theorem 4.5, M equals $2m$ or $2m+1$.

Step 4. Let us remark that we can multiply χ_m by any formal series $f \in Z(\mathfrak{g})[[u^{-1}]]$. Now choose f as indicated in (4.1). Then it turns out that the $M \times M$ matrix with entries

$$\phi_m(s_{ij}(u)) := \chi_m(s_{ij}(u)) \cdot f(u), \quad -m \leq i, j \leq m,$$

satisfies the condition $\delta(u) = 1$ of theorem 2.18 or, which is the same, the symmetry condition (2.3). This is verified by rather long calculations using the Perelomov-Popov formulas for the orthogonal and symplectic Lie algebras (about these formulas, see e.g. Želobenko [18]).

As a corollary we obtain that the collection $\phi := (\phi_m)$ defines a homomorphism of $Y^{\pm}(\infty)$ into $A(\mathfrak{g})$.

Step 5. It remains to show that $\ker \phi = \{0\}$ and that the statements (ii) and (iii) of the theorem hold. This is proved in the same manner as the corresponding results for $\mathfrak{g} = \mathfrak{gl}(\infty)$ (see [14] and [15]): here the main idea is to reduce all the problems to the graded commutative algebra $\mathrm{gr}\, A_m$ whose structure can be determined by making use of the classical invariant theory.

References

[1] I.V.Cherednik, *Factorizing particles on a half-line and root systems*, Theor. Math. Phys. **16** no. 1 (1984), 35–44.

[2] I.V.Cherednik, *Quantum groups as hidden symmetries of classical representation theory*, Diff. Geom. Meth. Math. Phys. (Proc. 17-th Intern. Conf.), World Scientific, Singapore, 1989, pp. 47–45

[3] J.Dixmier, *Algèbres enveloppantes*, Gauthier-Villars, Paris, 1974.

[4] V.G.Drinfeld, *Hopf algebras and the quantum Yang-Baxter equation*, Dokl. Akad. Nauk SSSR **283** (1985), 1060–1064; Soviet Math. Dokl. **32** no. 1 (1985), 245–258, English transl..

[5] V.G.Drinfeld, *Degenerated affine Hecke algebras and Yangians*, Funct. Anal. Appl. **20** no. 1 (1986)

[6] V.G.Drinfeld, *New realization of the Yangians and the quantized affine algebras*, Dokl. Akad. Nauk SSSR **296** (1987), 13–17; Soviet Math. Dokl. **36** (1988), 212–216, English transl..

[7] V.G.Drinfeld, *Quantum groups*, Proc. ICM-86, vol. 1, Berkeley, 1987, pp. 789–820.

[8] B.L.Feigin and D.B.Fuks, *Casimir operators in modules over the Virasoro algebra*, Dokl. Akad. Nauk SSSR **269** (1984), 1060–1064; Soviet Math. Dokl. **32** (1985), 1057–1060, English transl..

[9] V.G.Kac, *Laplace operators of infinite-dimensional Lie algebras and theta functions*, Proc. Nat. Acad. Sci. USA 81 (1984), 645–647.

[10] A.N.Kirillov and N.Yu.Reshetikhin, *The Yangians, Bethe ansatz and combinatorics*, Lett. Math. Phys. **12** (1986), 199–208.

[11] M.L.Nazarov, *Quantum Berezinian and the classical Capelli identity*, Lett. Math. Phys. **21** (1991) 123–131.

[12] M.L.Nazarov, *Yangians of the "strange" Lie superalgebras*, Published in this volume.

[13] G.I.Olshanskiĭ, *Extension of the algebra $U(\mathfrak{g})$ for the infinite-dimensional classical Lie algebras $\mathfrak{g}$ and the Yangians $Y(\mathfrak{gl}(m))$*, Dokl. Akad. Nauk SSSR **297** (1987), 1050–1054; Soviet Math. Dokl. **36** (1988), 569–573, English transl..

[14] G.I.Olshanskiĭ, *Yangians and universal enveloping algebras*, Diff. Geometry, Lie Groups and Mechanics. IX. Zap. Nauchn. Semin. LOMI **164** (1987), 142-150, Leningrad; J. Soviet Math. **47** no. 2 (1989), 2466–2473, English translation:.

[15] G.I.Olshanskiĭ, *Representations of infinite-dimensional classical groups, limits of enveloping algebras and Yangians*, Preprint (September, 1990), 61, To be published in a collection of the series "Advances in Soviet Mathematics", Amer. Math. Soc..

[16] N.Yu.Reshetikhin and M.A.Semenov-Tian-Shansky, *Central extensions of quantum groups*, Lett. Math. Phys. **19** (1990), 133–142.

[17] E.K.Sklyanin, *Boundary conditions for integrable quantum systems*, J. Phys. **A21** (1988), 2375-2389.

[18] D.P.Želobenko, *Compact Lie groups and their representations*, Transl. Math. Monogr. no. 40 (1973), Amer. Math. Soc., Providence, R.I..

INSTITUTE OF GEOGRAPHY OF THE USSR ACADEMY OF SCIENCES, STAROMONETNY 29, MOSCOW 109017 USSR.

DIFFERENTIAL GRADED LIE ALGEBRAS, QUASI–HOPF ALGEBRAS AND HIGHER HOMOTOPY ALGEBRAS

JIM STASHEFF

University of North Carolina at Chapel Hill,
Mathematics Department

Прежде всего я хотел бы поблагодарить организаторов этой конференции. Очень приятно встретиться с советскими математиками и физиками. Я надеюсь, что в будущем многие из вас смогут посетить Чепл Хил.

Сегодня я хотел бы обсудить связи между нашими работами на западе; работами японских физииков на Дальнем Востоке и вашиими работами в Советском Союзе; особенно работами Владимира Гершоновича Дринфельда. Я думаю, что здесь я возможно буду рассказывать вам, как использовать самовар.

Каждый любит язык своей страны. Поэтому я буду говорить по английски.

One of the wonders of contemporary developments in mathematical physics is the intertwining of many seemingly unrelated topics in pure mathematics. Today I would like to emphasize such interrelations rather than results in any one field.

Recently, Drinfel'd has introduced a notion of quasi-Hopf algebra in which associativity of the diagonal is modified in a way in which the pentagon condition plays a dominant role, analogous to the hexagonal Yang-Baxter equation replacement for commutativity. This pentagon condition and its Lie analog are cropping up in many places throughout theoretical physics: in Conformal Field Theory, in higher spin algebras, in String Field Theory, both open and closed, etc. The Lie analog is also directly relevant to deformation theory in all its manifestations, beginning with its implicit appearance in Douady's treatment of deformations of complex structure[Dou].

Since there will be many subtopics in my report, associative and Lie algebras and various 'quasi' or homotopy analogs, a chart showing their relationship might be helpful.

	ASSOCIATIVE	LIE (JACOBI)
STRICT		d.g. Lie Algebra
HOMOTOPY or "CUBIC"	quasi-bialgebra (Drinfel'd) Open SFT (HIKKO)	
STRONGLY HOMOTOPY or "NON-POLYNOMIAL"	Chains on loop space Homological Perturbation Theory	Small models for deformations Closed SFT (SZ/KKS)

Fig. 1

Let us begin with deformation theory. My point of view on deformation theory, leveloped in collaboration with Mike Schlessinger, is that the deformation theory of an lgebraic object is controlled by a differential graded algebra. Consider the case of the leformation theory of an associative algebra A.

In his pioneering work on deformation theory of associative algebras, Gerstenhaber G] created a bracket on the Hochschild cohomology $Hoch(A, A)$ which with respect to he Hochschild grading took $Hoch^p \otimes Hoch^q$ to $Hoch^{p+q-1}$. An infinitesimal deformaion θ was an element of $Hoch^2(A, A)$ and the primary obstruction to extending it to full deformation was given by $[\theta, \theta] \in Hoch^3(A, A)$. There exist Lie-Massey brackets , ,] such that if $[\theta_i, \theta_j] = 0$ for $i, j \in \{1, 2, 3\}$, then $[\theta_1, \theta_2, \theta_3]$ is defined in a uotient of $Hoch^{p+q+r-3}$ and similarly for $n > 3$ factors. Higher order obstructions o integrability of an infinitesimal θ are given by $[\theta, \theta, \dots, \theta] \in Hoch^3(A, A)$. (For a letailed discussion of Lie-Massey brackets, see Retakh [R].)

If we start with $Hoch(A)$ as the basic object, it is naturally a graded Lie algebra vith the additional structure of the higher order Massey brackets, but no differential. Iow can we interpret this as differential graded Lie algebra or something close to it?

Before answering this question, let us look at related phenomena which are simpler n two ways: they involve associative algebras rather than Lie algebras and they will nvolve only trilinear "corrections" to the basic law of associativity.

First, carrying my samovar to Tula, let me consider Drinfel'd's quasi-Hopf algebras D1,4,5].

Quasi-Hopf algebras

Recall that modules over an algebra admit a tensor operation if the algebra has a ompatible diagonal: using Heyneman-Sweedler notation

$$\Delta a = \Sigma a_{(1)} \otimes a_{(2)},$$

he A-module structure on $V \otimes W$ is defined by

$$a(v \otimes w) = \Sigma a_{(1)} v \otimes a_{(2)} w.$$

The usual tensor product of vector spaces is associative; we have a specific isomorphism

$$(U \otimes V) \otimes W \approx U \otimes (V \otimes W).$$

For this to hold as A-modules, it is sufficient (but NOT necessary) for Δ to be associative: $(\Delta \otimes 1)\Delta = (1 \otimes \Delta)\Delta$. More generally, we ask only for an invertible natural transformation

$$\alpha : (U \otimes V) \otimes W \approx U \otimes (V \otimes W).$$

(Mnemonic: α has a direction; it reconstructs the triple tensor product - call it "perestroika".)

The pentagon condition is the commutativity of Fig. 2:

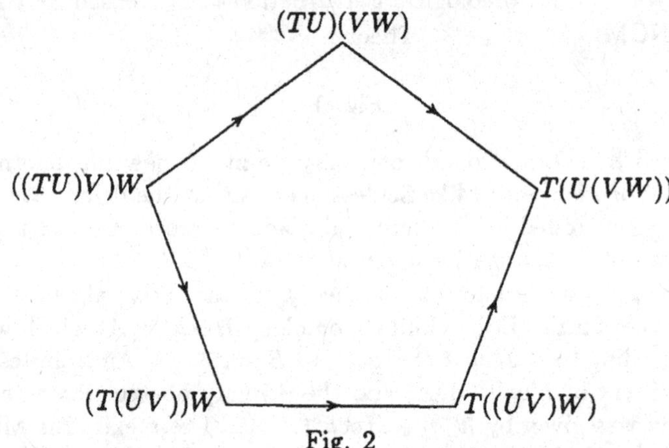

Fig. 2

For those who know the classical yoga of Racah or 6j-coefficients for angular momentum, this pentagon corresponds to the Biedenharn-Elliot identities [B], [E], [B-L], [Maj], the latter for a mathematical treatment.

Drinfel'd considers a not necessarily coassociative Hopf algebra in which α is realized in a special way:

Definition (Drinfel'd). A *quasi-Hopf algebra* $(A, \Delta, \epsilon, \phi)$ is a not necessarily coassociative Hopf algebra (A, Δ, ϵ) together with an invertible element ϕ of $A \otimes A \otimes A$ such that

$$(1 \otimes \Delta)\Delta(a) = \Phi[(\Delta \otimes 1)\Delta(a)]\Phi^{-1}$$

$$(1 \otimes 1 \otimes \Delta)(\Phi)(\Delta \otimes 1 \otimes 1)(\Phi) = (1 \otimes \Phi)(1 \otimes \Delta \otimes 1)(\Phi)(\Phi \otimes 1)$$

satisfying some additional axioms involving units, counits and antipodes.

If we think of A as $Map(X, k)$ for some space X and field k with Δ given by $\Delta(a)(x, y) = a(xy)$ for some multiplication $X \times X \to X$, then the pentagon condition can be rewritten:

$$\Phi(w, x, yz)\Phi(wx, y, z) = \Phi(x, y, z)\Phi(w, xy, z)\Phi(w, x, y).$$

The pentagon condition implies the pentagon condition on the tensor product of A-modules, as in the definition of tensor category [MacL].

For those to whom the classifying space BC of a category C is familiar, this means the category C of A-modules has a classifying space BC with a multiplication $BC \times BC \to BC$ which is, corresponding to 1), homotopy associative. The pentagon condition, being atisfied identically and not up to further conjugation, implies that BC has in turn a lassifying space BBC.

N.B. Even if Δ is associative, it is possible to have a non-trivial Φ, i.e. one that fixes he triple diagonal.

Another analogy takes place in the differential graded category.

Differential graded homotopy associative algebras

Definition. A *differential graded (associative) algebra* A (over a commutative ring R) onsists of a graded R-module $A = \{A^p\}$ together with an associative multiplication $A^p \times A^q \to A^{p+q}$ respecting the grading and a graded derivation $d: A^p \to A^{p+1}$ with $d^2 = 0$.

An obvious example is the deRham complex of differential forms. The differential ould equally well decrease degree by 1, examples being cubical singular chains on topoogical groups or multiplicative resolutions in homological algebra. In fact, historically, he strict associativity was relaxed in favor of homotopy associativity first for such chain omplexes of spaces of the homotopy type of a topological group or monoid [S4].

Definition. A *homotopy associative algebra* A (over a commutative ring R) consists of a raded R-module $A = \{A^p\}$ together with a multiplication $A^p \times A^q \to A^{p+q}$ respecting he grading, a graded derivation $d: A^p \to A^{p+1}$ with $d^2 = 0$ and an *associating homotopy* $h: A^p \times A^q \times A^r \to A^{p+q+r-1}$ such that

$$(ab)c - a(bc) = dh(a,b,c) + h(da,b,c) + (-1)^p h(a,db,c) + (-1)^{p+q} h(a,b,dc).$$

The algebra is called *pentagonal* if h satisfies the pentagon condition:

$$ah(b,c,e) - h(ab,c,e) + h(a,bc,e) - h(a,b,ce) + h(a,b,c)e = 0.$$

Remarkably, pentagonal algebras have recently been discovered in physics, namely, n open string field theory according to Kaku [K] and the Kyoto group: Hata, Itoh, Kugo, Kunitomo and Ogawa [HIKKO].

The pentagonal algebra of open string field theory

One version of an open string in physics is a geometric object - something like an oriented arc in a Riemannian manifold (M,g). One of the subtleties of string theory is the use of parametrized strings $X : [0,r] \to M$ to obtain results that are independent of the choice of parameterization . String interactions are handled by "joining"two strings to form a third; this is often pictured in one of three ways:

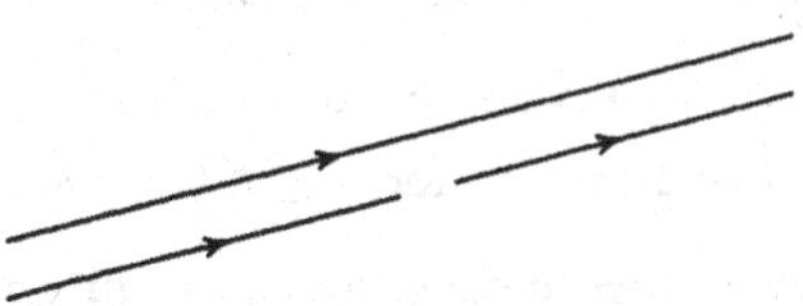

Fig. 3. E: endpoint interaction

Here we are dealing with a space of maps $Map(I, M)$ where $I = [0,1]$. (Physicists prefer $[0, \pi]$.) The endpoint joining E requires a reparameterization; define $X + Y : [0,1] \to M$ by

$$t \mapsto X(2t) \text{ for } 0 \leq 2t \leq 1$$
$$t \mapsto Y(2t-1) \text{ for } 1 \leq 2t \leq 2.$$

This operation failed to have units, inverses or associativity, though all were present "up to homotopy".

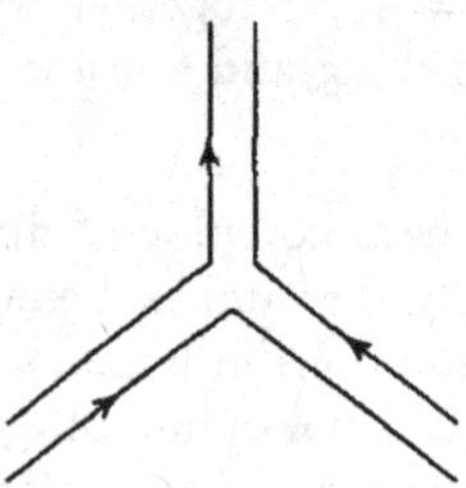

Fig. 3. M: midpoint or half overlap interaction

Here when the latter half of one path is the same as the first half of the second path but with reversed orientation, the interaction produces a third path by 'cancelling' the overlapping halves. The picture is symmetric with respect to cyclic permutations and is so interpreted.

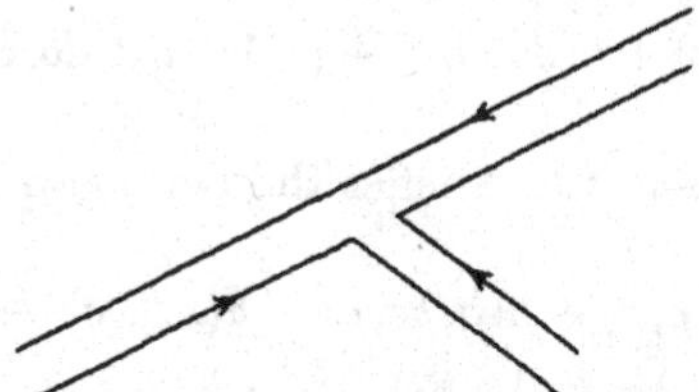

Fig. 3. V: variable overlap interaction

The endpoint case E is familiar in mathematics as far back as the study of the fundamental group; the midpoint case M was considered by Witten in 1986 [W] for string field theory, although in fact it had been considered independently in mathematics by Lashof [L] in 1956; the variable case V seems to have occurred first in physics in the work of Kaku [K], although it follows from an interpretation of the HIKKO interaction [HIKKO], which was considered independently in the study of loop spaces by Moore in 1956 [M].

So as to restore associativity, we follow Moore [M] and [HIKKO] and consider the space $\mathcal{P}M$ of parametrized paths $X : [0, r] \to M$. (Fixing r can be regarded as a (partial) choice of gauge.) Now we define the **endpoint** joining operation $X \vee Y$ for X,Y in $\mathcal{P}M$ such that X(r) = Y(0) with $X \vee Y : [0, r+s] \to M$ by

$$t \mapsto X(t) \text{ for } 0 \leq t \leq r$$
$$t \mapsto Y(t-r) \text{ for } r \leq t \leq r+s.$$

This operation is associative where defined and has units $m : [0,0] \to M$, but has inverses only up to homotopy.

From the physicist's symmetrical point of view, this picture also gives inverses in the endpoint gauge by reversing orientation and cancellation of a portion of one string by all of another. Consideration of multiple products then implies the variable overlap interaction and to a hidden destruction of associativity in what HIKKO refer to as "horn" diagrams.

The variable overlap joining V corresponds to the operation $X * Y$ defined as follows: Let $u = max(r$ such that $Y(t) = X(r-t)$ for $0 \leq t \leq r)$, then $X * Y : [0, r+s-2u] \to M$ by

$$t \mapsto X(t) \text{ for } 0 \leq t \leq r-u$$
$$t \mapsto Y(t-r+2u) \text{ for } r-u \leq t \leq r+s-2u.$$

Again $m : [0,0] \to M$ is a unit, and the reversal $\overline{X}(t) = X(r-t)$ provides an inverse, but now $*$ occasionally fails to be associative; for example in the configuration of Fig. 4.

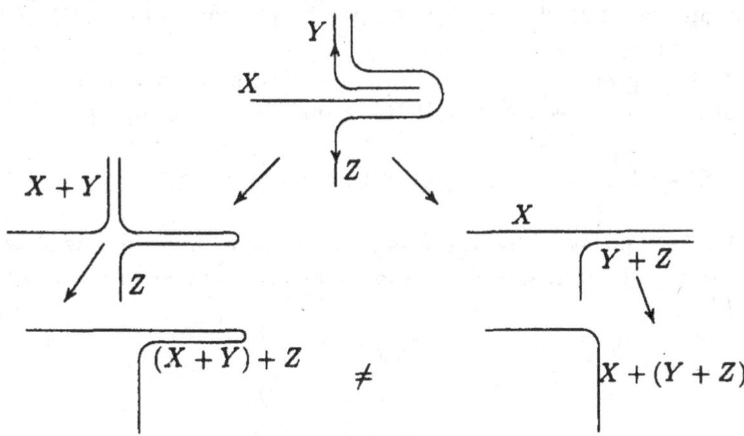

Fig. 4

On the other hand, $(X * Y) * Z$ and $X * (Y * Z)$ are clearly homotopic - just gradually shrink the back-and-forth part of $(X * Y) * Z$ - and hence $*$ is homotopy associative. From our point of view, it is this homotopy which gives rise to the "4-string vertex" in physics.

Convolution algebras

String field theory includes consideration of functionals $\Phi, \Psi : \mathcal{P}M \to \mathbf{R}$ (or $\mathbf{C}$). If $\mathcal{P}M$ were a compact group G, we could define the convolution $\Phi * \Psi$ by

$$(\Phi * \Psi)(g) = \int_G \Phi(gh)\Psi(h^{-1})d\mu$$

where $d\mu$ is an invariant measure on G with $\int_G d\mu = 1$. Instead we define

$$(\Phi * \Psi)(Z) = \int_{X*Y=Z} \Phi(X)\Psi(Y)d\mu$$

where $d\mu$ is an "appropriate measure" on $\mathcal{P}M$. Notice that since we are decomposing Z in all possible ways under $*$, the integration is over all paths W that have their source at a point of Z.

Fig. 5

There are considerable subtleties in making sense out of such an integral, in particular, making sense of the 'measure' and this generalized path integration; these issues have been addressed recently in detail by Wiesbrock [Wies]. He finds that this generalized path integration necessarily involves a metric on M.

(Un)fortunately, $*$ is not associative for paths and this failure carries over to $*$ for functionals. In fact, the best measure has a similar failure of associativity [Wies]. This gives rise to an operation $\Phi \circ \Psi \circ \Lambda$ for three functionals, which plays the role of an associating homotopy.

Without thinking of $\Phi \circ \Psi \circ \Lambda$ as an associating homotopy and certainly without knowledge of A_4-structures [S2,4], [HIKKO] consider the appropriate combination

$$\Phi * (\Psi \circ \Lambda \circ \Sigma) - (\Phi \circ \Psi \circ \Lambda) * \Sigma - (\Phi * \Psi) \circ \Lambda \circ \Sigma + \Phi \circ (\Psi * \Lambda) \circ \Sigma - \Phi \circ \Psi \circ (\Lambda * \Sigma)$$

and show that it vanishes. The way I look at this strict vanishing is shown in Fig. 6 where the horizontal (dashed and bottom) lines are mapped to the common value of the two ends.

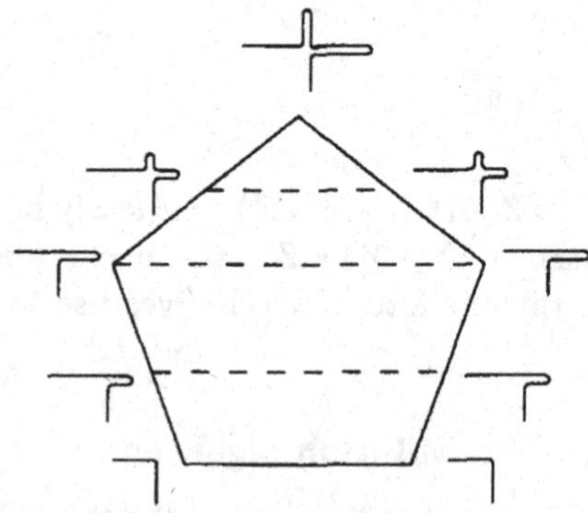

Fig. 6

Kaku also arrives at "4-point vertices" which correspond to homotopy associativity of * and, like the Kyoto group, asserts that higher order terms are unnecessary.

Strongly homotopy associative algebras

Instead of requiring that the associating homotopy h satisfy the strict pentagon condition, one could instead require that the pentagon condition be satisfied only modulo d and then consider higher order conditions in which the pentagon is replaced by the *associahedra* K_n. The index n refers to the number of variables in the iterated associativity

For $n = 4$, K_4 is the pentagon. The K_5-polytope also occurs in analyzing a certain obstruction in Drinfel'd's proof of Kohno's Theorem [Ko] [D1,5] [S2]; its Schlaegel diagram (stereographic projection) appears in Fig. 7, and John Harer has recently shown me an attractive 3-dimensional view shown in Fig. 8.

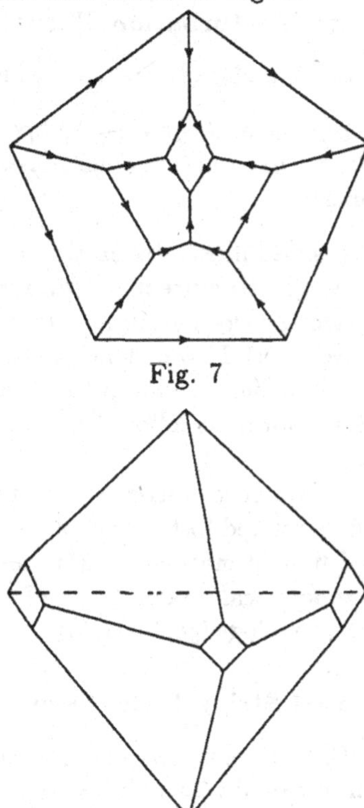

Fig. 7

Fig. 8

[The K_n-polytopes have recently received renewed attention in the combinatorics literature where they are referred to as *associahedra* [Ass].]

Such algebras are called A_∞-algebras or sha-algebras (strongly homotopy associative).

Definition. A *strongly homotopy associative (sha) algebra A* (over a commutative ring R)[S3,5] consists of a differential graded R-module $(A = \{A^p\}, d)$ together with module maps $m_i : A^{\otimes i} \to A$ of degree $i - 2$ such that

$$dm_i - (-1)^i m_i d_\otimes = \Sigma_{j+k=i+1} \Sigma_q \pm m_j(1 \otimes \ldots m_k \otimes \ldots 1).$$

Thus m_2 is a graded multiplication with respect to which d is a derivation and m_3 is an associating homotopy. (The formulas can be simplified further by setting $m_1 = d$.)

An augmented d.g. associative algebra A permits the construction of a d.g. coalgebra BA, known as the bar construction of A. In fact, A is completely characterized by BA. Similarly, for d of degree -1, the definition of an augmented sha-algebra can be incorporated in a corresponding d.g. coalgebra. Let sA denote the differential graded R-module such that $(sA)^p = A^{p-1}$ except that $(sA)^1 = \ker \epsilon : A^0 \to R$.

Theorem. *The structure of an sha-algebra on a differential graded module (A, d) is equivalent to a coderivation D on the tensor coalgebra $T^c sA$ such that $D|sA = sd$ and $D^2 = 0$.*

Homological Perturbation Theory [HPT]

A rich and significant class of examples of sha-algebras is provided by the following:

Theorem. *If (M, d) is a d.g. module and chain homotopy equivalent to an associative d.g. algebra (A, d), then M admits the structure of an sha-algebra (homotopy equivalent to A in the sense of sha-algebras).*

The significance of the result is well illustrated in the case $M = H(A)$ in which, even though $d = 0$ on $H(A)$, there are still structure maps m_i, reflecting the Massey products in $H(A)$. In this case, M is a strong deformation retract of A. Much of the early work in homological perturbation theory [HPT] was done in the latter context. The further generality of the theorem as stated here is due in one version to Heubschmann and Kadeishvili [HK], and has been shown to follow from the SDR case by Barnes and Lambe [BL].

Having described higher order homotopy associativity, it is not difficult to conceive of higher order homotopy structures related to the Jacobi identity, strongly homotopy Lie algebras. Again these occurred first in mathematics, though this time in an algebraic context directly, without any topological analog in mind. It was left to the physicists to reinvent them in terms of the topology/geometry of the free loop space.

Closed String Field Theory

In closed string field theory (CSFT), a closed string is pictured as a closed curve in a (Riemannian) manifold M. Since closed strings have no initial or terminal point, there are additional subtleties in making the space of closed strings S into a groupoid-like object. Subsequent work of several physicists, especially Saadi and Zwiebach [SZ] and Kugo, Kunitomo and Suehiro [KKS], leads to an 'algebra' of string fields which includes a binary operation but is NOT generated by that operation; rather it exhibits the structure of an sh-Lie algebra (defined below), which first appeared in the deformation theory of (differential graded) commutative algebras [SS].

Again we encounter the subtlety that the physics is described in terms of a parameterization of such a curve, e.g. a map of the circle S^1 into M but the physics should not depend on the parameterization. Thus the space of closed strings S can be described as the space of equivalence classes (under reparameterization) of maps of the circle into M, i.e. the quotient of $LM = Map(S^1, M)$ by the reparameterization group $Diff^+ S^1$. Fields then refer to functions on S or sections of some bundle over S. When particles are considered to be mathematical points, their interaction leads to consideration of the algebra of fields being given in terms of point-wise multiplication. For open strings, as I have mentioned above, interactions are pictured in terms of contact between strings (Figures 3.E, M, V) and an algebra of fields which is given by convolution [S1]. For closed strings, there is in [SZ] [KKS] a similar convolution operation, but there is more to the algebra than that, namely N-ary operations corresponding to simultaneous interactions among (N + 1) strings which do NOT decompose in terms of the binary operation. Instead, the operations for various N are related via the so-called BRST operator in a relation which has precisely the form of an sh-Lie algebra, the Lie analog of a strongly homotopy associative algebra (sha-algebra). [This is also the dual of the structure that appears on the indecomposables of a (non-minimal) Sullivan model [Su].]

This time we view the joining of two strings to form a third by extending the method due to Lashof [L] in the case of based loops and to Witten [W] for strings. The idea is that two closed strings Y and Z join to form a third $Y * Z$ if a semi-circle of one agrees with a reverse oriented semi-circle of the other. (Notice this avoids Witten's marking of the circle.) The join $Y * Z$ is formed from the complementary semi-circles of each. To be more precise, consider the configuration of three arcs $A_i, i = 0,1,2$ with the three initial points identified and the three terminal points identified, as in a circle together with a diameter.

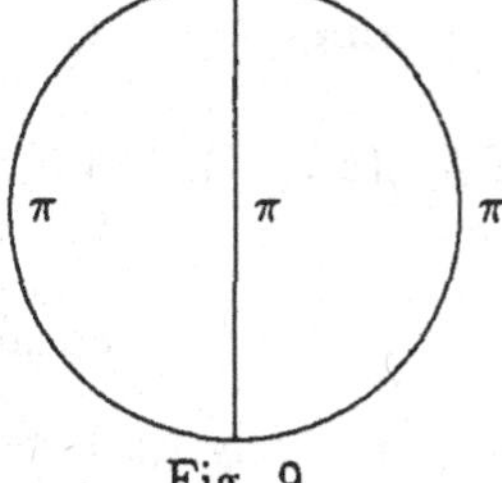

Fig. 9

(One is tempted to call this a (theta) Θ curve, but string field theory is likely to involve theta functions in the sense of number theory; there's enough confusion of terminology already! so we will refer to it as a **trihedron**.) To emphasize the symmetry and to fix parameterizations, consider the arcs to be great semi-circles on the unit 2-sphere in $\mathbf{R}^3$ parametrized by arc length from north pole to south pole. Denote the union of the three arcs by Θ. Denote by $\bar{A}_i$ the arc parametrized in the reverse direction. Let C_i denote any isometry $C_i : S^1 \hookrightarrow \Theta$ which agrees with A_j on one semi-circle and with $\bar{A}_k$ on the other for a cyclic permutation (i,j,k) of $(0,1,2)$. (Up to rotation, C_i is A_j followed by $\bar{A}_k$.) Given any map $X : \Theta \to M$, let $X_i = X \circ C_i$.

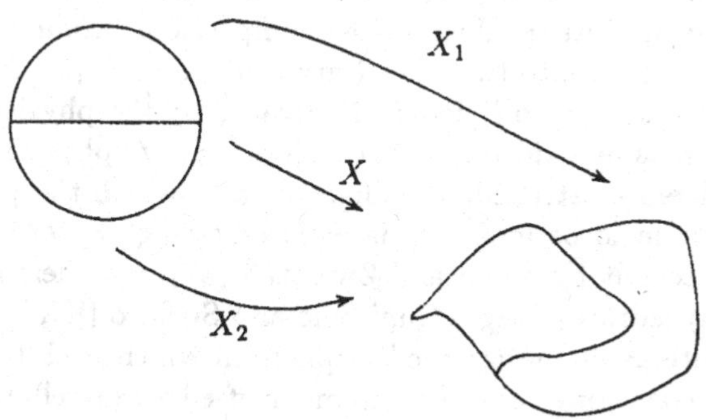

Fig. 10

Now let ϕ, ψ be function(al)s on S, the space of strings. Define the **convolution product** $\phi * \psi$ as follows:

$$(\phi * \psi)(X_0) = \int \phi(X_1)\psi(X_2)d\mu$$

where the integral is over all isometries C_i and all maps $X : \Theta \to M$ such that $X_0 = X \circ C_0$. (Thus $\phi * \psi$ depends on all ways of **decomposing** X_0 into two loops X_1 and X_2.) Again there are some really subtle measure theoretic questions here, which have been addressed by Wiesbrock [Wies].

In accordance with the usual paradigm, KKS desire an "action" of the form

$$S = \phi \cdot Q_B \phi + \Sigma_{N=3}^{\infty}(\phi \ldots \phi) \qquad (N \text{ times})$$

invariant with respect to a variation

$$\delta\phi = \delta_\Lambda \phi = Q_B \Lambda + \Sigma_{N=1}^{\infty}\lceil \phi \ldots \phi\Lambda \rceil \qquad (\phi \text{ repeated } N \text{ times}).$$

We can simplify the formulas by setting $\lceil \Lambda \rceil := Q_B \Lambda$ and $(\phi_0 \phi_1) := \phi_0 \cdot Q_B \phi_1$. The formulas then become

$$S = \Sigma_{N=2}^{\infty}(\phi \ldots \phi) \qquad (N \text{ times})$$

and

$$\delta\phi = \Sigma_{N=0}^{\infty}\lceil \phi \ldots \phi\Lambda \rceil \qquad (\phi \text{ repeated } N \text{ times}).$$

The operator Q_B is referred to in the physics literature as a BRST-operator. We interpret this to mean that ϕ_i is a some sort of differential form on the space of free loops and Q_B is the exterior derivative d along the orbits of the reparameterization group. Henceforth we will write d instead of Q_B. If the formula for $\delta\phi$ is to be homogeneous, ϕ must be of degree 3 and Λ of degree 2.

But what sort of differential forms? One possibility might be the semi-infinite forms of Feigin [F] for the reparameterization algebra $diff^+ \; S^1$ with coefficients in the ring of functions on the space of free loops.

Restricted polyhedra

Now we are ready to suggest a slightly revisionist, primarily topological interpretation of the N-ary operations of [SZ] and [KKS] which will include a modified form of the $*$-product of Kaku and of Witten. Just as the composition product $*$ was based on the decomposition of one loop into two, the N-ary operations $\lceil \phi_1 \ldots \phi_N \rceil$ are based on the decomposition of one loop into N loops. Motivated by the physical interpretation which sees the trihedron as imbedded in a world sheet, several physicists, starting with Kaku [K], have considered a tetrahedral configuration in which the perimeter of each face is regarded as a circle to be mapped via a closed string. Extension to polyhedra with 5 faces was worked out by Saadi and Zwiebach [SZ] and their lead was carried through to general polyhedra by Kugo, Kunitomo and Suehiro [KKS]. Here polyhedra refer to cell decompositions of the oriented 2-sphere in which each face (=2-cell) has boundary (perimeter) consisting of a finite number of edges (1-cells). Each face and hence its perimeter carries the orientation induced from S^2. The polyhedra are restricted geometrically in that each edge is assigned a length such that:

1) Saadi and Zwiebach: the perimeter of each face has length 2π (this implies each edge has length $\leq \pi$), and
2) Kugo, Kunitomo and Suehiro: any simple closed edge path has length $\geq 2\pi$.

[A. S. Schwartz suggested that the 1-skeleta of such polyhedra can be regarded as the limiting case of a Riemann sphere with $N+1$ holes of maximal size so that only the 1-skeleton is left; there is only 'boundary' and no interior.]

There is only one restricted trihedron, Θ.

For tetrahedra, the restrictions are precisely that opposite edges have equal lengths, say (a_1, a_2, a_3), with $0 \leq a_i \leq \pi$ and $\Sigma a_i = 2\pi$.

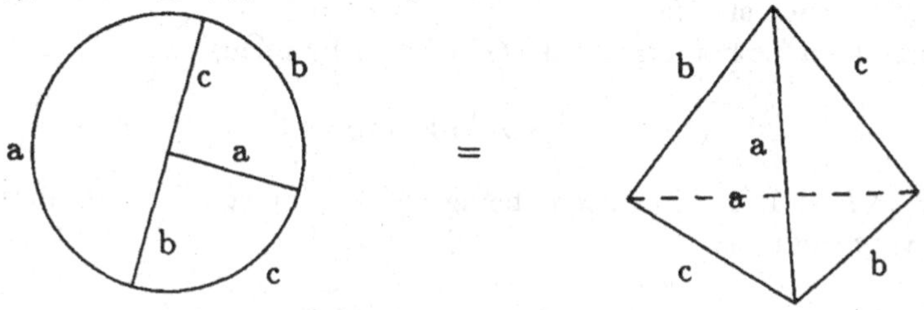

Fig. 11

In other words, the "moduli" space of restricted tetrahedra is given by the union of two 2-simplices (one for each orientation of the tetrahedron) with vertices in common.

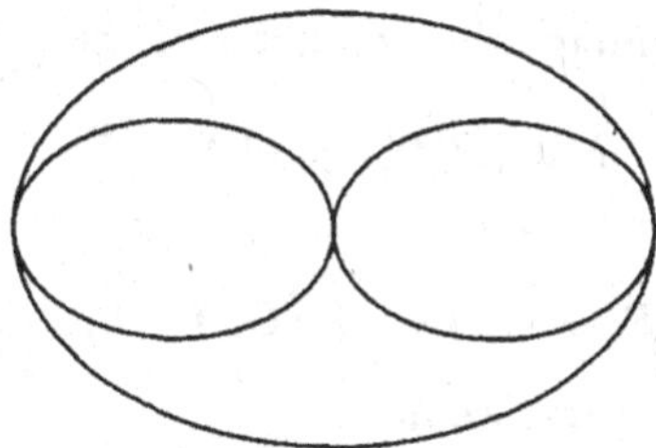

Fig. 12

More generally, let $\mathcal{M}_N$ be the "moduli" space of all restricted $(N+1)$-hedra P with an arbitrary ordering of the faces from 0 to N. As a space, $\mathcal{M}_N$ is given the topology of the local coordinates which are the edge lengths. As P varies by varying the metric on the underlying ordered $(N+1)$-hedron $|P|$, we have a cell, $\mathcal{M}_{|P|}$. The cells of maximal dimension correspond to 3-valent polyhedra and the dimension of these cells is $2N-4$, with faces corresponding to $(N+1)$-hedra with one 4-valent vertex, etc. when an edge length goes to zero.

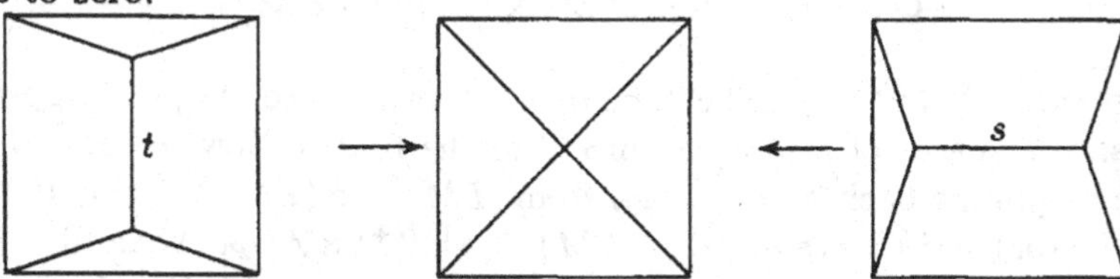

Fig. 13

As $|P|$ varies over all $(N+1)$-hedra, the $\mathcal{M}_{|P|}$ fit together to form $\mathcal{M}_N$. Thus $\mathcal{M}_N$ is described as a finite cell complex in which each cell has boundary composed of a finite number of cells of lower dimension.

The cell complex $\mathcal{M}_N$ has a "boundary" corresponding to saturation of the inequalities 2). KKS refer to such polyhedra as **critical**, i.e. if there is an edge path enclosing at least two faces and of length precisely 2π. Separating the polyhedron P along this edge path produces two restricted polyhedra Q and R.

The separation can be regarded as giving a partition of the set $\{0,\ldots,N\}$ or, if the faces of P are ordered, as an **unshuffle** giving induced orderings on Q and R.

Conversely, if we have two restricted polyhedra Q and R, we can form a twisted connected sum $Q\#R$ by deleting two faces $F_Q \in Q$ and $F_R \in R$ and identifying their perimeters by an arbitrary orientation reversing isometry. The result will be a restricted polyhedron P. If the faces of Q and R are ordered, we can define such a twisted connected sum for each shuffle.

Thus we can describe boundary "facets" of $\mathcal{M}_N$ by inclusions

$$(*) \qquad S^1 \times \mathcal{M}_Q \times \mathcal{M}_R \hookrightarrow \mathcal{M}_N$$

where $Q + R = N + 1$, the inclusions being indexed by (Q, R)-unshuffles and the S^1 coordinate giving the twist.

The N-ary operations

Given forms ϕ_i on the space of free loops LM, define $\lceil \phi_1 \ldots \phi_N \rceil$, a form on LM of dimension $\Sigma_1^N q_i - 3(N-1)$, as follows: For any restricted $(N+1)$-hedron P with isometries $C_i : S^1 \hookrightarrow P$ with image the perimeter of the i-th face , consider

$$Map(P, M) \rightarrow (LM)^{N+1}$$

by

$$(X : P \rightarrow M) \mapsto (X \circ C_i, i = 0, \ldots, N).$$

For fixed $P \in \mathcal{M}_N$, as the C_i vary, we have

$$Map(P, M) \times (S^1)^{N+1} \mapsto (LM)^{N+1}$$

by

$$(X : P \rightarrow M, \theta_0, \ldots, \theta_N) \rightarrow (X \circ C_i \circ \theta_i),$$

interpreting $\theta \in S^1$ as a rotation $\theta \in Aut(S^1)$. As P varies over all $(N+1)$-hedra, the above families fit together to form a "bundle" over $\mathcal{M}_N$

$$\mathcal{Q}_N = \cup_{|P|} Map(P, M) \times (S^1)^{N+1} \times \mathcal{M}_{|P|}$$

with a map into $(LM)^{N+1}$. (The boundary of each piece $\mathcal{M}_{|P|}$ is approached via isometries so the spaces of maps pass nicely to the limit.) Now define $\lfloor \phi_1 \ldots \phi_N \rfloor$ on $Map(P, M)$ by pulling back $\phi_1 \times \cdots \times \phi_N$ from $(LM)^N$ to $(LM)^{N+1}$ (via the projection off the 0-th factor) and thence to $Map(P, M) \times (S^1)^{N+1} \times \mathcal{M}_{|P|}$, then fibre integrating over $(S^1)^{N+1} \times \mathcal{M}_{|P|}$. Now let $Map_{X_0}(|P|, M)$ denote the space of maps $X \rightarrow M$ with specified X_0. This subspace can be considered as the "fibre" of the projection $Map(P, M) \rightarrow LM$ given by composition with C_0. (Suitably defined) generalized path integration over this fibre and summation over ordered topological types $|P|$ will produce the desired form $\lceil \phi_1 \ldots \phi_N \rceil_{X_0}$ evaluated at the loop X_0. (To be somewhat more in keeping with the physics literature, we can denote this fibre integration as

$$\int_{X_0} \lfloor \phi_1 \ldots \phi_N \rfloor DX$$

to mean an integral over all $X : P \to M$ such that $X \circ C_0 = X_0, P \in \mathcal{M}_{N-1}$. There are considerable subtleties in making sense out of such an integral, in particular, making sense of the 'measure' and this generalized path integration; these issues have been addressed recently in detail by Wiesbrock [W]. He finds that this generalized path integration necessarily involves a metric on M.)

As KKS assert, the invariance of S is a consequence of

$$(16.N) \quad d\lceil\phi_1 \ldots \phi_N\rceil + \Sigma_1^N \lceil\phi_1 \ldots d\phi_i \ldots \phi_N\rceil = \Sigma_2^{N-1}\lceil\lceil\phi_{i_1} \ldots \phi_{i_Q}\rceil\phi_{i_{Q+1}} \cdots \phi_{i_N}\rceil$$

where the sum is over all unshuffles of $1, \ldots, N$ (signs and factorial fudge factors being ignored for now). This relation holds because of the combinatorial/topological structure of the parameter spaces over which integration takes place in defining the operations. The moduli space $\mathcal{M}_N$ of all restricted $(N+1)$-hedra as defined is a cell complex with a cellular boundary. The verification of the crucial relation (16.N) is a matter of a careful application of Stokes' Theorem.

But what do these relations signify? In particular, the binary operation $\lceil\phi_1\phi_2\rceil$ has the graded symmetry

$$\lceil\phi_1\phi_2\rceil = (-1)^{(q_1-1)(q_2-1)}\lceil\phi_2\phi_1\rceil,$$

but need not satisfy the Jacobi identity. Instead (16.3) asserts (writing d for Q_B and ignoring signs):

$$d\lceil\phi_1\phi_2\phi_3\rceil + \lceil d\phi_1\phi_2\phi_3\rceil + \lceil\phi_1 d\phi_2\phi_3\rceil + \lceil\phi_1\phi_2 d\phi_3\rceil = \Sigma\lceil\lceil\phi_{i_1}\phi_{i_2}\rceil\phi_{i_3}\rceil,$$

the sum again being over unshuffles of 1, 2, 3. If the LHS were zero, we would have a form of the Jacobi identity. In the language of homotopy theory, $\lceil\phi_1\phi_2\rceil$ satisfies the Jacobi identity **up to homotopy** . The conditions (16.N) specify that the **homotopy** $\lceil\phi_1\phi_2\phi_3\rceil$ satisfies higher order homotopy conditions - to all orders. These are precisely the conditions which in mathematics are summarized in the name "strongly homotopy Lie algebra", abbreviated "sh Lie algebra" [SS]. This is essentially related to the fact that δ generates a differential of square zero on a suitable module.

Strongly homotopy Lie algebras

First recall the definition of a differential graded Lie algebra.

Definition. A *differential graded Lie algebra* A (over a commutative ring R) consists of a graded R-module $A = \{A^p\}$ together with a bracket $A^p \times A^q \to A^{p+q}$ respecting the grading, satisfying the graded analogs of the usual skew-commutativity and Jacobi identities, together with a graded derivation $d : A^p \to A^{p+1}$ with $d^2 = 0$.

If we look at formula (16.N) formally, we see (up to sign and factorial fudge factors), the signature equation of :

Definition. An *sh-Lie algebra (strongly homotopy Lie algebra)* consists of a d.g. module (L, d) together with module maps $\lambda_i : \Lambda^i L \to L$ of degree $i - 1$ such that

$$d\lambda_n + \lambda_n d = \Sigma_{p+q=n}\Sigma_{(p,q)-shuffles}(-1)^\sigma \lambda_{p+1}(\lambda_q \otimes 1)\sigma$$

where σ is the (p, q)-shuffle.

Here we regard $\Lambda^i L$ as the sub-module of $L^{\otimes i}$ invariant under the signed permutations, except that the signs are those corresponding to $(sL)^{\otimes i}$. Thus just as in the associative case, the definition is equivalent to one on the graded commutative *coalgebra* $\Lambda^c L$ as follows (what follows is joint work with Tom Lada [LaS] extending work of Elizabeth Jones [J]):

Proposition. *An sh-Lie algebra structure on a d.g. module (L, d) is equivalent to a differential $D \in Coder \Lambda^c L$ such that $D|sL = sd$ and $D^2 = 0$.*

The application of Homological Perturbation Theory is more subtle in this situation since the proof of [GS] can not be symmetrized. The result however is the direct analog.

Theorem. *If (M, d) is a d.g. module and a strong deformation retract of a d.g. Lie algebra (L, d), then M admits the structure of an sh-Lie algebra (homotopy equivalent in the sense of sh-Lie algebras to L).*

Again the significance of the result is well illustrated in the case $M = H(L)$ in which, even though $d = 0$ on $H(L)$, there are still structure maps λ_i, reflecting the Lie-Massey brackets in $H(L)$ [R]. These are structures of great relevance to deformation theory as was shown in joint work with Mike Schlessinger [SS].

Deformation theory of algebras

In his pioneering work on deformation theory of associative algebras, Gerstenhaber [G1] created a bracket on the Hochschild cohomology $Hoch(A, A)$, but this bracket was not induced from a differential graded Lie algebra structure on the underlying complex. In Schlessinger and Stasheff [SS] , we did construct a differential graded Lie algebra structure on a complex giving the Harrison cohomology $Harr(A, A)$ of a commutative algebra A in characteristic 0. The corresponding results for associative algebras are given in [S6] where the essential idea may be revealed more transparently.

[While this exposition was being developed, I received copies of preprints by Lecomte et al [LR], [LMS] which emphasize and generalize the analogous results for Lie algebras, subject to conditions of finite dimensionality, thus avoiding the use of coalgebras and coderivations.]

The original definition of $Hoch(A, A)$ used the standard (bar) construction [C], [EM]. Ignoring its differential, the bar construction BA is in fact the cofree connected graded coalgebra cogenerated by A [Mi]. The (normalized) Hochschild complex for $Hoch(A, A)$ is $Hom(BA, A)$ with coboundary δ given by

$$(\delta h)[\bar{a}_0|\dots|\bar{a}_n] = hd[\bar{a}_0|\dots|\bar{a}_n] + \bar{a}_0 h[\bar{a}_1|\dots|\bar{a}_n] + (-1)^n h[\bar{a}_0|\dots|\bar{a}_{n-1}]\bar{a}_n.$$

Our essential observation is:

Proposition. *As a differential graded module, $(Hom(BA, A), \delta)$ is isomorphic to $(Coder(BA), D)$.*

Here $Coder(BA) \subset Hom(BA, BA)$ consists of graded coderivations, i.e. those graded k-linear maps h such that

$$\Delta h[\bar{a}_1|\dots|\bar{a}_n] = \Sigma h[\bar{a}_1|\dots|\bar{a}_p] \otimes [\bar{a}_{p+1}|\dots|\bar{a}_n] + (-1)^{p|h|}[\bar{a}_1|\dots|\bar{a}_p] \otimes h[\bar{a}_{p+1}|\dots|\bar{a}_n]$$

where $|h|$ =deg h is defined by $h : A^{\otimes k} \to A^{\otimes k-|h|}$.

It is easy to check that $Hom(BA, BA)$ is a differential graded algebra under composition with $Dh = d_B h \pm h d_B$ and that $Coder(BA)$ is a subcomplex. On the other hand, graded coderivations form a graded Lie algebra under graded commutator of compositions, i.e.$[\theta, \phi] := \theta \circ \phi - (-1)^{|\theta||\phi|}\phi \circ \theta$ and D is a derivation with respect to this bracket. Thus $(Coder(BA), [\ ,\ ], D)$ is a differential graded Lie algebra and $Hoch(A, A)$, being isomorphic to $H(Coder(BA), D)$, inherits the structure of a graded Lie algebra. The composition bracket translates, up to sign, to the Gerstenhaber bracket on $Hom(BA, A)$.

Exactly the same approach works on other deformation theories which are controlled by cochains in a complex of the form $Hom(CA, A)$ where CA is a suitably cofree coalgebra cogenerated by A. The deformation theory of bialgebras is controlled by $Hom(BA, \Omega A)$, where ΩA is the cobar construction [A] on A considered as a coalgebra. At the time (summer 1989) I began to write [S6], this was not known to be isomorphic to any $Coder(CA)$, although the beginnings of a bracket are evidenced in the work of Gerstenhaber and Schack [GS]. The paper of Lecomte and Roger [LR], however, does provided a differential graded Lie algebra controlling the deformation theory of Lie bialgebras, explicating Drinfel'd's complex in his ICM talk [D3]. Working under suitable finiteness restrictions, their d.g. Lie algebra is of the form $Hom(\Lambda^c E, \Lambda E) \approx \Lambda(E^* \oplus E)$, the latter inheriting the structure of a graded Poisson algebra from the canonical symplectic form on $E^* \oplus E$. The obvious extension of this approach to $Hom(BA, \Omega A)$ should provide the appropriate d.g. Lie algebra controlling bialgebra deformations.

Drinfel'd's proof of Kohno's theorem [D1,4,5] is a deformation theoretic argument except that there is currently no formal deformation theory for his quasi-Hopf algebras. The cobar construction does feature prominently in the proof, but, at a crucial point, he has resort to the associahedron K_5. This step could be avoided if we could imbed the argument in a complex controlling the deformation theory of quasi-Hopf algebras. Recent work of Markl [Mar], developed in another context, may do the job. It should also be possible to define analogous complexes for the deformation theory of triangular or quasi-triangular, strict or quasi-Hopf algebras. "We've only just begun."

References

[A] J.F.Adams, *On the cobar construction*, Colloque de topologie algebrique Louvain, 1956, pp. 82–87.

[BL] D.Barnes and L.A.Lambe, *A fixed point approach to homological perturbation theory*, Proc. AMS (to appear).

[Ass] L.J. Billera, P. Filliman and B. Sturmfels, *Construction and complexity of secondary polytopes*, Adv. Math. **83** (1990), 155–179.

C.W. Lee, *The associahedron and triangulations of the n-gon*, Europ. J. Comb. **10** (1989), 551–560.

C.W. Lee, *Regular triangulations of convex polytopes*, preprint DIMACS Tech. Rep. 90-16.

[B] L.C.Biedenharn, *An identity satisfied by the Racah coefficients*, J. Math. Phys. **31** (1953), 287–293.

[BL] L.C.Biedenharn and J.D.Louck, *Angular Momentum in Quantum Physics*, Addison-Wesley, 1981.

[CE] C. Chevalley and S. Eilenberg, *Cohomology theory of Lie groups and Lie algebras*, Trans. Amer. Math. Soc. **63** (1948), 85–124.

[Dou] A.Douady, *Obstruction primaire à la déformation, exposé 4*, Seminaire Henri CARTAN, 1960/61.

[D1] V.G.Drinfel'd, Alg. Anal. 1 no. 6 (1989), 114–148; *Quasi-Hopf algebras*, Leningrad Math. J. **1** (1990), 1419–1457.

[D2] V.G.Drinfel'd, Alg. Anal. 1 no. 2 (1989), 30–46 (in Russian); *On the concept of cocommutative Hopf algebras*, Leningrad Math. J. **1** (1990).

[D3] V.G.Drinfel'd, *Quantum groups*, Proc. ICM-86 (Berkeley), vol. 1, AMS, 1987, pp. 798–820.

[D4] V.G.Drinfel'd, *Quasi-Hopf algebras and Knizhnik-Zamolodchikov equations*, preprint ITP-89-43E (1989).

[D5] V.G.Drinfel'd, Alg. Anal. **2** no. 4 (1990), 149–181. (in Russain)

[EM] S.Eilenberg and S.MacLane, *On the groups $H(\Pi, n).I$*, Ann. of Math. **58** (1953), 55–106.

[E] J.P.Elliott, *Theoretical studies in nuclear structure V*, Proc. Roy. Soc. **A218** (1953), 370.

[F] B.L.Feigin, *The semi-infinite homology of Kac-Moody and Virasoro Lie algebras*, Russian Math. Surveys **39** (1984), 155–156; *Russian original*, Usp. Mat. Nauk **39** (1984), 195–196.

[G1] M. Gerstenhaber, *The cohomology structure of an associative ring*, Ann. of Math. **78** (1963), 267–288.

[G2] M. Gerstenhaber, *On the deformation of rings and algebras*, Ann. of Math. **79** (1964,), 59–103; *On the deformation of rings and algebras III*, Ann. of Math. **88** (1968), 1–34.

[GS] M.Gerstenhaber and S.D.Schack, *Bialgebra cohomology, deformations, and quantum groups*, PNAS, USA **87** (1990), 478–481.

M.Gerstenhaber and S.D.Schack, *Algebras, bialgebras, quantum groups and algebraic deformations*, Proc. Conference on Deformation Theory and Quantization with Applications to Physics, Amherst, June 1990, Contemporary Math to appear.

[HIKKO] H.Hata, K.Itoh, T.Kugo, H.Kunitomo and K.Ogawa, Phys. Rev. D **34** (1986), 2360–2429.

H.Hata, K.Itoh, T.Kugo, H.Kunitomo and K.Ogawa, Phys. Rev. D **35** (1987), 1318–1355.

T.Kugo, *String Field Theory*, Lectures delivered at 25th Course of the International School of Subnuclear Physics on "The SuperWorld II", Erice, August 6-14, (1987).

[HK] J.Huebschmann and T.Kadeishvili, *Minimal models for chain algebras over a local ring*, Math. Zeitschrift (to appear).

[HPT] V.K.A.M. Gugenheim, *On the chain complex of a fibration*, Ill. J. Math. **3** (1972), 398–414.

V.K.A.M.Gugenheim, *On a perturbation theory for the homology of the loop-space*, J. Pure & Appl. Alg. **25** (1982), 197–205.

V.K.A.M.Gugenheim and L.Lambe, *Applications of perturbation theory in differential homological algebra I*, Ill. J. Math. **33** (1989).

V.K.A.M.Gugenheim, L.Lambe and J.Stasheff, *Algebraic aspects of Chen's twisting cochain*, Ill. J. Math. **34** (1990), 485–502.

V.K.A.M.Gugenheim, L.Lambe and J.Stasheff, *Perturbation theory in differential homological algebra II*, Ill. J. Math. (to appear).

V.K.A.M.Gugenheim and J.Stasheff, *On perturbations and A_∞-structures*, Bull. Soc. Math. de Belg. **38** (1986), 237–246.

L.Lambe, *Homological Perturbation Theory - Hochschild Homology and Formal Groups*, Proc. Conference on Deformation Theory and Quantization with Applications to Physics, Amherst, June 1990, AMS, to appear.

L.Lambe and J.D.Stasheff, *Applications of perturbation theory to iterated fibrations*, Manuscripta Math. **58** (1987), 363–376..

[J] E.Jones, *A study of Lie and associative algebras from a homotopy point of view*, Master's Project, NCSU (1990).

[K] M.Kaku, *Why are there two BRST string field theories?*, Phys. Lett. B **200** (1988), 22–30.

M.Kaku, *Deriving the four-string interaction from geometric string field theory*, preprint, CCNY-HEP88/5.

M.Kaku, *Geometric derivation of string field theory from first principles: Closed strings and modular invariance*, preprint, CCNY-HEP-88/6.

M.Kaku, *Introduction to Superstrings*, Springer-Verlag, 1988.

M.Kaku and J.Lykken, *Modular invariant closed string field theory*, preprint, CCNY-HEP-88/7.

[Ko] T.Kohno, *Monodromy representations of braid groups and Yang-Baxter equations*, Ann. Inst. Fourier **37.4** (1987), 139–160.

T.Kohno, *Linear representations of braid groups and classical Yang-Baxter equations*, Proc. Conf. on Artin's Braid Groups, Santa Cruz , 1986, vol. 78, AMS, 1988, pp. 339–363.

T.Kohno, *Quantized universal enveloping algebras and monodromy of braid groups*, Nagoya preprint (1988).

[KKS] T.Kugo, H.Kunitomo and K.Suehiro, *Non-polynomial closed string field theory*, Phys. Lett. **226B** (1989), 48–54.

T.Kugo and K.Suehiro, *Nonpolynomial closed string field theory: Action and gauge invariance*, Nucl. Phys. B **337** (1990), 434–466.

[L] R.Lashof, *Classification of fibre bundles by the loop space of the base*, Annals of Math. **64** (1956), 436–446.

[LMS] P.A.B. Lecomte, P.W. Michor and H. Schicketanz, *The multigraded Nijenhuis-Richardson algebra, its universal property and applications*, JPAA (to appear).

[LR] P.A.B. Lecomte and C. Roger, *Modules et cohomologies des bigèbres de Lie*, C.R.A.S, Paris **310** (1990), 405–410.

[MacL] S.MacLane, *Categories for the working mathematician*, Springer-Verlag, 1971.

S.MacLane, *Natural associativity and commutativity*, Rice Univ. Studies **49** (1963), 28–46.

[Maj] S.Majid, *Quasitriangular Hopf algebras and Yang-Baxter equations*, Int. J. Mod. Phys. A **5** (1990), 1-91.

[Mar] M.Markl, *A cohomology theory for A(m)-algebras*, JPAA volume in honour of Alex Heller.

[Mi] W. Michaelis, *Lie coalgebras*, Adv. in Math. **38** (1980), 1–54.

[M] J.C.Moore, *The double suspension and p-primary components of the homotopy groups of spheres*, Bol. Soc. Math. Mex. **1** (1956), 28–37.

[R] V.S.Retakh, *Lie-Massey brackets and n-homotopically multiplicative maps of DG-Lie algebras*, JPAA volume in honour of Alex Heller.

V.S. Retakh, *Massey operations in the cohomology of Lie superalgebras and deformations of complex analytical algebras*, Funct. Anal. Appl **11** no. 4 (1977), 88–89.

V.S. Retakh, *Massey operations in Lie superalgebras and differentials in Quillen spectral sequences*, Funct. Anal. Appl **12** no. 4 (1978), 91–92.

V.S. Retakh, *Massey operations in Lie superalgebras and differentials in Quillen spectral sequences*, Colloquim Mathematicum no. 50 (1985), 81–94. (Russian)

[SZ] M.Saadi and B.Zwiebach, *Closed string field theory from polyhedra*, Ann. Phys. (N.Y.) **192** (1989), 213–227.

[SS] M. Schlessinger and J. D. Stasheff, *The Lie algebra structure of tangent cohomology and deformation theory*, J. of Pure and Appl. Algebra **38** (1985), 313–322.

M. Schlessinger and J. D. Stasheff, *Deformation theory and rational homotopy type*, Publ. Math. IHES (to appear - eventually).

[S1] J. Stasheff, *An almost groupoid structure for the space of (open) strings and implications for string field theory*, Advances in Homotopy Theory, LMS Lect. Note Series 139, 1989, pp. 165–172.

[S2] J. Stasheff, *Drinfel'd's quasi-Hopf algebras and beyond*, Proc. Conference on Deformation Theory and Quantization with Applications to Physics, Amherst, June 1990, AMS, to appear.

[S3] J.D. Stasheff, *H-spaces from a homotopy point of view*, LNM 161, Springer-Verlag, 1970.

[S4] J.D. Stasheff, *On the homotopy associativity of H-spaces I*, Trans. AMS **108** (1963), 275–292.

[S5] J.D. Stasheff, *On the homotopy associativity of H-spaces II*, Trans. AMS **108** (1963), 293–312.

[S6] J. Stasheff, *The intrinsic bracket on the deformation complex of an associative algebra*, JPAA volume in honour of Alex Heller.

[Su] D.Sullivan, *Infinitesimal computations in topology*, Publ. Math. IHES **47** (1977), 269–331.

[Wies] H.-W. Wiesbrock, *A note on the construction of the C^*-Algebra of bosonic strings*, preprint, FUB-HEP-90/.

H.-W. Wiesbrock, *The quantum algebra of bosonic strings*, preprint, FUB-HEP/89-9.

H.-W. Wiesbrock, *The mathematics of the string algebra*, preprint, DESY 90-003.

[W] E. Witten, *Non-commutative geometry and string field theory*, Nuclear Physics B **268** (1986), 253–294.

E. Witten, *Interacting field theory of open strings*, Nuclear Physics B **276** (1986), 291-324.

University of North Carolina at Chapel Hill, Mathematics Department, CB# 3250, Phillips Hall Chapel Hill, NC 27599-3250, USA

QUANTUM DEFORMATION OF THE FLAG VARIETY

EARL J. TAFT

Department of Mathematics at New Brunswick, Rutgers University

In [6], the author and J. Towber proposed a quantum deformation of flag schemes and Grassman schemes for $GL(n)$. In this note, we illustrate and clarify the nature of our construction.

We briefly recall the construction of [6]. Let k be a commutative ring with $1, n$ a positive integer. Let $\Lambda(n)$ be the exterior algebra over k with basis $1, f(i_1, ..., i_r) = x_{i_1} \wedge x_{i_2} \wedge ... \wedge x_{i_r}$ for $1 \leq r \leq n, 1 \leq i_1 < i_2 .. < i_r \leq n$. These 2^n elements are taken as commutative generators for the commutative algebra $\Lambda^+(n)$ with the usual relations on the symbols $\wedge$, together with the Young symmetry relations (using the numbering of [6]):

$$\sum_{1 \leq \lambda_1 < ... < \lambda_r \leq t+r} (-1)^{\lambda_1 + ... + \lambda_r} f(i_1, ..., \hat{i}_{\lambda_1}, ..., \hat{i}_{\lambda_r}, ..., i_{t+r}) \times f(i_{\lambda_1}, ..., i_{\lambda_r}, j_1, ..., j_{s-r}) = 0 \tag{1.2c}$$

whenever $1 \leq r \leq s \leq t \leq n$, and for all choices $1 \leq i_1 < ... < i_{t+r} \leq n$ and $1 \leq j_1 < ... < j_{s-r} \leq n$. For a fixed $t, 1 \leq t \leq n$, the subalgebra of $\Lambda^+(n)$ generated by all $f(i_1, ..., i_t)$ is the Grassman algebra $Gr^t(n)$. $\Lambda^+(n)$ is graded by the "shapes" $\alpha = (a_1, ..., a_s)$, where $n \geq a_1 \geq ... \geq a_s > 0$ for any non-negative integer s.

Let q be generic over k, i.e. we work over the ring $k[q, q^{-1}]$. Our deformation $\Lambda_q^+(n)$ of $\Lambda^+(n)$ is the non-commutative $k[q, q^{-1}]$-algebra, also generated by $x_{i_1} \wedge ... \wedge x_{i_r} = f_q(i_1, ..., i_r)$, where now $f_q(i_1, ..., i_r) = 0$ if two i's coincide, and $f_q(i_{\sigma_1}, ..., i_{\sigma_r}) = (-q)^{-I(\sigma)} f_q(i_1, ..., i_r)$ for σ in S_r, where $I(\sigma)$ is the number of inversions in σ. The relations (1.2c) are here replaced by similar relations (3.2c) where sign $(-1)^{\lambda_1 + \cdots + \lambda_r}$ is replaced by $(-q)^{-I(i_1, ..., \hat{i}_{\lambda_1}, ..., \hat{i}_{\lambda_r}, ..., i_{t+r}, i_{\lambda_1}, ..., i_{\lambda_r})}$. There are also additional "strainghtening" relations (3.2d) for $1 \leq r < s \leq n, 1 \leq i_1 < ... < i_s \leq n$ and $1 \leq j_1 < ... < j_r \leq n$:

$$f_q(j_1, ..., j_r) f_q(i_1, ..., i_s) = \sum_{1 \leq \lambda_1 < ... < \lambda_s \leq s} (-q)^{I(\lambda_1, ..., \lambda_r, 1, ..., \hat{\lambda}_1, ..., \hat{\lambda}_r, ..., s)} \times f_q(j_1, ..., j_r, i_1, .., \hat{i}_{\lambda_1}, ..., \hat{i}_{\lambda_r}, .., i_s) \cdot f_q(i_{\lambda_1}, ..., i_{\lambda_r}). \tag{3.2d}$$

$\Lambda_q^+(n)$ is also graded by the shapes α. We recall that $\Lambda_q^+(n)$ is a left comodule algebra over the quantum matrix algebra $M_q(n)$ (and the quantum groups $GL_q(n)$ and $SL_q(n)$), with the comodule structure map involving certain quantum minors in $M_q(n)$, and that $\Lambda_q^+(n)$ is thus isomorphic (as left comodule algebra) to certain subalgebra of $M_q(n)$

which is also a left coideal. Finally the graded pieces $\Lambda_q^\alpha(n)$ are subcomodules, which are free finite rank $k[q,q^{-1}]$ - modules with standard basis of straightened monomials corresponding to the shape α (i.e. the Poincaré-Birkhoff-Witt property holds).

As an example of (1.2c), let $n=3, r=s=1, t=2, i_1=1, i_2=2, i_3=3$. Then $(x_1\wedge x_2)x_3-(x_1\wedge x_3)x_2+(x_2\wedge x_3)x_1=0$ in the commutative algebra $\Lambda^+(3)$. The same data for (3.2c) give the relation $(x_1\wedge x_2)x_3+(-q)^{-1}(x_1\wedge x_3)x_2+(-q)^{-2}(x_2\wedge x_3)x_1=0$ in the non-commutative algebra $\Lambda_q^+(3)$. As an example of (3.2d), let $n=3, r=1, s=2, i_1=1, i_2=2, j_1=3$. Then we get the straightening relation $x_3(x_1\wedge x_2)=-q^{-1}(x_2\wedge x_3)x_1+(x_1\wedge x_3)x_2$ in $\Lambda_q^+(2)$. However, we also note that for $n=2, r=s=t=1$, (3.2c) gives $x_2x_1=qx_1x_2$ in $\Lambda_q^+(2)$, which is to be regarded as a straightening relation in the quantum Grassman algebra $Gr_q^1(2)$.

We will comment later about the fact that our relations (3.2c) are deformations of the Plucker relations. We illustrate this now with $Gr^2(4)$ for $k=\mathbb{C}$, i.e., 2 planes in $\mathbb{C}^4$. This is generated by $f(i,j)=x_i\wedge x_j, 1\le i<j\le 4$, with Plucker relation (commutative) $f(12)f(34)-f(13)f(24)+f(14)f(23)=0$. The p-th graded piece ($\alpha = 4>2\ge 2\ge 2\ge \ldots \ge 2$, with p 2's) is spanned by $f_T=f(i_1j_1)\ldots f(i_pj_p), 1\le i_k\le j_k\le 4, i_1\le i_2\le \ldots\le i_p, j_1\le j_2\le\ldots\le j_p$, where T corresponds to the row-standard Young tableau

$$T=\begin{array}{|cc|}\hline i_1 & j_1\\ \hline i_2 & j_2\\ \hline \cdot & \cdot\\ \cdot & \cdot\\ \cdot & \cdot\\ \hline i_p & j_p\\ \hline\end{array}$$

Let us consider 5 cases of (3.2c) with $n=4, \quad s=t=2$.

	r	$i_1,\ldots,i_{t+r}$	$j_1,\ldots,j_{s-r}$
1.	1	1,2,3	4
2.	1	1,2,4	3
3.	1	1,3,4	2
4.	1	2,3,4	1
5.	2	1,2,3,4	$\emptyset$

These yield five relations $T_1,T_2,T_3,T_4,T_5=0$. (We write f_{ij} for $f_q(i,j)$).

$$T_1=f_{12}f_{34}-q^{-1}f_{13}f_{24}+q^{-2}f_{23}f_{14}$$

$$T_2^*=qT_2=f_{12}f_{34}+f_{14}f_{23}-q^{-1}f_{24}f_{13}$$

$$T_3^*=f_{13}f_{24}-q^{-1}f_{14}f_{23}-q^{-1}f_{34}f_{12}$$

$$T_4^*=f_{23}f_{14}-q^{-1}f_{24}f_{13}+q^{-2}f_{34}f_{12}$$

$$T_5=f_{12}f_{34}-q^{-1}f_{13}f_{24}+q^{-2}f_{14}f_{23}+q^{-2}f_{23}f_{14}-q^{-3}f_{24}f_{13}+q^{-4}f_{34}f_{12}$$

We indicate how the six products $f_{ij}\quad f_{kl},\quad i<j, k<l$, with i,j,k,l mutually distinct, can be written in terms of the standard tableaux $\begin{array}{|cc|}\hline 1 & 2\\ \hline 3 & 4\\ \hline\end{array}$ and $\begin{array}{|cc|}\hline 1 & 3\\ \hline 2 & 4\\ \hline\end{array}$

corresponding to $G = f_{12}f_{34}$ and $H = f_{13}f_{24}$ respectively. First note that $0 = q^2(T_5 - T_1) = f_{14}f_{23} - q^{-1}f_{24}f_{13} + q^{-2}f_{34}f_{12}$. Comparison with T_4^* yields $f_{14}f_{23} = f_{23}f_{14}$. Then $T_1 = 0$ yields $f_{14}f_{23} = f_{23}f_{14} = q^{-2}G + qH$. Then $T_2^* = 0$ gives $f_{24}f_{13} = (q - q^3)G + q^2H$. Note that when $q = 1$, this becomes $f_{24}f_{13} = f_{13}f_{24}$. Finally, T_3^* gives $f_{34}f_{12} = q^2G = q^2f_{12}f_{34}$. We also note that in the first four rows of the preceding table, different choices of j_1 give relations like $f_{13}f_{12} = qf_{12}f_{13}$. So it is clear that at $q = 1$, we recover the commutative algebra $GR^2(4)$.

We also give an example of (3.2d) in $\Lambda_q^+(4)$. Take $r = 1, s = 3, i_1 = 2, i_2 = 3, i_3 = 4$ and $j_1 = 1$. Then $x_1(x_2 \wedge x_3 \wedge x_4) = (x_1 \wedge x_3 \wedge x_4)x_2 + (-q)^{-1}(x_1 \wedge x_2 \wedge x_4)x_3 + (-q)^{-2}(x_1 \wedge x_2 \wedge x_3)x_4$.

We next comment on the commutative relations being deformed. The relations (1.2c) are Plucker relations. They are of the form

$$T(K';I,J) = \sum_{\substack{K \subseteq I, L \subseteq K' \\ \#K = \#K'}} (-1)^{\#K} \operatorname{In}\{(K,L) \circ f(I)f(J)\}$$

for all $I = \{i_1, ..., i_t\}, 1 \leq i_1 < i_2 ... < i_t \leq n$, (all sets are contained in $\{1,2,...,n\}$ and written in increasing order),$K' \subseteq J, \#J \leq \#I, K' \neq \emptyset$, where $f(I) = x_{i_1} \wedge ... \wedge x_{i_t}$, and $Int(K,K') \circ f(I)f(J)$ interchanges k_1 with k_1', etc. For example, $Int(\{2,6\},\{1,3\}) \circ (x_2 \wedge x_4 \wedge x_6)(x_1 \wedge x_3) = (x_1 \wedge x_4 \wedge x_3)(x_2 \wedge x_6) = -(x_1 \wedge x_3 \wedge x_4)(x_2 \wedge x_6)$. In the earlier papers of Towber [7,8] in which relations on the flag manifold were discussed, a different variation $Y(K';I,J)$ was used, where $Y(K';I,J) = -f(I)f(J) + \sum_{\substack{K \subseteq I \\ \#K = \#K'}} Int(K,K') \circ f(I)f(J)$ (here K' may be empty, in which case $Y(\emptyset; I,J) = 0$). The two sets of relations are equivalent, since each T is an integral linear combination of the Y's, and vice-versa. Specifically (with $K' \neq \emptyset$), $T(K';I,J) = \sum_{V \subseteq K'} (-1)^{\#V} Y(V;I,J)$ and $Y(K';I'J) = \sum_{\substack{V \subseteq K \\ V \neq \emptyset}} (-1)^{\#V} T(V;I,J)$. Our relations (3.2c) for $\Lambda_q^+(n)$ are quantum deformations $T_q(K';I,J)$ of $T(K';I,J)$, rather than of $Y(K';I,J)$, which we did not see how to deform directly. In our proof of the standard basis theorem for the graded pieces $\Lambda_q^\alpha(n)$, the spanning argument is analogous to the one given in [7] for the $q = 1$ case using the relations $Y(K';I,J)$. Since these relations are equivalent to (1.2c), the straightened monomials span $\Lambda_q^\alpha(n)$. The linear indepedence argument is given by supposing a dependence relation (over $k[q,q^{-1}]$) and obtaining a contradiction by letting $q \to 1$.

We also note the existence of the Garnir relations, or full Plucker relations $0 = G(i_1, ..., i_s; j_1, ... j_t; l; k_1, ..., k_u) = \sum'_{\pi \in S_t} (sign\, \pi)(x_{i_1} \wedge ... \wedge x_{i_s} \wedge x_{j_{\pi 1}} \wedge ... \wedge x_{j_{\pi l}}) \times (x_{j_{\pi(l+1)}} \wedge ... \wedge x_{j_{\pi t}} \wedge x_{k_1} \wedge ... \wedge x_{k_u})$ with $t \geq s + l \geq u + (t - l)$, all i, j, k's between 1 and n where $\sum'$ is over all π in S_t such that $\pi 1 < ... < \pi l$ and $\pi(l+1) < ... < \pi t$. These include (1.2c) ($s = 0$). Conversely, let $G_q(i_1, ..., i_s; j_1, ... j_t; l; k_1, ... k_u)$ be as in G, but with sign π replaced by $(-q)^{-l(\pi)}$, where we may as well assume that $1 \leq i_1 < ... < i_s < j_1 < ... < j_t \leq n$ and $1 \leq i_1 < ... < i_s < k_1 < ... < k_u \leq n$.

Then we note that G_q is a $k[q,q^{-1}]$ - linear combination of the T_q of relation (3.2c), and for $q = 1$, each G is an integral linear combination of the relations T of (1.2c). This is clear if $s = 0$, and follows by induction on $s \geq 1$ by noting that $G_q(i_1, ..., i_s; j_1, ... j_t; l; k_1, ... k_u) = G_q(i_1, ... i_{s-1}; i_s, j_1, ..., j_t; l+1; k_1, ..., k_u) - (-q)^{-(l+1)} \times G_q(i_1, ..., i_{s-1}; j_1, ..., j_t; l+1; i_s, k_1, ..., k_u)$.

Finally, we note that since the submission of our preprint [6] in January 1989, similar constructions using relations T_q or G_q were given in [1] and [4]. In [4], the standard basis theorem for $\Lambda_q^+(n)$ is proved by use of the Bergman diamond lemma. More general quantum constructions for flag varieties and Schubert varieties for some of the classical groups (or other Lie algebras) have been given in [2], [3] and [5]. Our construction is for type A_n. [2], for example, treats the types A_n, B_n, C_n, D_n, G_2 and E_6.

References

[1] M.Hashimoto and T.Hayashi, *Quantum multilinear algebra*, Preprint (1989).

[2] V.Lakshmibai and N.Reshetikhin, *Quantum deformations of flag and Schubert schemes*, Preprint (1990).

[3] S.Levendorskii and Ya.Soibelman, *Representation theory for the algebras of functions on compact quantum groups and some applications*, Preprint (1990).

[4] M.Noumi, H.Yamada and K.Mimachi, *Finite dimensional representations of the quantum group $GL_q(n, \mathbb{C})$ and the zonal spherical functions on $U_q(n-1) \setminus U_q(n)$*, Preprint (1990).

[5] Ya.Soibelman, *Representations of $\mathbb{C}_h(K)$ and Schubert cells*, Rostov University preprint (1989).

[6] E.Taft and J.Towber, *Quantum deformation of flag schemes and Grassman schemes I. A q-deformation of the shape algebra for $GL(n)$*, Preprint (1989); J. Algebra (to appear).

[7] J.Towber, *Two new functions from modules to algebras*, J. Algebra **47** (1977), 80–104.

[8] J.Towber, *Young Symmetry, the flag manifold, and representations of $GL(n)$*, J. Algebra **49** (1979), 414–462.

Department of Mathematics at New Brunswick, Rutgers University, New Brunswick, New Jersey, 08903, USA

ZONAL SPHERICAL FUNCTIONS ON QUANTUM SYMMETRIC SPACES AND MACDONALD'S SYMMETRIC POLYNOMIALS

KIMIO UENO AND TADAYOSHI TAKEBAYASHI

Department of Mathematics, Waseda University, Tokyo

ABSTRACT. We will study zonal spherical functions on quantum symmetric space $GL_q(N+1)/\ O_q(N+1))$, and will show that those for the case $N=2$ are given by Macdonald's polynomials of the A_2 type. Some $q-$analogues of hypergeometric series associated with the quantum symmetric spaces will be discussed.

Introduction. In [3], I.G.Macdonald proposed a wide class of orthogonal polynomials which are associated with root systems, and include some parameters. Let R be a reduced root system, and P the weight lattice determined by R. Macdonald's polynomials are parametrized by dominant integral weights λ, and are written as $P_\lambda(e^\mu;q,t)$, where e^μ ($\mu \in P$) denotes an element of the group-ring of P, and q,t are parameters. The importance of Macdonald's polynomials is that they connect zonal spherical polynomials on symmetric spaces G/K (G is a real semisimple Lie group, and K is a maximal compact subgroup of G) with zonal spherical functions on G/K (G is a p-adic semisimple Lie group). The former case is realized when $q=1$ (the classical limit), and the latter case is so when $q=0$ (the crystalization).

Macdonald proposed, at the same time in [3], a quite fascinating problem.

"Does there exist a group-like object which gives an account of Macdonald's polynomials from a geometrical viewpoint?"

In this paper, we will make an attempt to solve this problem. The answer is that the group-like object is " quantum groups", however our answer is incomplete in many aspects. There are many problems which should be investigated in the future.

We give some comments on notations used in the text.

Throughout this paper, we assume that $|q|<1$. Macdonald's polynomials in section 3 are of the type A_N, and the symbol $(a;q)_r (0 \le r \le \infty)$ denotes the $q-$shifted factorial: $(a;q)_r = \prod_{k=1}^{r}(1-aq^{k-1}), (a;q)_0 = 1$.

1. Let $\mathcal{U}_q(g\ell(N+1))$ be the quantum enveloping algebra introduced in Jimbo [1]. The Chevalley generators are denoted by e_j, f_j $(1 \le j \le N)$, and $q^{\pm\varepsilon_i/2}$ $(0 \le i \le N)$ with

Key words and phrases. Zonal spherical functions, quantum symmetric spaces, Macdonald's symmetric polynomials.

satisfying the defining relations

$$q^{\epsilon_i/2}e_jq^{-\epsilon_i/2} = \begin{cases} q^{1/2}e_j & i=j-1 \\ q^{-1/2}e_j & i=j \\ e_j & i\neq j-1,j. \end{cases}$$

$$q^{\epsilon_i/2}f_jq^{-\epsilon_i/2} = \begin{cases} q^{-1/2}f_j & i=j-1 \\ q^{1/2}f_j & i=j \\ f_j & i\neq j-1,j. \end{cases} \tag{1.1}$$

$$[e_i, f_j] = \delta_{ij}\frac{k_j^2 - k_j^{-2}}{q-q^{-1}} \qquad (k_j = q^{(\epsilon_{j-1}-\epsilon_j)/2}),$$

$$e_j^2 e_{j\pm1} - (q+q^{-1})e_je_{j\pm1}e_j + e_{j\pm1}e_j^2 = 0,$$

$$f_j^2 f_{j\pm1} - (q+q^{-1})f_jf_{j\pm1}f_j + f_{j\pm1}f_j^2 = 0.$$

This algebra is a non-commutative, non-cocommutative Hopf algebra with a comultiplication defined by

$$\begin{gathered}\Delta(q^{\pm\epsilon_i/2}) = q^{\pm\epsilon_i/2}\otimes q^{\pm\epsilon_i/2}, \\ \Delta(e_j) = e_j\otimes k_j + k_j^{-1}\otimes e_j, \qquad \Delta(f_j) = f_j\otimes k_j + k_j^{-1}\otimes f_j.\end{gathered} \tag{1.2}$$

As a dual Hopf algebra, one can introduce the coordinate ring of the quantum group $GL_q(N+1)$, $A(GL_q(N+1))$. Let x_{ij} $(0\leq i,j\leq N)$ be the coordinate variables of the quantum group, which satisfy, for $i<j$, $k<\ell$, the following relations:

$$\begin{gathered} qx_{ik}x_{i\ell} = x_{i\ell}x_{ik}, \qquad qx_{ik}x_{jk} = x_{jk}x_{ik}, \\ qx_{i\ell}x_{j\ell} = x_{j\ell}x_{i\ell}, \qquad qx_{jk}x_{j\ell} = x_{j\ell}x_{jk}, \\ x_{i\ell}x_{jk} = x_{jk}x_{i\ell}, \\ x_{ik}x_{j\ell} - q^{-1}x_{i\ell}x_{jk} = x_{j\ell}x_{ik} - qx_{jk}x_{i\ell}. \end{gathered} \tag{1.3}$$

We denote by $A(\mathrm{Mat}_q(N+1))$ an associative algebra generated by these coordinate variables with defining relations (1.3). Then the coordinate ring $A(GL_q(N+1))$ is precisely defined to be the localization of $A(\mathrm{Mat}_q(N+1))$ at $(\det_q)^{-1}$, where $\det_q$ denotes the quantum determinant:

$$\det_q = \sum_{w\in W}(-q^{-1})^{\ell(w)}x_{0,w(0)}\cdots x_{N,w(N)} \tag{1.4}$$

which is a central element of $A(\mathrm{Mat}_q(N+1))$. W denotes the symmetric group of degree $N+1$.

The comultiplication of this Hopf algebra is determined by

$$\Delta(x_{ij}) = \sum_k x_{ik}\otimes x_{kj}. \tag{1.5}$$

The duality between $\mathcal{U}_q(\mathfrak{gl}(n+1))$ and $A(GL_q(N+1))$ induces two representations, $dR : \mathcal{U}_q(\mathfrak{gl}(n+1)) \longrightarrow \mathrm{End}_{\mathbf{C}}(A(GL_q(N+1)))$ and $dL : \mathcal{U}_q(\mathfrak{gl}(N+1)) \longrightarrow \mathrm{End}_{\mathbf{C}}(A(GL_q(N+1)))$. The action of the generators in the former case is prescribed by

$$
\begin{gathered}
dR(q^{e_i/2})(x_{k\ell}) = q^{\delta_{i\ell}(1/2)} x_{k\ell}, \\
dR(e_j)(x_{k\ell}) = \delta_{j\ell} x_{k,\ell-1}, \qquad dR(f_j)(x_{k\ell}) = \delta_{j-1,\ell} x_{k,\ell+1},
\end{gathered}
\tag{1.6}
$$

and extends to the whole as an algebra homomorphism with the twisted derivation rule (1.2). The latter one is an anti-algebra homomorphism whose action is given by

$$
\begin{gathered}
dL(q^{e_i/2})(x_{k\ell}) = q^{\delta_{ik}(1/2)} x_{k\ell}, \\
dL(e_j)(x_{k\ell}) = \delta_{j-1,k} x_{k+1,\ell}, \qquad dL(f_j)(x_{k\ell}) = \delta_{j,k} x_{k-1,\ell}.
\end{gathered}
\tag{1.7}
$$

We call these representations the differential representations of $\mathcal{U}_q(\mathfrak{gl}(N+1))$.

2. Set $G = GL_q(N+1)$ and $K = O_q(N+1)$. In what follows, we shall consider zonal spherical functions on the quantum symmetric space G/K. Here we should be careful for the notion G/K, because K is not a quantum subgroup of G.

We can define the coordinate ring $A(G/K)$ only in terms of the differential representations: $A(G/K) = \{\varphi \in A(G); dR(D_j)(\varphi) = 0 (1 \leq j \leq N)\}$, where we have set

$$
D_j = \sqrt{-1}(q^{-1/2} k_j e_j - q^{1/2} k_j f_j). \tag{2.1}
$$

This ring is a left $A(G)$-comodule. The subring $A(K\backslash G/K)$ of bi-K-invariant elements is defined to be

$$
A(K\backslash G/K) = \{\varphi \in A(G); dR(D_j)(\varphi) = dL(D_j)(\varphi) = 0 (1 \leq j \leq N)\} \tag{2.2}
$$

Now we determine the class 1 representations with respect to K.

Let $(V(q^{\Lambda/2}), \pi)$ be an irreducible $\mathcal{U}_q(\mathfrak{gl}(N+1))$-representation of finite dimensions with highest weight $q^{\Lambda/2} = (q^{\lambda_0/2}, \ldots, q^{\lambda_N/2})$, where λ_j are integers, $\lambda_0 \geq \lambda_1 \geq \cdots \geq \lambda_N$(see [1], [2]).Define

$$
V(q^{\Lambda/2})_K = \{v \in V(q^{\Lambda/2}); \pi(D_j)v = 0, 1 \leq j \leq N\} \tag{2.3}
$$

Using the Gelfand-Tsetlin basis [1] [2], we obtain

Theorem 1.

(i) $\dim V(q^{\Lambda/2}) \leq 1$.
(ii) *If* $\dim V(q^{\Lambda/2}) = 1$, *then* $\lambda_{j-1} - \lambda_j$ $(1 \leq j \leq N)$ *are even integers.*

From this theorem, we can reveal the structure of the ring $A(K\backslash G/K)$.

Theorem 2. *The ring* $A(K\backslash G/K)$ *is commutative:*

$$
A(K\backslash G/K) = \mathbf{C}[z_1, \ldots, z_N, \det_q, (\det_q)^{-1}],
$$

where

$$z_m = \sum_{\#I=\#J=m} q^{|J|-|I|}(\xi_I^J)^2. \tag{2.4}$$

The above summation ranges over all the increasing sequences I and $J = (j_1, \dots, j_m)$ of length m, and $|J| = j_1 + \cdots + j_m$. ξ_I^J stands for a quantum minor determinant with label (I, J).

Let $A(H)$ be the coordinate ring of the maximal diagonal subgroup $H : A(H) = \mathbb{C}[t_0^{\pm 1}, \dots, t_N^{\pm 1}]$, which is a commutative Hopf algebra with a comultiplication $\Delta(t_i^{\pm 1}) = t_i^{\pm 1} \otimes t_i^{\pm 1}$. Let P denote the projection of $A(G)$ onto $A(H)$ defined by $P(x_{ij}) = \delta_{ij} t_i$, which is a Hopf algebra homomorphism. P induces the following isomorphism of rings:

$$P : \mathbb{C}[z_1, \dots, z_N, z_{N+1}, z_{N+1}^{-1}] \xrightarrow{\sim} \mathbb{C}[\tau_0^{\pm 1}, \dots, \tau_N^{\pm 1}]^W, \tag{2.5}$$

where $z_{N+1} = (\det_q)^2$ and $\tau_i = t_i^2$. The correspondence is given by $P(z_j) = s_j(\tau_0, \dots, \tau_N)$ (the fundamental symmetric polynomial of order j).

Generators of the center of $\mathcal{U}_q(\mathfrak{gl}(N+1))$ was found by Jimbo [1]: The generating functional expression for those is

$$Z(q^\mu) = \sum_{w \in W} (-q)^{\ell(w)} E_{N,w(N)}(q^{\mu-N}) \dots E_{0,w(0)}(q^\mu) \tag{2.6}$$

where μ is a complex parameter, and $E_{ij}(q^\mu)$ is defined to be

$$E_{ij}(q^\mu) = \begin{cases} q^{\mp(\mu + (\epsilon_i + \epsilon_j - 1)/2)} E_{ij} & (i \lessgtr j) \\ \dfrac{q^{e_i + \mu} - q^{-e_i - \mu}}{q - q^{-1}} & (i = j) \end{cases} \tag{2.7}$$

and E_{ij} is given recursively by $E_{i-1,i} = e_i$, $E_{i,i-1} = f_i$ and $E_{ij} = E_{ik}E_{kj} - q^{\pm 1} E_{kj} E_{ik}$ $(i \lessgtr k \lessgtr j)$.

Taking the isomorphism (2.5) into account, let us set

$$S(q^\mu) = P \circ dR(Z(q^\mu)) \circ P^{-1} \in \operatorname{End}_{\mathbb{C}}(\mathbb{C}[\tau_0^{\pm 1}, \dots, \tau_N^{\pm 1}]^W). \tag{2.8}$$

Let T_{ϵ_i} be a q-shift operators defined by

$$T_{\epsilon_i} \varphi(\tau_0, \dots, \tau_i, \dots, \tau_N) = \varphi(\tau_0, \dots, q^2 \tau_i, \dots, \tau_N), \tag{2.9}$$

and $\Delta_+(\tau_0, \dots, \tau_N) = \prod_{i<j} (\tau_i - \tau_j)$, the Vandermonde determinant.

Direct computation shows the following assertion:

Proposition 3. *For $N = 2$, the operator $S(q^\mu)$ is written in the form*

$$S(q^\mu) = \frac{1}{(q - q^{-1})^3} (S_3 \cdot q^{3\mu - 3} - S_2 \cdot q^{\mu - 2} + S_1 \cdot q^{-\mu + 2} - S_0 \cdot q^{-3\mu + 3}) \tag{2.10}$$

where the coefficients in the above are q-difference operators given by

$$S_0 = T_{-\varepsilon_0-\varepsilon_1-\varepsilon_2}, \qquad S_3 = T_{\varepsilon_0+\varepsilon_1+\varepsilon_2},$$

$$S_1 = \frac{\Delta_+(q\tau_0, q^{-1}\tau_1, q^{-1}\tau_2)}{\Delta_+(\tau_0,\tau_1,\tau_2)} T_{\varepsilon_0-\varepsilon_1-\varepsilon_2} + \frac{\Delta_+(q^{-1}\tau_0, q\tau_1, q^{-1}\tau_2)}{\Delta_+(\tau_0,\tau_1,\tau_2)} \times$$
$$\times T_{-\varepsilon_0+\varepsilon_1-\varepsilon_2} + \frac{\Delta_+(q^{-1}\tau_0, q^{-1}\tau_1, q\tau_2)}{\Delta_+(\tau_0,\tau_1,\tau_2)} T_{-\varepsilon_0-\varepsilon_1+\varepsilon_2} \tag{2.11}$$

$$S_2 = \frac{\Delta_+(q\tau_0, q\tau_1, q^{-1}\tau_2)}{\Delta_+(\tau_0,\tau_1,\tau_2)} T_{\varepsilon_0+\varepsilon_1-\varepsilon_2} + \frac{\Delta_+(q\tau_0, q^{-1}\tau_1, q\tau_2)}{\Delta_+(\tau_0,\tau_1,\tau_2)} T_{\varepsilon_0-\varepsilon_1+\varepsilon_2} +$$
$$+ \frac{\Delta_+(q^{-1}\tau_0, q\tau_1, q\tau_2)}{\Delta_+(\tau_0,\tau_1,\tau_2)} T_{-\varepsilon_0+\varepsilon_1+\varepsilon_2}.$$

Zonal spherical functions on $GL_q(3)/O_q(3)$ is by definition, simultaneous eigenfunctions in $\mathbb{C}[\tau_0^{\pm 1}, \tau_1^{\pm 1}, \tau_2^{\pm 1}]^W$ for the operators S_m ($m = 0, 1, 2, 3$).

These operators S_m immediately generalize as follows.
Let k be a complex parameter. Set

$$S_m(k) = \sum_{+\dots m} \frac{\Delta_+(q^{\pm 2k}\tau_0, \dots, q^{\pm 2k}\tau_N)}{\Delta_+} T_{\pm\varepsilon_0 \pm \dots \pm \varepsilon_N} \tag{2.12}$$

where the summation ranges over the set of sequences consisting of (+) and (−) in which (+) appears m times.

Let us propose conjectures or problems concerning these operators.

Problem 4.

(i) *Show that, for arbitrary N, the coefficients of $S(q^\mu)$ is given by $S_m(1/2)$ ($0 \le m \le N$).*

(ii) *Give the operators $S_m(k)$ a geometrical interpretation as in Proposition 3.*

3. In this section, we shall consider more on the operators $S_m(k)$. Apparently, these operators are W-invariant, and provide endomorphism of $\mathbb{C}[\tau_0^{\pm 1}, \dots, \tau_N^{\pm 1}]^W$. Moreover we see that these operators are triangular with respect to the basis $\{m_\lambda\}_{\lambda;partitions}$ of $\mathbb{C}[\tau_0^{\pm 1}, \dots, \tau_N^{\pm 1}]^W$, where m_λ is a monomial symmetric polynomial associated with a partition λ (Macdonald [4]); $m_\lambda = \sum_{\mu \in W\cdot\lambda} \tau^\mu$. For a fixed k, one can introduce a positive definite hermitian inner product $\langle\cdot,\cdot\rangle_k$ on $\mathbb{C}[\tau_0^{\pm 1}, \dots, \tau_N^{\pm 1}]^W$:

$$\langle\varphi, \psi\rangle_k = \frac{1}{(N+1)!} \operatorname{Res}_{\tau=0}\{\varphi(\tau)\psi^*(\tau) w(\tau,k) \frac{d\tau_0 \wedge \cdots \wedge d\tau_N}{\tau_0 \dots \tau_N}\}, \tag{3.1}$$

where the weight function $w(\tau, k)$ is

$$w(\tau,k) = \prod_{i \ne j} \frac{(\tau_i/\tau_j; q^4)_\infty}{(q^{4k}\tau_i/\tau_j; q^4)_\infty} \tag{3.2}$$

and for $\psi(\tau) = \sum_\alpha c_\alpha \tau^\alpha$, we have set $\psi^*(\tau) = \sum_\alpha \bar{c}_\alpha \tau^{-\alpha}$. (As for this inner product, refer to Macdonald [4]). Then we see that the operators $S_m(k)$ are hermitian under this inner product.

Thus we have

Proposition 5. *The operators $S_m(k)$ mutually commute, and simultaneous eigenfunctions in $\mathbb{C}[\tau_0, \ldots, \tau_N]^W$ of these operators are given by the Macdonald polynomials $\{P_\lambda(\tau; q^4, q^{4k})\}_{\lambda;partitions}$.*

This proposition and Proposition 3 leads us to

Theorem 6. *Zonal spherical polynomials on the quantum symmetric space $GL_q(3)/O_q(3)$ are given by the Macdonald polynomials $(P_\lambda(\tau_0, \tau_1, \tau_2; q^4, q^2))_{\lambda;partitions}$.*

Next we discuss a q-analogue of the Lauricella function (Jackson [5]):

$$\Phi_D\left(\begin{matrix} q^\alpha, q^{\beta_1}, \ldots, q^{\beta_k} \\ q^\gamma \end{matrix}; q, x_1, \ldots, x_k\right) = \\ = \sum_{m_1, \ldots, m_k = 0}^{\infty} \frac{(q^\alpha; q)_{m_1+\cdots+m_k}(q^{\beta_1}; q)_{m_1} \cdots (q^{\beta_k}; q)_{m_k}}{(q^\gamma; q)_{m_1+\cdots+m_k}(q; q)_{m_1} \cdots (q, q)_{m_k}} x_1^{m_1} \ldots x_k^{m_k}. \tag{3.3}$$

This series is absolutely convergent for $|x_j| < 1$ $(1 \leq j \leq k)$, and has an integral representation of the Pochhammer type for $\mathrm{Re}(\alpha) > 0$;

$$\Phi_D\left(\begin{matrix} q^\alpha, q^{\beta_1}, \ldots, q^{\beta_k} \\ q^\gamma \end{matrix}; q, x_1, \ldots, x_k\right) \\ = \frac{\Gamma_q(\gamma)}{\Gamma_q(\alpha)\Gamma_q(\gamma-\alpha)} \int_0^1 u^{\alpha-1} \frac{(qu; q)_\infty \prod_{j=1}^k (q^{\beta_j} x_j u; q)_\infty}{(q^{\gamma-\alpha}u; q)_\infty \prod_{j=1}^k (x_j u; q)_\infty} d_q u \tag{3.4}$$

where $\int \cdot d_q u$ stands for the Jackson integral, and $\Gamma q(x)$ is the q-gamma function [6];

$$\Gamma_q(x) = \frac{(q; q)_\infty}{(q^x; q)_\infty}(1-q)^{1-x}$$

Using this integral representation, we can show the following.

Proposition 7. *Let $\zeta_j = \tau_{j-1}/\tau_N$ $(1 \leq j \leq N)$ and ν be a complex parameter. Then*

$$\Psi(\zeta, \nu) = (\zeta_1 \ldots \zeta_N)^\nu \Phi_D\left(\begin{matrix} q^{4(N+1)\nu}, q^{4k}, \ldots, q^{4k} \\ q^{4(N+1)\nu+4-4k} \end{matrix}; q^4, q^{4-4k}\zeta_1, \ldots, q^{4-4k}\zeta_N\right) \tag{3.5}$$

is a simultaneous eigenfunction of the operators $S_m(k)$.

References

[1] Jimbo, M., *A q-analogue of $U_q(\mathfrak{gl}(N+1))$, Hecke algebra, and the Yang-Baxter equation.*, Lett. Math.Phys. **11** (1986), 247.

[2] Ueno, K., Takebayashi, T. and Shibukawa, Y., *Construction of Gelfand-Tsetlin basis for $U_q(\mathfrak{gl}(n+1))$-modules*, Publ.of RIMS **26** no. 1 (1990), 667.

[3] Macdonald, I.G., *Orthogonal polynomials associated with root systems*, Preprint.

[4] Macdonald, I.G., *Symmetric functions and Hall polynomials*, Oxford University Press, 1979.

[5] Jackson, F.H., *On basic double hypergeometric functions.*, Quart.J.Math. **13** (1942), 69.

[6] Gasper, G. and Rahman, M., *Basic Hypergeometric Series*, Encyclopedia of Mathematics and its Applications **34** (1990), Cambridge University Press, Cambridge.

DEPARTMENT OF MATHEMATICS, WASEDA UNIVERSITY, 3-4-1 OHKUBO SHIJUKU-KU, TOKYO 169, JAPAN

HIDDEN QUANTUM GROUPS INSIDE KAC–MOODY ALGEBRAS

A. ALEKSEEV, L. FADDEEV AND M. SEMENOV-TIAN-SHANSKY

Leningrad Branch of Steklov Mathematical Institute

1. Introduction

Fascinating links between Conformal Field Theory and Quantum groups discovered recently suggest that Quantum groups also have a direct bearing on the representation theory of Kac-Moody algebras. It is the purpose of the present note to trace down this hidden quantum group symmetry in the framework of Kac-Moody algebras. Our main result is that the monodromy of quantum Kac-Moody current when properly regularized satisfies the commutation relations of the quantized universal enveloping algebra $U_q(\mathfrak{g})$ with q related to the central charge k via $q = \exp(\frac{\pi i}{k+n})$. The regularized definition of the monodromy is based in its turn on a lattice version of the current algebra which we also describe in this paper. This algebra associated with a periodic 1-dimensional lattice is already quantum (i.e. incorporates parameter q; in fact, it is defined for any q, not only for roots of unity) and also takes into account the central charge. It may be regarded as a nontrivial deformation of $U_q(\mathfrak{g})^{\otimes N}$ (The very existence of such deformations is a typically quantum phenomenon. Indeed, it is well known that classical semi-simple Lie groups and Lie algebras are rigid. By contrast, quantum universal enveloping algebras admit certain deformations which may be regarded as finite-dimensional counterparts of central extensions of current algebras). Our first key result is the monodromy theorem for this lattice algebra which asserts that the monodromy satisfies the commutation relations of $U_q(\mathfrak{g})$. In the scaling limit the lattice current algebra becomes the ordinary classical current algebra with fixed central charge, while the monodromy remains quantum. This result has several important corollaries. First , we are able to describe the regularized Casimir operators for the Kac-Moody algebra and to relate them to the quantum Casimir operators for $U_q(\mathfrak{g})$. Second, we define for the lattice algebra the structure of a $U_q(\mathfrak{g})$ - comodule. This structure reproduces the well known tensor product properties of $U_q(\mathfrak{g})$. Remarkably, it essentially coincides with the fusion rules in Conformal Field Theory.

The present note represents a part of the research program now in progress. Our approach combines ideas from different independent sources. We should mention the papers [1] and [2] which lead to the definition of lattice current algebras and the papers [3-7] on quantum exchange algebras. Some aspects of our results were reported in [8]. While for the lattice current algebras our results are completely rigorous, the details of our construction in the scaling limit are still to be worked out. In this note we content ourselves with the physical level of rigour. Complete proofs and the mathematical background of the construction of lattice current algebras will be given in a subsequent paper.

2. Current algebras: a brief reminder

Let $\mathfrak{g}$ be a finite dimensional simple Lie algebra, $\tilde{\mathfrak{g}}$ the associated current algebra on the circle. For simplicity, we may assume that $\mathfrak{g} = sl(2,\mathbb{C})$. By definition, the current algebra with central charge k is a Lie algebra with generators $J^a(x)$, $0 \le x \le 2\pi$, associated with an orthogonal basis $\{\sigma^a\}$ in $\mathfrak{g}$, satisfying the commutation relations

$$[J^a(x), J^b(y)] = f_c^{ab} J^c(x)\delta(x-y) + \frac{ik}{2\pi}\delta^{ab}\delta'(x-y) \tag{1}$$

where f_c^{ab} are the structure constants of $\mathfrak{g}$ with respect to the basis $\{\sigma^a\}$. To get the more familiar generators one has to perform the formal Fourier transform

$$J^a(x) = \frac{1}{2\pi}\sum_{k\in\mathbb{Z}} e^{ikx} J_k^a.$$

Then (1) is equivalent to

$$[J_m^a, J_n^b] = f_c^{ab} J_{m+n}^c + k\delta^{ab}\delta_{m+n,0}. \tag{1'}$$

We are interested in representations of the algebra (1) which may be integrated to projective unitary representations of the corresponding compact loop group. This means that if the generators $J^a(x)$ correspond to the orthogonal basis of the compact real form of $\mathfrak{g}$, the constant ik should be integer and purely imaginary. We shall see below that passing to the compact real form of the lattice current algebra defined in the next section presents some difficulties, so we have to deal with complex algebras while still keeping the natural condition on the central charge. It is convenient to suppress the Lie algebra indices in (1). Let us put

$$J(x) = J^a(x)\sigma_a, \qquad C = \sigma_a \otimes \sigma_a. \tag{2}$$

Then (1) is rewritten as

$$[J_1(x), J_2(y)] = \frac{1}{2}[C, J_1(x) - J_2(y)]\delta(x-y) + \frac{ik}{2\pi}C\delta'(x-y). \tag{3}$$

where the indices 1,2 refer to two copies of the auxiliary space $\mathfrak{g}$ and $J_i(x)$ is regarded as an element of $\tilde{\mathfrak{g}}\otimes\mathfrak{g}_i \subset \tilde{\mathfrak{g}}\otimes\mathfrak{g}_1\otimes\mathfrak{g}_2$. The commutation relations (3) are invariant with respect to the action of gauge transformations

$$J(x) \mapsto g(x)J(x)g(x)^{-1} + \frac{ik}{2\pi}\partial_x g\cdot g^{-1}(x) \tag{4}$$

and in particular with respect to global gauge transformations

$$J(x) \mapsto gJ(x)g^{-1} \tag{5}$$

The monodromy of the current $J(x)$ is formally defined by

$$M = P\exp(\frac{2\pi i}{k}\int_0^{2\pi} J(x)dx). \tag{6}$$

Since the commutation relations for $J(x)$ are highly singular, it is not easy to derive the commutation relations for M. One of our main assertions is that when properly regularized M has the commutation relations of the quantum universal enveloping algebra $U_q(\mathfrak{g})$ (with q related to central charge k by $q = \exp(\frac{\pi i}{k+n})$). This may be regarded as a new type of anomaly which leads to the spontaneous breakdown of the global gauge symmetry (5). (There is some evidence that this anomaly is associated with the cohomology class in $H^3(\mathfrak{g})$ which also generates, via transgression, the Schwinger anomaly in the commutation relations (3) - see discussion in Section 5 below).

The usual definition of the quantized universal enveloping algebra $U_q(\mathfrak{g})$ is by means of generators and relations (For $\mathfrak{g} = sl(2)$ it is due to Kulish, Reshetikhin and Sklyanin, while the general case was worked out by Drinfeld and Jimbo)(see [9]). As an algebra , $U_q(sl(2))$ is generated by the elements H, X_+, X_- which satisfy the following relations

$$\begin{aligned} [H, X_\pm] &= \pm 2X_\pm, \\ [X_+, X_-] &= \frac{\sinh(hH)}{\sinh(h)}. \end{aligned} \tag{7}$$

The coalgebra structure which determines the tensor product of the representations is given by

$$\begin{aligned} \Delta H &= H \otimes 1 + 1 \otimes H, \\ \Delta X_\pm &= e^{\frac{hH}{2}} \otimes X_\pm + X_\pm \otimes e^{-\frac{hH}{2}}. \end{aligned} \tag{8}$$

There is a different way to express these commutation relations, due to Faddeev, Reshetikhin and Takhtajan [10]. Let $R = R(q)$ be the quantum R-matrix associated with the standard representation of $U_q(sl(2))$ in $\mathbb{C}^2, R \in \mathrm{End}(\mathbb{C}^2 \otimes \mathbb{C}^2)$:

$$R = \begin{pmatrix} q & 0 & 0 & 0 \\ 0 & 1 & 0 & 0 \\ 0 & q-q^{-1} & 1 & 0 \\ 0 & 0 & 0 & q \end{pmatrix}, q = e^h \tag{9}$$

The R-matrix (9) satisfies the quantum Yang-Baxter identity

$$R_{12}R_{13}R_{23} = R_{23}R_{13}R_{12}. \tag{10}$$

Along with R we shall also need the matrixes

$$R^+ = PRP, \qquad R^- = R^{-1} \tag{11}$$

(Here $P \in \mathrm{End}(\mathbb{C}^2 \otimes \mathbb{C}^2)$ is the permutation matrix defined by $(P(x \otimes y) = y \otimes x)$. Consider the 2×2-matrices L^+, L^- whose coefficients are generators of $U_q(sl(2))$:

$$L^+ = \begin{pmatrix} q^{-H} & -(q-q^{-1})X_+ \\ 0 & q^H \end{pmatrix}, \begin{pmatrix} q^H & 0 \\ -(q-q^{-1})X_- & q^{-H} \end{pmatrix}. \tag{12}$$

Then the following commutation are satisfied

$$L_1^{\pm} L_2^{\pm} R^+ = R^+ L_2^{\pm} L_1^{\pm}, \qquad L_1^+ L_2^- R^+ = R^+ L_2^- L_1^+. \tag{13}$$

It is convenient to combine L^+, L^- into a single matrix $L = L^+(L^-)^{-1}$. The commutation relations for L are

$$(R^-)^{-1} L_1 R^- L_2 = L_2 (R^+)^{-1} L_1 R^+ \tag{14}$$

One can show that a free associative algebra generated by the matrix coefficients of $L = (L_{ij})$ satisfying the relations (14) is isomorphic to $U_q(sl(2))$. The rearrangement of simple commutation relations (7) into the rather complicated form (13,14) may seem artificial. However, there are profound mathematical reasons to prefer this form of commutation relations. In a sense, passing from (7) to (14) or (13) is similar to passing from a Lie algebra to the corresponding Lie group. We shall not go into further discussion of this point (although a complete exposition of the underlying theory is still lacking in the existing literature), but simply notice that the commutation relations for the monodromy (6) have exactly the form (14). Another important fact is that relations (14) serve as a starting point to define the lattice version of the current algebra.

Let us finally discuss the coproduct structure and the tensor products in terms of the generators $L^{\pm}, L$. Assume that there are two representations of $U_q(\mathfrak{g})$ in linear spaces $\mathcal{H}', \mathcal{H}''$. Let L', L'' be the matrices of generators acting in $\mathcal{H}', \mathcal{H}''$ respectively. Put

$$\Delta L = (L^+)' L'' ((L^-)^{-1})'. \tag{15}$$

We may regard ΔL as a 2×2-matrix whose coefficients are linear operators acting in $\mathcal{H}' \otimes \mathcal{H}''$. Then ΔL also satisfies commutation relations (14) and hence defines a representation of $U_q(\mathfrak{g})$ in $\mathcal{H}' \otimes \mathcal{H}''$. One can show that this construction agrees with the standard definition of the tensor product for $U_q(\mathfrak{g})$ based on (8).

3. Lattice version of current algebras

As already explained in the Introduction, our strategy will be as follows. First, we describe the lattice version of commutation relation (3) and state the corresponding monodromy theorem. Then we discuss the continuous limit.

By definition, the lattice current algebra $\mathcal{A}_{LC}$ is a free algebra generated by matrix coefficients of 2×2-matrices $L^i, i = 1, 2, \ldots, N$, satisfying the following relations

$$\begin{aligned} {L_1}^i {L_2}^i &= R^+ {L_2}^i {L_1}^i R^-, \qquad {L_1}^i R^- {L_2}^{i-1} = {L_2}^{i-1} {L_1}^i, \\ {L_1}^i {L_2}^j &= {L_2}^j {L_1}^i \qquad \text{for } |i-j| \geq 2\,. \end{aligned} \tag{16}$$

We assume that $i + N \equiv i$, i.e. that the matrices L^i are associated with a periodic 1-dimensional lattice. To make these relations more transparent let us introduce the R-matrix on the lattice

$$R^{\pm}_{i-j} = \begin{cases} R^{\pm}, & i = j, \\ I, & i \neq j. \end{cases} \tag{17}$$

Commutation relations (14) for $U_q(\mathfrak{g})^{\otimes N}$ may be rewritten in the form

$$(R^-_{i-j})^{-1} L_1{}^i R^-_{i-j} L_2{}^j = L_2{}^j (R^+_{i-j})^{-1} L_1{}^i R^+_{i-j}.$$

The relations (16) have the form

$$(R^-_{i-j})^{-1} L_1{}^i R^-_{i-j+1} L_2{}^j = L_2{}^j (R^+_{i-j-1})^{-1} L_1{}^i R^+_{i-j}\ . \tag{18}$$

In other words, the perturbation of the relations (14) consists in replacing conjugation by $R^\pm$ with lattice gauge transformations (cf. [2] where similar commutation relation are derived for central extensions of quantum Kac-Moody algebras). We omit the mathematical background which serves as a motivation for this definition (Its classical counterpart is already present in [1]). Relations (18) are not ultralocal in that the neighbouring matrices L^i, $L^{i\pm 1}$ do not commute with each other. In [8] a change of variables is described which replaces (18) with local relations. This change of variables is important for the study of representations of $\mathcal{A}_{LC}$. We are planning to discuss these questions in a separate publication. The main property of the lattice current algebra is given by the following theorem.

Theorem. *Put*

$$M = L_N L_{N-1} \ldots L_2 L_1. \tag{19}$$

Then the monodromy matrix M satisfies relations (14).

The commutation relations of M with L_i is given by

$$M_1 R^-_i L_2^i (R^-_i)^{-1} = L_2^i R^+_{i-1} M_1 (R^+_{i-1})^{-1}. \tag{20}$$

One may notice that the definition of M depends on the initial point $k = 1$. In a similar way we may define monodromies

$$M_k = \prod_{i=k}^{\overset{\curvearrowleft}{k+N-1}} L_i \tag{21}$$

which satisfy (14) and also

$$M_1^k R^-_{i-k+1} L_2^i (R^-_{i-k+1})^{-1} = L_2^i (R^+_{i-k})^{-1} M_1^k R^+_{i-k}. \tag{22}$$

Commutation relations (18) admit certain quantum automorphisms. To describe them let us first recall the definition of the quantum group (as opposed to the quantum universal algebra). By definition, the algebra $A = \mathrm{Fun}_q(G)$ is a free algebra with generators $T = (T_{ij})$ and relations

$$T_1 T_2 R^+ = R^+ T_2 T_1. \tag{23}$$

For various classical groups these relations are supplemented by the symmetry relations for T and the condition on its determinant. For instance, for $G = SL(2)$ the latter relation has the form

$$\det{}_q T = T_{11} T_{22} - q T_{12} T_{21} = 1. \tag{24}$$

One can show that as a Hopf algebra A is the dual of $U_q(\mathfrak{g})$. The duality is set up by

$$\langle L^{\pm}, T\rangle = (R^{\pm})^{-1}. \tag{25}$$

Now let $A^{\otimes N} = A \otimes A \otimes \cdots \otimes A$ and put

$$\tilde{L}^i = T^{i+1} L^i (T^i)^{-1} , \tag{26}$$

where we assume that L^i and T^j commute with each other. Then $\tilde{L}^i$ satisfies the same relations (18). (In more mathematical terms, (26) defines the structure of an A^N-comodule on $\mathcal{A}_{LC}$).

Another interesting question is the structure of the Casimir operators of $\mathcal{A}_{LC}$.

Theorem. *For generic q and N odd the center of $\mathcal{A}_{LC}$ is a free algebra generated by 1 and the elements*

$$C_k = \operatorname{tr} q^{2\rho} M^k , \qquad k = 1, 2, \ldots, \operatorname{rank}\ \mathfrak{g} \tag{27}$$

It is well known that C_k generate the center of $U_q(\mathfrak{g})$ (This was proved by [10], cf. also [11], [12]).

In more mathematical terms we may regard the monodromy map as an embedding

$$M^* \ : \ U_q(\mathfrak{g}) \hookrightarrow \mathcal{A}_{LC}.$$

The theorem then asserts that the extension

$$\operatorname{cent} U_q(\mathfrak{g}) \hookrightarrow U_q(\mathfrak{g}) \hookrightarrow \mathcal{A}_{LC}$$

is central. The semiclassical version of this theorem was proved in [1]. Along with the monodromy (19) we shall also need the wave functions

$$u^i = \prod_{1 \le k \le i}^{\curvearrowleft} L_i \tag{28}$$

which satisfy the following exchange relations

$$u_1^i u_2^j = u_2^j u_1^i R^+, \ i > j, \ \ u_1^i u_2^j = u_2^j u_1^i R^-, \ i < j \ . \tag{29}$$

The exchange algebra (29) will be used in Sec.5 to derive the scaling limit. Relations at one point have the form

$$u_1^i u_2^i = R^+ u_2^i u_1^i R^-. \tag{30}$$

4. Lattice current algebra as a comodule

It is possible to define an action of $\mathcal{A}_{LC}$ in tensor products of representation spaces of $\mathcal{A}_{LC}$ with representation spaces of $U_q(\mathfrak{g})$. Let $\mathcal{H}$ be a representation space of $\mathcal{A}_{LC}$ and V a representation space for $U_q(\mathfrak{g})$. We denote by $(L^i)'$ the matrices of generators of $\mathcal{A}_{LC}$ acting in $\mathcal{H}$ and by $(N^{\pm})''$ the matrices of generators of $U_q(\mathfrak{g})$ satisfying relations (9) which act in V.

We define matrices $\tilde{L}^i$ acting in $\mathcal{H} \otimes V$ by

$$\begin{aligned} \tilde{L}^i &= (L^i)', \qquad i \neq 1, N, \\ \tilde{L}^1 &= (L^1)'((N^-)^{-1})'', \qquad \tilde{L}^N = (N^+)''(L^N)' . \end{aligned} \tag{31}$$

It is easy to check that matrices $\tilde{L}^i$ satisfy relations (18). In more mathematical terms, we may regard the mapping $C \ : \ L^i \ \to \ \tilde{L}^i$ as a homomorphism from $\mathcal{A}_{LC}$ into $\mathcal{A}_{LC} \otimes U_q(\mathfrak{g})$. Clearly, it satisfies

$$(C \otimes \mathrm{id})C = (\mathrm{id} \otimes \Delta)C \tag{32}$$

and hence is consistent with the coproduct Δ in $U_q(\mathfrak{g})$. The monodromy $M(L)$ is mapped into $(N^+)''M(L)'((N^-)^{-1})''$ and hence we have a commutative diagram

$$\begin{array}{ccc} \mathcal{A}_{LC} & \xrightarrow{C} & \mathcal{A}_{LC} \otimes U_q(\mathfrak{g}) \\ {\scriptstyle M^*}\uparrow & & \uparrow{\scriptstyle M^*\mathrm{id}} \\ U_q(\mathfrak{g}) & \xrightarrow{\Delta} & U_q(\mathfrak{g}) \otimes U_q(\mathfrak{g}) \end{array}$$

which expresses consistency with the monodromy map.

The comodule structure of $\mathcal{A}_{LC}$ suggest simple tensor properties of its representation. First of all, since $\mathcal{A}_{LC}$ and $U_q(\mathfrak{g})$ have common Casimir operators, it seems plausible that an irreducible representation $\mathcal{H}_s$ of $\mathcal{A}_{LC}$ is uniquely specified by its "spin" s. Moreover, as a $U_q(\mathfrak{g})$-module the space $\mathcal{H}_s$ decomposes into direct sum of "isotypical" components with the same spin s. Thus

$$\mathcal{H}_s = \mathcal{H} \otimes V_s$$

where the monodromy acts trivially in $\mathcal{H}$. We then have

$$\mathcal{H}_s \otimes V_j \simeq (\mathcal{H} \otimes V_s) \otimes V_j \simeq (\mathcal{H} \otimes V_j) \otimes V_s$$

and hence (by the uniqueness assumption)

$$\mathcal{H}_s \otimes V_j \simeq \mathcal{H}_j \otimes V_s \, .$$

Moreover, the decomposition into irreducible representations of $\mathcal{H}_s \otimes V_j$ and its braiding properties are completely determined by the corresponding properties of $U_q(\mathfrak{g})$. Clearly, this picture resembles the fusion algebra in Conformal Field Theory. The tensor algebra still makes sense if q is a root of unity (the latter condition being imposed by the scaling properties of our algebra, as explained in the next section).

5. The scaling limit

Let us now discuss the continuous limit of the lattice current algebra. We shall argue that the commutation relations for currents become those of the ordinary Kac-Moody algebra, while the commutation relations for the monodromy remain quantum. As mentioned in the Introduction our arguments will be not completely rigorous. The starting point is the exchange algebra (29). If we replace the lattice variable i with the continuous variable x we get the following exchange algebra

$$u_1(x)u_2(y) = u_2(y)u_1(x)R(x-y) \tag{33}$$

where

$$R(x-y) = R^+\theta(x-y) + R^-\theta(y-x), \qquad \theta(x) = \begin{cases} 1, & x>0, \\ 0, & x<0. \end{cases} \tag{34}$$

For $x=y$ we get formally

$$u_1(x)u_2(x) = R^+u_2(x)u_1(x)R^-$$

but this relation has of course a dubious status since it does not survive when u_1 and u_2 are smoothed down by averaging with test functions.

The current $J(x)$ is formally the derivative of $u(x)$

$$J(x) = \partial_x u \cdot u^{-1}. \tag{35}$$

However, since the commutator of u's has singularities at coinciding points we must take care to extract them first. Thus we write

$$\frac{1}{\varepsilon}u(x+\varepsilon)u(x)^{-1} = \frac{A}{\varepsilon} + J(x) + O(\varepsilon). \tag{36}$$

By dimension count the singular term A/ε is the only one possible. For symmetry reasons, the matrix A is scalar, $A = aI$. We may also write

$$u'(x+\varepsilon)u(x)^{-1} = \frac{B}{\varepsilon} + \text{regular terms}, \tag{37}$$

where B is a matrix in the tensor square. Again for symmetry reasons, $B_{ijkl} = b\delta_{jk}\delta_{il}$. A simple calculation shows that $a = nb$ where n is the size of our matrices.

The relation of the current J to the lattice variables is formally

$$L^i = P\exp\int_x^{x+\varepsilon} J(y)dy. \tag{38}$$

Thus we expect that in the scaling limit lattice variables are close to identity. A more thorough inspection shows, however, that this is true only for normal ordered exponentials

$$:L^i: = :P\exp\int_x^{x+\varepsilon} J(y)dy: \xrightarrow[\varepsilon\to 0]{} I\,. \tag{39}$$

The lattice variables L^i differ from these normal ordered exponentials by a finite factor

$$L^i = e^a\ :L^i: \tag{40}$$

where a is the same constant as in (36). Below we shall actually compute this constant using the operator expansion. The computation of the commutation relations for currents which follow from the exchange algebra (33) is based on the following general formula for the R-matrices

$$R^{\pm} = F_{12} q^{\pm P} F_{21}^{-1} \tag{41}$$

where P is the permutation matrix, F_{12} an invertible matrix in $\mathbb{C}^n \otimes \mathbb{C}^n$ and $F_{21} = PF_{12}P$. Hence we have for the R-matrix $R(x-y)$:

$$R(x-y) = F_{12} q^{P \,\mathrm{sign}(x-y)} F_{21}^{-1}. \tag{42}$$

Formulae (41),(42) are based on the general philosophy of 'quasi-Hopf algebras', due to Drinfeld [13]. Before proceeding to the actual computation it is worth to comment on the meaning of these formulae and their general implications. As already noted, the regularized commutation relations for the monodromy are the manifestation of symmetry breaking which introduces quantum R-matrices into the basic commutation relations. Now, for higher rank groups the quantization and hence the R-matrices are certainly not unique, and one might wonder why the particular R-matrix which defines the theory is preferred to all the others. According to the general philosophy of Drinfeld [13], the uniqueness of quantization is restored, modulo some natural equivalence relation, in a larger class of algebras for which the coassociativity constraint is replaced by a milder assumption. Formula (41) shows, in particular, that any quantum R-matrix is 'gauge equivalent' to the standard R-matrix

$$R_0 = q^P. \tag{43}$$

(More precisely, this is true for a certain class of R-matrices which define the so-called quasi-triangular Hopf algebras; all physically relevant R-matrices fall within this class).

The permutation operator P, or, more generally, the Casimir operator C, is stable i.e. it does not depend on the particular choice of quantization (One might show that this stability has a cohomological nature: the Casimir operator defines an element in $H^3(\mathfrak{g})$ which may be regarded as the cohomology class of quantization. Incidently, this same cohomology class also determines, via transgression, the cohomology class of the central extension of the current algebra.) Of course, it would be conceptually more simple to work directly with the R-matrix (43), but as explained in [13] this leads to complications with associativity.

Our general conclusion resolving the difficulty referred to above is that one may take any R-matrix in the given 'gauge class' and this will lead to essentially equivalent results (Choosing an R-matrix means roughly to specify some particular regularization of the theory). The independence of the choice of R is manifested in the computation of the Schwinger commutation relation for currents: as we shall see, the answer depends only on P, and the 'gauge matrix' F_{12} is cancelled out.

Let us now proceed to the actual computation. We shall compute the commutation relations between J and u in two ways, which will allow to fix the so far unspecified constant a. We have for $X > x - \varepsilon > y$:

$$u_1(x)u_1(x-\varepsilon)^{-1}u_2(y) = u_1(x)u_2(y)R^{+}(x-y-\varepsilon)^{-1}u_1(x-\varepsilon)^{-1}. \tag{44}$$

Using the expansion (36) and the formula

$$R^+(x-y-\varepsilon)^{-1} = R^+(x-y)^{-1} + 2\varepsilon \ln q \cdot F_{12} P F_{21}^{-1} \delta(x-y) + o(\varepsilon)$$

we get, after simple calculation,

$$[J_1(x), u_2(y)] = 2 \ln q P u_2(y) \delta(x-y). \tag{45}$$

On the other hand, the expansion (37) implies that

$$J_1(x) u_2(y) = -\frac{b}{x-y} P u_2(y) + \text{regular terms}. \tag{46}$$

The operator expansion of the form

$$A(x)B(y) = \frac{1}{x-y} C(y) + \text{regular terms}$$

implies, by the Sokhotski-Plemelj formula, that

$$[A(x), B(y)] = 2\pi i C(y) \delta(x-y).$$

Thus a comparison of (45) and (46) yields

$$b = -\frac{\ln q}{\pi i} \tag{47}$$

and hence

$$a = -\frac{n \ln q}{\pi i}. \tag{48}$$

Now we are ready to calculate the commutator of currents. We have

$$\begin{aligned} &[J_1(x), \frac{1}{\varepsilon} u_2(y+\varepsilon) u_2(y)^{-1}] \\ &= \frac{2\ln q}{\varepsilon} P u_2(y+\varepsilon) u_2(y)^{-1} \delta(x-y-\varepsilon) \\ &- \frac{2\ln q}{\varepsilon} u_2(y+\varepsilon) u_2(y)^{-1} P \delta(x-y) + O(\varepsilon) \\ &= 2\ln q [P, J_2(y)] \delta(x-y) - 2 \ln q (1+a) P \delta'(x-y) + O(\varepsilon). \end{aligned}$$

and hence

$$[J_1(x), J_2(y)] = 2\ln q [P, J_2(y)] \delta(x-y) - 2\ln q (1+a) P \delta'(x-y). \tag{49}$$

Assume that $\ln q$ is purely imaginary,

$$\ln q = \frac{\pi i}{k+n}. \tag{50}$$

Then

$$[J_1(x), J_2(y)] = \frac{2\pi i}{k+n}[P, J_2(y)]\delta(x-y) - 2\frac{2\pi i k}{(k+n)^2}P\delta'(x-y). \tag{51}$$

If we rescale the current by setting

$$\tilde{J}(x) = \frac{i(k+n)}{2\pi}J,$$

we get

$$[\tilde{J}_1(x), \tilde{J}_2(y)] = \frac{1}{2}[P, \tilde{J}_1(x) - \tilde{J}_2(y)]\delta(x-y) + \frac{ik}{2\pi}P\delta'(x-y). \tag{52}$$

Thus we get

$$\tilde{J}(x) = \frac{i(k+n)}{2\pi}\partial_x u \cdot u^{-1}. \tag{53}$$

Let us notice that the presence of $i = \sqrt{-1}$ in the formula (53) for the rescaled current causes difficulties in dealing with real forms of our algebras. Indeed, (53) implies that $\tilde{J}(x)$ is anti-Hermitian (i.e. lies in the compact form of the current algebra) if $u(x)$ is Hermitian. But in that case the monodromy

$$M = P\exp\{-\frac{2\pi i}{k+n}\int_o^{2\pi}\tilde{J}(x)dx\}$$

does not satisfy the expected unitarity condition. The same trouble is also reflected by the fact that the quantization parameter $q = \exp(\frac{\pi i}{k+n})$ is a root of unity while a bona fide $U_q(su(n))$ is defined for real q. Of course, the emergence of quantum groups $U_q(\mathfrak{g})$ for q a root of unity is highly typical for the problems of Conformal Field Theory.

References

[1] M.A.Semenov-Tian-Shansky, Publ.RIMS **21(no.6)** (1985), 1237–1260, Kyoto Univ.

[2] N.Yu.Reshetikhin and M.A.Semenov-Tian-Shansky, Lett.Math.Phys. **19** (1990), 133–142.

[3] O.Babelon, Phys.Lett. **B215** (1988), 523-527.

[4] B.Blok, Tel-Aviv University preprint (1989).

[5] A.Alekseev and S.Shatashvili, Commun.Math.Phys. **133** (1990), 353–368.

[6] L.D.Faddeev, Commun.Math.Phys. **132** (1990), 131-138.

[7] L.D.Faddeev, Lectures given in Schladming (1989).

[8] A.Alekseev,L.D.Faddeev,M.A.Semenov-Tian-Shansky and A.Volkov, *The unravelling of the quantum group structure in the WZNW theory*, Preprint CERN-TH-5981/91 (1991).

[9] V.Drinfeld, *Quantum groups*, Proc.ICM-86 Berkeley,California,USA,1986, 1987, pp. 798–820.

[10] L.D.Faddeev, N.Yu.Reshetikhin and L.A.Takhtajan, Alg. Anal. **1** no. 1 (1989), 178–206. (in Russian)

[11] V.Drinfeld, Alg. Anal. **1** no. 2 (1989), 30–47. (in Russian)

[12] N.Yu.Reshetikhin, Alg. Anal. **1** no. 2 (1989), 169–189. (in Russian)

[13] V.Drinfeld, Alg. Anal. **1** no. 6 (1989), 114–149. (in Russian)

Leningrad Branch of Steklov Mathematical Institute, Fontanka, 27, Leningrad, 191011, USSR

LIOUVILLE THEORY ON THE LATTICE AND UNIVERSAL EXCHANGE ALGEBRA FOR BLOCH WAVES

O.BABELON

Laboratoire de Physique Théorique et Hautes Energies, Paris

ABSTRACT. We review some aspects of Liouville theory and the relation between its integrable and conformal structures. We emphasis its lattice version which exhibits the role of quantum groups.

1. Introduction

The discovery by Polyakov [1] of the role played by Liouville's equation in string theory revived our interest in this model. What makes Liouville's theory so fascinating are its multiple relations to the theory of Riemann surfaces, conformal field theory, integrable systems... So, the investigation of this very simple model is likely to bring into light unexpected relations between Riemann surfaces, infinite dimensional Lie algebras (the Virasoro algebra), quantum groups, etc... Concerning Riemann surfaces, we will simply quote the results of Zograf and Takhtadjan [2]. The study of Liouville theory, based on its conformal structure, was performed by Gervais and Neveu [3]. The integrable side of the model was considered by Faddeev and Takhtajan [4]. A synthesis between these two aspects was realized in [5], thereby obtaining a simple explanation of the occurrence of quantum groups in conformal field theory. We present here a short review of the subject, organized around the exchange algebra and the introduction of a lattice regulator. Besides providing the necessary regularizations, this exhibits interesting algebraic structures like non-ultralocal commutation relations (see also [6,7,8]) or a lattice deformation of the Virasoro algebra.

2. Continuous theory

Let $z_\pm = x \pm t$ denote the light cone coordinates. Liouville's equation reads

$$\partial_z \partial_{\bar{z}} \phi = e^{2\phi} . \tag{1}$$

One remarkable feature of this equation is that it can be written as a zero curvature condition. Let E_+, E_-, H be the three generators of the Lie algebra sl_2

$$\begin{aligned} [H, E_\pm] &= \pm 2 E_\pm \\ [E_+, E_-] &= H . \end{aligned}$$

Following the general construction of Toda systems, we introduce the field Φ with values in the Cartan subalgebra of sl_2

$$\Phi = \frac{1}{2}\phi H$$

and define

$$A_{z_+} = \partial_{z_+}\Phi + e^{ad\Phi}E_+ \tag{2}$$

$$A_{z_-} = -\partial_{z_-}\Phi + e^{-ad\Phi}E_- . \tag{3}$$

The zero curvature condition

$$F_{z_+z_-} = \partial_{z_+}A_{z_-} - \partial_{z_-}A_{z_+} + [A_{z_+}, A_{z_-}] = 0 \tag{4}$$

is equivalent to the equation of motion (1).

As a first application of this result, let us construct the general solution of eq. (1). We follow the analysis of Leznov and Saveliev [9]. Since the zero curvature condition holds, we write $A_{z_\pm}$ as a pure gauge

$$A_{z\pm} = -\partial_{z_\pm}T \cdot T^{-1} \tag{5}$$

where T belongs to the group SL_2. We write T in two different ways

$$T = e^{-\Phi}G_1 = e^{\Phi}G_2 \tag{6}$$

and we introduce the Gauss decompositions

$$G_1 = N_+^{-1}Q_-^{-1} \qquad G_2 = N_-Q_+ . \tag{7}$$

The plus and minus signs refer to upper or lower triangular matrices respectively. We assume moreover that $N_\pm$ are strictly triangular (the identity on the diagonal). The linear system eq. (5) reduces to the conditions

$$\partial_{z_-}Q_+ = 0 \qquad \partial_{z_+}Q_- = 0 \tag{8}$$

and $Q_\pm$ are solutions of linear systems of the form

$$\partial_{z_+}Q_+ = (P(z_+)H - E_+)Q_+ \tag{9}$$

$$\partial_{z_-}Q_- = -Q_-(\overline{P}(z_-)H - E_-). \tag{10}$$

Here $P(z_+)$ and $\overline{P}(z_-)$ are two chiral fields which will parametrize the general solution of eq. (1). To construct it, we consider representations of sl_2 with highest weight $\lambda^{(j)}_{max}$. To each highest weight vector $|\lambda^{(j)}_{max}>$ we associate one line vector $\xi^{(j)}$ and one column vector $\overline{\xi}^{(j)}$ as follows

$$\xi^{(j)} =< \lambda^{(j)}_{max}|G_2 =< \lambda^{(j)}_{max}|Q_+ \tag{11}$$

$$\overline{\xi}^{(j)} = G_1^{-1}|\lambda^{(j)}_{max} >= Q_-|\lambda^{(j)}_{max} > . \tag{12}$$

Eq. (8) imply that they are chiral vectors, functions of P and $\overline{P}$ respectively. Noticing that we can also write

$$\xi^{(j)}(x) =< \lambda^{(j)}_{max}|e^{-\Phi}T(x) \tag{13}$$

$$\overline{\xi}^{(j)}(x) = T^{-1}(x)e^{-\Phi}|\lambda^{(j)}_{max} > \tag{14}$$

we get the general solution of eq. (1) in the form of a scalar product

$$e^{-2\lambda^{(j)}_{max}(\Phi)} = \xi^{(j)} \cdot \overline{\xi}^{(j)} .$$

The above solution exhibits a splitting of chiralities, it is therefore intimately related to the conformal structure of the theory. To see this, we take H and $E_\pm$ in the spin 1/2 representation,

$$H = \begin{pmatrix} 1 & 0 \\ 0 & -1 \end{pmatrix} \qquad E_+ = \begin{pmatrix} 0 & 1 \\ 0 & 0 \end{pmatrix} \qquad E_- = \begin{pmatrix} 0 & 0 \\ 1 & 0 \end{pmatrix}$$

the matrix elements of $\xi^{(1/2)}(x)$ will satisfy a second order differential equation of the form

$$(\partial_x^2 - \mathcal{U})\xi^{(1/2)}(x) = 0 . \tag{15}$$

Using eq. (13), we find

$$\mathcal{U} = (\partial_{z_+}\phi)^2 - \partial^2_{z_+}\phi \tag{16}$$

while using eq. (11), we get

$$\mathcal{U}(x) = P^2(x) + P'(x) . \tag{17}$$

We recognize in eq. (16) the chiral energy momentum tensor of Liouville theory. The passage to the P variable is just the Miura transformation.

In a spin j representation the analog of eq. (15) is a differential equation of order $2j+1$.

$$(\partial^{2j+1}_{z_+} + \sum_{i=0}^{2j-1} \mathcal{U}_i \partial^i_{z_+})\xi^{(2j+1)} . \tag{18}$$

As it is well known, these equations are the classical limits of the decoupling condition for null vectors in conformal field theory [10,11]. So, the splitting of chiralities which occurs naturally in the method of solution of Leznov and Saveliev also exhibits the conformal structure of the theory.

To talk about the integrable structure of Liouville theory, we introduce Poisson brackets. Let $\pi = \dot{\phi}$ be the conjugate variable of ϕ

$$\{\pi(x), \phi(y)\} = 2\gamma\delta(x-y) .$$

In eq. (16) we use the equations of motion to eliminate the higher order time derivatives in favor of π. Then we can calculate the Poisson brackets of $\mathcal{U}$

$$\{\mathcal{U}(x), \mathcal{U}(y)\} = 2\gamma(2\mathcal{U}(x)\partial_x + \mathcal{U}'(x) - \frac{1}{2}\partial_x^3)\delta(x-y) . \tag{19}$$

This is the Virasoro algebra. Alternatively, one could use eq. (17) to reach the same result provided

$$\{P(x), P(y)\} = \gamma\delta'(x-y). \tag{20}$$

The main result of this integrable approach is that one can compute the Poisson brackets of T in eq. (5). More precisely, consider at fixed time the linear system

$$(\partial_x + A_x)T = 0$$

where $A_x = A_{x_+} + A_{x_-}$. Take the solution normalized by

$$T(0) = 1 \tag{21}$$

then we have [12,4]

$$\{T(x) \otimes T(x)\} = -\frac{\gamma}{2}[r^{\pm}, T(x) \otimes T(x)]. \tag{22}$$

The matrices $r^{\pm}$ are solutions of the classical Yang-Baxter equation and are given by

$$r^{+} = H \otimes H + 4E_{+} \otimes E_{-} \tag{23}$$

$$r^{-} = -H \otimes H - 4E_{-} \otimes E_{+}. \tag{24}$$

The combination of the integrable and the conformal structures is obtained when we compute the Poisson brackets of ξ and $\overline{\xi}$. Starting from eq. (13, 14) and eq. (22), we immediately find [5]

$$\{\xi^{(j)}(x) \otimes \xi^{(j')}(y)\} = \frac{\gamma}{2}\xi^{(j)}(x) \otimes \xi^{(j')}(y)[\theta(x-y)r^{+} + \theta(y-x)r^{-}] \tag{25}$$

$$\{\overline{\xi}^{(j)}(x) \otimes \overline{\xi}^{(j')}(y)\} = \frac{\gamma}{2}[\theta(x-y)r^{-} + \theta(y-x)r^{+}]\overline{\xi}^{(j)}(x) \otimes \overline{\xi}^{(j')}(y) \tag{26}$$

$$\{\xi^{(j)}(x) \otimes \overline{\xi}^{(j')}(y)\} = -\frac{\gamma}{2}\xi^{(j)}(x) \otimes 1 \cdot r^{-} \cdot 1 \otimes \overline{\xi}^{(j')}(y) \tag{27}$$

$$\{\overline{\xi}^{(j)}(x) \otimes \xi^{(j')}(y)\} = -\frac{\gamma}{2}\overline{\xi}^{(j)}(x) \otimes 1 \cdot r^{+} \cdot 1 \otimes \xi^{(j')}(y). \tag{28}$$

This exchange algebra which is expressed in terms of the classical r-matrix also encodes the conformal structure of the theory. Indeed from the ξ's one can reconstruct the energy momentum tensor $\mathcal{U}(x)$. It suffices to remark that if $\xi^1(x)$ and $\xi^2(x)$ are linearly independent solutions of eq. (15). We can rewrite this equation as

$$\det\begin{pmatrix} \xi & \xi^1 & \xi^2 \\ \xi' & \xi^{1\prime} & \xi^{2\prime} \\ \xi'' & \xi^{1\prime\prime} & \xi^{2\prime\prime} \end{pmatrix} = 0.$$

Thus $\mathcal{U}$ is given in terms of ξ^1 and ξ^2 by a Wronskian type expression and one can compute the Poisson bracket of $\mathcal{U}$ from eq. (25). Of course, we recover eq. (19).

The exchange algebra is left invariant by the transformation

$$\xi \longrightarrow \xi\, g_-^{-1} \qquad \overline{\xi} \longrightarrow g_+\, \overline{\xi} \tag{29}$$

provided we have non trivial Poisson brackets between $g_\pm$

$$\begin{aligned}
\{g_+ \overset{\otimes}{,} g_+\} &= \frac{\gamma}{2}[r^\pm, g_+ \otimes g_+] \\
\{g_- \overset{\otimes}{,} g_-\} &= \frac{\gamma}{2}[r^\pm, g_- \otimes g_-] \\
\{g_- \overset{\otimes}{,} g_+\} &= \frac{\gamma}{2}[r^-, g_- \otimes g_+] \\
\{g_+ \overset{\otimes}{,} g_-\} &= \frac{\gamma}{2}[r^+, g_+ \otimes g_-]
\end{aligned} \tag{30}$$

and we assume that $g_\pm$ commute with $\xi, \overline{\xi}$. It was shown in [13] that the transformations eq. (29) are nothing else than the dressing transformations, written on the chiral objects ξ and $\overline{\xi}$. Eq. (30) express the Lie Poisson action of these transformations [14]. When quantized, this becomes the quantum group symmetries we encounter in conformal field theory.

As seen from these considerations the basis where the exchange algebra assumes the simple form eq. (25) is not unique. Moreover there are basis where it may look quite different.

Since the coefficients in eq. (18) are periodic functions of x there is one basis which is very natural and plays an important role, it is the basis of quasi periodic solutions (Bloch waves)

$$\psi^{(j)}(x + 2\pi) = \psi^{(j)}(x) e^{2\pi P_0 H}$$

where P_0 is the Bloch momentum. It is an important problem is to find the exchange algebra in the ψ basis. To do so, we start from the linear system eq. (9). Consider its solution normalized by the condition $Q(0) = 1$. The monodromy matrix S is defined by

$$Q(x + 2\pi) = Q(x) S \qquad S = Q(2\pi)\,.$$

To construct the Bloch waves we have to diagonalize S. Let

$$S = g^{-1} e^{2\pi P_0 H} g$$

the matrix g is uniquely determined if we require it to be strictly upper triangular. Then we set

$$\psi^{(j)}(x) = < \lambda^{(j)}_{max} | Q(x) g^{-1} \rho \tag{31}$$

where ρ is a diagonal matrix. These $\psi^{(j)}(x)$ have diagonal monodromy

$$\psi^{(j)}(x + 2\pi) = \psi^{(j)}(x) e^{2\pi P_0 H}\,.$$

The matrix ρ is chosen as follows

$$\rho = e^{-K_+ H} e^{\theta H}\,. \tag{32}$$

The constant K_+ is defined by

$$\int_0^x P(y)dy = K_+ + P_0 x + \sum_{n\neq 0} \frac{P_n}{in} e^{inx}$$

where we have introduced the Fourier decomposition of $P(x)$

$$P(x) = \sum_n P_n e^{inx} .$$

The role of K_+ is to eliminate from $\psi^{(j)}(x)$ all the remaining dependence in the normalization point $x = 0$. The variable θ is the conjugate variable of P_0

$$\{\theta, P_0\} = \frac{\gamma}{2\pi} .$$

With these choices, $\psi^{(j)}(x)$ is a conformal object of weight $-j$ and the exchange algebra is

$$\begin{aligned}\{\psi^{(j)}(x) \otimes_{,} \psi^{(j')}(y)\} = -\frac{\gamma}{4}\psi^{(j)}(x) \otimes \psi^{(j')}(y)\Big[&(r^+ - r^-)\epsilon(x-y) \\ &- \frac{(A+D)}{(A-D)}(r^+ + r^-)\Big]\end{aligned} \tag{33}$$

where $A = e^{2\pi P_0}$, $D = e^{-2\pi P_0}$. Detailed proofs are given in [15].

3. Going to the lattice

To generalize all these results to the quantum case one has to define a suitable ordering for operators. A natural choice would be normal ordering. It has the advantage of preserving conformal invariance [3].

Another possibility, the one we will follow, is to go to the lattice and use the lattice ordering [16]. Conformal invariance is lost, in fact it is only deformed (see below), but this does not matter as far as the exchange algebra is concerned. It has the advantage that the quantum group structure is made explicit from the very beginning.

To go to the lattice, the first step is to discretize eq. (9). As usual, we set

$$Q_n = L_n Q_{n-1} . \tag{34}$$

The problem is to define a suitable Poisson bracket for the L_n's. We set

$$\begin{aligned}\{L_n \otimes_{,} L_m\} &= \frac{\gamma}{2}\delta_{nm}[r, L_n \otimes L_m] \\ &- \frac{\gamma}{2}\delta_{n,m+1} L_n \otimes 1\,(H \otimes H)\,1 \otimes L_m \\ &+ \frac{\gamma}{2}\delta_{n,m-1} 1 \otimes L_m\,(H \otimes H)\,L_n \otimes 1 .\end{aligned} \tag{35}$$

This may be considered as a lattice version of eq. (20) since if we let

$$L_n = 1 + \Delta(P_n - E_+) + O(\Delta^2)$$

eq. (35) gives back eq. (20) when the lattice spacing Δ tends to zero.

As explained in [16] the motivation to introduce eq. (35) is that it produces the straightforward lattice generalization of the exchange algebra eq. (25). In fact, let

$$Q_n = L_n L_{n-1} \cdots L_1 \qquad n < N$$

by straightforward calculations, we get

$$\begin{aligned}\{Q_n \overset{\otimes}{,} Q_m\} = \frac{\gamma}{2} Q_n \otimes Q_m \cdot \\ \cdot\Big\{\theta(n-m)\big[-r + Q_m^{-1} \otimes Q_m^{-1}(r - H\otimes H)Q_m \otimes Q_m\big] \\ +\theta(m-n)\big[-r + Q_n^{-1} \otimes Q_n^{-1}(r + H\otimes H)Q_n \otimes Q_n\big]\Big\}.\end{aligned} \tag{36}$$

Projecting on highest weights

$$\sigma_n^{(j)} = < \lambda_{max}^{(j)} | Q_n$$

we obtain

$$\{\sigma_n^{(j)} \overset{\otimes}{,} \sigma_m^{(j')}\} = -\frac{\gamma}{2} \sigma_n^{(j)} \otimes \sigma_m^{(j')} [\theta(n-m) r^+ + \theta(m-n) r^-].$$

The relations eq. (35) constitute a non-ultralocal generalization of the usual ultralocal formulae which are at the basis of the Hamiltonian approach to the Inverse Scattering Method [17].

Next we consider a lattice with N sites and periodic boundary conditions. We define the monodromy matrix S

$$S = L_N L_{N-1} \cdots L_1$$

so that

$$Q_{N+n} = Q_n S .$$

Remembering that $\{L_1 \overset{\otimes}{,} L_N\} \neq 0$ (in contrast to the ultralocal case), one can show that

$$\begin{aligned}\{Q_n \overset{\otimes}{,} S\} = \frac{\gamma}{2} Q_n \otimes S\Big[-r + Q_n^{-1} \otimes Q_n^{-1}(r + H\otimes H)Q_n \otimes Q_n \\ - H \otimes S^{-1} H S\Big]\end{aligned} \tag{37}$$

$$\begin{aligned}\{S \overset{\otimes}{,} S\} = \frac{\gamma}{2} S \otimes S\Big[-r + S^{-1} \otimes S^{-1} r S \otimes S \\ + S^{-1} H S \otimes H - H \otimes S^{-1} H S\Big].\end{aligned} \tag{38}$$

We will need also the Poisson brackets of the matrix ρ

$$\{\rho \overset{\otimes}{,} \rho\} = 0 \tag{39}$$

$$\{Q_n \overset{\otimes}{,} \rho\} = -\frac{\gamma}{2} Q_n \otimes \rho (H \otimes H) \qquad n < N \tag{40}$$

$$\{S \overset{\otimes}{,} \rho\} = -\frac{\gamma}{2}\Big[S \otimes \rho H \otimes H + (H \otimes H) S \otimes \rho\Big]. \tag{41}$$

One can repeat on the lattice the same analysis as the one performed in [15] and we get the lattice version of eq. (33)

$$\{\psi_n^{(j)} \otimes_, \psi_m^{(j')}\} = -\frac{\gamma}{4}\psi_n^{(j)} \otimes \psi_m^{(j')}\Big[(r^+ - r^-)\epsilon(n-m) - \frac{(A+D)}{(A-D)}(r^+ + r^-)\Big].$$

So the exchange algebra is the same on the lattice and in the continuum, as expected.

4. Quantum theory on the lattice

To quantize the theory, all we have to do is to give the quantum version of eq. (35). Introducing the notation

$$L_{1n} = L_n \otimes I \qquad L_{2n} = I \otimes L_n$$

a natural quantum generalization of eq. (35) is

$$R_{12}L_{1n}L_{2n} = L_{2n}L_{1n}R_{12} \tag{42}$$

$$L_{1n}L_{2,n+1} = L_{2,n+1}A_{12}L_{1n} \tag{43}$$

where

$$R_{12} = 1 - \frac{i\gamma}{2}r + O(\gamma^2)$$

$$A_{12} = 1 + \frac{i\gamma}{2}H \otimes H + O(\gamma^2).$$

We will define the quantum theory by requiring that R_{12} is a solution of the quantum Yang-Baxter equation

$$R_{12}R_{13}R_{23} = R_{23}R_{13}R_{12}$$

and we take

$$A_{12} = q^{-\frac{1}{2}H\otimes H} \qquad q = e^{-i\gamma}.$$

Let us call $R_{12}^{\pm}$ the two solutions of the Yang-Baxter equation whose classical limits are $r_{12}^{\pm}$ respectively. Applying the automorphism $\sigma(x \otimes y) = y \otimes x$ to eqs. (42, 43) we also get

$$[R_{21}]^{-1}L_{1n}L_{2n} = L_{2n}L_{1n}[R_{21}]^{-1}$$

$$L_{2n}L_{1,n+1} = L_{1,n+1}A_{12}L_{2n}.$$

Since $R_{12}^- = [R_{21}^+]^{-1}$, we see that we may use indifferently R_{12}^+ or R_{12}^- in eq. (42). Their universal form is [19].

$$R_{12}^+(q) = q^{\frac{1}{2}H\otimes H}\sum_{i=0}^{\infty}(q-q^{-1})^i \frac{q^{-\frac{i(i+1)}{2}}}{[i]!}q^{\frac{i}{2}H}E_+^i \otimes q^{-\frac{i}{2}H}E_-^i$$

$$R_{12}^-(q) = q^{-\frac{1}{2}H\otimes H}\sum_{i=0}^{\infty}(-1)^i(q-q^{-1})^i \frac{q^{\frac{i(i+1)}{2}}}{[i]!}q^{\frac{i}{2}H}E_-^i \otimes q^{-\frac{i}{2}H}E_+^i$$

where now $H,\ E_\pm$ denote the generators of the quantum group $\mathcal{U}_q(sl_2)$ [18,19]

$$[H, E_\pm] = \pm 2E_\pm$$
$$[E_+, E_-] = \frac{q^H - q^{-H}}{q - q^{-1}}\,.$$

As in the ultralocal case, the crucial property of eqs. (42, 43) is that they can be integrated. For this however, one has to change slightly the definition of Q_n. We define

$$Q_n = L_n B L_{n-1} B \cdots B L_1 \tag{44}$$

where B is the diagonal matrix

$$B = q^{-\frac{1}{4}H^2}\,. \tag{45}$$

We also introduce the monodromy matrix S

$$S = L_N B L_{N-1} \cdots B L_1 \tag{46}$$

so that if the lattice is periodic with N sites we have

$$Q_{n+N} = Q_n B S\,. \tag{47}$$

With all these definitions at hand we get [16]

1) If $n = m < N$

$$R_{12} Q_{1n} Q_{2n} = Q_{2n} Q_{1n} R_{12} \tag{48}$$

2) If $n > m$, $(n, m < N)$

$$Q_{1n} Q_{1m}^{-1} A_{21} R_{12} Q_{1m} Q_{2m} = Q_{2m} Q_{1n} R_{12} \tag{49}$$

3) If $n = N,\ m < N$

$$S_1 Q_{1m}^{-1} A_{12} R_{12} Q_{1m} Q_{2m} = Q_{2m} A_{12} S_1 R_{12} \tag{50}$$

4) If $n = m = N$

$$R_{12} S_1 A_{12} S_2 = S_2 A_{12} S_1 R_{12} \tag{51}$$

where by definition

$$Q_n Q_m^{-1} = L_n B L_{n-1} B \cdots B L_{m+1} B\,.$$

Projecting eq. (49) on highest weight vectors

$$\sigma_n^{(j)} = < \lambda_{max}^{(j)} | Q_n$$

we have

If $n \neq m$

$$\sigma_{1n}^{(j)}\ \sigma_{2m}^{(j')} = \sigma_{2m}^{(j')}\ \sigma_{1n}^{(j)}\ R_{12}^{\pm}(q), \qquad \pm = \epsilon(n-m) \tag{52}$$

If $n = m$

$$q^{\frac{1}{2}\lambda_{max}^{(j)} \cdot \lambda_{max}^{(j')}}\ \sigma_{1n}^{(j)}\ \sigma_{2n}^{(j')} = \sigma_{2n}^{(j')}\ \sigma_{1n}^{(j)}\ R_{12}^{+}(q)\,. \tag{53}$$

Finally, we will need the diagonal matrix ρ with the following properties

$$\rho_1\rho_2 = \rho_2\rho_1 \tag{54}$$

$$A_{12}^{2(1-\alpha)} S_1\rho_2 A_{12}^{2\alpha} = \rho_2 S_1 \tag{55}$$

$$Q_{1n}\rho_2 A_{12}^{2\alpha} = \rho_2 Q_{1n} \,. \tag{56}$$

These commutation relations are straightforward quantum generalizations of the classical formulae eqs. (39, 40, 41). We have introduced a parameter α. It is a freedom we have also at the classical level [15]. Eqs. (39, 40, 41) correspond to the choice $\alpha = 1/2$. As we will see this parameter drops out in the final formulae. The choice $\alpha = 0$ is simpler in many respects.

5. Universal Bloch wave algebra

At the classical level, the monodromy matrix is an element of the group SL_2 and it is upper triangular. Therefore we can write

$$S = e^{2\pi P_0 H} e^{ZE_+} \,.$$

From eq. (38) we find that

$$\{P_0, Z\} = 0 \,.$$

At the quantum level we look similarly for an upper triangular solution of eq. (51) depending on two commuting operators P_0 and Z. The answer is

$$S = e^{2\pi P_0 H} B \sum_{i=0}^{\infty} \frac{Z^i}{[i]!} q^{\frac{i}{2} H} E_+^i \tag{57}$$

where we used the notation

$$[i] = \frac{q^i - q^{-i}}{q - q^{-1}} \qquad [i]! = [1][2]\cdots[i] \,.$$

Now that S is defined, we can diagonalize it. We look for a matrix g^{-1} such that

$$BSg^{-1} = g^{-1} e^{2\pi P_0 H} B^2 \,. \tag{58}$$

Of course g^{-1} is determined only up to a diagonal matrix, and we choose it to be strictly upper triangular. We have

$$g^{-1} = \sum_{n=0}^{\infty} Z^n \, X_n(H) E_+^n \tag{59}$$

$$X_n(H) = (-1)^n \frac{1}{[n]!} \frac{q^{\frac{n(n+1)}{2}} A^n q^{-\frac{n}{2} H}}{\prod_{\nu=n}^{2n-1} (Aq^{-H+\nu} - Dq^{H-\nu})} \,. \tag{60}$$

The Bloch waves are defined as

$$\psi_n^{(j)} = \sigma_n^{(j)} g^{-1} \rho B^{2\alpha} \,. \tag{61}$$

They have diagonal monodromy

$$\psi^{(j)}(x+2\pi) = \psi^{(j)}(x) e^{2\pi P_0 H} q^{\frac{1}{2}H^2}.$$

In eq. (61) the matrix ρ and the parameter α are those defined in eq. (54, 55, 56).

The exchange algebra for the Bloch waves is [20]

$$\psi^{(j)}_{1n}\psi^{(j')}_{2m} = \psi^{(j')}_{2m}\psi^{(j)}_{1n}\ \mathcal{R}^{\pm}_{12}(P_0)$$

where

$$\mathcal{R}^{\pm}_{12}(P_0) = F^{-1}_{21}(P_0) R^{\pm}_{12}(q) F_{12}(P_0)$$

and

$$F_{12}(P_0) = \sum_{k=0}^{\infty} (q-q^{-1})^k \frac{(-1)^k}{[k]!} \frac{A^k}{\prod_{\nu=k}^{2k-1}(Aq^{H_2+\nu} - Dq^{-H_2-\nu})} q^{\frac{k}{2}(H_1+H_2)} E^k_+ \otimes E^k_- .$$

In these formulae, P_0 is the quantum Bloch momentum. As in all this paper $A = e^{2\pi P_0}$ and $D = e^{-2\pi P_0}$. For $j = j' = 1/2$ this exchange algebra was first obtained in [3].

A similar analysis can be performed for the other chirality. Bloch waves have diagonal monodromy

$$\overline{\psi}^{(j)}_{n+N} = q^{\frac{1}{2}H^2} e^{-2\pi \overline{P}_0 H} \overline{\psi}^{(j)}_n$$

and their exchange algebra reads

$$\overline{\psi}^{(j)}_{1n}\overline{\psi}^{(j')}_{2m} = \overline{\mathcal{R}}^{\pm}_{12}(\overline{P}_0)\overline{\psi}^{(j')}_{2m}\overline{\psi}^{(j)}_{1n}$$

$$\overline{\mathcal{R}}^{\pm}_{12}(\overline{P}_0) = \overline{F}_{12}(\overline{P}_0) R^{\mp}_{12}(q) \overline{F}^{-1}_{21}(\overline{P}_0)$$

where

$$\overline{F}_{12}(\overline{P}_0) = \sum_{k=0}^{\infty} (q-q^{-1})^k \frac{(-1)^k}{[k]!} \frac{\overline{A}^k}{\prod_{\nu=1}^{k}(\overline{A}q^{H_1-\nu} - \overline{D}q^{-H_1+\nu})} q^{\frac{k}{2}(H_1+H_2)} E^k_+ \otimes E^k_-$$

and

$$\overline{A} = e^{-2\pi \overline{P}_0} \qquad \overline{D} = e^{2\pi \overline{P}_0}.$$

6. The ξ-basis

Now that we have some control on the Bloch waves, we can try to construct other basis from this one. In particular, can one find a basis where the exchange algebra is the obvious quantum analog of eq. (25)

$$\xi^{(j)}_1(x)\xi^{(j')}_2(y) = \xi^{(j')}_2(y)\xi^{(j)}_1(x) R^{\pm}_{12}(q) \qquad \pm = \epsilon(x-y). \tag{62}$$

We already have the $\sigma^{(j)}(x)$ basis with this property. However, due to the normalization condition $Q(0) = 1$, the $\sigma^{(j)}$'s are not conformal covariant objects. We have shown in [5] that for $j = 1/2$ there exists a change of basis depending only on P_0

$$\xi^{(j)}(x) = \psi^{(j)}(x) M(P_0)$$

such that $\xi^{(1/2)}(x)$ does satisfy eq. (62). This result was later generalized to all j in [21]. We give here the universal form of $M(P_0)$ [20]. Since M depends only on P_0 and since P_0 commutes with the Virasoro generators, $\xi^{(j)}(x)$ has the same good conformal properties as $\psi^{(j)}(x)$.

The equation one has to solve reads (we write $M(A,D)$ instead of $M(P_0)$)

$$R^{\pm}_{12}F_{12}M_1(Aq^{H_2},Dq^{-H_2})M_2(A,D) = F_{21}M_2(Aq^{H_1},Dq^{-H_1})M_1(A,D)R^{\pm}_{12}\,. \qquad (63)$$

Its solution is [20]

$$M(P_0) = \sum_{n,m=0}^{\infty} \frac{(-1)^m}{[n]![m]!}\frac{A^m q^{\frac{1}{2}n(n-1)+m(n-m)}}{\prod_{\nu=1}^{n}(Aq^{\nu}-Dq^{-\nu})}E_+^n E_-^m q^{\frac{1}{2}(n+m)H}\,.$$

This result is itself a direct consequence of the following formula

$$F_{12}(P_0)M_1(Aq^{H_2},Dq^{-H_2})M_2(A,D) = \Delta_q[M(A,D)]$$

and we recall that

$$\Delta_q(E_\pm) = E_\pm \otimes q^{\frac{1}{2}H} + q^{-\frac{1}{2}H}\otimes E_\pm$$
$$\Delta_q(H) = H\otimes 1 + 1\otimes H\,.$$

7. Local fields

In this section we combine the two chiralities to construct a set of periodic and local fields [20]. In order to do it we have to impose the constraint $P_0 = \overline{P}_0$. We choose to work with independent zero modes P_0 and $\overline{P}_0$. Their conjugate variables are also considered to be independent so that $\psi_n^{(j)}$ and $\overline{\psi}_m^{(j')}$ commute

$$\psi_{1n}^{(j)}\overline{\psi}_{2m}^{(j')} = \overline{\psi}_{2m}^{(j')}\psi_{1n}^{(j)}\,.$$

The constraint $P_0 = \overline{P}_0$ is then imposed on the states. If $\mathcal{H}$ and $\overline{\mathcal{H}}$ are the representation spaces of the two chiral halves of the theory, the physical space $\mathcal{F}\subset\mathcal{H}\otimes\overline{\mathcal{H}}$ is defined by the condition

$$(P_0 - \overline{P}_0)\mathcal{F} = 0$$

i.e. we have the decomposition

$$\mathcal{F} = \bigoplus_j \mathcal{H}_{\omega_j}\otimes\overline{\mathcal{H}}_{\omega_j}$$

where ω_j denote the eigenvalues of P_0. The fields

$$\Phi_n^{(j)} = \psi_n^{(j)}B^{-1}\,\overline{\psi}_n^{(j)}$$

1) Admit a restriction to the subspace $\mathcal{F}$ where $P_0 = \overline{P}_0$.
2) Their restriction to $\mathcal{F}$ is periodic.
3) Their restriction to $\mathcal{F}$ is local.

8. Free field representation

Let φ_n be a periodic free field on the lattice and π_n its conjugate momentum so that

$$[\pi_n, \varphi_m] = i\gamma \frac{\delta_{nm}}{\Delta}$$

where Δ is the lattice spacing. A realization of eq. (42, 43) is given by

$$L_n = e^{\Delta \pi_n H} \left\{ B \sum_{i=0}^{\infty} \frac{\Delta^i}{[i]!} q^{\frac{i}{2}H} E_+^i \right\} e^{\frac{1}{2}(\varphi_{n-1}-\varphi_n)H} .$$

Moreover the constant ρ satisfying eqs. (54, 55, 56) (with $\alpha = 0$) is

$$\rho = e^{-\varphi_N H} .$$

For spin $\frac{1}{2}$ representations, we have ($q = e^{-i\gamma}$)

$$R_{12}^+ = \begin{pmatrix} q^{\frac{1}{2}} & 0 & 0 & 0 \\ 0 & q^{-\frac{1}{2}} & q^{\frac{1}{2}} - q^{-\frac{3}{2}} & 0 \\ 0 & 0 & q^{-\frac{1}{2}} & 0 \\ 0 & 0 & 0 & q^{\frac{1}{2}} \end{pmatrix} \qquad R_{12}^- = \begin{pmatrix} q^{-\frac{1}{2}} & 0 & 0 & 0 \\ 0 & q^{\frac{1}{2}} & 0 & 0 \\ 0 & q^{-\frac{1}{2}} - q^{\frac{3}{2}} & q^{\frac{1}{2}} & 0 \\ 0 & 0 & 0 & q^{-\frac{1}{2}} \end{pmatrix}$$

$$A_{12} = \begin{pmatrix} q^{-\frac{1}{2}} & 0 & 0 & 0 \\ 0 & q^{\frac{1}{2}} & 0 & 0 \\ 0 & 0 & q^{\frac{1}{2}} & 0 \\ 0 & 0 & 0 & q^{-\frac{1}{2}} \end{pmatrix} .$$

Notice also that $H^2 = 1$ and so $B = q^{-\frac{1}{4}}$. The following L_n is an exact realization of eqs. (42, 43)

$$L_n = \begin{pmatrix} e^{\Delta \hat{p}_n} & -\Delta q^{-\frac{1}{2}} e^{\Delta \overline{p}_n} \\ 0 & e^{-\Delta \hat{p}_n} \end{pmatrix} \tag{64}$$

where

$$\hat{p}_n = \pi_n - \frac{1}{2\Delta}(\varphi_n - \varphi_{n-1}) \qquad \overline{p}_n = \pi_n + \frac{1}{2\Delta}(\varphi_n - \varphi_{n-1}) .$$

We will need also

$$p_n = \pi_n - \frac{1}{2\Delta}(\varphi_{n+1} - \varphi_n)$$

9. Lattice deformation of the Virasoro algebra

A natural question which arises is: What happens to the Virasoro algebra? As we have seen, in the continuous theory one can reconstruct the Virasoro generators from the ξ or the ψ fields. One can apply the same construction on the lattice and we get an interesting lattice deformation of the Virasoro algebra [4,23,24,25].

First we find the lattice analog of eq. (15). We simply write that ξ_n^1 and ξ_n^2 are the two independent solutions of the linear recursion relation

$$\det \begin{pmatrix} \xi_n & \xi_n^1 & \xi_n^2 \\ \xi_{n+1} & \xi_{n+1}^1 & \xi_{n+1}^2 \\ \xi_{n+2} & \xi_{n+2}^1 & \xi_{n+2}^2 \end{pmatrix} = 0$$

or

$$W_n^{(1)}\xi_{n+2} - W_n^{(2)}\xi_{n+1} + W_{n+1}^{(1)}\xi_n = 0 \tag{65}$$

where

$$W_n^{(p)} = \xi_n^1\,\xi_{n+p}^2 - \xi_n^2\,\xi_{n+p}^1 .$$

Notice that since these are spin zero objects one can express them in the σ-basis as well. Setting

$$\xi_n = \sqrt{\frac{W_{n-1}^{(1)}W_n^{(1)}}{W_{n-1}^{(2)}}}\;\phi_n$$

eq. (65) becomes

$$(L-2)\phi = 0 \tag{66}$$

where L is the symmetric matrix

$$L_{nm} = \sqrt{S_n}\,\delta_{m,n+1} + \sqrt{S_m}\,\delta_{n,m+1}$$

and

$$S_n = 4\frac{W_{n+1}^{(1)}W_{n-1}^{(1)}}{W_n^{(2)}W_{n-1}^{(2)}} .$$

Eq. (66) is the lattice version of eq. (15).

From eq. (25) we may calculate the Poisson brackets of the W's. We find that they form a closed algebra only if $p=1$ or 2.

$$\begin{aligned}
\{W_n^{(1)},W_m^{(1)}\} &= -\frac{\gamma}{2}\,W_n^{(1)}\,W_m^{(1)}\,(\delta_{n,m-1}-\delta_{n,m+1})\\
\{W_n^{(1)},W_m^{(2)}\} &= -\frac{\gamma}{2}\,W_n^{(1)}\,W_m^{(2)}\,(-\delta_{n,m+2}+\delta_{n,m+1}-\delta_{nm}+\delta_{n,m-1})\\
\{W_n^{(2)},W_m^{(2)}\} &= -\frac{\gamma}{2}\,W_n^{(2)}\,W_m^{(2)}\,(\delta_{n,m-2}-\delta_{n,m+2}+2\delta_{n,m+1}-2\delta_{n,m-1})\\
&\quad + 2\gamma\,W_{n-1}^{(1)}\,W_{n+1}^{(1)}\,\delta_{n,m+1} - 2\gamma\,W_{m-1}^{(1)}\,W_{m+1}^{(1)}\,\delta_{n,m-1} .
\end{aligned}$$

Alternatively, we can use the generators $W_n^{(1)}$ and S_n

$$\begin{aligned}
\{W_n^{(1)},W_m^{(1)}\} &= -\frac{\gamma}{2}W_n^{(1)}\,W_m^{(1)}\,(\delta_{n,m-1}-\delta_{n,m+1})\\
\{W_n^{(1)},S_m\} &= 0\\
\{S_n,S_m\} &= \frac{\gamma}{2}S_nS_m[(4-S_n-S_m)(\delta_{n,m-1}-\delta_{n,m+1})\\
&\quad + S_{n-1}\delta_{n,m+2} - S_{m-1}\delta_{n,m-2}] .
\end{aligned} \tag{67}$$

Equation (67) was first obtained in [4]. It may be considered as a lattice deformation of the Virasoro algebra, since if we let

$$S_n \longrightarrow 1-\Delta^2\mathcal{U}(x)$$

we recover eq. (19) in the continuum limit $\Delta \longrightarrow 0$.

The Poisson structure eq. (67) is intimately related to integrability. One way to see this is to rewrite it in terms of the L-operator. We consider a lattice with an even number of sites N and we assume open boundary conditions for simplicity. Let X be a symmetric $(N+1)\times(N+1)$ matrix and let $R(X)$ be the classical r-matrix of the $sl(N+1)$ Toda chain

$$R(X) = \sum_{j>i} E_{ij}\, tr(E_{ji}X) - E_{ji}\, tr(E_{ij}X). \tag{68}$$

Let $L(X) = tr(LX)$ then we have [26]

$$\{L(X), L(Y)\} = L([X,Y]_D)$$

where

$$[X,Y]_D = [D(X),Y] + [X, D(Y)]$$

and

$$D(X) = \frac{\gamma}{2} R(LX + XL) - \frac{\gamma}{8} R(L^3X + XL^3).$$

From this result it follows immediately that $I_k(S_n) = tr(L^k)$ are in involution and the equation of motion with Hamiltonian I_k takes the Lax form. L is the Lax operator.

As the L-operator is really the important notion in the classical system, we first look for its quantum generalization. Let us define

$$W_n^{(p)} = \xi_n^1\, \xi_{n+p}^2 - q^{-1}\xi_n^2\, \xi_{n+p}^1 \qquad p = 1,2$$

and let us impose

$$W_n^{(0)} = \xi_n^1\, \xi_n^2 - q^{-1}\xi_n^2\, \xi_n^1 = 0$$

which is the only possible relation compatible with the commutation relations with $\xi_m, m \neq n$. From the definition of $W_n^{(0)}$, $W_n^{(1)}$ and $W_n^{(2)}$ we easily check that

$$q^{-\frac{1}{2}}\, W_{n+1}^{(1)}\, \xi_n - W_n^{(2)}\, \xi_{n+1} + q^{\frac{1}{2}}\, W_n^{(1)}\, \xi_{n+2} = 0$$

when $\xi_n = (\xi_n^1, \xi_n^2)$.

As in the classical situation the algebra closes only if $p = 1,2$.

$$\begin{aligned}
W_n^{(1)}\, W_m^{(1)} &= q^{\frac{1}{2}(\delta_{n,m-1}-\delta_{n,m+1})}\, W_m^{(1)}\, W_n^{(1)} \\
W_n^{(1)}\, W_m^{(2)} &= q^{\frac{1}{2}(-\delta_{n,m+2}+\delta_{n,m+1}-\delta_{nm}+\delta_{n,m-1})}\, W_m^{(2)}\, W_n^{(1)} \\
W_n^{(2)}\, W_m^{(2)} &= q^{\frac{1}{2}(\delta_{n,m-2}-\delta_{n,m+2}+2\delta_{n,m+1}-2\delta_{n,m-1})}\, W_m^{(2)}\, W_n^{(2)} \\
&\quad + (q^{-\frac{1}{2}} - q^{\frac{3}{2}})\, W_{n-1}^{(1)}\, W_{n+1}^{(1)}\, \delta_{n,m+1} + (q^{\frac{1}{2}} - q^{-\frac{3}{2}})\, W_{m-1}^{(1)}\, W_{m+1}^{(1)}\, \delta_{n,m-1}.
\end{aligned}$$

Notice that if $q^* = q^{-1}$, that is $q = e^{-i\gamma}$ with γ real, these relations are compatible with the reality condition

$$W_n^{(1)+} = W_n^{(1)},\ W_n^{(2)+} = W_n^{(2)}. \tag{69}$$

We introduce a new variable ϕ_n by

$$\xi_n = q^{-\frac{5}{8}n} \sqrt{W^{(1)}_{n-1}} \sqrt{W^{(1)}_n} \frac{1}{\sqrt{W^{(2)}_{n-1}}} \phi_n$$

to obtain

$$\Sigma_n \phi_n - 2\phi_{n+1} + \Sigma^+_{n+1} \phi_{n+2} = 0 \tag{70}$$

with

$$\Sigma_n = 2\frac{1}{\sqrt{W^{(2)}_n}} \sqrt{W^{(1)}_{n+1}} \sqrt{W^{(1)}_{n-1}} \frac{1}{\sqrt{W^{(2)}_{n-1}}} . \tag{71}$$

Equations (70, 71) define the quantum version of the Schroedinger equation on the lattice.

From this expression for the L-operator, a natural definition of S_n as a positive and self-adjoint operator is

$$S_n = \Sigma^+_n \Sigma_n .$$

In this form it is not difficult to derive the commutation relations of S_n^{-1}. We find

$$\begin{aligned}
[W^{(1)}_n, S^{-1}_m] &= 0 \\
[S^{-1}_n, S^{-1}_m] &= 0 \qquad \text{if } |n-m| > 2 \\
qS^{-1}_n S^{-1}_{n+1} &= q^{-1} S^{-1}_{n+1} S^{-1}_n + \frac{1}{4}(q - q^{-1})(S^{-1}_n + S^{-1}_{n+1}) \\
[S^{-1}_n, S^{-1}_{n+2}] &= \frac{1}{4}(q^3 - q)[S^{-1}_{n+2} + \frac{1}{4}(q^{-1} - 1)] \cdot \\
&\qquad \cdot [S^{-1}_{n+1} + \frac{1}{4}(q-1)]^{-1} \cdot [S^{-1}_n + \frac{1}{4}(q^{-1} - 1)] .
\end{aligned} \tag{72}$$

In the representation eq. (64), we get

$$W^{(1)}_n = -\Delta q^{-\frac{1}{4}} e^{\Delta \bar{p}_{n+1}}$$

$$W^{(2)}_n = -\Delta q^{-\frac{1}{2}} [e^{\Delta \bar{p}_{n+2}} e^{-\Delta \dot{p}_{n+1}} + e^{\Delta \dot{p}_{n+2}} e^{\Delta \bar{p}_{n+1}}]$$

so that finally

$$S^{-1}_n = \frac{1}{4}[1 + e^{-2\Delta p_n} + e^{2\Delta p_{n+1}} + q e^{2\Delta p_{n+1}} e^{-2\Delta p_n}] .$$

This is the quantum Miura transformation on the lattice [23,24,16].

References

[1] A.M.Polyakov, *Quantum Geometry of Bosonic Strings*, Phys.Lett. **103B** (1981), 207.

[2] P.Zograf, L.Takhtajan, *Action of Liouville equation is a generating function for the accessory parameters and the potential of the Weil-Petersson metric on the Teichmuller space*, Func. Anal. Appl. **19** (1985), 219; Math. USSR Sbornik **60** (1988), 143, 297.

[3] J.L.Gervais, A.Neveu, *Novel triangle relation and absence of tachyons in Liouville string field theory*, Nucl. Phys. **B238** (1984), 125.

[4] L.D.Faddeev, L.Takhtajan, *Liouville model on the lattice*, Springer Lecture Notes in Physics **246** (1986), 166.

[5] O.Babelon, *Extended conformal algebra and the Yang-Baxter equation*, Phys. Lett. **B215** (1988), 523.

[6] J.M.Maillet, *New integrable canonical structures in two-dimensional models*, Nucl. Phys. **B269** (1986), 54.

[7] A.Alekseev, L.Faddeev, M.Semenov-Tian-Shansky, A.Volkov, *The unravelling of the quantum group structure in the WZNW theory*, Preprint CERN Th.5981/91.

[8] F.Nijhoff, V.Papageorgiou, H.Capel, *Integrable Time-discrete systems: Lattices and Mappings*, Preprint Clarkson University INS 166 (1990); *Integrable Quantum Mappings*, Preprint Clarkson University INS 168 (1991).

[9] A.N.Leznov, M.V.Saveliev, *Representation of zero curvature for the system of non-linear partial differential equations* $x_{\alpha,z\bar{z}} = \exp(kx)_\alpha$ *and its integrability*, Lett. Math. Phys. **3** (1979), 489.

[10] A.A.Belavin, A.M.Polyakov, A.B.Zamolodchikov, *Infinite conformal symmetry in two dimensional quantum field theory*, Nucl. Phys. **B241** (1984), 333.

[11] J.L.Gervais, A.Neveu, *Dimension shifting operators and null states in 2D conformally invariant field theories*, Nucl. Phys. **B264** (1986), 557.

[12] E.Sklyanin, *On the complete integrability of the Landau-Lifshitz equation*, Preprint LOMI E-3-79, Leningrad.

[13] O.Babelon, D.Bernard, *Dressing Transformations and the Origin of the Quantum Group Symmetries*, Preprint SPhT 91/016, PAR LPTHE 91/15.

[14] M.Semenov-Tian-Shansky, *Dressing Transformations and Poisson Lie Group Actions*, Pub. RIMS **21** (1985), 1237.

[15] O.Babelon, L.Bonora, F.Toppan, *Exchange algebra and the Drinfeld-Sokolov theorem*, Preprint SISSA 65/90/EP; Commun. Math. Phys., (to appear).

[16] O.Babelon, L.Bonora, *Quantum Toda theory*, Phys. Lett. **253B** (1991), 365.

[17] L.D.Faddeev, *Integrable Models in* 1+1 *Dimensional Quantum Field Theory*, Les Houches Lectures, 1982; Elsevier, Amsterdam, 1984.

[18] M.Jimbo, *A q-Difference analogue of* $U(G)$ *and the Yang-Baxter equation*, Lett. Math. Phys. **10** (1985), 63.

[19] V.G.Drinfeld, *Quantum Groups*, Proc. of the ICM-86, vol. 1, Berkeley, 1987, p. 798.

[20] O.Babelon, *Universal exchange algebra for Bloch waves and Liouville theory*, Preprint PAR LPTHE 91/11; Commun. Math. Phys., (to appear).

[21] J.L.Gervais, *The quantum group structure of 2D gravity and minimal models*, Commun. Math. Phys. **130** (1990), 257.

[22] J.L.Gervais, *Solving the strongly coupled 2D gravity: 1. Unitary truncation and quantum group structure*, Preprint LPTENS 90/13.

[23] A.Volkov, Zapiski Nauch. Semin. LOMI **150** (1986), 17; Zapiski Nauch. Semin. LOMI **151** (1987), 24, (To be translated in Sov. Jour. Math.).

[24] A.Volkov, *Miura transformation on the lattice*, Theor. Math. Phys. **74** (1988), 135.

[25] O.Babelon, *Exchange formula and lattice deformation of the Virasoro algebra*, Phys. Lett. **238B** (1990), 234.

[26] O.Babelon, *Integrable systems associated to the lattice version of the Virasoro algebra. I. The classical open chain*, Talk given at the Workshop "Integrable systems and quantum groups" Pavia March 1-2 1990. Preprint PAR LPTHE 90/5.

Laboratoire de Physique Théorique et Hautes Energies, Université Pierre et Marie Curie, Tour 16, 1er étage, 4 Place Jussieu, 75252 Paris CEDEX 05-FRANCE

NON–LOCAL CURRENTS IN 2D QFT: AN ALTERNATIVE TO THE QUANTUM INVERSE SCATTERING METHOD

DENIS BERNARD[(1)] AND ANDRÉ LECLAIR[(2)]

[(1)]Service de Physique Théorique de Saclay, Gif-sur-Yvette
[(2)]Institute of Theoretical Physics, University of California, Santa Barbara

The formalism based on non-local charges that we propose provides an alternative to the quantum inverse scattering method for solving integrable quantum field theories in 2D. The content of the paper is:

1. Introduction: *historical background.*
2. The NLC approach to 2D QFT: *a summary.*
3. Exchange algebras and on-shell conservation laws: *why non-local charges are useful.*
4. The lattice construction: *the geometrical origin of non-local conserved currents.*
5. The continuum construction: *how to deal with non-local conserved currents.*
6. Examples: *Yangian and quantum group currents.*
7. Conclusions: *open problems.*

1. Introduction

We propose a non-perturbative approach to 2D massive quantum field theories which is based on quantum symmetries.

Symmetry in quantum field theory is widely recognized as being of fundamental importance. In $3+1$ spacetime dimensions, the likely symmetries of the S- matrix are subject to the severe limitations of the Coleman-Mandula theorem [1]. The theorem states that any symmetry group is necessarily locally isomorphic to the direct product of an internal symmetry group and the Poincaré group. These possible symmetries are normally not restrictive enough to allow a non-perturbative solution of the theory.

In lower-dimensional quantum field theory, some of the postulates of the Coleman-Mandula theorem may be relaxed in non-trivial way. Apart from the structure of the Poincaré group itself, one of the main assumptions of the theorem is that the symmetry group acts on multi-particle states as if they were tensor products of one-particle states. More specifically, let Q be a generator of the symmetry group and W denote the Hilbert space of one- particle states. The multi-particle states are spanned by $W\otimes\cdots\otimes W$. The action of the generator on a multi-particle state is an operator from $W\otimes\cdots\otimes W$ into itself, and will be referred to as the comultiplication $\Delta(Q)$. The proof of the Coleman-Mandula theorem assumed the trivial comultiplication:

$$\Delta(Q)=Q\otimes 1\otimes\cdots\otimes 1+1\otimes Q\otimes 1\otimes\cdots\otimes 1+\ldots 1\otimes\cdots\otimes Q \tag{1.1}$$

Supersymmetry in any number of dimensions is of course a well known example of how the hypothesis (1.1) can be circumvented.

The subject of integrable quantum field theory has a long history, with many impressive results. The Leningrad school developed the quantum inverse scattering method (QISM) to quantize integrable systems, including the sine-Gordon model [2][3]. It was this investigation that led to the theory of the Quantum Group. The QISM is an algebraization of the Bethe ansatz methods. In quantum field theory the QISM was developed as a way to quantize the theory in a manner that preserves the infinite number of commuting integrals of motion which exist in the classical theory. The Bethe ansatz has had many successes, but many of its features (such as the introduction of pseudo-vacuum and the entailed complicated filling of the Dirac sea) are rather unpleasant for the typical quantum field theorist.

This paper is primarily devoted to the description of some new quantum symmetries that exist in a wide variety of known integrable quantum field theories in $1+1$ dimensions. The conserved currents that generate the symmetries are non-local and further characterized by non-trivial equal-time commutation, or braiding, relations. These exceptional properties of the currents are responsible for the non-trivial comultiplication of the charges when acting on multi-particle states. That is, the assumption (1.1) of the Coleman-Mandula theorem is violated. Contrary to the QISM, the non-local charge framework deals with conserved charges that are not in involution. The efficiency of the method resides in the non-abelian nature and in the non-commutativity of the algebras generated by the non-local charges. Though the same algebraic structure of the Quantum Group appears in both the QISM and the theory of non-local charges,their physical content is different.

The existence of a non-trivial comultiplication for the non-local conserved charges implies that they belong to the algebraic framework of Hopf algebras, notably the Quantum Groups and the Yangian symmetries [4] [5] [6]. Our work thus provides explicit currents that generate the Quantum Group symmetries. The non-local charges for the Yangian symmetry were studied in [7] [8]. The Yangians are hidden symmetries of 2D massive current algebras. The Quantum Group symmetries are generated by currents having fractional Lorentz spin [9]. They provide a non-Abelian extension of the 2D Poincaré algebra. A typical example is the sine-Gordon theory for which the hidden quantum symmetry algebra is the quantum $sl(2)$ loop algebra. It is an infinite dimensional symmetry. The algebra generated by the non-local charges allows for a non-perturbative treatment of the theories.

2. The NLC approach to 2D QFT

The integrability of two dimensional quantum field theories requires local integrals of motion in involution. Only one integral of motion besides the trivial ones is required for factorization of the S-matrices. But, in general the Abelian character of algebra generated by these integrals of motion prevent them from providing significant non-perturbative results. For example, solving the two-dimensional conformal field theories only from their integrals of motion which are in involution would have been difficult. The non-local currents (NLC) approach is a proposal to avoid these difficulties. It is based on a formulation of the quantum field theories which relies on non-local quantum symmetries. The basic points of this approach can be summarized as follows:

•) **Definition and construction of the non-local conserved currents.** As we will illustrate in the following, there are at least three or four approaches to this constructive problem. Solving this problem requires characterizing (at least in part) the operator algebra of the models. Being non-local the currents will be characterized by non-trivial braiding relations among them and with the fields of the theories.

•) **The action of the charges on the fields.** The models invariant under non-local quantum symmetries are distinguished by their field content and the way the conserved charges act on them. In section 5, we will review the techniques involved in the description of these actions. Because of their non-locality, the charges act on the fields as generalized graded commutators and have non-trivial comultiplications. They generate a quantum algebra and are naturally interpreted as quantum Lie derivatives.

•) **The conservation laws and the S-matrices.** By their non-commutativity and non-cocommutativity the non-local charges provide non-perturbative information on the S-matrices. This information is hidden in the on-shell conservation laws for the non-local charges which turn out to be the exchange relations for the quantum algebra generated by non-local charges. In general, they determine the S-matrices.

•) **The Ward identities for the correlation functions.** Besides giving information on the S-matrices, the non-local charges also impose constraints on the correlation functions through the Ward identities.

There are a priori four approaches for defining non-local conserved currents. The first one consists in defining the theory on a lattice. In section 4, we will use this method to reveal the geometrical òrigin of the non-local currents. This approach requires analyzing how the non-local currents behave in the scaling limit. The other approaches apply directly to the continuum. Either one defines the quantum theory by specifying its operator algebra or one defines it as a relevant perturbation of its ultraviolet limit. We will present examples of both methods in section 6. Finally, another approach consists in defining the quantum field theory from its symmetries. In a way analogous to the standard approach to two-dimensional conformal field theories, the massive quantum field theories and their operator algebras are defined as the minimal models invariant under the quantum symmetries.

3. Exchange algebras and on-shell conservation laws

The simplest way to realize that non-local conserved charges provide non-perturbative information on the S-matrices consists in going backwards. Namely, in this section we assume that the S-matrices are known and we look at the conserved charges we can define using this knowledge.

Let us suppose that we have multiplets of asymptotic particles. We denote by W_A the vector spaces spanned by the particles of the species A. The two-particle to two-particle S-matrix is an operator S_{AB}:

$$S_{AB} : W_A \otimes W_B \rightarrow W_B \otimes W_A. \tag{3.1}$$

By Lorentz invariance, S only depends on the difference θ_{12} of the rapidities: $\theta_{12} = \theta_1 - \theta_2$. The rapidities parametrize the energy-momentum of the particles: ($E =$

$m\cosh\theta; p = m\sinh\theta$). Factorization requires $S(\theta_{12})$ to satisfy the Yang-Baxter equation:

$$(S_{BC}\otimes 1)(1\otimes S_{AC})(S_{AB}\otimes 1) = (1\otimes S_{AB})(S_{AC}\otimes 1)(1\otimes S_{BC}). \tag{3.2}$$

Unitarity and crossing symmetry further constrain the S-matrix.

As is well known, to any solution of the Yang-Baxter equation we can associate an exchange algebra. The algebra is generated by a matrix valued operator $T(\theta) = (T_{ij}(\theta))$. We denote by $T_A(\theta)$ the operators acting on W_A. They satisfy:

$$S_{AB}(\theta_{12})[T_A(\theta_1)\cdot T_B(\theta_2)] = [T_B(\theta_2)\cdot T_A(\theta_1)]S_{AB}(\theta_{12}). \tag{3.3}$$

A representation of the algebra (3.3) is provided by the S-matrix itself. Namely to any species C we may associate the generator $T^{(C)}(\theta)$ whose action on W_A is defined by $T^{(C)}_A(\theta) = S_{AC}(\theta)$. They form a representation of the algebra (3.3) because S satisfies the Yang-Baxter equation. Since $S(\theta_{12})$ depends only on the difference of the rapidities the exchange algebra (3.3) possesses an automorphism: $\theta \to \theta - \lambda$. Therefore we can define a one-parameter family of generators $\tau^\lambda(\theta)$ by:

$$\tau^\lambda(\theta) = T(\theta-\lambda). \tag{3.4}$$

They also satisfy the exchange algebra (3.3). The operators we are interested in are not the generators τ^λ but their logarithmic derivatives, denoted Q^λ:

$$Q^\lambda = \left(\frac{\partial}{\partial\lambda}\tau^\lambda\right)(\tau^\lambda)^{-1}. \tag{3.5}$$

On-shell the charges Q^λ act on the particles of type A and of fixed rapidities θ as $Q^\lambda_A(\theta) = (\frac{\partial}{\partial\lambda}\,\tau^\lambda_A(\theta))\,\tau^\lambda_A(\theta)^{-1}$.

Taking logarithmic derivatives transforms the multiplicative form of the exchange algebra (3.3) into additive form. Denote by $Q^\lambda_{A;ij}$ the matrix elements of Q^λ_A. The exchange algebra (3.3) is then equivalent to:

$$S_{AB}[Q^\lambda_{A;ij}\otimes 1 + \sum_{kl}\Theta^{\lambda;kl}_{A;ij}\otimes Q^\lambda_{B;kl}] = [Q^\lambda_{B;ij}\otimes 1 + \sum_{kl}\Theta^{\lambda;kl}_{B;ij}\otimes Q^\lambda_{A;kl}]S_{AB} \tag{3.6}$$

with $\Theta^{\lambda;kl}_{A;ij} = (\tau^\lambda_A)_{ik}(\tau^\lambda_A)^{-1}_{lj}$. Note that the braiding matrices Θ^λ_A act on the charges Q^λ_B by an adjoint action: $\sum_{kl}\Theta^{\lambda;kl}_{A;ij}Q^\lambda_{B;kl} = (\tau^\lambda_A\cdot Q^\lambda_B\cdot\tau^{\lambda-1}_A)_{ij}$. The equations (3.6) can alternatively be written as follows:

$$S_{AB}\left(\Delta_{A;B}(Q^\lambda_{ij})\right) = \left(\Delta_{B;A}(Q^\lambda_{ij})\right)\ S_{AB} \tag{3.7a}$$

$$\Delta_{A;B}(Q^\lambda_{ij}) = Q^\lambda_{A;ij}\otimes 1 + \sum_{kl}\Theta^{\lambda;kl}_{A;ij}\otimes Q^\lambda_{B;kl}. \tag{3.7b}$$

Equations (3.7) are naturally interpreted as conservation laws for the on-shell conserved charges Q^λ_{ij}. They act on the asymptotic particles of the species A and with rapidities θ as $Q^\lambda_{A;ij}(\theta)$. The generators Q^λ_{ij} are formal power series in λ; there is thus an infinite number of conserved charges. They are non-local because they have non-trivial comultiplications. The relations between non-locality and non-trivial comultiplications will be clarified later.

Finally let us show how we recover local conserved charges from the above construction. The local charges q^λ are given by the trace of the non-local charges:

$$q^\lambda = Tr(Q^\lambda) = Tr\left(\left(\frac{\partial}{\partial\lambda}\tau^\lambda\right)(\tau^\lambda)^{-1}\right). \tag{3.8}$$

They are local because the trace kills the Θ^λ factors in the comultiplications. They are obviously conserved:

$$S_{AB}(q_A^\lambda \otimes 1 + 1 \otimes q_B^\lambda) = (q_B^\lambda \otimes 1 + 1 \otimes q_A^\lambda)S_{AB}. \tag{3.9}$$

It is easy to check that the definitions (3.8) and (3.5) of the local and non- local charges are compatible with the bootstrap conditions.

In summary, *the exchange algebras reflect the conservation laws of on-shell non-local charges.* The NLC approach consists in developing this in the opposite direction by constructing the non-local charges in the quantum field theory and then by deriving the S-matrix using the exchange algebras, alias the conservation laws.

4. The lattice construction

Non-local conserved currents can be defined in a lattice formulation of models[1]. The lattice approach reveals the geometrical origin of the non-local conserved currents in a simple way: *Non-local conserved currents originate in the quantum group invariance of the local hamiltonian.* In other words, if the lattice model possesses a quantum group invariance there are non-local conserved currents associated to it.

To illustrate this point, let us consider a vertex model on square lattice. Each vertex of the lattice corresponds to a space-time point (x,t). In our convention the time t goes up and x-coordinate increases from the left to the right. On each link of the lattice there is a copy of a vector space V, called the configuration space, to which the generalized spins of the model belong. The vertex model is defined by the data of the configuration space V and of the R-matrix of the Boltzmann weights; the matrix R acts on $V \otimes V$. Graphically we will represent it by a crossing:

$$R: V \otimes V \to V \otimes V \equiv \quad + \tag{4.1}$$

4a) Lie group invariance and local conserved currents.

Before dealing with quantum groups, let us warm up with Lie groups by proving the physically obvious statement that Lie group invariance of the R-matrix implies the existence of local conserved currents taking values in the Lie algebra of the group.

Suppose that the configuration space V carries a representation of a Lie group G. We denote this representation by ρ. We suppose that the R-matrix is G-invariant; this means that:

$$R\,(\rho(g)\otimes\rho(g)) = (\rho(g)\otimes\rho(g))\,R \quad ; \quad \forall g \in G. \tag{4.2}$$

[1] The result presented in this section have been obtained in collaboration with G.Felder [10].

Taking Lie derivatives of equation (4.2) gives its infinitesimal form. If $t^a, a = 1, \dots, \dim \mathcal{G}$, form a basis of the Lie algebra $\mathcal{G}$ of G, the equation (4.2) implies:

$$R(T^a \otimes 1 + 1 \otimes T^a) = (1 \otimes T^a + T^a \otimes 1)\, R, \tag{4.3}$$

where $T^a = \rho(t^a)$ are the matrices representing the Lie algebra $\mathcal{G}$ in the configuration space V. Equation (4.3) admits the following simple graphical representation:

(4.4)

Here the crosses drawn on a link represent insertions of the matrices T^a acting on the copy of the configuration space V located in this link.

We now introduce currents, denoted by $J^a_\mu(x,t)$, taking values in $\mathcal{G}$. Their components, $J^a_x(x,t)$ and $J^a_t(x,t)$, are defined by inserting a matrix T^a acting on the copy of the configuration space V located on the link ending at the point (x,t) and respectively oriented in the x- or t-direction. This definition is better summarized by the following picture:

(x,t)

$J^a_x(x,t) =$ 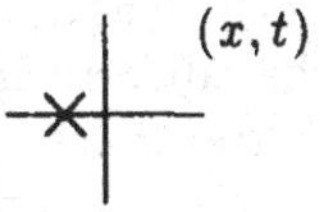 (4.5a)

(x,t)

$J^a_t(x,t) =$ 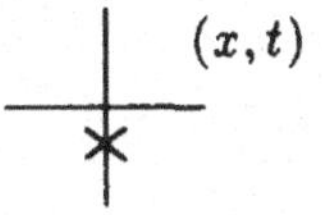 (4.5b)

The currents (4.5) are obviously local. The equations (4.3) or (4.4) are the lattice conservation laws for the currents (4.5):

$$\partial_\mu J^a_\mu(x,t) = 0. \tag{4.6}$$

This proves the standard statement: *Local conserved currents reflect Lie group invariance.*

4b) Quantum group invariance and non-local conserved currents.

Our aim is now to reproduce the arguments of the previous section with the Lie group replaced by a quantum group. The outcome will be the definition of non-local currents which are conserved due to the quantum group invariance of the R-matrix of Boltzmann weights.

Suppose, that the configuration space V carries a co-representation of the Hopf algebra $F(G_q)$ of 'functions over a quantum group'. By abuse of notation we denote the matrix elements of this co-representation by $\rho(g)$, $\rho(g) \in F(G_q)$. Quantum group invariance then means that the R-matrix is an intertwiner for $F(G_q)$;

$$R(\rho(g) \otimes 1)(1 \otimes \rho(g)) = (1 \otimes \rho(g))(\rho(g) \otimes 1)\, R. \tag{4.7}$$

Equation (4.7) is analogous to the equation (4.2) but is not quite the same because in the quantum group case, we have to be careful with the order of the tensor product since the matrix elements $\rho(g)$ do not commute.

As in classical case the infinitesimal form of the equation (4.7) is deduced by acting on it with Lie derivatives. This requires developing a differential calculus on the quantum group $F(G_q)$. We refer to [11][12] for the details concerning its formulation. The net outcome is the following infinitesimal form of (4.7):

$$R\,(T^a \otimes 1 + \Theta^a_b \otimes T^b) = (1 \otimes T^a + T^b \otimes \Theta^a_b)\,R. \tag{4.8}$$

Summation over repeated indices is implicit. The matrices T^a and Θ^a_b form a representation of the algebra $\mathcal{U}_q(\mathcal{G})$ of the quantum Lie derivatives acting on $F(G_q)$ in the configuration space V. The matrices Θ^a_b, which define a generalized grading, arise when trying to commute the quantum Lie derivatives through the quantum group elements. In other words, they define the generalized grading of the quantum Liebnitz rules. A graphical representation for the equation (4.8) is the following:

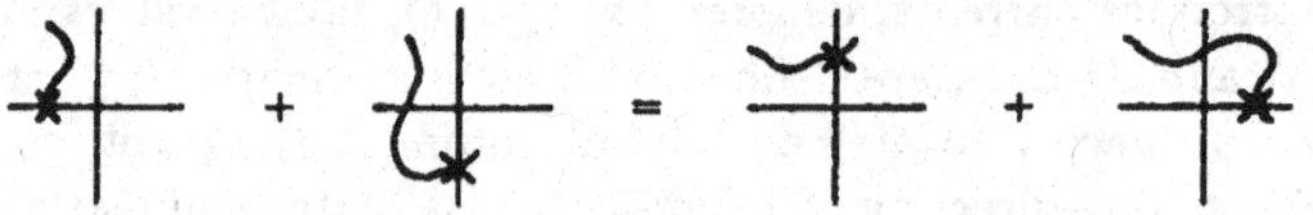

(4.9)

As before the crosses denote insertions of the matrices T^a but the wavy lines denote insertions of the matrices Θ^a_b. The fact that the lines and the crosses are touching indicates a summation of the adjacent indices.

The components of the currents are now defined by the following graphical representations:

$J^a_x(x,t)$ = (x,t) (4.10a)

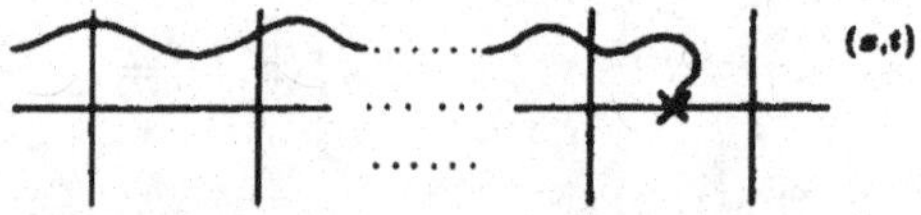

$J^a_t(x,t)$ = (x,t) (4.10b)

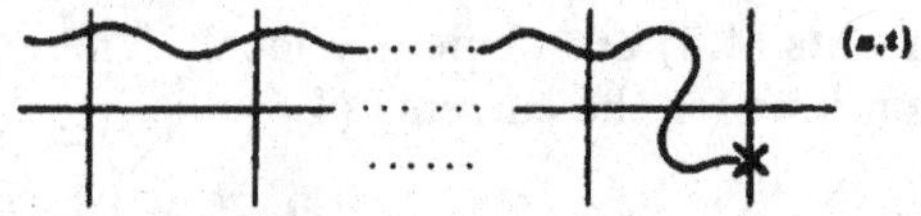

The definition (4.10) can be easily translated into precise equations but their description is cumbersome. In (4.10) the wavy lines represent the insertion of a collection of matrices Θ^a_b with an implicit summation over adjacent indices. The precise location of the string attached to the currents is irrelevant up to topological obstructions. The matrices Θ^a_b can be thought of as defining a quantum flat connection on the lattice. The definition (4.10) is analogous to the lattice definition of parafermions. Parafermions are products of disorder operators by spin fields; in (4.10) the crosses represent the analogue of the spin fields whereas the wavy strings are the analogue of the disorder lines. The non-locality of the local currents (4.10) is transparent in their definition. Moreover, the equations (4.8) and (4.9) are the conservation laws for them:

$$\partial_\mu J^a_\mu(x,t) = 0. \tag{4.11}$$

Hence we proved our claim: *Non-local currents reflect quantum group invariance.*

Further properties of the lattice non-local conserved currents will be described in [10] in collaboration with G.Felder. There, the connection between non-local conserved currents, differential calculus on quantum groups and their quasi-triangular quantum Lie algebras introduced in [12], as well as examples, will be developed.

5. The continuum construction

In this section, we first review the basic techniques for dealing with non-local currents. This framework was used in [8] and [9]. Examples will be given in section 6.

Due to the possibility of fields with non-trivial braiding relations, quantum field theories in two spacetime dimensions may have non-local conserved currents. The currents, which we denote by $J^a_\mu(x,t)$, are localized at the spacetime points (x,t). Their precise definition (e.g. from lattice construction we described in section 4 or directly in the continuum as we will describe it in section 6) requires attaching to the currents a one-dimensional curve going from $-\infty$ to the point (x,t). The precise location of the string attached to the currents is irrelevant except when topological obstructions are encountered. The non-locality of the currents is encoded in their equal-time braiding relations:

$$J^a_\mu(x,t)\ J^b_\nu(y,t)\ =\ R^{ab}_{cd}\ J^c_\nu(y,t)\ J^d_\mu(x,t)\quad ;\quad \text{for } x>y. \tag{5.1}$$

The above equation is implicitly time-ordered to the left, e.g. $J^a_\mu(x,t)J^b_\nu(y,t) \equiv J^a_\mu(x,t+\epsilon)J^b_\nu(y,t)$, for ϵ small and positive. The braiding relations (5.1) originate in the topological obstructions encountered while trying to move the string attached to the currents, and are displayed in fig.1.

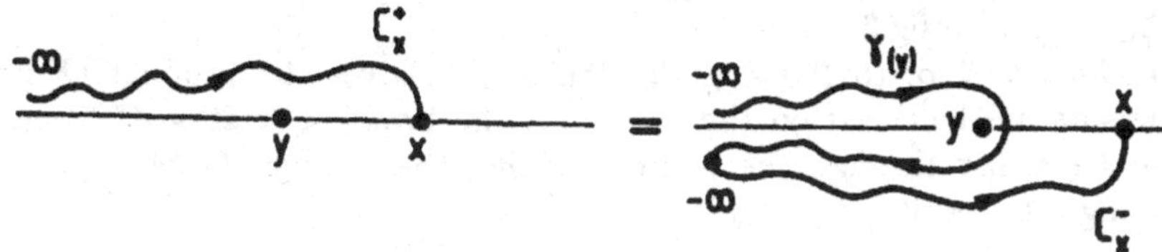

Fig.1: graphical representation of braiding

In this figure, time increases upward, and the positions of the strings are dictated by the time-ordering. Associativity of the operator algebra requires the matrix R^{ab}_{cd} to be a solution of the Yang-Baxter equation. A more complete discussion of braiding relations in 2D quantum field theories can be found in [13] [14].

For conserved currents, $\partial_\mu J^a_\mu(x,t) = 0$, the global conserved charges Q^a acting on the physical Hilbert space are $Q^a = \frac{1}{2\pi i}\int_t dx\ J^a_t(x,t)$. The charges $\hat{Q}^a$ acting on the fields $\Phi^k(y)$ are not defined by integrating the currents along an equal-time slice but along a path $\gamma(y)$ from $-\infty$ to $-\infty$ surrounding the point y:

$$\widehat{Q}\left(\Phi^k(y)\right)\ =\ \frac{1}{2\pi i}\int_{\gamma(y)} dz_\nu\ \epsilon^{\nu\mu}\ J^a_\mu(z)\ \Phi^k(y). \tag{5.2}$$

The contour $\gamma(y)$ is drawn in fig.2. There we have drawn the string attached to the currents in the position specified by the order of the fields in (5.2). The exact shape of the contour $\gamma(y)$ is irrelevant due to the conservation of the currents.

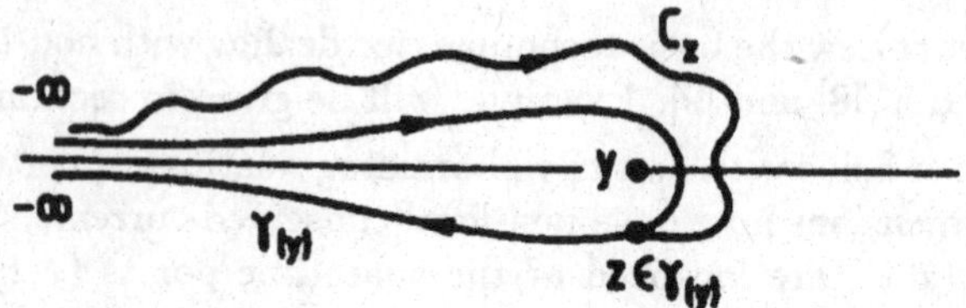

Fig.2: the contour of the integration for the action of non-local charges on fields

The action of the charges on the fields (5.2) can be expressed in terms of generalized commutators, as we now describe. The fields of the theory can be classified into multiplets according to their braiding relations with the currents. We suppose the following braiding relations

$$J^a_\mu(x,t)\ \Phi^k(y,t)\ =\ \Theta^{ak}_{bl}\ \Phi^l(y,t)\ J^b_\mu(x,t)\quad;\quad \text{for } x>y. \tag{5.3}$$

As in (5.1) these braiding relations arise from the obstructions for moving the string attached to the fields. To express (5.2) as generalized braided commutators, let us decompose the contour of integration $\gamma(y)$ into the difference of two contours γ_+ and γ_-, $\gamma(y)=\gamma_+-\gamma_-$, as in fig.3.

Integrating the r.h.s. of (5.2) along the contour γ_+ gives the product $Q^a\Phi^k(y)$. When the currents are localized on the curve γ_-, the braiding relations(5.3) can be used to move the string through the point y, giving the product $\Theta^{ak}_{bl}\Phi^l(y)Q^b$. We gather everything into result:

$$\hat{Q}\left(\Phi^k(y)\right)\ =\ Q^a\ \Phi^k(y)-\Theta^{ak}_{bl}\ \Phi^l(y)\ Q^b. \tag{5.4}$$

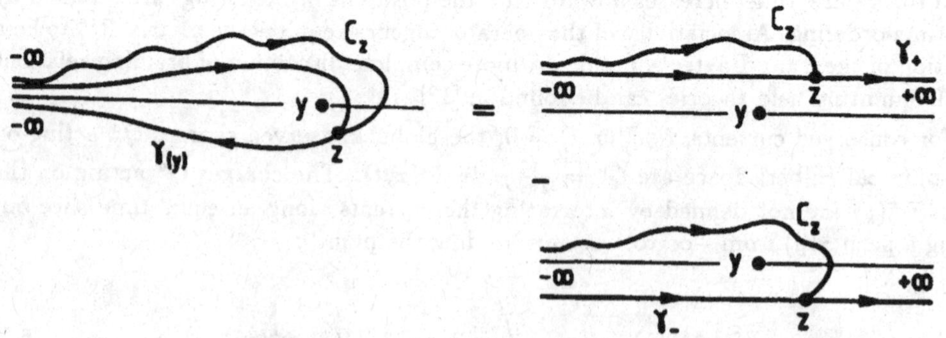

Fig.3: decomposition of the contour of fig.2

In particular if we consider $\Phi(y)$ as the time-component of the currents, (5.4) shows that the global conserved charges Q^a satisfy braided communication relations:

$$\mathrm{Adj}(Q^a)(Q^b)\ \equiv\ \hat{Q}^a(Q^b)\ =\ Q^a\ Q^b\ -\ R^{ab}_{cd}\ Q^c\ Q^d. \tag{5.5}$$

This defines a quantum adjoint action.

Next we consider the action of the charges on a product of fields. This will define for us the comultiplication $\Delta(Q^a)$ of the charges. For simplicity, consider the action on a product of two fields. This is defined as

$$\widehat{Q}^a\left(\Phi^k(y_1)\Phi^n(y_2)\right) = \frac{1}{2\pi i}\int_{\gamma_{12}} dz_\nu \epsilon^{\nu\mu} J^a_\mu(z)\, \Phi^k(y_1)\, \Phi^n(y_2). \tag{5.6}$$

The contour γ_{12} encloses both space-time points y_1 and y_2. The contour γ_{12} can be decomposed into the sum of two contours, $\gamma_{12} = \gamma_1 + \gamma_2$, as in fig.4.

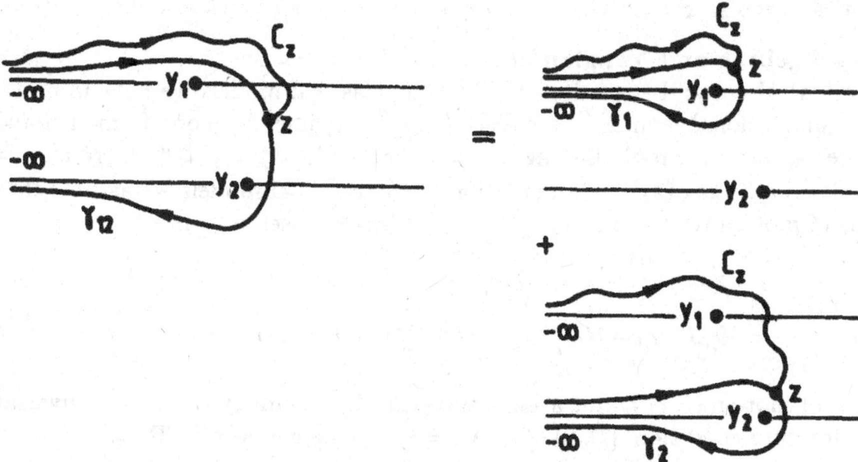

Fig.4: the contour defining the action of non-local charges on a product of fields.

The integration over the contour γ_1 gives the action of the charges on $\Phi^k(y_1)$. After having taken into account the braiding relations between the currents and the fields $\Phi^k(y_1)$, the integration over the contour γ_2 gives the action of the charges on $\Phi^n(y_2)$. Thus we obtain:

$$\widehat{Q}^a\left(\Phi^k(y_1)\Phi^n(y_2)\right) = \widehat{Q}^a\left(\Phi^k(y_1)\right)\Phi^n(y_2) + \left(\Theta^{ak}_{bl}\Phi^l(y_1)\right)\widehat{Q}^b\left(\Phi^n(y_2)\right). \tag{5.7}$$

Let us arrange the quantum numbers of the fields $\Phi(y_1)$ ($\Phi(y_2)$) into a vector space W_1 (W_2). The charges $\hat{Q}^a$ on the product of two fields is then an operator on $W_1 \otimes W_2$, which is denoted by $\Delta(\hat{Q}^a)$, and defines the comultiplication. In this compact notation(5.7) becomes

$$\Delta\left(\widehat{Q}^a\right) = \widehat{Q}^a \otimes 1 + \Theta^a_b \otimes \widehat{Q}^b \tag{5.8}$$

where Θ^a_b is the braiding operator and is a matrix acting on the vector space W_1. More specifically, $\langle l|\Theta^a_b|k\rangle = \Theta^{ak}_{bl}$. The same comultiplication holds for the global charges Q^a.

Hence, by simple contour manipulations we have shown: *The non-locality of the currents and their non-trivial braiding relations leads to generalized braided commutators and to non-trivial comultiplications.*

6. Examples

As we mentioned in section 2, there are at least three possible approaches for constructing non-local conserved currents in a non-perturbative way. The first one consists in defining the models on the lattice; we used this approach in section 4 to illustrate the geometrical origin of the non-local conserved currents. The two examples that we now describe illustrate two other approaches. The first example deals with current algebras in two dimensions; we show how it is possible to define the quantum theories by specifying the operators algebras. The second example deals with massive theories defined as perturbations of their ultraviolet limits. In both cases it leads us to non-perturbative construction of non-local conserved currents. We refer to the original papers [8] [9] for the details of the proofs.

6a) Current algebras and Yangian currents.

•) The classical theory: Let us first describe the classical models we have in mind. These are two-dimensional geometrical models. They are defined by a one-form, denoted by $J(x)$, valued in a semi-simple Lie algebra $\mathcal{G}$: $J(x)=\sum_a J^a_\mu(x)t^a dx^\mu$ where $t^a, a=1,\dots,\dim\mathcal{G}$, form a basis of $\mathcal{G}$[2] . By definition of the classical models we assume that the equations of motion require the $J^a_\mu(x)$ to be curl-free conserved currents:

$$\begin{aligned}\partial_\mu J^a_\mu(x)=0\\ \partial_\mu J^a_\nu(x)-\partial_\nu J^a_\mu(x)+f^{abc}J^b_\mu(x)J^c_\nu(x)=0.\end{aligned} \tag{6.1}$$

The equations of motion (6.1) admit a Lax representation. Namely they are equivalent to the zero curvature condition, $[\mathcal{D}_\mu(\lambda),\mathcal{D}_\nu(\lambda)]=0$, for the connection $\mathcal{D}_\mu(\lambda)$,

$$\mathcal{D}_\mu(\lambda) \;=\; \partial_\mu+\frac{\lambda^2}{\lambda^2-1}J_\mu(x)+\frac{\lambda}{\lambda^2-1}\epsilon_{\mu\nu}J_\nu(x). \tag{6.2}$$

The geometrical character of the equation (6.1) leads to the definition of an infinite set of conserved charges. These conserved charges are non-local. They can be defined in two different but equivalent ways: either one uses the transfer matrix defined from the connection $\mathcal{D}_\mu(\lambda)$ as a generating function for the non-local conserved charges as in ref. [15]; or one uses a recursive definition of the conserved currents as explained in ref. [16]. In the following we only need the two first conserved currents. The first one is the local one-form $J^{(0)^a}_\mu(x)=J^a_\mu(x)$. The second one, denoted by $J^{(1)}(x)=\sum_a J^{(1)^a}_\mu(x)t^a dx^\mu$, is defined by:

$$\begin{aligned}J^{(1)^a}_\mu(x) \;&=\; \epsilon_{\mu\nu}J^a_\nu(x)+\frac{1}{2}f^{abc}J^b_\mu(x)\phi^c(x)\\ \text{with } \partial_\mu\phi^c(x) \;&=\; \epsilon_{\mu\nu}J^c_\nu(x).\end{aligned} \tag{6.3}$$

The currents $J^{(1)^a}$ are non-local because $\phi^c(x)$ is non-local: $\phi^c(x)=\int_{C_x}\star J^c$ where C_x is a curve ending at the point x. Using eq. (6.1) it is easy to check that the currents $J^{(1)^a}_\mu(x)$ are conserved: $\partial_\mu J^{(1)^a}_\mu(x)=0$.

[2] We suppose the t^a orthonormalized. We use the convention: $[t^a,t^b]=f^{abc}t^c$ where f^{abc} denote the structure constants of $\mathcal{G}$.

•) The quantum theory: The currents $J_\mu^{(1)^a}(x)$ are the ones we want to quantize. A way to define the quantum models consists in imposing constraints on the operator algebra. This approach was described in ref. [8]. It is formulated by specifying the operator algebra generated by the quantum currents $J_\mu^a(x,t)$. This mainly consists in requiring that the operator product expansions of the currents $J_\mu^a(x,t)$ close on the currents and their descendants. It leads us to formulate an algebraic definition of these massive current algebras in two dimensions. A hidden consequence of this definition is that the massive current algebras actually describe perturbations of $\mathcal{G}^{(1)}$ affine Kac-Moody algebras by the perturbing fields $\Phi_{pert.}(x)=\sum_a J_\mu^a(x)J_\mu^a(x)$.

The properties of the 2D massive current algebras that we define in [8] ensure that the quantum currents $J_\mu^a(x,t)$ still satisfy the geometrical quantum equation of motion (6.3):

$$\begin{aligned} \partial_\mu J_\mu^a(x,t) &= 0 \\ \partial_\mu J_\nu^a(x) - \partial_\nu J_\mu^a(x) + f^{abc} \ : J_\mu^b(x) J_\nu^c(x) : &= 0 \end{aligned} \tag{6.4}$$

where the double dots denote an appropriate normal order. The proof of eq. (6.4) relies on a slight extension of a theorem due to Lüsher [7] which describes the operator product expansions between currents in a massive current algebra.

Having proved that the quantum conserved currents satisfy the quantum form (6.4) of the equations of motion (6.1), it is easy to define the quantum conserved currents $J^{(1)}(x,t)$. They are defined by a point splitting regularization ($\delta>0$):

$$\begin{aligned} J_\mu^{(1)^a}(x,t) &= \lim_{\delta\to 0^+} J_\mu^{(1)^a}(x,t|\delta) \\ J_\mu^{(1)^a}(x,t|\delta) &= Z(\delta)\epsilon_{\mu\nu} J_\nu^a(x,t) + \frac{1}{2} f^{abc} J_\mu^b(x,t)\phi^c(x-\delta,t) \end{aligned} \tag{6.5}$$

where $\phi^c(x,t)$, which satisfies $d\phi^c = \star J^c$, is defined by:

$$\phi^c(x,t) = \int_{C_x} \star J^c \tag{6.6}$$

The contour of integration C_x is a curve from ∞ to x. It is analogue of the wavy strings involved in the lattice definition of the non-local conserved currents described in section 4. The field $\phi(x,t)$ depends weakly on the contour C_x because $\star J$ is closed: $d\star J=0$.

In equation (6.5), $Z(\delta)$ is a renormalization constant which is determined by requiring that $J_\mu^{(1)^a}(x,t)$ are finite and conserved. The constant $Z(\delta)$ is logarithmically divergent in δ. The existence of the renormalization constant $Z(\delta)$ which, at the same time, ensures the finiteness and the conservation of the currents $J_\mu^{(1)^a}(x,t)$ is a non-trivial consequence of our definition of 2D massive current algebras.

•) The braiding relations and the algebra of conserved charges: The non-local character of the currents $J_\mu^{(1)^a}(x,t)$ is encoded in their braiding relations and in their Lorentz transformation law.

The former can be summarized as follows. Let $\Phi(y,t)$ be a quantum field local with respect to the currents $J_\mu^a(x,t)$. Then it satisfies the following equal-time braiding relations:

$$J_\mu^a(x,t)\ \Phi(y,t) = \Phi(y,t)\ J_\mu^a(x,t) \quad ; \quad \text{for } x \lessgtr y \tag{6.7a}$$

$$J_\mu^{(1)^a}(x,t)\Phi(y,t) = \Phi(y,t)J_\mu^{(1)^a}(x,t) \quad ; \quad \text{for } x < y \tag{6.7b}$$

$$J_\mu^{(1)^a}(x,t)\Phi(y,t) = \Phi(y,t)J_\mu^{(1)^a}(x,t) - \frac{1}{2}f^{abc}\hat{Q}_0^b\left(\Phi(y,t)\right)J_\mu^c(x,t) \quad ; \quad \text{for } x > y \tag{6.7c}$$

where $\hat{Q}_0^b$ are the global charges associated with the local conserved current $J_\mu^b(x,t)$. The first relation (6.7a) is simply the definition of the mutual locality of $J_\mu^a(x,t)$ and $\Phi(y,t)$. The proof of the braiding relations (6.7b,c) is the same as the proof of the braiding relations for disorder fields. It only relies on the manner of deforming the contour C_x entering in the definition of the currents $J_\mu^{(1)^a}(x,t)$.

Besides non-trivial braiding relations, the non-locality of the currents also implies that the non-local conserved charges generate a non-abelian extension of the two-dimensional Lorentz algebra. As explained in the introduction, this is possible since the Coleman-Mandula theorem [1] breaks down in two dimensions because the comultiplication of the symmetry algebra could be non-trivial. In other words our examples provide explicit counter-examples of the Coleman-Mandula theorem in two dimensions. In two dimensions the Poincaré algebra is generated by the momentum operators P_μ and Lorentz boost L is abelian. The momentum operators P_μ are the global charges associated with the conserved stress-tensor $T_{\mu\nu}(x) : \partial_\mu T_{\mu\nu}(x) = 0$. The Lorentz boost L is the global charge associated with the conserved boost current:

$$L_\mu(x) = \frac{1}{2}\epsilon^{\rho\sigma}\left(x_\rho T_{\mu\sigma}(x) - x_\sigma T_{\mu\rho}(x)\right). \tag{6.8}$$

On the local fields L acts as a commutator: $L(\Phi(y)) = L\,\Phi(y) - \Phi(y)\,L$. The currents $J_\mu^a(x)$, which are one-forms, transform covariantly under Lorentz boosts; the light-cone components $J_\pm^a(x,t)$ have Lorentz spin ± 1.

Let Q_0^b and Q_1^a denote the global charges respectively associated to the currents $J_\mu^a(x,t)$ and $J_\mu^{(1)^a}(x,t)$. They satisfy the following algebraic relations

$$\begin{aligned} [Q_0^a\,,\,Q_0^b] &= f^{abc}Q_0^c \quad ; \quad [L\,,\,Q_0^a] = 0 \\ [Q_0^a\,,\,Q_1^b] &= f^{abc}Q_1^c \quad ; \quad [L\,,\,Q_1^a] = -\frac{C_{Adj}}{4\pi i}Q_0^a \end{aligned} \tag{6.9}$$

where C_{Adj} is the Casimir of $\mathcal{G}$ in the adjoint representation if the normalization is defined by equation (6.9). The relations (6.9) are part of the defining relations of the semi-direct product of the Yangians $Y(\mathcal{G})$ by the Poincaré algebra. Only the Serre relations are missing. (See ref. [4] for more details on $Y(\mathcal{G})$.) It is worth noting that the normalization coefficients in the equations (6.9) are not arbitrary: They are the normalizations which ensure the crossing symmetry of the $Y(\mathcal{G})$-invariant S-matrices. The proof of the commutation relations (6.9) was given in ref. [8].

•) **Action on the fields and the comultiplications:** The action of the non-local charges on the quantum fields and their comultiplication follow from the general theory reviewed in section 5. Let $\Phi(y,t)$ be a quantum field local with respect to the conserved currents $J_\mu^a(x,t)$. Because the braiding relations between this field and the conserved currents are trivial for the currents $J_\mu^a(x,t)$ but non-trivial for the non-local currents

$J_\mu^{(1)^a}(x,t)$, the charges act as pure commutators in the first case but as generalized braided commutators in the second case:

$$\begin{aligned}\hat{Q}_0^a(\Phi(y)) &= Q_0^a\Phi(y) - \Phi(y)Q_0^a \\ \hat{Q}_1^a(\Phi(y)) &= Q_1^a\Phi(y) - \Phi(y)Q_1^a + \frac{1}{2}f^{abc}\hat{Q}_0^b(\Phi(y))\,Q_0^c.\end{aligned} \tag{6.10}$$

The equations (6.10) are proved by deforming the contour of integration $\gamma(y)$ and by using the braiding relations.

Let us now describe the comultiplications. The comultiplications just encode how the charges act on a product of fields, say $\Phi_1(y_1)\Phi_2(y_2)\ldots$. As explained in section 5, for the fields $\Phi_n(y_n)$ which are local with respect to the currents $J_\mu^a(x)$ the comultiplication can be deduced by deforming the contours of fig.4 and by using the braiding relations (6.7). For the charges Q_0^a associated to the local conserved currents $J_\mu^a(x,t)$, we find

$$\begin{aligned}\hat{Q}_0^a(\Phi_1(y_1)\,\Phi_2(y_2)) &= \hat{Q}_0^a(\Phi_1(y_1))\,\Phi_2(y_2) + \Phi_1(y_1)\hat{Q}_0^a(\Phi_2(y_2)) \\ \Delta\hat{Q}_0^a &= \hat{Q}_0^a\otimes 1 + 1\otimes\hat{Q}_0^a,\end{aligned} \tag{6.11}$$

for the charges Q_1^a associated to the non-local conserved currents $J_\mu^{(1)^a}(x,t)$ we have

$$\begin{aligned}\hat{Q}_1^a(\Phi_1(y_1)\Phi_2(y_2)) &= \hat{Q}_1^a(\Phi_1(y_1))\,\Phi_2(y_2) + \Phi_1(y_1)\,\hat{Q}_1^a(\Phi_2(y_2)) - \\ &\quad -\frac{1}{2}f^{abc}\hat{Q}_0^b(\Phi_1(y_1))\,\hat{Q}_0^c(\Phi_2(y_2)) \\ \Delta\hat{Q}_1^a &= \hat{Q}_1^a\otimes 1 + 1\otimes\hat{Q}_1^a - \frac{1}{2}f^{abc}\hat{Q}_0^b\otimes\hat{Q}_0^c.\end{aligned} \tag{6.12}$$

Equation (6.11) is the standard Lie algebra comultiplication as it should be. Equation (6.11) together with the equation (6.12) are defining the relations of the non-cocommutative comultiplication of the Yangians $Y(\mathcal{G})$.

•) **Action on the asymptotic states and the S-matrices:** We now describe how to obtain non-perturbative results for the S-matrices by examining the action of the non-local charges on the asymptotic states. The constraints of the S-matrices we obtain arise by requiring that they commute with the non-local conserved charges. In the cases we are dealing with, these commutation relations will simply be the exchange algebras for the Yangians $Y(\mathcal{G})$. The constructions we explained are model-independent. Different models are distinguished by their field content and spectrum of massive particles. The models may also be invariant under other symmetries. They will then not be completely specified by the Yangian invariance; in particular the quantum fields will carry other quantum numbers besides the $Y(\mathcal{G})$ indices. In these cases the S-matrices will admit a factorization whose elementary factors are separately invariant under various quantum symmetries.

The way to deduce the action of Q_1^a on the asymptotic states is clear. It is enough to compute the action on the one-particle states because we already know the comultiplication. First we should identify the (asymptotic) fields, called Φ_Λ, which create the asymptotic particles. Because of the global $\mathcal{G}$-invariance the fields Φ_Λ belong to some representation Λ of the Lie algebra $\mathcal{G}$. The antiparticles belong to the conjugated representation Λ^* and they are asymptotically created by the fields Φ_{Λ^*}. Secondly because the local conserved currents $J^a_\mu(x)$ appear in the operator product expansions between Φ_Λ and Φ_{Λ^*} they can be written as (generalized) normal ordered products of the two conjugated fields Φ_Λ and Φ_{Λ^*}. Finally once we have the expression of the currents in terms of Φ_Λ we can insert it into the definition (6.5) of $J^{(1)^a}_\mu(x)$ and use it to compute the action of Q^a_1.

By global $\mathcal{G}$-invariance, the asymptotic particles gather into multiplets. Each multiplet forms a representation ρ of the Lie algebra $\mathcal{G}$. The global charges Q^a_0 are represented by $\rho(Q^a_0) = T^a$ on the multiplets. The Lorentz transformation laws (6.9) of the charges implies that on-shell non-local charges depend linearly on the rapidity of the asymptotic particles: $\rho(Q^a_1) = \tau^a - \frac{\theta C_{Adj}}{4\pi i}T^a$. This follows because on-shell $L \equiv \partial/\partial\theta$. The conservation law for the two-body S-matrices are then:

$$(T^a \otimes 1 + 1 \otimes T^a)\, S(\theta_{12}) \;=\; S(\theta_{12})(T^a \otimes 1 + 1 \otimes T^a) \tag{6.13}$$

$$\left((\tau^a - \frac{\theta_2 C_{Adj}}{4\pi i}T^a) \otimes 1 + 1 \otimes (\tau^a - \frac{\theta_1 C_{Adj}}{4\pi i}T^a) - \frac{1}{2}f^{abc}T^b \otimes T^c\right)\, S(\theta_{12})$$
$$= S(\theta_{12})\left((\tau^a - \frac{\theta_1 C_{Adj}}{4\pi i}T^a) \otimes 1 + 1 \otimes (\tau^a - \frac{\theta_2 C_{Adj}}{4\pi i}T^a) - \frac{1}{2}f^{abc}T^b \otimes T^c\right).$$

with $\theta_{12} = \theta_1 - \theta_2$. These equations are the exchange relations for the Yangians $Y(\mathcal{G})$. They determine (up to scalar factors) the S-matrices.

6b) The sine-Gordon Model and $\widehat{sl_q(2)}$ currents.

In this section we describe the non-local charges that characterize the sine-Gordon theory. We will show that these charges generate the quantum $sl(2)$ loop algebra. Our analysis provides a new simple derivation of the soliton S-matrix.

•) The sine-Gordon theory: The quantum sine-Gordon theory is described by the Euclidean action

$$S \;=\; \frac{1}{4\pi}\int d^2z\, \partial_z\Phi\partial_{\bar z}\Phi \;+\; \frac{\lambda}{\pi}\int d^2z\; :\; \cos(\hat\beta\Phi)\; :\; . \tag{6.14}$$

The parameter $\hat\beta$ is a coupling constant; it is related to the conventionally normalized coupling by $\hat\beta = \beta/\sqrt{4\pi}$. The values of the coupling $\hat\beta = 1$ and $\hat\beta = \sqrt{2}$ are known to correspond to a free Dirac fermion and to the $SU(2)$ Gross-Neveu model respectively. The parameter λ defines the mass scale of the model; in the deep ultra-violet it is zero. For $\hat\beta \leq \sqrt{2}$ the action can be renormalized by normal-ordering the $\cos(\hat\beta\Phi)$ interaction and absorbing the infinities into λ; the coupling constant $\hat\beta$ is thereby unrenormalized.

We will treat the action (6.14) as a perturbation of a conformal theory in the sense developed by Zamolodchikov [17]. Namely, we treat the $\lambda\cos(\hat\beta\Phi)$ term as a perturbation of the conformal field theory corresponding to a single free boson. A more careful analysis shows, that it corresponds to a free boson compactified on a circle of radius $R = \hat\beta/2$. For $\hat\beta < \sqrt{2}$, the perturbing field is relevant: its (holomorphic,anti-holomorphic) anomalous dimensions are less than one. Following Zamolodchikov, this allows us to assume that the space of fields has not been drastically modified by the perturbation. In particular it allows us to suppose that all the operators $\mathcal{O}(x,t)$ of the sine-Gordon theory have a smooth ultra-violet limit and that they are in correspondence with the fields of the ultra-violet conformal field theory. We can thus label in a unique way the fields of the sine-Gordon theory by the corresponding fields in ultra-violet limit. In the massless limit, the free boson can be expanded as $\Phi(z,\overline{z}) = \phi(z) + \overline{\phi}(\overline{z})$ with $\langle\phi(w)\phi(z)\rangle = -\log(z-w)$ and similarly for $\overline{\phi}$. The fields of the ultra-violet conformal field theory are products of the chiral vertex operators $\exp(i\alpha\phi(z))$ and $\exp(i\alpha\overline{\phi}(\overline{z}))$ and of their Virasoro descendants. The anomalous (holomorphic, antiholomorphic) dimensions $(\Delta,\overline{\Delta})$ of these exponential operators are: $\Delta(\exp i\alpha\phi(z)) = \overline{\Delta}(\exp i\alpha\overline{\phi}(\overline{z})) = \frac{\alpha^2}{2}$. The perturbing operator $\cos(\hat{\beta}\Phi)$ is thus relevant for $\hat{\beta} < \sqrt{2}$ as indicated above.

In the deep ultra-violet limit the (anti)-chiral components $\phi(x,t)$ and $\overline{\phi}(x,t)$ can be expressed in a non-local way in terms of the sine-Gordon field $\Phi(x,t)$. The relations are:

$$\begin{aligned}\phi(x,t) &= \frac{1}{2}\left(\Phi(x,t) + \int_{-\infty}^{x} dy\ \partial_t\Phi(y,t)\right)\\ \overline{\phi}(x,t) &= \frac{1}{2}\left(\Phi(x,t) - \int_{-\infty}^{x} dy\ \partial_t\Phi(y,t)\right).\end{aligned} \tag{6.15}$$

Though the above non-local expressions (6.15) can only be derived in the massless limit, we can take them to define the chiral components ϕ and $\overline{\phi}$ in exponential operators in the massive theory also, because of the correspondence mentioned above. The vertex operators $\exp(i\alpha\phi(z))$ and $\exp(i\alpha\overline{\phi}(\overline{z}))$ are the Mandelstam-like operators.

The sine-Gordon theory has a well known topological current: $J^\mu(x,t) = \frac{\hat{\beta}}{2\pi}\epsilon^{\mu\nu}\partial_\nu\Phi(x,t)$ where $\epsilon^{\mu\nu} = -\epsilon^{\nu\mu}$. We take the convention $\epsilon^{01} = 1$. The topological charge is:

$$\tau = \frac{\hat{\beta}}{2\pi}\int_{-\infty}^{+\infty} dx\ \partial_x\Phi = \frac{\hat{\beta}}{2\pi}\left(\Phi(x=\infty) - \Phi(x=-\infty)\right). \tag{6.16}$$

The normalization of the topological current is fixed by the periodicity of the $\cos(\hat{\beta}\Phi)$ potential. More specifically, the topological solitons that correspond to single particles in the quantum theory are described classically by field configurations with $\tau = \pm 1$. These solitons are kinks that connect two neighboring vacua in the $\cos(\hat{\beta}\Phi)$ potential. In the quantum theory the topological charge $\tau(\mathcal{O})$ of an operator $\mathcal{O}$ is defined by the commutation relation $[\tau,\mathcal{O}] = \tau(\mathcal{O})\ \mathcal{O}$. The topological charge of the vertex operators is: $\tau(\exp(i\alpha\phi + i\overline{\alpha}\overline{\phi})) = \hat{\beta}(\alpha-\overline{\alpha})$.

•) **The non-local conserved charges and their algebra:** As proved in ref. [9] the sine-Gordon model admits four non-local conserved currents:

$$\partial_\mu J^\pm_\mu(x,t) = \partial_\mu\overline{J}^\pm_\mu(x,t) = 0. \tag{6.17}$$

The Lorentz spin s of the currents $J^\pm$ ($\overline{J}^\pm$) is $s = \frac{2}{\hat{\beta}^2}$ $(-\frac{2}{\hat{\beta}^2})$. The conservation law (6.17) can first be proved to first order in perturbation theory. Then, for $\hat{\beta}^2$ irrational, simple scaling arguments are enough to show that the first order is exact. Therefore the currents are conserved to all orders. From these conserved currents we define four conserved charges, $Q_\pm$ and $\overline{Q}_\pm$, respectively associated to the currents $J^\pm_\mu(x,t)$ and $\overline{J}^\pm_\mu(x,t)$. The Lorentz spin of the conserved charges is:

$$\frac{1}{\gamma} \equiv \mathrm{spin}(Q_\pm) = -\,\mathrm{spin}(\overline{Q}_\pm) = \frac{2}{\hat{\beta}^2} - 1. \tag{6.18}$$

The conserved currents whose exact expressions are given in ref. [9] are Mandelstam-like vertex operators. The conserved charges $Q_\pm$ and $\overline{Q}_\pm$ are thus non-local due to the fact that the (anti)-chiral components, ϕ and $\overline{\phi}$, of the sine-Gordon field Φ are non-local. This non-locality is reflected in the relations (6.15) which manifest the strings attached to the currents that were refereed to in section 4. The braiding relations arising from the non-locality are independent of the scale; thus they can be described in the ultra-violet limit without loss of information. For the currents of interest, the braiding relations are:

$$\begin{aligned} J_\mu^\pm(x,t)\overline{J}_\nu^\mp(y,t) &= q^{-2}\overline{J}_\nu^\mp(y,t)J_\mu^\pm(x,t) \quad ;\forall x,y \\ J_\mu^\pm(x,t)\overline{J}_\nu^\pm(y,t) &= q^{2}\overline{J}_\nu^\pm(y,t)J_\mu^\pm(x,t) \quad ;\forall x,y \end{aligned} \tag{6.19}$$

where $q=\exp(-2\pi i/\hat{\beta}^2)=-\exp(-i\pi/\gamma)$.

The algebra of the non-local charges is first obtained to lowest non-trivial order in perturbation theory. Simple scaling arguments then show that it is exact to all orders for $\hat{\beta}^2$ irrational. The results are:

$$Q_\pm\overline{Q}_\pm - q^2\overline{Q}_\pm Q_\pm = 0 \tag{6.20a}$$

$$Q_\pm\overline{Q}_\mp - q^{-2}\overline{Q}_\mp Q_\pm = a\left(1-q^{\pm 2\tau}\right) \tag{6.20b}$$

$$[\tau, Q_\pm] = \pm 2\, Q_\pm \tag{6.20c}$$

$$[\tau, \overline{Q}_\pm] = \pm 2\overline{Q}_\pm \tag{6.20d}$$

where a is some constant. The NLC approach consists in taking it as the non-perturbative definition of the theory. The algebra (6.20) is a known infinite dimensional algebra, namely the q-deformation of the $sl(2)$ affine Kac-Moody algebra, denoted $\widehat{sl_q(2)}$, with zero center [4][5]. Only the Serre relations for $\widehat{sl_q(2)}$ are missing in (6.20).

•) The action on the soliton fields: We now determine the manner in which the non-local charges are represented on the fields and on the asymptotic soliton states. We first construct the fundamental quantum field that create sine-Gordon solitons out of the vacuum. They must have topological charge ± 1. There are large families of operators with topological charge ± 1. These operators differ by products with local fields, and in general differ in Lorentz spin. Among them, there are four fields which generate these families. These fields, which we call the soliton fields, are defined by:

$$\Psi_\pm(x,t) = \exp\left(\pm\frac{i}{\hat{\beta}}\phi(x,t)\right), \quad \overline{\Psi}_\pm(x,t) = \exp\left(\mp\frac{i}{\hat{\beta}}\overline{\phi}(x,t)\right). \tag{6.21}$$

They have topological charges ± 1. The soliton fields in (6.21) are also characterized by non-trivial Lorentz spin: $\text{spin}(\Psi_\pm) = -\,\text{spin}(\overline{\Psi}_\pm) = \frac{1}{2\hat{\beta}^2}$.

The soliton fields transform covariantly under the action of the non-local charges. It can be proved that:

$$\begin{aligned} Q_\pm(\overline{\Psi}_\pm) &= 0,\ Q_\pm(\overline{\Psi}_\mp) = \lambda\ :\Psi_\pm(\chi_\pm\overline{\chi}_\mp):\equiv \lambda\hat{\Psi}_\pm \\ \overline{Q}_\pm(\Psi_\pm) &= 0,\ \overline{Q}_\pm(\Psi_\mp) = \lambda\ :\overline{\Psi}_\pm(\overline{\chi}_\pm\chi_\mp):\equiv \lambda\hat{\overline{\Psi}}_\pm. \end{aligned} \tag{6.22}$$

The fields $\widehat{\Psi}_\pm$ (or $\widehat{\overline{\Psi}}_\pm$), implicitly defined in (6.22), have topological charge ±1. They differ from the fields $\Psi_\pm$ (or $\overline{\Psi}_\pm$) by the local operators $\chi_\pm\overline{\chi}_\mp$ (or $\overline{\chi}_\pm\chi_\mp$). They also create solitons asymptotically. An easy computation shows that spin$(\widehat{\Psi}_\pm)$ = spin$(\overline{\Psi}_\mp)+\frac{1}{\gamma}$. Therefore the action (6.22) is consistent with the Lorentz spin$\pm\frac{1}{\gamma}$ of the non-local charges, as it must be.

The comultiplication of the charges defines their action on product of soliton fields. As explained in section 5, this comultiplication follows from the braiding of the currents with the soliton fields. The required braiding relations are:

$$\begin{aligned} J^\pm_\mu(x,t)\overline{\Psi}_\tau(y,t) &= q^{\pm\tau}\overline{\Psi}_\tau(y,t)J^\pm_\mu(x,t) \quad ; \quad \forall x,y \\ \overline{J}^\pm_\mu(x,t)\Psi_\tau(y,t) &= q^{\mp\tau}\Psi_\tau(y,t)\overline{J}^\pm_\mu(x,t) \quad ; \quad \forall x,y\,. \end{aligned} \tag{6.23}$$

Therefore, the comultiplications for the non-local charges $Q_\pm$ and $\overline{Q}_\pm$ are:

$$\Delta(Q_\pm) = Q_\pm\otimes 1 + q^{\pm H}\otimes Q_\pm \tag{6.24a}$$

$$\Delta(\overline{Q}_\pm) = \overline{Q}_\pm\otimes 1 + q^{\mp H}\otimes \overline{Q}_\pm \tag{6.24b}$$

$$\Delta(H) = H\otimes 1 + H\otimes 1. \tag{6.24c}$$

The last relation follows from the additivity of the topological charge.

Let us now determine the manner in which the non-local charges are represented on asymptotic (in the sense of scattering theory) multi-soliton states. Denote by $|\alpha=\pm\frac{1}{2},\theta\rangle$ a single-soliton state with topological charge $\tau=2\alpha=\pm1$ and rapidity θ. The vector space of single soliton states of fixed rapidity will be referred to as $W=$ vect.$\{|+\frac{1}{2}\rangle,|-\frac{1}{2}\rangle\}$.

Consider first the action on single-soliton states. The action of the charges on such states must form a representation of the algebra. This representation can be deduced as follows. We suppose that the fields (6.21) create the solitons. Taking into account the topological charges of the soliton fields, we have the following non-vanishing matrix elements,

$$\langle 0|\Psi_\pm(x,t)|\mp1/2,\theta\rangle \neq 0,\ \langle 0|\overline{\Psi}_\pm(x,t)|\mp1/2,\theta\rangle \neq 0, \tag{6.25}$$

as $t\to\pm\infty$. Analogous non-vanishing matrix elements exist for any operator having topological charge ±1 and which differs from the soliton fields by multiplication with a local operator. Thus we can take either fields of the family generated by the soliton fields $\Psi_\pm$ or $\overline{\Psi}_\pm$ to create the soliton state $|\pm\frac{1}{2},\theta\rangle$ asymptotically. From the equation (6.22), one infers that the charges Q_+ and $\overline{Q}_+$ will transform anti-solitons to solitons and vice versa for the charges Q_- and $\overline{Q}_-$. Moreover, in rapidity space, a Lorentz boost is represented as a shift of θ: $\theta\to\theta-\alpha$. The on-shell operators $\exp(\pm\theta/\gamma)$ have Lorentz spin$\pm\frac{1}{\gamma}$. Taking all these facts together, we find the following representation of the charges on the asymptotic solitons,

$$\begin{aligned} Q_\pm &= c\,e^{\theta/\gamma}E_\pm q^{\pm H/2} \\ \overline{Q}_\pm &= c\,e^{-\theta/\gamma}E_\pm q^{\mp H/2} \\ \tau &= H \end{aligned} \tag{6.26}$$

where c is a constant, $H = \mathrm{diag}(+1,-1)$ and $E_\pm$ are the Pauli spin matrices $\sigma_\pm$. The action on multi-soliton states is deduced from the comultiplication (6.24).

It is instructive to compare the above results with known structure of the $\widehat{sl_q(2)}$ loop algebra. The isomorphism of the representation (6.26) to the $\widehat{sl_q(2)}$ representation is made explicit by identifying the spectral parameter x in the principal gradation with $\exp(\theta/\gamma)$. The comultiplication (6.24) that we derived in the quantum field theory can be compared with the known comultiplication of $\widehat{sl_q(2)}$; they are equivalent as they must be. In particular this implies that the comultiplication provides a representation of the algebra (6.20) on asymptotic states with an arbitrary number of particles. This fact is important in establishing the non-local charges as true symmetries of the theory.

•) **The S-matrix from the non-local charges:** We now demonstrate how one can use the non-local charges to obtain non-perturbative information about the sine-Gordon theory by providing a derivation of the soliton S-matrix. The integrability of the sine-Gordon theory implies that the set of in-coming and out-going momenta are the same. Let $W_1 \otimes W_2$ denote the Hilbert space of two-soliton states of fixed rapidities, i.e. $W_1 \otimes W_2$ is spanned by the states $|\alpha_1 = \pm\frac{1}{2}, \theta_1\rangle \otimes |\alpha_2 = \pm\frac{1}{2}, \theta_2\rangle$. The two-particle to two-particle S-matrix is an operator, $\check{S} : W_1 \otimes W_2 \to W_2 \otimes W_1$. By Lorentz invariance $\check{S}$ depends only on the combination $\theta_1 - \theta_2$. Apart from the rapidity dependence, $\check{S}$ depends on the coupling $\hat{\beta}$. In order to keep this dependence in mind we denote the two-body S-matrix by $\check{S}(\frac{x_1}{x_2}; q)$ with $x_i = \exp(\theta_i/\gamma)$ and $q = -\exp(-i\pi/\gamma)$.

The S-matrix must commute with the action of the non-local charges since they are symmetries of the theory:

$$[\check{S}\ ,\ \Delta(H)] = [\check{S}\ ,\ \Delta(Q_\pm)] = [\check{S}\ ,\ \Delta(\overline{Q}_\pm)] = 0\ . \tag{6.27}$$

Let us rewrite (6.27) in a slightly different form. Representing the charges as in (6.26) and multiplying both sides of (6.27) by $q^{\tau/2} \otimes q^{\tau/2}$ or $q^{-\tau/2} \otimes q^{-\tau/2}$ wherever appropriate, we find:

$$[\check{S}(\frac{x_1}{x_2}; q), H \otimes 1 + 1 \otimes H] = 0 \tag{6.28a}$$

$$\begin{aligned} &\check{S}(\frac{x_1}{x_2}; q)\ \left(x_1 E_\pm \otimes q^{\mp H/2} + q^{\pm H/2} \otimes x_2 E_\pm\right) \\ &= \left(x_2 E_\pm \otimes q^{\mp H/2} + q^{\pm H/2} \otimes x_1 E_\pm\right)\ \check{S}(\frac{x_1}{x_2}; q) \end{aligned} \tag{6.28b}$$

$$\begin{aligned} &\check{S}(\frac{x_1}{x_2}; q)\ \left(x_1^{-1} E_\pm \otimes q^{\pm H/2} + q^{\mp H/2} \otimes x_2^{-1} E_\pm\right) \\ &= \left(x_2^{-1} E_\pm \otimes q^{\pm H/2} + q^{\mp H/2} \otimes x_1^{-1} E_\pm\right)\ \check{S}(\frac{x_1}{x_2}; q) \end{aligned} \tag{6.28c}$$

Jimbo [5] has proven that the solution $\check{S}(x; q)$ to the above equation is unique up to an overall scalar function $v(\theta_{12} \equiv \theta_1 - \theta_2)$. He showed that the solution automatically satisfies the Yang-Baxter equation, which is required for factorization of the multiparticle S-matrix. Constraints on $v(\theta_2)$ can be found by imposing crossing and unitarity. The minimal solution is the known sine-Gordon S-matrix [18].

7. Conclusions

It is now clear that the NLC approach to two dimensional integrable quantum field theories is very effective. Because it deals directly with the physical vacuum and with the quantum operator algebra it avoids some of the unpleasant features of the quantum inverse scattering method.

In section 6, we only presented the basic examples: the Yangian symmetry of the 2D current algebras and the $\widehat{sl_q(2)}$ symmetry of the sine-Gordon model. Many more examples can be and should be developed. Already, the NLC approach was successfully used to discover the general structure of the S-matrices for new massive theories. This was done in ref. [19] [20] for the restricted and unrestricted models and for any value of the central charges of the underlying Kac-Moody algebras. The S-matrices which were discovered this way have the following tensorial structure:

$$S(\theta) \;=\; \mathbf{X}_{CDD}(\theta)\;\left(S_{\mathcal{G}}(\theta) \otimes S^{frac}_{susy}(\theta)\right) \tag{7.1}$$

Each factor in (7.1) reflects a quantum symmetry. $\mathbf{X}_{CDD}(\theta)$ are Toda-like CDD factors. $S^{frac}_{susy}(\theta)$ are S-matrices, generally in the RSOS form, which are invariant under fractional supersymmetries. The S-matrices $S_{\mathcal{G}}(\theta)$ are $Y(\mathcal{G})$-invariant or $\mathcal{U}_q(\widehat{\mathcal{G}})$-invariant depending on the nature of the conformal field theories and of their perturbations. For the restricted models the $S_{\mathcal{G}}(\theta)$ factors are in the RSOS form. It remains to study the consequences of this general structure by developing many more examples.

Another interesting problem concerns the computation of the correlation functions. Very few results in this direction are known. The differential equations satisfied by the correlation functions are known only for the massive Ising model. However Smirnov's approach [21] to correlation functions expresses them in terms of form factors, and these form factors are in turn deduced from the S-matrices. Since the non-local currents characterize the S-matrix, it seems that they will determine, at least in part, the correlation functions. However, all the consequences of the Ward identities for the non-local currents have not been fully explored up to now.

There exists another approach to the correlation function problem. As we learned in conformal field theory, to constrain the correlation functions we need symmetry algebras which possess only infinite dimensional representations. Those which we described in this paper have finite dimensional representations; we used these representations to find S-matrices. To restrict ourselves to infinite dimensional representations it is natural to look for central extensions of the NLC algebras. This direction leads to hierarchies of symmetry algebras with the corresponding hierarchies of integrable models. The simplest examples are the hierarchies of Lie algebras, the finite Lie algebra $\mathcal{G}$, the loop algebra $\widehat{\mathcal{G}}$, the affine Kac-Moody algebra $\mathcal{G}^{(1)}$ and their quantum deformations, and the corresponding hierarchies of Toda models over $\mathcal{G}$, $\widehat{\mathcal{G}}$ or $\mathcal{G}^{(1)}$. The last Toda theories are singular conformal theories [22] invariant under the quantum affine algebra $\mathcal{U}_q(\mathcal{G}^{(1)})$ with center [9]. The approach to 2D quantum field theories based on the compatibility between these two symmetries seems very promising.

Acknowledgements. We thank G.Felder for his interest in this subject and for his collaboration. D.B. wishes to thank all the members of the Leningrad Branch of the Steklov Mathematical Institute for their cordial welcome. A.L. thanks the Institute in

Santa Barbara for hospitality. This work was partially supported by the US National Science Foundation under Grant No. PHY89-04035.

References

[1] S.Coleman, J.Mandula, Phys. Rev. **159** (1967), 1251.
[2] E.K.Sklyanin, L.A.Takhtajan, L.D.Faddeev, Theor. Math. Phys. **40** (1980), 688.
[3] L.D.Faddeev, *Les Houches Lectures 1982*, Elsevier Science Publisher, 1984.
[4] V.G.Drinfeld, Sov.Math.Dokl. **32** (1985), 254; Sov.Math.Dokl. **36** (1988), 212.
[5] M.Jimbo, Lett. Math. Phys. **10** (1985), 63; Lett. Math. Phys. **11** (1986), 247; Commun. Math. Phys. **102** (1986), 537.
[6] N.Y.Reshetikhin, L.A.Takhtajan, L.D.Faddeev, Leningrad Math. J. **1** (1990), 193.
[7] M.Lüscher, Nucl.Phys. **B135** (1978), 1.
[8] D.Bernard, *Hidden Yangians in 2D massive current algebras*, Saclay Preprint no. SPhT-90/109; Commun. Math. Phys. (to appear).
[9] D.Bernard, A.Leclair, *Quantum group symmetries and non-local currents in 2D QFT*, Preprint no. CLNS-90/1027 no. SPhT-90/144.
[10] D.Bernard, A.Leclair, *Non-local currents in 2D: the lattice approach* (to appear).
[11] S.L.Woronowicz, Commun. Math. Phys. **122** (1989), 125.
[12] D.Bernard, *A propos du calcul differential sur les groupes quantiques*, Preprint no. SPhT-90/119; a paraitre dans C.R.Acad.Sci., Paris.
[13] J.Fröhlich, *Cargese Lectures 1987, G 'tHooft et al (eds)*, Plenum, NY.
[14] K.Fredenhagen, K.H.Rehren, B.Schroer, Commun. Math. Phys. **125** (1989), 201.
[15] M.Lüscher, K.Pohlmeyer, Nucl. Phys. **B137** (1978), 46–54.
[16] E.Brezin et al., Phys. Lett. **82B** (1979), 442–444.
[17] A.B.Zamolodchikov, Adv. Studies Pure Math. **19** (1990), 641.
[18] A.Zamolodchikov, Al.Zamolodchikov, Annals Phys. **120** (1979), 252.
[19] D.Bernard, A.Leclair, Phys. Lett. **B247** (1990), 309.
[20] C.Ahn, D.Bernard, A.Leclair, Nucl. Phys. **B346** (1990), 409.
[21] F.A.Smirnov, Theor. Math. Phys. **60** (1987), 356; J.Phys. **A19** (1986), L575.
[22] O.Babelon, L.Bonora, Phys. Lett. **B244** (1990), 220.

(1) SERVICE DE PHYSIQUE THÉORIQUE DE SACLAY, F-91191 GIF-SUR- YVETTE, FRANCE

(2) INSTITUTE OF THEORETICAL PHYSICS, UNIVERSITY OF CALIFORNIA, SANTA BARBARA, CALIFORNIA 93106 AND NEWMANN LAB. OF NUCLEAR STUDIES, CORNELL UNIVERSITY, ITHACA NY 14853, USA

INDUCED REPRESENTATIONS AND TENSOR OPERATORS FOR QUANTUM GROUPS*

L.C. BIEDENHARN AND M. A. LOHE

Department of Physics, Duke University, Durham

ABSTRACT. The analog of the Borel-Weil construction of irreducible representations as holomorphic sections of holomorphic line bundles is constructed for quantum groups and applied to $U_q(2)$ and $U_q(3)$. The concept of a tensor operator for a quantum group and the corresponding q-analog to the generalized Wigner-Eckart theorem are developed and discussed with examples.

1. Introduction

Quantum groups are Hopf algebras [1-3] which are neither commutative nor cocommutative. This new mathematical structure first appeared explicitly in research at Leningrad as an abstraction from physical problems arising in the quantum inverse scattering method [1,4] and in integrable models from statistical mechanics and quantum field theory [2,5] . Since this initial introduction, interest in quantum groups and related applications has grown tremendously in both mathematics and physics.

A quantum group may be considered alternatively as a *deformation* of the universal enveloping algebra of an underlying classical Lie group. The continuous deformation parameter, q, can be suggestively written as $q = e^{\hbar}$ so that for $\hbar \rightarrow 0$ we obtain 'classical', undeformed Lie algebra; this is the sense in which "quantized" is to be understood.

We will focus in this paper on two related aspects of quantum groups: the construction of the algebra of tensor operators of a given quantum group and the extension of the Frobenius-Wigner-Mackey method of induced representations to obtain all unitary irreducible representations (irreps) of a compact quantum group (with generic q). Our explicit results will be restricted, for simplicity, to the quantum group $U_q(n)$ although the q-generalization to all classical Lie groups exists. The results we will present on q-tensor operators are largely expository [6], most results having been published previously [7], but our results on induced q-irreps are, we believe, new.

*Supported, in part, by the National Science Foundation, PHY-9008007.

2. The Construction of Unitary Irreps

There exists an elegant geometric procedure for constructing all unitary irreps of a compact classical Lie groups, namely, the Borel-Weil (BW) construction which realizes irreps as sections of line bundles over Kähler manifolds. Our objective in this section is to describe the quantum group analog of this construction for $U_q(n)$, treating, however, only the $U_q(2)$ case in full detail. In order to bring out the essential ideas most clearly we will begin with the q-boson construction of irreps of $U_q(n)$, and then extend these techniques to the q-analog of the BW procedure.

Jimbo [5] has given the defining algebraic relations for the quantum group $SU_q(n)$ corresponding to a deformation of the classical simple Lie algebra A_{n-1}. Denote the q-generators in the Chevalley-Weyl basis as $\{E_{i,i+1}, E_{i+1,i}, H_i\}$, $i = 1,\ldots,n-1$ (where to simplify the notation we also write $E_{i,i+1}$ as E_i^+ and $E_{i+1,i}$ as E_i^-), then the algebra is defined by:

$$[H_i, H_j] = 0, \quad [E_i^\pm, E_j^\pm] = 0, \quad [H_i, E_j^\pm] = \pm k_{(i,j)} E_j^\pm \tag{2.1}$$

where

$$k_{(i,j)} = \begin{cases} 1 & i = j \\ -\dfrac{1}{2} & i = j+1 \\ 0 & \text{otherwise} \end{cases}$$

and

$$[E_i^+, E_j^-] = \delta_{ij} \frac{q^{H_i} - q^{-H_i}}{q^{\frac{1}{2}} - q^{-\frac{1}{2}}} \tag{2.2}$$

The commutation relations are completed by the q-analog of the Serre relations:

$$\sum_{\nu=0}^{2} (-1)^\nu \begin{bmatrix} 2 \\ \nu \end{bmatrix}_q (E_i^\pm)^{2-\nu} E_j^\pm (E_i^\pm)^\nu = 0, \qquad (i \neq j) \tag{2.3}$$

where the q-binomial coefficient is defined by

$$\begin{bmatrix} n \\ m \end{bmatrix}_q \equiv \frac{[n]!}{[m]![n-m]!}, \qquad [n]! \equiv [n][n-1]\ldots[1] \tag{2.4}$$

We use the notation $[n]$ for the q-number

$$[n] \equiv \frac{q^{\frac{n}{2}} - q^{-\frac{n}{2}}}{q^{\frac{1}{2}} - q^{-\frac{1}{2}}} = q^{\frac{n-1}{2}} + q^{\frac{n-3}{2}} + \cdots + q^{-\frac{(n-1)}{2}}, n \in \mathbf{Z},\ q \in \mathbf{R}^+. \tag{2.5}$$

Remark. Note that $[n]$ is symmetric under $q \to q^{-1}$, and has precisely n terms whose powers decrease by steps of *unity*. Just as for angular momentum theory, these requirements force the use of half-integers, here integral powers of $q^{1/2}$. The literature is not uniform, however, and q often appears for $q^{1/2}$, with corresponding steps of *two*.

The Hopf algebra operations take the form:

$$\begin{gathered} \Delta(H_i) = H_i \otimes 1 + 1 \otimes H_i, \quad \Delta(E_i^\pm) = E_i^\pm \otimes q^{H_i/4} + q^{-H_i/4} \otimes E_i^\pm, \\ \varepsilon(1) = 1, \quad \varepsilon(E_i^\pm) = \varepsilon(H_i) = 0, \quad \gamma(E_i^\pm) = -q^{\mp\frac{1}{2}} E_i^\pm, \quad \gamma(H_i) = -H_i. \end{gathered} \tag{2.6}$$

The quantum group $U_q(n)$ is defined by the generators above plus the additional generator H_n which commutes with all other generators. The Hopf algebra operations for H_n are the same as those for the other H_i.

Let us now consider realizations of these algebras, using an approach which stems from the physical literature. This approach (for classical Lie groups) uses the boson operator, a, and its Hermitian conjugate, the destruction operator, $\overline{a}$, obeying the Heisenberg algebra $[\overline{a}, a] = 1$, with the cyclic "vacuum" ket-vector $|0\rangle$ defined by $\overline{a}|0\rangle = 0$. Equivalently, we can construct basis vectors as functions of the complex variable z and regard the operators $a, \overline{a}$ as multiplication by z and differentiation $\partial/\partial z$ respectively, with the vacuum ket omitted. Explicit basis vectors can be constructed by repeated application of the boson creation operator to the vacuum state, and can equivalently be regarded as polynomials in z.

Let us now construct q-analogs to the boson operators [8-14]. Introduce the $q-$ *creation operator* a^q, its Hermitian conjugate the $q-$ *destruction operator* $\overline{a}^q$, and the *boson vacuum* $|0\rangle = 0$ defined by $\overline{a}^q|0\rangle = 0$. Instead of the Heisenberg algebra, postulate the algebraic relation:

$$\overline{a}^q a^q - q^{1/2} a^q \overline{a}^q = q^{-N/2}, \tag{2.7}$$

where N is the (Hermitian) *number operator* satisfying

$$[N, a^q] = a^q, \qquad [N, \overline{a}^q] = -\overline{a}^q, \text{ with } N|0\rangle \equiv 0. \tag{2.8}$$

This algebra is a deformation of the Heisenberg algebra, which is obtained for $q \to 1$.

As in the case of the boson operators, we can equivalently regard the operator a^q as effecting multiplication by the complex variable z, but now $\overline{a}^q$ is represented not by differentiation, but by the finite difference operator D_z defined by

$$D_z f(z) = \frac{f(zq^{\frac{1}{2}}) - f(zq^{-\frac{1}{2}})}{z(q^{\frac{1}{2}} - q^{-\frac{1}{2}})} \tag{2.9}$$

for suitable functions $f(z)$, and N is given by

$$N f(z) = z\frac{\partial f(z)}{\partial z} \tag{2.10}$$

Evidently D_z acts as a finite difference operator and for $q \to 1$ becomes differentiation $\partial/\partial z$. The commutation relations (2.7, 2.8) are satisfied. A useful property is

$$D_z z^n = [n] z^{n-1} \tag{2.11}$$

where $[n]$ is defined by (2.5). Let us define the q-exponential function $\exp_q$ by

$$\exp_q z = \sum_{n=0}^{\infty} \frac{z^n}{[n]!} \tag{2.12}$$

then it follows from (2.11) that $D_z \exp_q(Az) = A\exp_q(Az)$ where A is a constant, or an operator independent of z. The q-exponential is a q-analog of the classical exponential function although as such it is not unique; it is however invariant under $q \leftrightarrow q^{-1}$. The operator D_z and q-extensions to classical functions are not new to quantum groups, having been studied long since by Jackson [15]. In fact the theory of q-functions dates back to Heine and has been developed extensively by Askey [16], Milne [17] and Andrews [18], who has provided a review of q-series; see also Gasper and Rahman [19] for a recent exposition.

The q-exponential and the operator D_z appear in the BW construction of $U_q(n)$ states, and several further properties will be used extensively there, in particular

$$zD_z = [N] = a^q \overline{a}^q, \qquad D_z z = [N+1] = \overline{a}^q a^q \tag{2.13}$$

which follow from the definitions (2.9,2.10), or equivalently (2.7,2.8).

It is now easy to define a q-analog for the algebra of the generators of the quantum group $SU_q(2)$. In the language of the q-boson operators, define a pair of mutually commuting q-bosons a_i^q and $\overline{a}_i^q$ for $i = 1, 2$. Then we have:

$$E_{12} = a_1^q \overline{a}_2^q, \qquad E_{21} = a_2^q \overline{a}_1^q, \qquad H_1 = \frac{1}{2}(N_1 - N_2). \tag{2.14}$$

It can be verified that these generators satisfy the quantum algebra of $SU_q(2)$.

Consider now for $SU_q(2)$ the basis vector defined by:

$$\left| \begin{smallmatrix} m_{12} & & 0 \\ & m_{11} & \end{smallmatrix} \right\rangle \equiv ([m_{12} - m_{11}]![m_{11} - m_{22}]!)^{-\frac{1}{2}} (a_1^q)^{m_{11} - m_{22}} (a_2^q)^{m_{12} - m_{11}} |0\rangle. \tag{2.15}$$

Notation : The vector $|(m)\rangle$ above is labelled using the Gel'fand-Weyl pattern (m). Each basis vector $|(m)\rangle$ in the flag manifold belonging to the unitary irrep $[m_{1n} \ldots m_{nn}]$ of $U(n)$ is labelled uniquely by a triangular pattern of integers:

$$(m) = \begin{pmatrix} m_{1n} & & m_{2n} & & \cdots & & m_{nn} \\ & m_{1,n-1} & & m_{2,n-1} & \cdots & m_{n-1,n-1} & \\ & & & \cdots & & & \\ & & & m_{11} & & & \end{pmatrix}$$

with $m_{ij} \geq m_{i,j-1} \geq m_{i+1,j}$. It follows from the Lusztig-Rosso theorem [20,21] that these labels are invariant under deformation and hence properly label the basis vectors for flag manifolds in the unitary irreps of quantum group $U_q(2)$. For the simple group $SU_q(n)$ the unitary irreps have $m_{nn} = 0$.

It is easily verified that the vectors in (2.15) form a basis for the unitary irrep $[m_{12}, 0]$ of $SU_q(2)$ labelled by the eigenvalue $(m_{11} - m_{12})/2$ of H_1. The vector space $\mathbf{V}_{[m_{12},0]}$ spanned by the $(m_{12}+1)$ basis vectors $\left\{ \left| \begin{smallmatrix} m_{12} & & 0 \\ & m_{11} & \end{smallmatrix} \right\rangle \right\}$, $m_{11} = 0, 1, \ldots m_{12}$, carries a unitary irrep of $SU_q(2)_q$ and *every* unitary irrep of $SU_q(2)$ is realized in this way. Let us denote the direct sum of these spaces by $\mathbf{V}$:

$$\mathbf{V} \equiv \sum_{m_{12}} \oplus \mathbf{V}_{[m_{12},0]}. \tag{2.16}$$

Let us now consider the quantum group $U_q(2)$ and its unitary irreps. The q-boson realization of the algebra of $U_q(2)$ is given by (2.14) with the adjunction of the operator $H_2 =$

$\frac{1}{2}(N_1+N_2)$, or equivalently by adjoining N_1 and N_2 independently. In order to construct all unitary irreps of $U_q(2)$ via a q-boson realization it is necessary to use two independent (commuting) pair of q-bosons, (a_1^q, a_2^q) and (b_1^q, b_2^q). Using the comultiplication operation, one finds for the generators (*dropping the $q-$ superscript label hence forth*):

$$\begin{aligned}
E_{12} &= a_1\bar{a}_2 \otimes q^{(N_1^b-N_2^b)/4} + q^{-(N_1^a-N_2^a)/4} \otimes b_1\bar{b}_2, \\
E_{21} &= a_2\bar{a}_1 \otimes q^{(N_1^b-N_2^b)/4} + q^{-(N_1^a-N_2^a)/4} \otimes b_2\bar{b}_1, \\
N_1 &= N_1^a \otimes 1 + 1 \otimes N_1^b \quad N_2 = N_2^a \otimes 1 + 1 \otimes N_2^b
\end{aligned} \tag{2.17}$$

where the subscripts a, b refer to the q-boson sets $\{a\}, \{b\}$ respectively. The generators in (2.17) act on the space $\mathbf{V} \otimes \mathbf{V}$ and we seek irreducible subspaces of $\mathbf{V} \otimes \mathbf{V}$ in which the states carry Gel'fand-Weyl labels (m) of $U_q(2)$.

Remarkably, the desired basis vectors can be compactly written in the operator form [14]

$$\left|\begin{smallmatrix} m_{12} & & m_{22} \\ & m_{11} & \end{smallmatrix}\right\rangle = M^{-\frac{1}{2}}(a_{12})^{m_{22}}(a_1)^{m_{11}-m_{22}}(a_2)^{m_{12}-m_{11}}|0\rangle, \tag{2.18}$$

where

$$M = \frac{[m_{12}+1]![m_{12}-m_{11}]![m_{11}-m_{22}]![m_{22}]!}{[m_{12}-m_{22}+1]!}$$

and a_{12} is the operator defined by:

$$a_{12} = q^{(N_2^a+N_1^b+1)/4}a_1b_2 - q^{-(N_1^a+N_2^b+1)/4}a_2b_1 \tag{2.19}$$

The proof, that the vectors (2.18) are indeed a realization of the orthogonal Gel'fand-Weyl basis, proceeds by expanding the operator a_{12} so that (2.18) becomes a sum of homogeneous polynomials in the operators $\{a_i, b_i\}$. In carrying out this expansion we use essentially the q-binomial theorem, which can be expressed elegantly in terms of non-commutating coordinates. In the problem at hand, these coordinates can for example be defined by

$$x = q^{(N_2^a+N_1^b+1)/4}a_1b_2 \qquad , y = q^{-(N_1^a+N_2^b+1)/4}a_2b_1$$

which satisfy $xy = qyx$ and so can be regarded as quantum coordinates [24]

Remark. Taking the inner product of (2.18) with product states (given by (2.15) for $\{a\}$ and $\{b\}$) defines the explicit Wigner-Clebsch-Gordan coefficient reducing $\mathbf{V} \otimes \mathbf{V}$.

We are now in a position to develop the q-analog to the Borel-Weil (BW) construction. Recall that the BW construction for a compact classical group G, taken for simplicity here to be $U(2)$, constructs irreps as holomorphic sections of a holomorphic line bundle. The homogeneous space $U(2)/T$, with $T \equiv U(1) \times U(1)$, can be made into a Kähler manifold. To every character of T one associates a holomorphic line bundle over $G/T \simeq S^2$, which carries a left G-action. The sections of this bundle are irreps of $U(2)$ with highest weight given by the character of T.

In order to carry this out explicitly, let $\left|\begin{smallmatrix} m_{12} & & m_{22} \\ & m_{11} & \end{smallmatrix}\right\rangle$ be the eigenvector of $T = U(1) \times U(1)$ whose eigenvalues are: $E_{11} \to m_{12}$, $E_{22} \to m_{22}$. (The notation implies by design that $|(m)\rangle$ is a highest weight vector in $U(2)$). The BW construction augments this vector by a function which satisfies: $f(gt^{-1}) = \chi_\lambda(t)f(g)$ where $t \in T$ and χ_λ is a character of T. Such functions are easily constructed and are given by:

$$f(g) \equiv \left\langle \begin{smallmatrix} m_{12} & & m_{22} \\ & m_{11} & \end{smallmatrix} \middle| e^{zE_{12}} \middle| \begin{smallmatrix} m_{12} & & m_{22} \\ & m_{11} & \end{smallmatrix} \right\rangle, \tag{2.20}$$

which are matrix elements in $U(2)$ of the operator $g = e^{zE_{12}}$. Note that the state on the right in the bracket denotes an arbitrary vector in the irrep $[m_{12}, m_{22}]$. Since E_{12} is a raising operator, (2.20) is actually a polynomial in z, and clearly holomorphic.

The explicit normalized vectors $|m)_{BW}$ of the irrep $[m_{12}, m_{22}]$ are given by:

$$\left| \begin{smallmatrix} m_{12} & & m_{22} \\ & m_{11} & \end{smallmatrix} \right\rangle_{BW} = \left\langle \begin{smallmatrix} m_{12} & & m_{22} \\ & m_{11} & \end{smallmatrix} \middle| e^{zE_{12}} \middle| \begin{smallmatrix} m_{12} & & m_{22} \\ & m_{11} & \end{smallmatrix} \right\rangle \left| \begin{smallmatrix} m_{12} & & m_{22} \\ & m_{11} & \end{smallmatrix} \right\rangle. \tag{2.21}$$

There are many features [23] to discuss for these irreps, (for example, the explicit left action), but we will omit these for brevity since such results will be clear from the explicit q-analog results below.

To proceed to the quantum group $U_q(2)$ is now straightforward. The q-exponential $\exp_q$ replaces the ordinary exponential in (2.21), and the carrier space for all unitary irreps of $U_q(2)$ is then given by the vectors:

$$\begin{aligned} \left| \begin{smallmatrix} m_{12} & & m_{22} \\ & m_{11} & \end{smallmatrix} \right\rangle_{q-BW} &= \left\langle \begin{smallmatrix} m_{12} & & m_{22} \\ & m_{11} & \end{smallmatrix} \middle| exp_q(zE^q_{12}) \middle| \begin{smallmatrix} m_{12} & & m_{22} \\ & m_{11} & \end{smallmatrix} \right\rangle \left| \begin{smallmatrix} m_{12} & & m_{22} \\ & m_{11} & \end{smallmatrix} \right\rangle \\ &= \left(\frac{[m_{12} - m_{22}]!}{[m_{11} - m_{22}]!} \right)^{\frac{1}{2}} \frac{z^{m_{12} - m_{11}}}{\sqrt{[m_{12} - m_{11}]!}} \left| \begin{smallmatrix} m_{12} & & m_{22} \\ & m_{11} & \end{smallmatrix} \right\rangle. \end{aligned} \tag{2.22}$$

The left action by the $U_q(2)$ generators $\mathbf{g} = \{E_{12}, E_{21}, N_1, N_2\}$, which we will denote by $\Gamma(\mathbf{g})$, (that is: $\mathbf{g} \to \Gamma(\mathbf{g})$) is given by:

$$\begin{aligned} \Gamma(N_1) &= \mathcal{N}_1 - N, \qquad & \Gamma(E_{12}) &= D_z, \\ \Gamma(N_2) &= \mathcal{N}_2 - N, \qquad & \Gamma(E_{21}) &= z[\mathcal{N}_1 - \mathcal{N}_2 - N], \end{aligned} \tag{2.23}$$

where D_z and N are given by (2.9,2.10), with $\mathcal{N}_1, \mathcal{N}_2$ denoting the action on $\left| \begin{smallmatrix} m_{12} & & m_{22} \\ & m_{11} & \end{smallmatrix} \right\rangle$ in (2.22) of the abstract generators N_1, N_2. By direct calculation we can prove:

Lemma. *The map* $\mathbf{g} \to \Gamma(\mathbf{g})$ *is an isomorphism, of the quantum group algebra* $\mathbf{g} \in U_q(2)$.

The "K_q-factor" defined by $K_q^2(m) = [m_{12} - m_{22}]!/[m_{12} - m_{11}]!$ has the significance of defining the metric in the Kähler manifold (S^2 in the classical case). We believe that this construction for the q-analog case allows one to define the metric for the analog to a "q-Kähler manifold for the quantum group $U_q(2)$.

It is significant that in this construction all irrep vectors of $U_q(2)$ appear as monomials, involving only the *single* complex variable z (or equivalently, a single q-boson), in sharp contrast to the *two* q-bosons involved in (2.14), or the *four* bosons in (2.17). This suggests that the explicit construction of all unitary irreps of $U_q(n)$ can be achieved with only $n-1$ q-bosons, and this is indeed correct; the corresponding recursive approach is a generalization of the Borel-Weil result. The generalization, for the classical Lie group $U(n)$, corresponds to a constructing holomorphic sections of a bundle whose fibers carry irreps of the sub-group $U(n-1)\times U(1)$. The base manifold is now $U(n)/(U(n-1)\times U(1))$ having $(n-1)$ complex dimensions.

The essentials of the quantum group construction of the q-Borel-Weil basis vectors for $U_q(n)$ irreps is mostly easily comprehended from the $U_q(3)$ example. First, let us define the operators E_{ij} from the generators $E_1^{\pm}, E_2^{\pm}$ given in the definition (2.1-2.4) of $U_q(3)$:

$$\begin{aligned} E_{12} &= E_1^+, \qquad E_{23} = E_2^+, \qquad E_{13} = q^{-E_{22}/2}(q^{1/2}E_{12}E_{23} - E_{23}E_{12}), \\ E_{21} &= E_1^-, \qquad E_{32} = E_2^-, \qquad E_{31} = q^{E_{22}/2}(q^{-1/2}E_{32}E_{21} - E_{21}E_{32}), \end{aligned} \tag{2.24}$$

The operators E_{13}, E_{31} are defined in this particular way in order to satisfy

$$[E_{13}, E_{12}] = 0 = [E_{13}, E_{23}] \text{ and } [E_{31}, E_{21}] = 0 = [E_{31}, E_{32}]$$

as follows from the relations (2.3).

We can now define the (normalized) q-Borel-Weil basis vectors for $U_q(3)$ by the explicit form:

$$\left|\begin{matrix} m_{13} & m_{23} & m_{33} \\ \multicolumn{3}{c}{m_{12} \quad m_{22}} \\ \multicolumn{3}{c}{m_{11}} \end{matrix}\right\rangle_{q-BW} = \sum_m \left\langle\begin{matrix} m_{13} & m_{23} & m_{33} \\ \multicolumn{3}{c}{m_{12} \quad m_{22}} \\ \multicolumn{3}{c}{m_{11}} \end{matrix}\right| e(z,E) \left|\begin{matrix} m_{13} & m_{23} & m_{33} \\ \multicolumn{3}{c}{m_{12} \quad m_{22}} \\ \multicolumn{3}{c}{m_{11}} \end{matrix}\right\rangle \left|\begin{matrix} m_{13} & m_{23} & m_{33} \\ \multicolumn{3}{c}{m_{12} \quad m_{22}} \\ \multicolumn{3}{c}{m_{11}} \end{matrix}\right\rangle \tag{2.25-2.26}$$

where

$$e(z,E) \equiv exp_q(z_1 E_{13}) exp_q(z_2 E_{23}). \tag{2.26}$$

It can be shown [24] that the left action by the $U_q(3)$ generators has the form

$$\begin{aligned} &\Gamma(N_1) = \mathcal{N}_1 - n_1, \qquad \Gamma(N_2) = \mathcal{N}_2 - N_2, \qquad \Gamma(N_3) = \mathcal{N}_3 + N_1 + N_2, \\ &\Gamma(E_{13}) = D_1, \qquad \Gamma(E_{23}) = D_2, \\ &\Gamma(E_{12}) = q^{N_2/2}\mathcal{E}_{12} - q^{\mathcal{N}_2/2} z_2 D_1, \qquad \Gamma(E_{21}) = q^{N_1/2}\mathcal{E}_{21} - q^{\mathcal{N}_1/2} z_1 D_2, \\ &\Gamma(E_{32}) = q^{(1-\mathcal{N}_3)/2} z_1 \mathcal{E}_{12} + z_2[\mathcal{N}_2 - \mathcal{N}_3 - N_1 - N_2], \\ &\Gamma(E_{31}) = q^{(\mathcal{N}_3-1)/2} z_2 \mathcal{E}_{21} + z_1[\mathcal{N}_1 - \mathcal{N}_3 - N_1 - N_2], \end{aligned} \tag{2.27}$$

where $D_i \equiv D_{z_i}$, $N_i = z_i\partial/\partial z_i$, and $\{\mathcal{E}_{12}, \mathcal{E}_{21}, \mathcal{N}_1, \mathcal{N}_2, \mathcal{N}_3\}$ denote the action of the $U_q(2)\times U(1)$ generators on the Gelfand-Weyl basis in (2.25). It can be checked directly that the defining relations of the $U_q(3)$ algebra are satisfied.

By expanding out the q-exponentials and taking matrix elements, the basis vectors in (2.27) can be given a more understandable form:

$$\left|\begin{matrix} m_{13} & m_{23} & m_{33} \\ \multicolumn{3}{c}{m_{12} \quad m_{22}} \\ \multicolumn{3}{c}{m_{11}} \end{matrix}\right\rangle_{q-BW} = K_q \sum_m \frac{z_1^{m-m_{11}}(-z_2)^{w+m_{11}-m}}{\sqrt{[m-m_{11}]![w+m_{11}-m]!}}\, {}_qC^{j_1 j_2 j}_{m_1 m_2 m_1+m_2} \left|\begin{matrix} m_{13} & m_{23} & m_{33} \\ \multicolumn{3}{c}{m_{12} \quad m_{22}} \\ \multicolumn{3}{c}{m} \end{matrix}\right\rangle \tag{2.28}$$

where

$$
\begin{aligned}
&w = m_{13} + m_{23} - m_{12} - m_{22}, \qquad j_1 = \frac{1}{2}(m_{13} - m_{23}), \qquad j_2 = \frac{1}{2}w \\
&j = \frac{1}{2}(m_{12} - m_{22}), \qquad m_1 = m - \frac{1}{2}(m_{13} + m_{23}), \qquad m_2 = m_{11} - m + \frac{1}{2}w.
\end{aligned}
\tag{2.29}
$$

Remark. Equation (2.28) shows the desired general form for the recursive $q - BW$ extension to $U_q(n)$. This form shows a q-boson irrep in $SU_q(n-1)$ q-WCG coupled to an abstract vector in $U_q(n-1) \times U(1)$ to give normalized irrep vectors in $U_q(n)$ with a suitable multiplicative constant K_q. *Note that all elements in this construction are well-defined recursively, including q-WCG coefficients discussed in the following section.*

3. Tensor Operators for Quantum Groups

Symmetry, as well known, plays a fundamental role in quantum physics and applications of symmetry techniques in physics have correspondently been a source of innovations in the mathematics of group theory.[1] A given symmetry group has the implications for physics not only of classifying possible state vectors and associated quantum numbers but also of classifying and partially determining physical transitions between these symmetry allowed states. This latter structure is determined by the *algebra of tensor operators*. Since this algebra is not well known in mathematics we must now give a precise definition.

Consider a compact Lie group G to be a physical symmetry; for definiteness, let G be $U(n)$. This symmetry allows one to classify a Hilbert space basis for physical states using the irreps of G. To discuss the problem of interactions, which induce transitions within and between irreps, it is useful to construct a *model space* [25], $\mathbf{M}$, defined to be the direct sum of vectors carrying unitary irreps of the group G, each equivalence class of irreps occurring *once and only once*. The operators on $\mathbf{M}$ are defined to belong to the linear space $\mathbf{T}$ with the action:

$$
\mathbf{T}: \quad \mathbf{M} \to \mathbf{M}. \tag{3.1}
$$

Using equivalence, the symmetry G can be exploited to give a partial classification of the operators belonging to $\mathbf{T}$, that is to say, if under the symmetry $g \in G$, $\mathbf{T}: \mathbf{M} \to \mathbf{U}(g)\mathbf{M}$ where $\mathbf{U}(g)$ is an operator whose action on $\mathbf{M}$ corresponds to a direct sum of irrep matrices, we have the equivalence condition:

$$
\mathbf{U}(g)\mathbf{T}_i\mathbf{U}^{-1}(g) = \sum_j \mathbf{T}_j D_{ji}(g) \tag{3.2}
$$

where $\mathbf{D}(g)$ is a representation, then the *tensor operators* $\{\mathbf{T}_i\}$ - sets of operators - may be classified by the irrep labels of G, with individual operators in the tensor operator set being labelled by Gelfand-Weyl patterns (for $G = U(n)$).

---

[1] For example the first "BW" construction for $U(2)$ in physics long preceded the papers of Borel and Weil.

At the Lie algebra level the action of the generator $X_\alpha \in \mathbf{g}$ on the irreducible tensor operator $\mathbf{T}_{(m)}$, with (m) a Gelfand-Weyl pattern, is the *adjoint action*:

$$[X_\alpha, \mathbf{T}_{(m)}] = \sum_{(m')} \langle (m')|X_\alpha|(m)\rangle T_{(m')}, \tag{3.3}$$

with $\sum_{(m')}\langle (m')|X_\alpha|(m)\rangle T_{(m')}$ the matrix of the generator X_α for the irrep $[m]$.

The tensor operator classification is not categoric; this is clear from noting that multiplying a given irreducible tensor operator labelled by $[m]$ by an invariant operator leaves the classification unchanged. To eliminate this freedom one can define *unit tensor operators* whose norm (see below) is unity. For $G = U(n)$ with $n \geq 3$, *distinct* unit tensor operators with the same irrep labels $[m_{1n} \dots m_{nn}]$ exist, but it has been shown [26] that a *canonical*(natural) labelling splitting this multiplicity exists for $n = 3$, and very likely for all $n > 3$

Applications of symmetry structures in quantum physics are heavily dependent on the fundamental theorem for tensor operators (the generalized Wigner-Eckart theorem). Such a theorem *need not exist* for a given symmetry structure, but depends rather upon the specific way in which the symmetry is realized [27]. The two required properties of the realization as given above are: equivalence and the adjoint action (or more generally the derivative property).

It is a consequence of these two properties that the generalized Wigner-Clebsch-Gordan(WCG) coefficients for a given symmetry group G occur in two logically distinct ways:

(a) as *coupling coefficients* for the Kronecker product of irreps carried by kinematically independent constituent systems, (*Clebsch-Gordan problem*) and,
(b) as *matrix elements* (up to a rotationally invariant scale factor) or physical transition operators, (the *Wigner-Eckart problem*).

Conversely, if equivariance and derivation are not valid for a given realization, then this latter results (b) fails.

It is not obvious that one can extend both (a) and (b) to quantum groups, particularly when one realizes that the derivative property corresponds to a *commutative co-product*, which is invalid for a quantum group. Moreover, the equivariance condition is problematic as well. For a Lie algebra, equivariance is effected by the adjoint action. This cannot work for quantum group, since the quantum group irrep corresponding to the adjoint representation is finite dimensional in contrast to the infinite number of linearly independent algebra elements obtained under commutation.

The resolution of the Clebsch-Gordan problem for $U_q(2)$ has been published by many authors [28-33] and is contained in section 2, as noted above. For the Wigner-Eckart problem, the resolution is given in [7,34]. We follow [7] below.

Let the Hilbert space on which the operators act be the model space, $\mathbf{M}$. The operators on $\mathbf{M}$ belong to the linear space $\mathbf{T}$ with the action: $\mathbf{T} : \mathbf{M} \rightarrow \mathbf{M}$. We introduce now the key concept of an *induced action*; this is mapping:

$$\mathbf{T} \otimes \mathbf{T} \rightarrow \mathbf{T}, \tag{3.4}$$

which we will denote $A(B) = C$, that is A acts on B to give C, with $A, B, C \in \mathbf{T}$. (One must carefully distinguish this action from the natural product action of operators on Hilbert space, which we denote as usual be juxtaposition: AB).

Using these concepts, we can now recognize that the standard setting for the generalized Wigner-Eckart theorem for classical Lie algebras consists first of the *induced co-product* (Δ):

$$\Delta(\mathbf{g})(A \otimes |m\rangle) = \mathbf{g}(A) \otimes \mathbf{1}|m\rangle + \mathbf{1}(A) \otimes \mathbf{g}|m\rangle, \tag{3.5}$$

where $\mathbf{g}$ is a generator, $A \in \mathbf{T}$ and $|m\rangle \in \mathbf{M}$, followed by a mapping c:

$$c: \ \mathbf{T} \otimes \mathbf{M} \to \mathbf{M}. \tag{3.6}$$

Applying this operation in the standard setting yields:

$$\mathbf{g}(A|m\rangle) = \mathbf{g}(A)|m\rangle + A \otimes \mathbf{g}|m\rangle \tag{3.7}$$

for which one uses the *induced action* $\mathbf{g}(A) \equiv [\mathbf{g}, A]$, that is, the adjoint action.

The novelly in this construction is that the co-product Δ acts, via (3.5), on the tensor product $\mathbf{T} \otimes \mathbf{M}$ of *different vector spaces* and, in addition, there is a further operation (the mapping c) which requires that the induced co-product and the induced action be compatible. The compatibility of these two structures is guaranteed for the standard (Lie group) case since the standard matrix action of operators in Hilbert space induces (by communication) a commutative co-product (a derivation) and the induced commutator action of the generators realizes, by equivariance, an irrep carried by the tensor operators (this compatibility is the content of the tensor operator theorem (generalized Wigner-Eckart theorem)).

The compatibility requirement can be expressed most succinctly by using the language of diagrams. Consider the following diagram (where E is a generator):

$$\begin{array}{ccc} \mathbf{T} \otimes \mathbf{M} & \xrightarrow{\Delta(E)} & \mathbf{T} \otimes \mathbf{M} \\ c\downarrow & & c\downarrow \\ \mathbf{M} & \xrightarrow{E} & \mathbf{M}. \end{array} \tag{3.8}$$

The requirement of compatibility is that the diagram above be commutative. Assuming a given co-product determines a compatible action (or vice-versa). Thus, for example, for a Lie group one uses the diagonal (commutative) co-product and determines the commutator to be the compatible induced action.

It should now be clear how to extend this structure to tensor operators for quantum groups (q-tensor operators). Let us formalize these considerations:

Definition [7]. Let $\mathbf{T}$ denote the vector space of operators mapping the *model space* $\mathbf{M}$ of the compact quantum group G_q into itself: $\mathbf{M} \xrightarrow{\mathbf{T}} \mathbf{M}$. An irreducible q-tensor operator in a set of operators, $\{\mathbf{t}_{\Xi,\xi}\} \in \mathbf{T}$ which carries a finite-dimensional irrep Ξ, with vectors ξ, of the quantum group G_q. That is:

$$E_\alpha(\mathbf{t}_{\Xi,\xi}) = \sum_{\xi'} \langle \Xi, \xi' | E_\alpha | \Xi, \xi \rangle \mathbf{t}_{\Xi,\xi'}, \tag{3.9}$$

where E_α is a generator of $G_q, E_\alpha(\mathbf{t}_{\Xi,\xi'})$ denotes as a action of E_α on $\mathbf{T}$, and $\langle \cdots \rangle$ denotes the matrices of the generators for the irrep Ξ. A q-tensor operator is accordingly a linear combination of irreducible q-tensor operators, with coefficients invariant under the q-group action.

Theorem [7]. *If $\{\mathbf{t}_{\Xi,\xi}\}$ is a q-tensor operator of the compact quantum group G_q such that the co-product of G_q is compatible with the action $E_\alpha(\mathbf{t}_{\Xi,\xi})$, that is, diagram (3.8) is commutative, then the matrix elements of $\{\mathbf{t}_{\Xi,\xi}\}$ in* **M** *are proportional to the q-WCG coefficients of G_q with the constant of proportionality an invariant.*

Conversely, if $\{\mathbf{t}_{\Xi,\xi}\}$ is a q-tensor operator and the matrix elements of $\{\mathbf{t}_{\Xi,\xi}\}$ are proportional to the q-WCG coefficients, then diagram (3.8) is commutative.

4. The Algebra of q-Tensor Operators

The q-WCG coefficients are the matrix elements of a map (denoted W) from $\mathbf{M}\otimes\mathbf{M}$ into $\mathbf{M}$. Expressing this diagrammatically we have the following *commutative* diagram:

$$\begin{array}{ccc} \mathbf{M}\otimes\mathbf{M} & \xrightarrow{\Delta(E)} & \mathbf{M}\otimes\mathbf{M} \\ W\big\downarrow & & \big\downarrow W \\ \mathbf{M} & \xrightarrow{E} & \mathbf{M}. \end{array} \tag{3.8}$$

Now, because of the way the action has be defined on the q-tensor operator $\{\mathbf{t}_{\Xi,\xi}\}$, for $\mathbf{t}_{\alpha,a}$ acting on the vector $\left|\begin{smallmatrix} b \\ \beta \end{smallmatrix}\right\rangle$ - an element of $\mathbf{T}\otimes\mathbf{M}$ - we have a well-defined compatible co-product action $\Delta(E)$, with $\mathbf{T}\otimes\mathbf{M}$ replacing $\mathbf{M}\otimes\mathbf{M}$. It follows from the diagram that the mapping c can be identified (to within an invariant factor) with the mapping W, that is, the q-WCG coefficients. Moreover the co-product Δ has a well-defined compatible action that extends to $\mathbf{T}\otimes\mathbf{T}$ thus yielding a further commuting diagram. It follows that we have as corollaries to the q-tensor operator theorem:

Corollaries.

(a) There exists an algebra of q-tensor operators, $\mathbf{T}\otimes\mathbf{T}\xrightarrow{W}\mathbf{T}$, carrying products of irreducible q-tensor operators into irreducible q-tensor operators.

(b) There exists a product carrying an irreducible q-tensor operators into an invariant(having the properties of an inner product). Thus a norm exists and irreducible *unit* q-tensor operators ($\hat{\mathbf{T}}$) are well-defined whose matrix elements, by the fundamental tensor operator theorem, are q-WCG coefficients.

(c) Denote the mapping in Cor. (a) by $\mathbf{T}\circledS\mathbf{T}\in\mathbf{T}$, and denote the invariant product in Cor. (b) by $\mathbf{T}\cdot\mathbf{T}$. Then the (6-j) operators are defined by: $\hat{\mathbf{T}}\cdot(\hat{\mathbf{T}}\circledS\hat{\mathbf{T}})$.

To clarify the meaning of these results let us give some examples.

Example 1: Define the operator pair by:

$$t_{\frac{1}{2},\frac{1}{2}}\equiv a_1^q q^{-N_2/4}, \qquad t_{\frac{1}{2},-\frac{1}{2}}\equiv a_2^q q^{-N_1/4}. \tag{4.2}$$

This pair is q-tensor operator of $SU_q(2)$, with the explicit induced action

$$J_\pm(t)=J_\pm t q^{-J_z/2}-q^{(-J_z\pm1)/2}tJ_\pm. \tag{4.3}$$

(Note that for $q\to1$ we recover the commutator action.) One verifies that this induced action on $t_{\frac{1}{2},\pm\frac{1}{2}}$ obeys the equivariance condition with $j=\frac{1}{2}$, that is:

$$J_\pm(t_{\frac{1}{2},\mp\frac{1}{2}})=t_{\frac{1}{2},\pm\frac{1}{2}},\ J_\pm(t_{\frac{1}{2},\pm\frac{1}{2}})=0,\ J_z(t_{\frac{1}{2},\pm\frac{1}{2}})=\pm\frac{1}{2}t_{\frac{1}{2},\pm\frac{1}{2}} \tag{4.4}$$

Remark. By using q-WCG operator coupling, this result for the spin-$\frac{1}{2}$ q-tensor operators extends *the induced action* to all q-tensor operators in $SU_q(2)$.

Example 2: It was remarked earlier that the generators of $SU_q(2)$ are *not* irreducible q-tensor operators. The q-tensor operator in $SU_q(2)$ carrying the three-dimensional (adjoint) irrep is denoted $\mathbf{T}_{1,m}$ with $m = \pm 1, 0$, and has the explicit boson operator realization:

$$\mathbf{T}_{1,1} = \sqrt{[2]}q^{(N_1^q+2-N_2^q)/4}a_1^q\overline{a}_2^q, \tag{4.5a}$$

$$\mathbf{T}_{1,0} = q^{-(N_2^q+1)/2}[N_1^q] - q^{(N_1^q+1))/2}[N_2^q], \tag{4.5b}$$

$$\mathbf{T}_{1,-1} = \sqrt{[2]}q^{(N_1^q-2-N_2^q)/4}a_2^q\overline{a}_1^q, \tag{4.5c}$$

The Casimir invariant (scalar product $\mathbf{T}_1 \cdot \mathbf{T}_1$) has the eigenvalue

$$\mathbf{T}_1 \cdot \mathbf{T}_1 \rightarrow [2J][2J+1] \,. \tag{4.6}$$

The unusual form for $\mathbf{T}_{1,0}$ in (4.5b) should be noted. One can verify that under the induced action $\mathbf{T}_{1,m}$ transforms properly as the adjoint q-irrep; for $q \rightarrow 1$ we recover the commutator action and the usual matrix elements of 2**J**.

It is interesting that the less systematic approaches yield forms [28,35] for the Casimir invariant different than (4.6) and distinct from each other, forms that introduce *fractional* q-numbers. (Although q-integers do not form a number field, not even a ring, algebra operations in $SU_q(2)$ do preserve the q-integer form, in particular, fractional q-numbers are never necessary.)

Remark. The algebra of tensor operators, and its q-tensor operator extension, are thoroughly understood for $SU(2)$ and $SU_q(2)$ and are almost as fully developed for $SU(3)$ but less so for $SU_q(3)$. *These operator algebras are characterized by a well-defined and fully explicit operator product expansion.* Little is known of their subalgebra structure although it is known that the universal enveloping algebra is a subalgebra and it is believed that some diffeomorphism group commutator algebras are also subalgebras.

Acknowledgements. We would like to express our thanks to Professor L.Faddeev for the invitation to participate in the inaugural conference opening the Euler International Mathematical Institute at Leningrad. We wish to thank Dr. Piotr P. Kulish for his help and for discussions at the conference.

References

[1] E. Sklyanin, L. Takhtajan, and L. Faddeev, Teor. Math. Phys. **40** (1979), 194.
[2] P. Kulish, N. Y. Reshetikhin, Zap. Nauch. Seminarov LOMI **101** (1981), 101; *ibid.*, J.Soviet Math. **23** (1983), 2435.
[3] V. G. Drinfeld, *Quantum Groups*, Proc. ICM-86 (Berkeley) **1** (1986), 798; Sov. Math. Dokl. **36** (1988), 212.
[4] L. Faddeev, Les Houches Lectures (1982) (1984), Amsterdam,Elsevier.
[5] M. Jimbo, Lett. Math. Phys. **10** (1985 2), 63.
[6] L. C. Biedenharn, *XVIIIth International Colloqium on Group Theoretical Methods in Physics*, to be published by Springer-Verlag ((Moscow, 4-9 June 1990)).
[7] L. C. Biedenharn and M. Tarlini, Lett.Math.Phys. **20** (1990), 271.
[8] L. C. Biedenharn, J.Phys.A.Math.Gen. **22** (1989), L873.
[9] A. J. Macfarlane, J.Phys.A.Math.Gen. **22** (1989), 4581.

[10] C.-P. Sun and H.-C Fu, J.Phys.A.Math.Gen. **22** (1989), L983.
[11] P. Kulish and E. Damaskinsky, J.Phys. **A23** (1990), L415.
[12] M. Chaichian and P. Kulish, Phys.Lett. **234B** (1990), 72.
[13] M. Chaichian, P. Kulish and J. Lukierski, Phys.Lett. **237B** (1990), 401.
[14] L. C. Biedenharn and M. A. Lohe, *invited paper at the Rochester Conference (honoring Prof. S.Okubo)*, Proceedings, World Scientific, Singapore (4-5 May 1990), (to appear).
[15] F. H. Jackson, Messenger Math. **38** (1909), 57–61; *ibid*, 62–64; Quart. J. Pure Appl. Math. **41** (1910), 192–203.
[16] R. Askey and J. A. Wilson, Memoirs Amer. Math. Soc. **319** (1985).
[17] S. C. Milne, Advances in Math. **72** (1988), 59–131.
[18] G. E. Andrews, *Reg. Conf. Series in Math. no.66*, (Providence, R.I., AMS).
[19] G. Gasper and M. Rahman, *"Basic Hypergeometric Series"*, Encyclopedia of Mathematics and Its Applications **35** (1990), (Cambridge University Press).
[20] G. Lusztig, Adv. in Math. **70** (1988), 237.
[21] M. Rosso, Commun. Math. Phys. **117** (1989), 581.
[22] Y. Manin, Commun. Math. Phys. **123** (1989), 163.
[23] R. LeBlanc and D.J.Rowe, J. Math. Phys. **29** (1988), 767 and references cited here.
[24] M. A. Lohe and L. C. Biedenharn, *"Inductive Construction of Representations of the Quantum Unitary Groups"*, in preparation.
[25] I. M. Gel'fand and A. V. Zelevinsky, Societé Math. de France, Astérisque, hors serie (1985), 117.
[26] L. C. Biedenharn, A. Giovannini and J. D. Louck, J. Math. Phys. **8** (1967), 691.
[27] L. C. Biedenharn and J. D. Louck, *Angular Momentum in Quantum Physics*, Encyclopedia of Mathematics and Its Applications **8** (1981), Addison Wesley; reprinted Cambridge University Press (1989).
[28] A. N. Kirillov and N. Yu. Reshetikhin, *Representations of the Algebra $U_q(s\ell(2))$, q-Orthogonal Polynomials and Invariants of Links*, Preprint, LOMI (1988).
[29] M. Nomura, J. Math. Phys. **30** (1988), 2397; J. Math. Soc. Jap. **58** (1989), 2694; ibid **59** (1990), 439.
[30] L. Vaksman, Sov. Math. Dokl. **39** (1989), 467.
[31] H. Ruegg, J. Math. Phys. **31** (1990), 1085.
[32] V. A. Groza, I.I. Kachurik and A. U. Klimyk, J. Math. Phys. **31** (1990), 2769.
[33] T. H. Koornwinder, Nederl. Acad. Wetensch. Proc. Ser. **A92** (1989), 97.
[34] A. N. Kirillov and N. Yu. Reshetikhin, Commun.Math.Phys. **134** (1991), 421–431.
[35] E. Witten, Nucl. Phys. **B330** (1990), 285.

DEPARTMENT OF PHYSICS, DUKE UNIVERSITY, DURHAM, NC 27706, USA

AFFINE TODA FIELD THEORY : S-MATRIX VS PERTURBATION

H.W.BRADEN[(1)], E.CORRIGAN[(2)], P.E.DOREY[(3)] AND R.SASAKI[(4)]

[(1)] Department of Mathematics, University of Edinburgh
[(2)] Department of Mathematical Sciences, University of Durham
[(3)] Service de Physique Théorique de Saclay, Gif-sur-Yvette
[(4)] Uji Research Center, Yukawa Institute for Theoretical Physics, Kyoto

1. Introduction

The starting point of the present research was Zamolodchikov's work [1,2] which revealed an integrable S-matrix theory containing eight massive scalar particles, starting from a certain deformation of minimal $c = 1/2$ Conformal Field Theory (CFT) , the Ising model at criticality. Later many authors, including Eguchi-Yang [3], Hollowood-Mansfield [4], Fateev-Zamolodchikov [5] related the mass spectrum, conserved quantities etc., to the coset construction of the CFT and thereby suggested a connection to Affine Toda Field Theory (ATFT). This connection is not straightforward, however , since the coupling constant is imaginary and takes only certain discrete values[1]. Affine Toda Field Theory can also be considered as an integrable solvable deformation of non-Affine Toda Field Theory which is solvable and conformal invariant. Both types of Toda Field Theories are well known examples of Lorentz-invariant solvable classical field theory. Thus we are led to study Affine Toda Field Theory in its own right as a solvable *bona fide* Lagrangian Quantum Field Theory. Besides its connection with CFT, the ATFT has another reason to become a good theoretical laboratory for investigating the structure of solvable quantum field theories. Previous study [8,11–16] suggests that the ATFT contains only those particles which are manifest in the Lagrangian. This is based on the S-matrix bootstrap which is built upon the assumption of the existence of bound states. So in the case of ATFT , all the particles are both elementary and bound states at the same time and they appear in the Lagrangian explicitly. This is in sharp contrast with the case of the Sine-Gordon theory (A_1 affine Toda Field Theory with imaginary coupling), another well known solvable quantum field theory. The Sine-Gordon has a rich spectrum of particles, the solitons and their bound states, the breathers. Whereas its Lagrangian contains only a single neutral scalar field. It is also well known that the bound states cannot be dealt with in the framework of perturbation theory, the only established means of calculation in field theory.

So we are in a position to explore ATFT from two fronts: to obtain the exact S-matrix using the bootstrap assumptions on the one hand and to calculate various scattering

[1] See for example [6–10]

amplitudes and Green's functions in the framework of Tomonaga-Schwinger-Feynman-Dyson perturbation theory starting from the Lagrangian. Either of them has its own deficiency; the former lacks clear first principles, so various assumptions made can never be proven. Only their self-consistency can be checked. The latter has a solid first principle but in practice only lower order perturbation can be calculated.

It is not a priori trivial or clear whether these two approaches are mutually consistent. It is a question to be checked by actual comparison of the outcome of both approaches. And the check cannot be conclusive due to the limitation of perturbation theory. With this limitation in mind we are going to answer the question in the affirmative during this talk. So for the time being both approaches can be regarded as an incomplete substitute of a complete formulation of solvable quantum field theory yet to be discovered. Another qualitative outcome of the present research is the marked universality of ATFT. Most of the quantities or statements can be expressed in such a way as to apply equally to any member of simply-laced ATFT, or any particular particle or scattering channel considered.

The plan of the talk is as follows: In the next section we introduce the Lagrangian of ATFT and extract necessary information to be used in the S-matrix theory and in the perturbative approach. In section 3 various properties of the discovered S-matrices are discussed. One novel feature of these S-matrices is the multiple pole structure [12,13,17]. For example the E_8 S-matrices have up to 12 order poles. In section 4 we check various properties of the above S-matrices, like mass renormalization, multiple pole residues etc., in the perturbation theory. As for the poles we could show that simple, double and cubic pole residues are correctly given by perturbation. The 4-th order poles and up to 12-th order poles are beyond the reach of our present ability.

2. (Semi) Classical Properties of Affine Toda Field Theory

Affine Toda field theory is defined by the action density

$$\mathcal{L} = \frac{1}{2}\partial_\mu\phi\cdot\partial^\mu\phi - V(\phi), \qquad V(\phi) = \frac{m^2}{\beta^2}\sum_{i=0}^{r} n_i e^{\beta\alpha_i\cdot\phi}. \tag{2.1}$$

Here ϕ^a $a = 1,\ldots,r$ are r-component real scalar fields in $1+1$ dimensions and β is a real coupling constant and m is a real parameter setting the mass scale of the theory. The vectors α_i, $i = 1,\ldots,r$ are the simple roots of any one of the rank r Lie algebras belonging to the series A, D or E[2]. The roots are chosen to have length $\sqrt{2}$. It is a classically integrable [18,19] field theory having an infinite number of conserved quantities but it has no soliton solutions. There may well be complex soliton solutions for imaginary β.

Expanding the potential $V(\phi)$ around the classical vacuum $\phi\equiv 0$ we get

$$V(\phi) = \frac{m^2}{\beta^2}h + \frac{1}{2}(M^2)^{ab}\phi^a\phi^b + \frac{1}{3!}c^{abc}\phi^a\phi^b\phi^c + \cdots,$$

[2]We restrict ourselves to the simply-laced theories for reasons given earlier [11,17] and $-\alpha_0 = \sum_1^r n_i\alpha_i$ is the highest root ($n_0 = 1$).

where $h = \sum_0^r n_i$ is the Coxeter number of g , M^2 is the mass matrix and c^{abc} is the 3-point coupling etc.,

$$(M^2)^{ab} = m^2 \sum_i n_i \alpha_i^a \alpha_i^b, \quad c^{abc} = m^2 \beta \sum_{i=0}^r n_i \alpha_i^a \alpha_i^b \alpha_i^c. \tag{2.2}$$

By diagonalising the mass matrix we get the mass spectrum, which may be found elsewhere [19,12,17,15,20,21,22] and will not be given here. The first universal result (as yet unexplained) is [12,23,17] that the set of masses for the theory of type (2.1) based on the simple roots of the Lie algebra g constitutes the components of the Perron-Frobenius eigenvector of the Cartan matrix[3] corresponding to g and given by:

$$C_{ij} = \alpha_i \cdot \alpha_j.$$

Indeed, if we write $\mathbf{m} = (m_1, m_2, \ldots, m_r)$ then

$$C\mathbf{m} = \mu_{\text{min}}\mathbf{m} \qquad \mu_{\text{min}} = 4\sin^2\frac{\pi}{2h}. \tag{2.3}$$

As a consequence of this observation, each mass m_a (or mass eigenstate) of the theory can be associated tentatively (but unambiguously) with a spot on the Dynkin diagram for g and hence with a fundamental representation (λ_a) of the Lie algebra g. The above relationship can be generalized to the eigenvalues of the conserved charges of the theories [8,25] ($\mathbf{q}_s = (q_s^1, q_s^2, \ldots, q_s^r)$)

$$C\mathbf{q}_s = \mu_s \mathbf{q}_s \quad \mu_s = 4\sin^2\frac{s\pi}{2h}, \tag{2.4}$$

where the spin s runs over the possible exponents [26] of the Lie algebra g. Here the conserved charge Q_s acts on single-particle mass eigenstates:

$$Q_s|p_a\rangle = q_s^a e^{s\theta_a}|p_a\rangle, \tag{2.5}$$

in which θ_a is the rapidity. This observation suggests that the Cartan matrix should play a fundamental role in the quantum theory.

The second universal result is the magnitude of the three-point couplings between the mass eigenstates. It has a nice geometrical interpretation [12,15,27] (whenever the coupling is non-zero) that,

$$|c^{abc}| = \frac{4\beta}{\sqrt{h}}\Delta^{abc}, \tag{2.6}$$

where Δ^{abc} is the area of the triangle whose sides are the masses of the three particles participating in the coupling, m_a, m_b, m_c. Furthermore, the masses form a triangle whose angles are integer multiples of π/h. The observation (2.6) plays a crucial role in verifying within perturbation theory properties of the exact S-matrices that have been conjectured for all of these theories.

[3] Shankar [24] suggested that the masses are related to the Cartan matrix in a different context.

Curiously, the above conjectured association of the mass m_a (mass eigenstates) with the fundamental representation λ_a respects the three-point couplings between the mass eigenstates. That is

$$c^{abc} \neq 0 \quad \Rightarrow \quad \lambda_a \otimes \lambda_b \otimes \lambda_c \supset (1), \tag{2.7}$$

where (1) denotes the identity representation. In other words $\lambda_{\bar{c}}$ belongs to the Clebsch-Gordan series [8,13] of $\lambda_a \otimes \lambda_b$. Actually the converse is true only for the cases of A_n and D_4. Namely in all the other cases the Lagrangian (2.1) requires that there should be 'holes' in the Clebsch-Gordan series. The above Clebsch-Gordan property, which is a property of the classical couplings, must be present at all orders in perturbation theory. We have checked to third order of perturbation that the 'holes' in the Clebsch-Gordan table are preserved for D_5 and E_6 [13,28], E_7 and E_8 [28]. This seems to us to be a non-trivial result for the following reason. Take the D_5 case. There, the only hole in the C-G table occurs for the particle corresponding to the adjoint representation, of dimension 45. Clearly, the tensor product of two adjoint representations contains itself. However, the affine Toda coupling table forbids this particle to self-couple. On the other hand, there is a three-point Green function in perturbation theory in which the forbidden coupling is constructed using loops and is therefore at least of order β^3. However, the sum of the contributing diagrams at order β^3 vanishes on-shell.

3. Exact S-Matrices for A, D, E series

In this chapter we outline the S-matrix approach. The basic assumption is that the infinite number of conserved quantities survive into the quantum theory. As a consequence only elastic scatterings are allowed and all multiparticle S-matrices are factorized into a product of two particle elastic S-matrices $S_{ab}(\theta)$, which depend on the rapidity difference θ of the two incoming particles a and b. They are merely 'phases' for real θ . The requirement of Unitarity and Crossing are

$$\begin{aligned} &\text{Unitarity} \quad S_{ab}(\theta)S_{ab}(-\theta) = 1, \\ &\text{Crossing} \quad S_{ab}(i\pi - \theta) = S_{a\bar{b}}(\theta). \end{aligned} \tag{3.1}$$

As a consequence S_{ab} has $2\pi i$ periodicity. Suppose $S_{ab}(\theta)$ has a pole at $\theta = i\theta^c_{ab}$ ($0 < \theta^c_{ab} < \pi$ and $\bar{\theta} = \pi - \theta$), then the bootstrap requires

$$\text{Bootstrap} \quad S_{cd}(\theta) = S_{ad}(\theta - i\bar{\theta}^b_{ac})S_{bd}(\theta + i\bar{\theta}^a_{bc}). \tag{3.2}$$

An additional requirement is its coupling dependence. We expect that the S-matrix becomes trivial in the limit of weak coupling; $\lim_{\beta \to 0} S_{ab}(\theta, \beta) = 1$. We also invoke the classical data—masses and couplings—to determine a solution to the bootstrap program for the full S-matrix of the quantized theory.

Here we list the two particle elastic S-matrices for the A_n and D_n affine Toda field theory obtained in this way. For A_n series ATFT we rediscovered [12,17] the S-matrix given by Arinshtein-Fateev-Zamolodchikov [29] some 10 years ago. The S-matrices for D_n series [12,17] and for E_6 [12,17] E_7 [30,12,17] and E_8 [16,12,17] are believed to be new. The basic building block of the S-matrix satisfying the periodicity and unitarity may be taken to be of the form

$$(x) = \frac{\sinh\left(\frac{\theta}{2} + \frac{i\pi x}{2h}\right)}{\sinh\left(\frac{\theta}{2} - \frac{i\pi x}{2h}\right)}, \tag{3.3}$$

where h is the Coxeter number. In order to incorporate the coupling constant dependence it is advantageous to introduce the modified building block

$$\{x\} = \frac{(x+1)(x-1)}{(x+1-B)(x-1+B)}, \tag{3.4}$$

where B is a function of β , $B(\beta) = \sum_{m=1}^{\infty} B_m \beta^{2m}$, which cannot be determined by Unitarity, Crossing or Bootstrap. Now $\{0\} = \{h\} = 1$, $\{x \pm 2h\} = \{x\}$, $\{-x\} = \{x\}^{-1}$ and the crossed version of $\{x\}$ is $\{h - x\}$. We also note the fundamental property

$$\{x\}_B = \{x\}_{-B+2}. \tag{3.5}$$

In terms of these we have the S-matrices:

$$a_n^{(1)[29,12,17]} \quad S_{ab} = \{a+b-1\}\{a+b-3\}\ldots\{|a-b|+1\} = \prod_{\substack{|a-b|+1\\ \text{step } 2}}^{a+b-1} \{p\}$$

$$d_n^{(1)[12,17]} \quad S_{ab} = \prod_{\substack{|a-b|+1\\ \text{step } 2}}^{a+b-1} \{p\}\{h-p\} \qquad a, b = 1, 2, \ldots n-2$$

$$S_{s'a} = S_{sa} = S_{\bar{s}a} = \prod_{\substack{0\\ \text{step } 2}}^{2a-2} \{n-a+p\} \tag{3.6}$$

$$d_{even}^{(1)} \quad S_{ss} = S_{s's'} = \prod_{\substack{1\\ \text{step } 4}}^{h-1} \{p\} \qquad S_{ss'} = \prod_{\substack{3\\ \text{step } 4}}^{h-3} \{p\}$$

$$d_{odd}^{(1)} \quad S_{ss} = S_{\bar{s}\bar{s}} = \prod_{\substack{1\\ \text{step } 4}}^{h-3} \{p\} \qquad S_{s\bar{s}} = \prod_{\substack{3\\ \text{step } 4}}^{h-1} \{p\}$$

One of the interesting features of the exact S-matrices is their multiple pole structure. For example, the S-matrices for A_n have simple and double poles, D_n have up to 4-th order poles, E_6 up to 6-th, E_7 up to 8-th and E_8 up to 12-th order poles. On examining the S-matrices we note that the lowest order term in the coefficient (we will refer to it as the pole residue) of a pth order pole as a function of β at the rapidity value $\theta = i\theta_0$ is given by the relation:

$$S_{ab}(\theta) \sim \left(\frac{\beta}{\sqrt{2h}}\right)^{2p} \frac{\kappa_p}{(\theta - i\theta_0)^p}, \tag{3.7}$$

where $\kappa_p = \pm i$ or 1 according to whether p is odd or even.

This is a universal form independent of the particular S-matrix element or the particular particle pair ab. Note, we have found that the odd order poles correspond to genuine bound states for the purpose of the bootstrap while those of even order not. The bound states appearing in any channel are of course supposed to be a subset of the r possible asymptotic states; for real β, as we mentioned above, there are conjectured to be no further states in the theory beyond those described by the fields appearing explicitly in the Lagrangian (2.1).

The requirement that there be no extra poles in the physical region can be expressed universally, *i.e.*, $0 < B < 2$. This together with the knowledge of the sine-Gordon S-matrix [31] led many authors to conjecture

$$B(\beta) = \frac{1}{2\pi}\frac{\beta^2}{1+\beta^2/4\pi}. \tag{3.8}$$

If this conjecture is true (3.5) entails the 'duality'between weak and strong coupling regime

$$S(\theta,\beta) = S(\theta,\frac{4\pi}{\beta}). \tag{3.9}$$

In the next section we show that these singularities together with the residues can be explained from the Lagrangian (2.1) as the Landau singularities [32] of the Feynman diagrams.

4. Field Theory Aspects

The mass spectrum (or mass ratios) of the exact S-matrices turn out to be the same as the classical mass spectrum obtained from the Lagrangian. This requires 'explanation'since mass gets renormalized in field theory. In two dimensions the only divergence comes from the tadpole diagrams and these can be removed by normal ordering. But still finite mass renormalization is necessary. By calculating the propagator bubble (self-energy part) to order β^2 we encounter again an elegant and universal result [12,17] for simply laced ATFT

$$\frac{\delta m_c^2}{m_c^2} = \frac{\beta^2}{4h}\cot\frac{\pi}{h}. \tag{4.1}$$

This justifies to order β^2 the assumption that the masses renormalize in such a way as to maintain the classical ratios. For non-simply laced ATFT the above property does not hold. However, in that context there are recent developments concerning theories based on super-algebras [33]. The fact that the classical mass ratios are not preserved is considered to be an evidence that the infinite number of conserved quantities develops anomaly in non-simply lased theories when quantized.

Next we turn to the problem of explaining the multiple poles and their residues from Feynman diagrams [14]. Given a particular on-shell elastic scattering process $ab \to ab$ characterized by the rapidity difference $\theta = \theta_a - \theta_b$, we seek Feynman diagrams singular as $\theta \to i\theta_0$ or equivalently, $s \to s_0$ where s_0 is below threshold. In particular, to obtain the leading contributions to the residue of the highest order pole in a particular channel, we seek the most complicated Feynman diagrams, constructed from the three-point couplings (2.2) between mass eigenstates, for which all the internal propagators may be simultaneously on-shell. Here, leading means of lowest order in the coupling constant β. If the pole is of order p, as it is in (3.7), then the number of internal propagators will be equal to $p + 2L$, where L is the number of loops in each of the contributing Feynman diagrams. Restricting ourselves to the leading terms has the advantage of removing the need for any discussion of renormalization beyond omitting diagrams containing propagator bubbles on internal or external legs.

Since p_a and p_b are the momenta of the external scattering particles we may write $s = (p_a + p_b)^2$, $p_a^2 = m_a^2$, $p_b^2 = m_b^2$. Further, let $p_a^{(0)}$ and $p_b^{(0)}$ denote a pair of values of these momenta for which $s = s_0$, the pole position. Given a Feynman diagram integrand for which the external momenta and all internal momenta are on-shell simultaneously, we may express the internal momenta $p_i^{(0)}$, $i = 1, \dots, P = p + 2L$, which therefore also satisfy ${p_i^{(0)}}^2 = m_i^2$ in the following unique way: $p_i^{(0)} = a_i p_a^{(0)} + b_i p_b^{(0)}$ $i = 1, \dots, P$, where the coefficients a_i and b_i will need to be determined in each specific case. If we allow p_a and p_b to drift from the singular configuration then, defining $p_i = a_i p_a + b_i p_b$ we maintain momentum conservation and we will have $p_i^2 - m_i^2 = a_i b_i (s - s_0) \equiv \epsilon_i (s - s_0)$. In order to introduce the loop momenta conveniently, we take the situation for which all propagators are on-shell to correspond to $l_1 = l_2 = \dots = l_L = 0$ where l_i is a loop momentum. Then, each propagator momentum P_i in general non-singular position can be written as a linear combination of p_i and the various loop momenta: $P_i = p_i + \sum_{j=1}^{L} \lambda_{ij} l_j \equiv p_i + k_i$. Thus, we find $P_i^2 - m_i^2 = \epsilon_i (s - s_0) + 2p \cdot k_i + k_i \cdot k_i$. Finally, we scale all loop momenta by a factor $(s - s_0)$, $l_i \to (s - s_0) l_i$, in which case the denominator of the typical propagator is scaled as follows,

$$P_i^2 - m_i^2 \to (s - s_0) \left[\epsilon_i + 2p_i \cdot k_i + (s - s_0) k_i \cdot k_i \right] .$$

Thus, taking all this into account, we find for a diagram containing P propagators and L loops

$$\int \prod_{\text{loops}} d^2 l_i \prod_{\text{propagators}} \frac{1}{P_i^2 - m_i^2 + i\epsilon}$$
$$\sim \frac{1}{(s - s_0)^{P-2L}} \int \prod_{\text{loops}} d^2 l_i \prod_{\text{propagators}} \frac{1}{2p_i^{(0)} \cdot k_i + \epsilon_i + i\epsilon} \tag{4.2}$$

explicitly demonstrating the pole of order $p = P - 2L$ with a residue which is given in terms of a multiple integral in which the denominators of the propagators are replaced by **linear** functions of the loop momenta. To evaluate (4.2) we use Cauchy's theorem.

In (4.2) we have omitted all the couplings and usual vertex and propagator factors. These will be re-instated once the integrals on the right hand side have been evaluated. Besides these factors, there will also be flux and normalization factors to yield a proper comparison with the S-matrix elements. For convenience, we list them all here: (a) For each vertex: $-4\beta i(2\pi)^2 (h)^{-1/2}$, (b) For each propagator: $i(2\pi)^{-2}$, (c) Flux and normalization factor: $-i(16\pi^2 m_a m_b \sin\theta_0)^{-1} = -i(32\pi^2 \Delta)^{-1}$, where $i\theta_0$ is the rapidity at which the singularity occurs. In addition, in (a) there will also be a further factor corresponding to a certain area (see (2.6)) and an additional phase. For future reference, we also note the relationship between the s and θ variables near the singularity:

$$s - s_0 = (\theta - i\theta_0) i 2 m_a m_b \sin\theta_0 = 4i\Delta(\theta - i\theta_0). \tag{4.3}$$

Wherever it appears, Δ is the area of the triangle with sides of length m_a and m_b with an angle θ_0 between them.

The types of diagrams we are seeking have been known for more than 20 years [32]: namely those Feynman diagrams whose *dual* diagrams are *planar*. The problem is essentially geometric : we seek *tilings* of a parallelogram (not necessarily exact) in terms of the basic triangles representing the classical three point couplings (2.6). The 'not necessarily exact' is required to cover the cases of crossed Feynman diagrams.

Let us first discuss the first order poles. To leading order in β, these arise from simple tree graphs of the following type

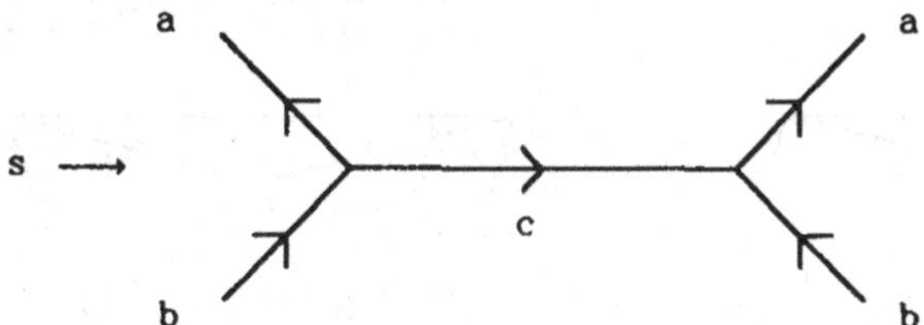

Fig. 1.

where the propagator provides the singularity. As a consequence of (2.6), each of the three-point couplings is proportional to the area Δ which therefore cancels out completely between the vertex parts in the numerator and the factors from the flux and (4.3). Gathering together all the other factors gives the universal result:

$$S(\theta) \sim \frac{i}{\theta - i\theta_0}\left(\frac{\beta^2}{2h}\right)$$

i.e. eq (3.7), for $p = 1$.

The leading order contribution for the second order poles is provided by a set of four box diagrams in general. A typical single box diagram, together with its on-shell dual diagram is

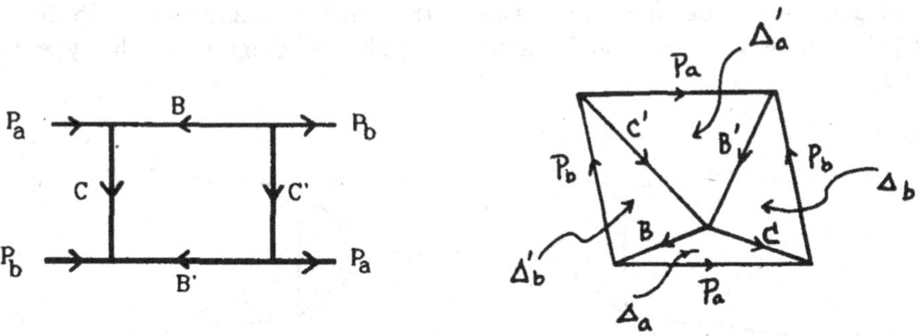

Fig. 2a.

where we label the internal (on-shell) momenta and also the areas of the four triangles $\Delta_a, \Delta_b, \Delta_a', \Delta_b'$, respectively, as shown (they satisfy $\Delta_a' + \Delta_a = \Delta = \Delta_b' + \Delta_b$). There are two such diagrams, interchanging primed and unprimed variables, but they have the same numerical value. Moreover there are two crossed box diagrams which are depicted in Fig.2b,c together with the dual diagrams.

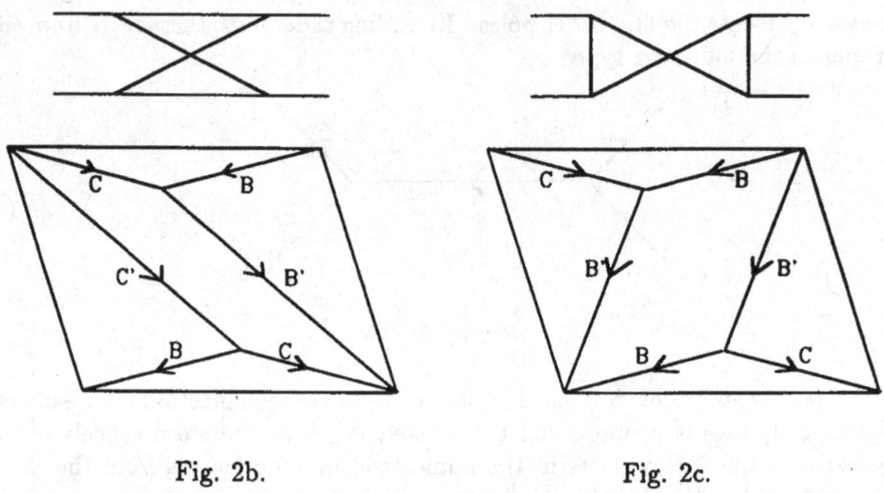

Fig. 2b. Fig. 2c.

The summing of diagrams also requires the phases to multiply up suitably. For the details of the actual evaluation of these diagrams, see ref.[14]. Here we simply give the result. Their contribution to the residue in the unit of $(\beta/\sqrt{2h})^4$ is (a) $2\Delta'_a/\Delta$, (b) $(\Delta_a\Delta_b - \Delta'_a\Delta'_b)\Delta^{-2}$, (c) $(\Delta_a\Delta'_b - \Delta'_a\Delta_b)\Delta^{-2}$. Thus summing them we find that the residue is $(\beta/\sqrt{2h})^4$, the desired result.

The situation regarding the third order poles is not so straightforward and the analysis depends upon several facts about affine Toda field theories that do not appear to have been noticed before. We have checked for all simply laced ATFT [14] that the leading order residues are correctly given by perturbation. Here we only mention the types of diagrams to be calculated. Among the singular diagrams contributing to the cubic poles, there are in general 9 one-particle reducible diagrams of the type depicted in Fig.3,

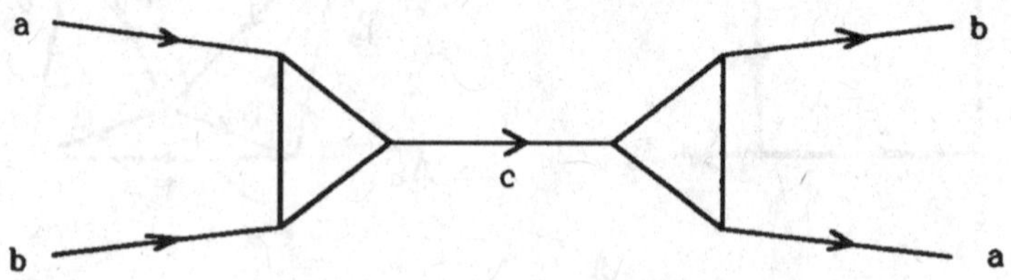

Fig. 3.

in which the middle propagator is on-shell at the pole position. There are also other singular diagrams of the double box type depicted in Fig.4 and other double boxes containing various crosses. Including multiplicities, there are 17 of these diagrams in total.

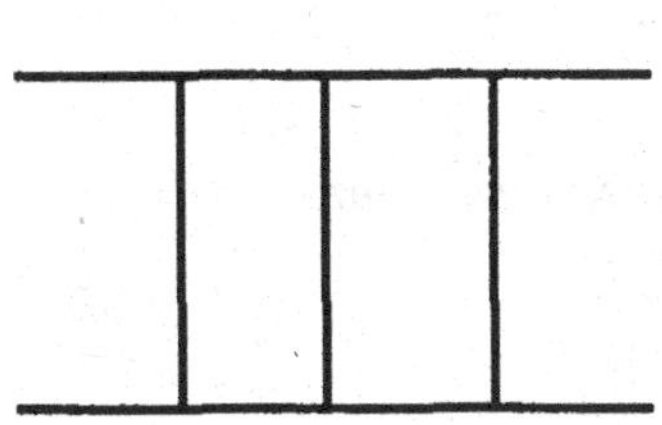

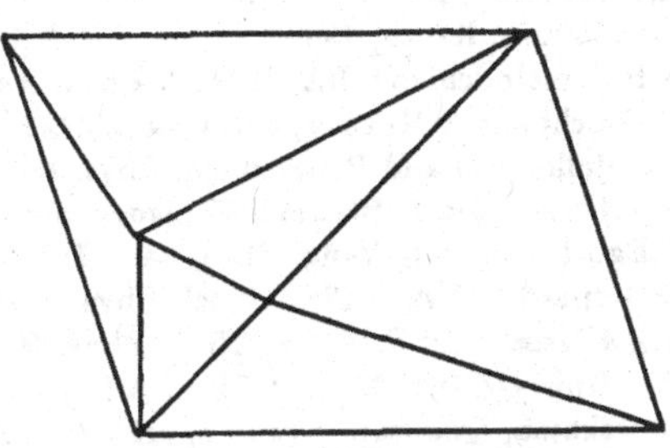

Fig. 4.

Continuing the above calculation for fourth order poles and beyond seems increasingly difficult. Even enumerating the contributing diagrams is an unsolved problem, let alone evaluating them. For example the number of diagrams contributing the 12-th order poles in E_8 theory well exceeds 70 million.

The last subject to be discussed is the coupling dependence of the S-matrix. The S-matrix elements obtained for an affine Toda field theory depend on the coupling constant β through one unknown function $B(\beta)$. As we have already mentioned this cannot be determined by unitarity, crossing or the bootstrap. Although there is no *a priori* reason, let us assume that these functions are identical for all simply laced ATFT: the lowest order (*i.e.* tree) perturbation result, $B(\beta) = \frac{1}{2\pi}\beta^2 + o(\beta^4)$, supports this hypothesis [17,20]. By calculating the simple pole residues to the next lowest order of perturbation, *i.e.* to β^4 , we found [34]

$$B(\beta) = \frac{\beta^2}{2\pi}(1 - \frac{\beta^2}{4\pi}) + o(\beta^6). \tag{4.4}$$

This supports the above hypothesis of universality and demonstrates the conjecture (3.8) to the required order. It should be stressed that the next leading order calculation inevitably contains the effects of renormalization mass and wavefunction.

We have given several pieces of evidence that perturbation theory manages to reproduce various salient features of the conjectured exact S-matrices of ATFT. At present, we do not see how to use perturbation theory to provide an efficient description of the quantum field theory; an alternative formulation may well be required in order to find a proper understanding of the conjectured S-matrices and other features such as the mass-renormalization and the Clebsch-Gordan property. Certainly, the knowledge from other approaches , for example, the Quantum Group approach to imaginary coupling ATFT, investigations of the Bethe-Salpeter equations for the bound states in ATFT and the algebraic Bethe ansatz method advocated for many years by Faddeev and others [35] would be helpful in the search for such a re-formulation.

Acknowledgements

One of us (RS) wishes to thank the organizers of the Workshop for warm hospitality and Yamada Science Foundation for a travel grant. We thank W. Nahm for informing us of Shankar's work

References

[1] A.B.Zamolodchikov, *Integrable Field Theory from Conformal Field Theory*, Proc. of the Taniguchi Symposium, Kyoto, 1988.
[2] A.B.Zamolodchikov, Int. J. Mod. Phys. **A4** (1989), 4235.
[3] T.Eguchi and S-K Yang, Phys. Lett. **B224** (1989), 373.
[4] T.J.Hollowood and P.Mansfield, Phys. Lett. **B226** (1989), 73.
[5] V.A.Fateev and A.B.Zamolodchikov, Int. J. Mod. Phys. **A5** (1990), 1025.
[6] T.Eguchi and S-K Yang, Phys. Lett. **B235** (1990), 282.
[7] D.Bernard and A.LeClair, Nucl. Phys. **B340** (1990), 721.
[8] T.R.Klassen and E.Melzer, Nucl. Phys. **B338** (1990), 485.
[9] F.A.Smirnov, Int. J. Mod. Phys. **A6** (1991), 1407.
[10] T.Nakatsu, *Quantum Group approach to Affine Toda field theory*, Tokyo preprint UT-567.
[11] H.W.Braden, E.Corrigan, P.E.Dorey and R.Sasaki, Phys. Lett **B227** (1989), 411.
[12] H.W.Braden, E.Corrigan, P.E.Dorey and R.Sasaki, *Aspects of perturbed conformal field theory, affine Toda field theory and exact S-matrices*, Proc. XVIII Inter. Conf. Differential Geometric Methods in Theoretical Physics, Lake Tahoe, USA 2-8 July 1989 (to appear).
[13] H.W.Braden, E.Corrigan, P.E.Dorey and R.Sasaki, *Aspects of affine Toda field theory*, Proceedings of the 10th Winter School on Geometry and Physics, Srni, Czechoslovakia; Integrable systems and Quantum Groups, Pavia, Italy; Spring Workshop on Quantum Groups, ANL, USA (to appear).
[14] H.W.Braden, E.Corrigan, P.E.Dorey and R.Sasaki, *Multiple Poles and Other Features of Affine Toda field theory*, preprint NSF-ITP-90-174, DTP-901-57, YITP/U-90-25; Nucl. Phys. **B**, in press.
[15] P.Christe and G.Mussardo, Nucl. Phys **B330** (1990), 465.
[16] C.Destri and H.J.de Vega, Phys. Lett. **B233** (1989), 336.
[17] H.W.Braden, E.Corrigan, P.E.Dorey and R.Sasaki, Nucl. Phys. **B338** (1990), 689.
[18] V.G.Drinfel'd and V.V.Sokolov, J. Sov. Math. **30** (1984), 1975.
G.Wilson, Ergod. Th. and Dynam. Sys. **1** (1981), 361.
D.I.Olive and N.Turok, Nucl. Phys. **B215** (1983), 470.
[19] A.V.Mikhailov, M.A.Olshanetsky and A.M.Perelomov, Comm. Math. Phys. **79** (1981), 473.
[20] P.Christe and G.Mussardo, *Elastic S-matrices in (1+1) dimensions and Toda field theories*, Int. J. Mod. Phys. (to appear).
[21] G.Mussardo and G.Sotkov, *Bootstrap program and minimal integrable models*, preprint UCSBTH--89-64/ISAS-117-89.
[22] G.Mussardo, *Away from criticality: some results from the S-matrix approach*, in [12].
M.Koca and G.Mussardo, *Mass formulae in Toda field theories*, preprint CERN-TH 5659/90.
[23] P.G.O.Freund, T.Klassen and E.Melzer, Phys. Lett. **B229** (1989), 243.
[24] R.Shankar, Phys. Lett. **B92** (1980), 333.
[25] P.E.Dorey,, *Root Systems and Purely Elastic S-Matrices*, Saclay preprint SPhT/90-169, Ph.D. thesis Durham, unpublished.
[26] S.Helgason, *Differential geometry, Lie groups and symmetric spaces*, Acad. Press, 1978.
[27] P.Christe, *S-matrices of the tri-critical Ising model and Toda systems*, in [12].
[28] H.W.Braden, E.Corrigan, and R.Sasaki (to appear).
[29] A.E.Arinshtein, V.A.Fateev and A.B.Zamolodchikov, Phys. Lett. **B87** (1979), 389.
[30] P.Christe and G.Mussardo, Nucl. Phys **B330** (1990), 465.
[31] A.B.Zamolodchikov and Al.B.Zamolodchikov, Ann. Phys. **120** (1979), 253.
[32] R.J.Eden et al, *The analytic S-matrix*, Cambridge University Press, 1966.
[33] M.T.Grisaru, S.Penati and D.Zanon, *Mass corrections in affine Toda theories based on Lie superalgebras*, preprint BRX-TH-304/IFUM-390-FT.
[34] H.W.Braden and R.Sasaki, *The S-matrix coupling dependence for a, d and e affine Toda field theory*, Phys. Lett. **B**, in press.
[35] L.Faddeev and L.Takhtajan, *Hamiltonian methods in the theory of solitons*, Springer, 1987.
E.K.Sklyanin, Nucl.Phys. **B326** (1989), 719.

UJI RESEARCH CENTER, YUKAWA INSTITUTE FOR THEORETICAL PHYSICS, KYOTO, UJI 611, JAPAN

CONTRACTIONS OF QUANTUM GROUPS

E.CELEGHINI[1,2], R.GIACHETTI[3,2], E.SORACE[2,1] AND M.TARLINI[2,1]

[1] Dipartimento di Fisica dell' Università, Firenze.
[2] I.N.F.N. Sezione di Firenze.
[3] Dipartimento di Matematica dell' Università, Bologna.

1. Introduction

In the study of Lie algebras the contraction [Gl] is an useful tool that, in our opinion, has often received less consideration than the due one. Indeed, although not completely trivial, contraction is a very simple technique: nevertheless it succeeds in reproducing practically any required results on non semisimple Lie algebras starting from semisimple ones.

The general theory of quantum groups has been developed at a considerably high level, see [Dr, F1, J1, Mn, W1] and for review [Mj], however except for simplest cases, it is well beyond the supply of concrete examples; indeed what we know about the Hopf algebras [Ab] arising from semisimple algebras and about their quantum deformations is quite a lot with respect to the few informations we have been able to gather when considering non semisimple structures.

It thus emerges very clearly the possibility of using the contraction technique in a constructive way by extending it to quantum groups in order to gain insight and produce worked out examples of "non semisimple" quantum groups from "semisimple" ones.

The general pattern we shall assume is the following: since it is possible to perform the so called "classical limit" on the q-parameter, going from a simple quantum group ($SQG \equiv q$-deformations of simple algebras) to a simple group (SG) and *after* a contraction to a contracted group (CG), we can try to interchange the order of operations and perform a contraction *before*, so to define a contracted quantum group (CQG) from the given SQG, reproducing then the same CG in the classical limit. Otherwise expressed we want to study the possibility of constructing the following commutative diagram

$$\begin{array}{ccccc} & & \varepsilon \to 0 & & \\ & SQG & \longrightarrow & CQG & \\ q \to 1 & \downarrow & & \downarrow & q \to 1 \\ & SG & \longrightarrow & CG & \\ & & \varepsilon \to 0 & & \end{array}$$

Here we are not so much interested in developing a complete theory, rather than in finding a number of examples easy to handle, on which calculations can be performed and ideas tested. We could say that our approach is somewhat constructive: therefore, we shall not be deeply involved in universal mathematical aspects, but we shall consider our results as examples giving strong indications about what should be the general theory of q–deformations of non semisimple structures.

Indeed, while it is straightforward to mimic the usual contraction both at the level of generators and representations for algebras and coalgebras – the only caution being an appropriate rescaling of the q-parameter also –, we find some technical difficulties in determining the contracted R-matrices: they may diverge and only in some examples we are allowed by peculiar situations to extract a finite contribution verifying the Yang-Baxter equation [Bx, J1]. The general theory for the R-matrix contractions is at the moment not completely clear and some problems are probably related to the nontrivial cohomology arising from the contraction. It would also be interesting to clarify the relation of the existence of a R–matrix to the possibility of directly performing the contraction on the C^*–algebra of the representative functions [W2].

We want to conclude this brief introduction by pointing out that for a given simple structure a large number of rescalings of the generators and the q-parameter can be given: however many of them are trivial and others are not allowed since they lead to unwanted singularities.

The plan of the following sections is:

2. the study of the easiest CQGs with physical relevance: $E(2)_q$, $H(1)_q$, $E(3)_q$ and $E(2,1)_q$, $G(2)_q$ at the level of algebra and coalgebra and, in some cases, at the level of representations;

3. an analysis of the R-matrices for $H(1)_q$, $E(3)_q$ and $G(2)_q$;

4. a short discussion of the matrix quantum groups $H(1)_q$ and $E(3)_q$;

5. some concluding remarks.

2. Quantum groups obtained by contractions

From a technical point of view we want again to stress the fact that, as in the Lie algebra case, there is not a unique way to perform a contraction. For the quantum groups the q-parameter must be included in the variables to be rescaled and care has to be taken in order not to fall directly into the classical situation.

Not to be overwhelmed by the technical aspects we shall directly consider the contractions of the simplest quantum groups, namely $SU(2)_q$, $SO(4)_q \simeq SU(2)_q \otimes SU(2)_q$ and their central extensions.

As it is well known for $SU(2)_q$ (letting $e^z = q$) the algebraic relations are [KR1]

$$[J_3, J_\pm] = \pm J_\pm, \ [J_+, J_-] = [2J_3]_q \equiv \frac{\mathrm{sh}(zJ_3)}{\mathrm{sh}(z/2)} \ . \tag{2.1}$$

The coalgebra is

$$\begin{aligned} \Delta J_\pm &= e^{-zJ_3/2} \otimes J_\pm + J_\pm \otimes e^{zJ_3/2} \ , \\ \Delta J_3 &= \mathbf{1} \otimes J_3 + J_3 \otimes \mathbf{1} \ , \end{aligned} \tag{2.2}$$

while counit and antipode read

$$\gamma(J_\pm) = -e^{zJ_3/2} \ J_\pm \ e^{-zJ_3/2} \ , \ \gamma(J_3) = -J_3 \tag{2.3}$$

$$\epsilon(J_\pm) \ = \ \epsilon(J_3) \ = \ 0 \ . \tag{2.4}$$

2.1 The q–deformations of $e(2)$ [C1]. Let us start to consider the "q–analog" of the contraction of $su(2)$ to the algebra of the rigid movements of the plane, $e(2)$. We find it more convenient to perform a change of basis and substitute the generators $\{J_+, J_-, J_3\}$ by $\{J_1, J_2, J_3\}$, where

$$J_1 = \frac{1}{2}(J_+ + J_-)\,, \qquad J_2 = -\frac{i}{2}(J_+ - J_-)\,.$$

It is then straightforward to see that, in this basis, equations (2.1) take the form

$$[J_1, J_2] = i\,\frac{\mathrm{sh}(zJ_3)}{2\,\mathrm{sh}(z/2)}, \qquad [J_2, J_3] = iJ_1, \qquad [J_3, J_1] = iJ_2\,. \tag{2.5}$$

Let us now consider the scaling

$${}^t(P_y, J, P_x, w) \;=\; \mathrm{diag}\{\varepsilon, 1, \varepsilon, \varepsilon^{-1}\}\; {}^t(J_1, J_2, J_3, z)\,.$$

Taking now the limit $\varepsilon \to 0$ of (2.5), a nontrivial quantum behaviour is maintained and the contracted relations read

$$[P_x, P_y] = 0, \qquad [J, P_x] = iP_y, \qquad [J, P_y] = -\frac{i}{w}\;\mathrm{sh}(wP_x)\,.$$

By contraction of (2.2-4) we obtain coproduct, antipode and counit, the Hopf algebra properties being of course preserved in the limit as can be proved by straightforward calculations:

$$\Delta J = e^{-wP_x/2} \otimes J + J \otimes e^{wP_x/2}, \; \Delta P_x = 1 \otimes P_x + P_x \otimes 1,$$
$$\Delta P_y = e^{-wP_x/2} \otimes P_y + P_y \otimes e^{wP_x/2},$$

$$\gamma(J) = -J + \frac{i}{2}\,wP_y, \; \gamma(P_x) = -P_x, \; \gamma(P_y) = -P_y,$$
$$\epsilon(J) = \epsilon(P_x) = \epsilon(P_y) = 0\,.$$

We call this Hopf algebra the Euclidean quantum group $E(2)_q$. The algebra $e(2)$ is obviously the limit $w \to 0$ and can be obtained also by a stronger rescaling of z directly from $SU(2)_q$ showing that some contractions afford to trivial results.

It is interesting to obtain by contraction also the Casimir of $E(2)_q$. We recall that the quantum group $SU(2)_q$ has two different expressions for the Casimir according to the fact that the representation has integer or half-odd j [Bd], that are respectively

$$J_-J_+ + [J_3]_q\,[J_3+1]_q, \qquad \text{and} \qquad J_-J_+ + \left([J_3 + \frac{1}{2}]_q\right)^2 .$$

It is easy to be seen that the contractions of both of them give the same result and we shall explicitly work out the integer j case. We have

$$J_-J_+ + [J_3]_q[J_3+1]_q = J_1^2 + J_2^2 - \frac{\mathrm{sh}(zJ_3)}{2\,\mathrm{sh}(z/2)} + \frac{\mathrm{sh}(zJ_3/2)\;\mathrm{sh}(z(J_3+1)/2)}{\mathrm{sh}^2(z/2)}$$
$$= \varepsilon^{-2}P_y^2 + J^2 - \frac{\mathrm{sh}(wP_x)}{2\,\mathrm{sh}(\varepsilon w/2)} + \frac{\mathrm{sh}(wP_x/2)\;\mathrm{sh}(w(P_x+\varepsilon)/2)}{\mathrm{sh}^2(\varepsilon w/2)}\,.$$

Taking the leading terms in the limit $\varepsilon \to 0$ we get the Casimir operator for $E(2)_q$,

$$C = P_y^2 + \frac{4}{w^2}\ \mathrm{sh}^2(wP_x/2)\ ,$$

which, for vanishing w, reproduces the classical $P_x^2 + P_y^2$.

According to the general scheme presented in section 2.2, we can obtain by contraction the representations of $E(2)_q$, that, in the momentum basis, are very similar to the usual ones since here also we diagonalize P_x and P_y. Of course the Casimir is conserved, which implies that we are not performing a rotation on a circle but on a less symmetrical curve of equation $C = const$. This is reflected by the expression of J which reads

$$J = i\ P_y\ \frac{\partial}{\partial P_x} - \frac{i}{w}\ \mathrm{sh}(wP_x)\ \frac{\partial}{\partial P_y}\ .$$

This is an example showing that quantum deformations allow to analyze situations with apparently lower symmetry - as the curve $C = const$ - by concepts and tools analogous to the completely symmetrical case. In this way it is possible to dismiss with continuity part of the symmetry while keeping all the other features unchanged.

To stress the relevance of both algebra and coalgebra aspects in quantum groups, we note that the following simple transformation in the enveloping algebra

$$Z_x = \frac{2}{w}\ \mathrm{sh}(wP_x/2),\ Z_y = P_y,\ W = \mathrm{sesh}(wP_x/2)\ J\ ,$$

permits to reproduce the commutation relations of the $e(2)$ algebra but the coproducts thus obtained are nontrivial, due to the intrinsic quantum character of the structure we found.

As noticed in the introduction, we shall now present a different deformation of $e(2)$ generated by a different contraction. The following transformation on the generators

$$^t(P_x, P_y, J, z)\ =\ \mathrm{diag}\{\varepsilon, \varepsilon, 1, 1\}\ ^t(J_1, J_2, J_3, z)$$

gives

$$[P_x, P_y] = 0,\ [J, P_x] = iP_y,\ [J, P_y] = -iP_x\ .$$

Moreover we get:

$$\Delta J = 1 \otimes J + J \otimes 1,\ \Delta P_i = e^{-zJ/2} \otimes P_i + P_i \otimes e^{zJ/2}\ ,$$

for the coproduct,

$$\gamma(J) = -J\ ,\ \gamma(P_x) = -e^{wJ/2} P_x e^{-wJ/2},\ \gamma(P_y) = -e^{wJ/2} P_y e^{-wJ/2},$$

for the antipode,

$$\epsilon(J) = \epsilon(P_x) = \epsilon(P_y) = 0,$$

for the counit. This is the structure described in [VK], which is again an $e(2)$ at the level of algebra but deformed as coalgebra.

2.2 $H(1)_q$ and the contractions of representations. Let us now turn to a number of contractions connected with the Heisenberg algebra. The Lie algebra we want to deform is the following:

$$[A, A^\dagger] = H,\ [N, A] = -A,\ [N, A^\dagger] = A^\dagger,\ [H, \cdot\,] = 0\ . \tag{2.6}$$

It can be obtained as a contraction of $u(2)$ as has been analyzed in [Gc]. According to the scheme previously proposed, we define the following transformation on the four generators of $U(2)_q \equiv SU(2)_q \otimes U(1)$ and on the parameter $z = \log q$.

$$\begin{pmatrix} A \\ A^\dagger \\ N \\ H \\ w \end{pmatrix} = \begin{pmatrix} \varepsilon & 0 & 0 & 0 & 0 \\ 0 & \varepsilon & 0 & 0 & 0 \\ 0 & 0 & -1 & \varepsilon^{-2} & 0 \\ 0 & 0 & 0 & 2 & 0 \\ 0 & 0 & 0 & 0 & \varepsilon^{-2} \end{pmatrix} \begin{pmatrix} J_+ \\ J_- \\ J_3 \\ K \\ z \end{pmatrix} \tag{2.7}$$

where K is the new $U(1)$ generator; this is exactly the singular transformation – beside the part related to z – giving (2.6) from $u(2)$. Taking the limit $\varepsilon \to 0$ we get [C2]

$$[A, A^\dagger] = \frac{\mathrm{sh}(wH/2)}{w/2},\ [N, A] = -A,\ [N, A^\dagger] = A^\dagger,\ [H, \cdot\,] = 0\ . \tag{2.8}$$

We use again the contraction procedure to determine coproduct, counit and antipode and we see that the Hopf algebra requirements are satisfied. From (2.2–4):

$$\Delta A = e^{-wH/4} \otimes A + A \otimes e^{wH/4}\ , \qquad \Delta A^\dagger = e^{-wH/4} \otimes A^\dagger + A^\dagger \otimes e^{wH/4}\ ,$$

while for H and N we have obviously

$$\Delta N = 1 \otimes N + N \otimes 1\ ,\ \Delta H = 1 \otimes H + H \otimes 1\ .$$

Here also there is the possibility of obtaining a q–deformation of the Heisenberg algebra different from (2.8) by using the following transformation on the generators of $SU(2)_q$:

$${}^t(H, X, P, w) \;=\; \mathrm{diag}\{\varepsilon^2, \varepsilon, \varepsilon, \varepsilon^{-1}\}\ {}^t(J_1, J_2, J_3, z)$$

from which

$$\begin{aligned} [H, X] \;&=\; 0,\quad [H, P] \;=\; 0,\quad [X, P] \;=\; iH\ , \\ \Delta H \;&=\; e^{-wP/2} \otimes H + H \otimes e^{wP/2}\ , \\ \Delta X \;&=\; e^{-wP/2} \otimes X + X \otimes e^{wP/2}\ , \\ \Delta P \;&=\; 1 \otimes P + P \otimes 1\ . \\ \gamma(X) \;&=\; -X + i\frac{w}{2}H,\quad \gamma(P) \;=\; -P,\quad \gamma(H) \;=\; -H\ . \end{aligned}$$

As in the case of the deformation of $e(2)$ described in [VK] the only non trivial part is the coalgebra. In the section devoted to the study of the R matrices we will see that in this case, in analogy to [VK], but contrary to what occurs for $H(1)_q$ previously described, it is not possible to find the R matrix by contraction.

We want to stress that none of the Heisenberg algebra deformations described here fulfill the algebraic relations of the q–deformed oscillators introduced in [Bd, Mf]. It is possible to recover that algebra through a contraction of $SU(2)_q$ [K1, Ng], but loosing the Hopf structure: on the other hand it has been shown that the superalgebra $osp(1|2)$ can be deformed to a Hopf superalgebra closing the q–oscillators defining relations; however the creation and annihilation operators have to be treated as odd generators [KR2, C3].

Recently it has been claimed [Ya] that the Hopf (bosonic) algebra of the q–oscillators has been obtained, but unfortunately that paper demonstrates only the necessity of a careful analysis of the actual content of a Hopf algebra to discriminate between a true new q–deformation and a mere redefinition of the generators. The Hopf algebra presented for the q-deformed oscillators is nothing else than a trivial rephrasing [K2] of the $SU(2)_q$ Hopf algebra for $q \neq 1$ and is not defined as a coalgebra for $q = 1$. Indeed it can be verified that the transformation

$$J_3 = N - \frac{i\alpha}{\log q}, \quad J_- = e^{i\alpha/2}\left(\frac{q+1}{q-1}\right)^{1/2} a_q, \quad J_+ = e^{i\alpha/2}\left(\frac{q+1}{q-1}\right)^{1/2} a_q^\dagger,$$

where $\alpha = \pi/2 + 2k\pi$ ($k \in \mathbf{Z}$, brings for $q \neq 1$ both the algebra and the coalgebra given in [Ya] to those of $SU(2)_q$. The transformation is singular for $q \to 1$ and this explains the appearance of the Heisenberg algebra in this limit together with the divergence of the coalgebra. Thus it is not a q–deformation of the harmonic oscillator Hopf algebra.

We shall now use the example of $H(1)_q$ to illustrate the extension of the contraction procedure to representations, performing directly the limit on the representations of the generators to obtain the corresponding representations of the contracted Hopf algebra. The technique is similar to the one defined in the Lie case [CT]: we start from the Hilbert space of the states of representations defined by a complete set of commuting observables, *e.g.* the Gelfand-Tseitlin variables, that reduce for $su(2)$ to $\{j, m\}$. In the quantum case for $|q| \neq 1$ the structure of the representations remains the same. We only have to give an appropriate treatment of the parameter q. To specify the representations means to write the (generalized) functions:

$${}_z\langle\, j, m' \mid J_i \mid j, m\,\rangle_z = f_i(j, m', m; z)\,.$$

We now perform a ε–dependent transformation:

$$\begin{pmatrix} p \\ n \\ w \end{pmatrix} = M(\varepsilon) \begin{pmatrix} j \\ m \\ z \end{pmatrix}.$$

This allows us to build (for $\varepsilon \neq 0$ where the transformation is regular) a one parameter family of Hilbert spaces isomorphic to the first one, where the basis

$$\mid p, n, \varepsilon\,\rangle_w = \mid j, m\,\rangle_z$$

s always orthonormal. On this we define the matrix elements g_i of the generators according to

$$_w\langle\, \varepsilon, p', n' \mid J_i \mid \varepsilon, p, n \,\rangle_w \;\equiv\; g_i(\varepsilon, p', n', p, n; w) = f_i(j', m', j, m; z)\,.$$

In most cases the contraction limit $\varepsilon \to 0$ diverges and must be renormalized: the renormalization factors ε^{n_i} are then used to define the generators of a contracted algebra according to $Y_i = \varepsilon^{n_i} J_i$. The corresponding representations result in:

$$\begin{aligned} _w\langle\, p', n' \mid Y_i \mid p, n \,\rangle_w &= \lim_{\varepsilon\to 0} \varepsilon^{n_i}\; _w\langle\, \varepsilon, p', n' \mid J_i \mid \varepsilon, p, n \,\rangle_w \\ &= \lim_{\varepsilon\to 0} \varepsilon^{n_i} g_i(\varepsilon, p', n', p, n; w)\,. \end{aligned}$$

The procedure sketched here is general, provided that some technical points are taken into account (see [CT] for details in the Lie case) and the unitarity of the representations can be preserved in the limit.

Let us now obtain the representations of $H(1)_q$ from those of $SU(2)_q$:

$$\begin{aligned} _z\langle\, j, m' \mid J_\pm \mid j, m \,\rangle_z &= \left([j \mp m]_q\, [j \pm m + 1]_q\right)^{1/2} \delta_{m', m\pm 1} \\ _z\langle\, j, m' \mid J_3 \mid j, m \,\rangle_z &= m\, \delta_{m', m}\,. \end{aligned} \tag{2.9}$$

The transformation we consider is

$$\begin{pmatrix} p \\ n \\ w \end{pmatrix} = \begin{pmatrix} 2\varepsilon^2 & 0 & 0 \\ 1 & -1 & 0 \\ 0 & 0 & \varepsilon^{-2} \end{pmatrix} \begin{pmatrix} j \\ m \\ z \end{pmatrix}$$

and equations (2.9) can be written in the form

$$\begin{aligned} _w\langle\, \varepsilon, p, n' \mid J_+ \mid \varepsilon, p, n \,\rangle_w &= \left([n]_s\, [\varepsilon^{-2} p - n + 1]_s\right)^{1/2} \delta_{n', n-1} \\ _w\langle\, \varepsilon, p, n' \mid J_- \mid \varepsilon, p, n \,\rangle_w &= \left([n+1]_s\, [\varepsilon^{-2} p - n]_s\right)^{1/2} \delta_{n', n+1} \\ _w\langle\, \varepsilon, p, n' \mid J_3 \mid \varepsilon, p, n \,\rangle_w &= (\varepsilon^{-2} p/2 - n)\delta_{n', n}\,, \end{aligned} \tag{2.10}$$

where $s = \exp\{\varepsilon w\}$. Considering the limit of (2.10) we have to define:

$$A = \varepsilon\, J_+\,, \qquad A^\dagger = \varepsilon\, J_-\,, \qquad H = 2\varepsilon^2\, J_3\,.$$

Eventually we find:

$$\begin{aligned} _w\langle\, p, n' \mid A \mid p, n \,\rangle_w &= \left(n\, \frac{\mathrm{sh}(wp/2)}{w/2} \right)^{1/2} \delta_{n', n-1} \\ _w\langle\, p, n' \mid A^+ \mid p, n \,\rangle_w &= \left((n+1)\, \frac{\mathrm{sh}(wp/2)}{w/2} \right)^{1/2} \delta_{n', n+1} \\ _w\langle\, p, n' \mid H \mid p, n \,\rangle_w &= p\, \delta_{n', n}\,. \end{aligned}$$

The relevance of the contraction of representations here described is shown in the third equation of (2.10). We see that H is central and the new generator N emerges naturally as the subleading term of J_3 [C2]: we recover in this way the relation $J_3 = \varepsilon^{-2} H/2 - N$ defined at the level of the contraction of the algebra $U(2)_q$.

2.3 The 3-dim Euclidean quantum group $E(3)_q$ [C4]. In this section we shall discuss the case of the semisimple group $SO(4)_q$. The Dynkin diagram of $so(4)$ is

$$\begin{matrix}\alpha_1 & \alpha_2 \\ \bigcirc & \bigcirc\end{matrix}$$

where α_1 and α_2 are the positive roots satisfying $<\alpha_k,\alpha_\ell>=\delta_{k\ell}$, $(k,\ell=1,2)$. In the Chevalley basis the commutation relations read

$$[H_{\alpha_k},(E_\pm)_{\alpha_\ell}]=\pm 2\;\delta_{k\ell}\;(E_\pm)_{\alpha_\ell}\;,\quad [(E_+)_{\alpha_k},(E_-)_{\alpha_\ell}]=\delta_{k\ell}\;H_{\alpha_\ell}\;,$$

which, under quantization $H_{\alpha_k}\to 2J_3^k$ and $(E_\pm)_{\alpha_k}\to J_\pm^k$, result in

$$\begin{aligned}[][J_3^k,J_\pm^\ell]&=\pm\;\delta_{k\ell}\;J_\pm^k\\ [J_+^k,J_-^\ell]&=\delta_{k\ell}\;\frac{\mathrm{sh}(zJ_3^k)}{\mathrm{sh}(z/2)}\;(e^z=q)\;.\end{aligned}\tag{2.11}$$

Remark. In principle the two disconnected components of $SO(4)$ could accommodate two unrelated deformation parameters z_1 and z_2. However, as we will see, the contraction is impossible at the coalgebra level if $z_1\neq -z_2$. This is also evidently related to what occurs in the tensor product of representations [BT, KS] so that we shall assume z and $-z$ for the deformation parameters of the two factors of $SO(4)$.

We now perform the usual change of basis

$$J_s=J_s^1+J_s^2\;,\;N_s=J_s^1-J_s^2\;,\;(s=\pm,3)$$

which, in the Lie case, would single out the diagonal $su(2)$ subalgebra. We find that the deformed commutators are the following ones:

$$\begin{aligned}[][J_+,J_-]=[N_+,N_-]&=2\;\mathrm{sh}(zJ_3/2)\;\mathrm{ch}(zN_3/2)\;/\;\mathrm{sh}(z/2)\;,\\ [J_\pm,N_\mp]&=\pm 2\;\mathrm{sh}(zN_3/2)\;\mathrm{ch}(zJ_3/2)\;/\;\mathrm{sh}(z/2)\;.\end{aligned}\tag{2.12}$$

Remark. From (2.12) it appears that no $SO(3)_q$ can be found in $SO(4)_q$, in agreement with the general fact that the chain of subgroup inclusions is of the form $SO(n-2)_q\subset SO(n)_q$.

Considering now the quantum comultiplication and antipode, we see that nothing changes for J_3 and N_3 with respect to the Lie case, while for $J_\pm$ and $N_\pm$ we have

$$\begin{aligned}\Delta J_\pm&=e^{-zJ_3^1/2}\otimes J_\pm^1+J_\pm^1\otimes e^{zJ_3^1/2}+e^{zJ_3^2/2}\otimes J_\pm^2+J_\pm^2\otimes e^{-zJ_3^2/2}\;,\\ \Delta N_\pm&=e^{-zJ_3^1/2}\otimes J_\pm^1+J_\pm^1\otimes e^{zJ_3^1/2}-e^{zJ_3^2/2}\otimes J_\pm^2-J_\pm^2\otimes e^{-zJ_3^2/2}\;,\end{aligned}\tag{2.13}$$

and

$$\begin{aligned}\gamma(J_\pm)&=-e^{zJ_3^1/2}\;J_\pm^1\;e^{-zJ_3^1/2}-e^{-zJ_3^2/2}\;J_\pm^2\;e^{zJ_3^2/2}\;,\\ \gamma(N_\pm)&=-e^{zJ_3^1/2}\;J_\pm^1\;e^{-zJ_3^1/2}+e^{-zJ_3^2/2}\;J_\pm^2\;e^{zJ_3^2/2}\;,\end{aligned}\tag{2.14}$$

where we want again to underline the choices of the positive and negative signs in front of the parameter z.

Let us now study the contraction of the Hopf algebra, by rescaling the generators and the parameter according to

$$^t(P_s, J_s, w) \;=\; \mathrm{diag}\{\varepsilon, 1, \varepsilon^{-1}\}\; ^t(N_s, J_s, z), \; (s = \pm, 3) \, .$$

We then get the commutation relations

$$\begin{aligned}
&[J_3, J_\pm] = \pm J_\pm \ , \\
&[J_3, P_\pm] = [P_3, J_\pm] = \pm P_\pm \ , \\
&[J_s, P_s] = [P_\ell, P_s] = 0 \ , \\
&[J_+, J_-] = 2\ J_3\ \ \mathrm{ch}(wP_3) \ , \\
&[J_\pm, P_\mp] = \pm(2/w)\ \mathrm{sh}(wP_3) \ .
\end{aligned} \tag{2.15}$$

The last two of the previous equations make evident the quantum character of the structure thus obtained.

Again the nontrivial coproducts are $\Delta P_\pm$ and $\Delta J_\pm$. The leading terms (of order ε^{-1}) are sufficient to produce the expressions

$$\Delta P_\pm = e^{-wP_3/2} \otimes P_\pm + P_\pm \otimes e^{wP_3/2} \, . \tag{2.16}$$

We see that the actual form of $\Delta J_\pm$ is

$$\begin{aligned}
\Delta J_\pm = 0 \exp\Big(-\frac{\varepsilon w}{2}(J_3 + \frac{1}{\varepsilon}P_3)\Big) \otimes \frac{1}{2}(J_\pm + \frac{1}{\varepsilon}P_\pm) + \frac{1}{2}(J_\pm + \frac{1}{\varepsilon}P_\pm) \otimes \\
\exp\Big(\frac{\varepsilon w}{2}(J_3 + \frac{1}{\varepsilon}P_3)\Big) + \exp\Big(\frac{\varepsilon w}{2}(J_3 - \frac{1}{\varepsilon}P_3)\Big) \otimes \frac{1}{2}(J_\pm - \frac{1}{\varepsilon}P_\pm) \\
+ \exp\Big(-\frac{\varepsilon w}{2}(J_3 - \frac{1}{\varepsilon}P_3)\Big) \otimes \frac{1}{2}(J_\pm + \frac{1}{\varepsilon}P_\pm) \ ,
\end{aligned}$$

so that, in the contraction limit a perfect cancellation of the divergent terms occurs, due to the choice of opposite signs of z in the two $SU(2)_q$ components, leading to the finite result

$$\Delta J_\pm = e^{-wP_3/2} \otimes J_\pm + J_\pm \otimes e^{wP_3/2} - \frac{w}{2}\left(e^{-wP_3/2}\ J_3 \otimes P_\pm - P_\pm \otimes e^{wP_3/2}\ J_3\right) \tag{2.17}$$

For the antipode we have

$$\begin{aligned}
\gamma(J_3) = -J_3 \ , \ \gamma(J_\pm) = -(J_\pm \pm wP_\pm) \ , \\
\gamma(P_s) = -P_s \ , \ (s = \pm, 3) \ .
\end{aligned} \tag{2.18}$$

Summing up the results so far obtained, we have

Proposition. *The relations* (2.15 – 18) *define a quantum group that will be called the three dimensional* Euclidean quantum group, $E(3)_q$.

Remark. Again there is no inclusion of $SO(3)_q$.

In order to obtain the Casimir operators of $E(3)_q$ we consider the Casimir of the two $SU(2)_q$ factors, namely

$$\left[\frac{J_3 \pm \varepsilon^{-1} P_3}{2}\right]_{e^{2\varepsilon w}} \left[\frac{J_3 \pm \varepsilon^{-1} P_3}{2} + 1\right]_{e^{2\varepsilon w}} + \left(\frac{J_- \pm \varepsilon^{-1} P_-}{2}\right)\left(\frac{J_+ \pm \varepsilon^{-1} P_+}{2}\right).$$

The terms of order ε^{-2} of this expression lead then to the first Casimir

$$C_1 = P_1^2 + P_2^2 + \frac{4}{w^2}\ \mathrm{sh}^2\left(\frac{wP_3}{2}\right), \tag{2.19}$$

while the terms of order ε^{-1} give the second one

$$C_2 = J_1 P_1 + J_2 P_2 + \frac{1}{w}\ J_3\ \mathrm{sh}(wP_3)\ . \tag{2.20}$$

The general procedure exposed in the previous section gives finally the representations of $E(3)_q$ by contraction.

Proposition. *On* $L^2(\mathbf{R}^3)$ *with coordinates* p_1, p_2, p_3 *the* P_i *are represented by the multiplication by* p_i, *while the* J_i *by*

$$\begin{aligned} J_1 &= -i\left(p_2\ \frac{\partial}{\partial p_3} - \frac{1}{w}\ \mathrm{sh}(wp_3)\ \frac{\partial}{\partial p_2}\right) + c_2\ \frac{p_1}{p_1^2 + p_2^2}\ , \\ J_2 &= -i\left(\frac{1}{w}\ \mathrm{sh}(wp_3)\ \frac{\partial}{\partial p_1} - p_1\ \frac{\partial}{\partial p_3}\right) + c_2\ \frac{p_2}{p_1^2 + p_2^2}\ , \\ J_3 &= -i\left(p_1\ \frac{\partial}{\partial p_2} - p_2\ \frac{\partial}{\partial p_1}\right). \end{aligned} \tag{2.21}$$

where c_2 *is the eigenvalue of* C_2.

Remark. As for $E(2)_q$ there exists a nonlinear change of variables that permits the linearization of the right hand sides of the algebra relations. Indeed, if

$$\begin{aligned} Z_1 &= P_1\ , & W_1 &= \mathrm{sech}(wP_3/2)\ J_1\ , \\ Z_2 &= P_2\ , & W_2 &= \mathrm{sech}(wP_3/2)\ J_2\ , \\ Z_3 &= (2/w)\ \mathrm{sh}(wP_3/2)\ , & W_3 &= J_3\ , \end{aligned}$$

then

$$[Z_i, Z_j] = 0\ ,\ [W_i, Z_j] = i\epsilon_{ijk} Z_k\ ,\ [W_i, W_j] = i\epsilon_{ijk} W_k\ .$$

Of course the coproducts have to be computed consequently and show that these elements are not primitive.

2.4 The two dimensional Galilei quantum group $G(2)_q$ and a realization of $E(2,1)_q$ [C4]. In this final subsection we are going to determine a Hopf algebra that deforms the 2-dimensional Galilei Lie algebra [Bc]. We define

$$^t(P_1, P_2, M, K_2, K_1, J, v) = \mathrm{diag}\{1, 1, \varepsilon, \varepsilon, -\varepsilon, 1, \varepsilon^{-1}\}\ ^t(P_1, P_2, P_3, , J_1, J_2, J_3, w) . \tag{2.22}$$

From (2.15), in the limit $\varepsilon \to 0$, we obtain

$$[J, K_\pm] = \pm K_\pm , \qquad [K_\pm, P_\mp] = (2i/v)\ \mathrm{sh}(vM) ,$$
$$[J, P_\pm] = \pm P_\pm , \qquad [K_+, K_-] = [P_+, P_-] = 0 ,$$

M being a central element. The coproducts read

$$\Delta J = J \otimes \mathbf{1} + \mathbf{1} \otimes J,\ \Delta M = M \otimes \mathbf{1} + \mathbf{1} \otimes M ,$$
$$\Delta P_\pm = e^{-vM/2} \otimes P_\pm + P_\pm \otimes e^{vM/2} ,$$
$$\Delta K_\pm = e^{-vM/2} \otimes K_\pm + K_\pm \otimes e^{vM/2} , \tag{2.23}$$

while the antipodes and counits are the same as for the Lie algebra $g(2)$.

Let us now determine the structure of the Casimir operators as given by the contraction. To this purpose we consider the ε expansion of the generator P_3 and of the Casimir C_1. As can be seen from the structure of the representations, these are of the form

$$P_3 = \varepsilon^{-1} M - \varepsilon T ,\ C_1 = \varepsilon^{-2} Q_1 + C ,\ C_2 = \varepsilon^{-1} Q_2 ,$$

where C is again a central element, due to the fact that invariance is maintained at each order of the ε-expansion. Equation (2.19) easily yields

$$Q_1 = (4/v^2)\ \mathrm{sh}^2(vM/2) ,\ C = P_1^2 + P_2^2 - (2/v)\ \mathrm{sh}(vM)\ T ,$$

from which it is evident that T represents the deformed Galileian energy with nonvanishing commutation relations

$$[T, K_\pm] = \mp i P_\pm .$$

In a similar way we can treat the Casimir C_2, obtaining for Q_2

$$Q_2 = K_2 P_1 - K_1 P_2 + (J/v)\ \mathrm{sh}(vM) ,$$

and it is evident that $S = vQ_2/\,\mathrm{sh}(vM)$ can be interpreted as the q-analogue of the spin operator of $g(2)$. An immediate consequence of the contraction procedure is that even in this case we can immediately establish the following result.

Proposition. *On the space $L^2(\mathbf{R}^2)$ with coordinates p_1, p_2, the P_i and M are diagonal, while the representation of K_i and J is given by the operators*

$$K_1 = \frac{i}{v}\ \mathrm{sh}(vM)\ \frac{\partial}{\partial p_1} - q_2\ \frac{p_2}{p_1^2 + p_2^2} ,$$
$$K_2 = \frac{i}{v}\ \mathrm{sh}(vM)\ \frac{\partial}{\partial p_2} + q_2\ \frac{p_1}{p_1^2 + p_2^2} , \tag{2.24}$$
$$J = -i \left(p_1\ \frac{\partial}{\partial p_2} - p_2\ \frac{\partial}{\partial p_1} \right) ,$$

q_2 being the eigenvalue of Q_2.

Having determined the quantum group $G(2)_q$, we shall show that, for the quantum case also, it is possible to use the generators of $G(2)_q$ for obtaining a quantum deformation of the canonical realizations of $E(3)$ as well as of the 2-dimensional Poincaré group $E(2,1)$. To this purpose we define

$$P_3 = \frac{2}{w}\ \text{arcsh}\left(\frac{w}{2}\left(\frac{4}{w^2}\ \text{sh}^2(\frac{wM}{2}) \mp p_1^2 \mp p_2^2\right)^{1/2}\right), \tag{2.25}$$

the minus sign referring to $E(3)$ and the plus sign to $E(2,1)$. Then, denoting by $\{,\}_+$ the anticommutator, a direct calculation shows that the operators

$$\begin{aligned} J_1 &= \frac{i}{2w}\ \{\text{sh}(wp_3), \frac{\partial}{\partial p_2}\}_+ + c_2\ \frac{p_1}{p_1^2+p_2^2}\ , \\ J_2 &= -\frac{i}{2w}\ \{\text{sh}(wp_3), \frac{\partial}{\partial p_1}\}_+ + c_2\ \frac{p_2}{p_1^2+p_2^2}\ , \\ J_3 &= -i\left(p_1\ \frac{\partial}{\partial p_2} - p_2\ \frac{\partial}{\partial p_1}\right) \end{aligned} \tag{2.26}$$

satisfy the commutation relations (2.15) and the similar for $E(2,1)_q$. Solving (2.24) for ∂_{p_i} with $q_2 = c_2$, $v = w$ and substituting into (2.26) we obtain the realization

$$\begin{aligned} J_1 &= \frac{1}{2\ \text{sh}(wM)}\ \{\text{sh}(wP_3), K_2\}_+ \pm S\ P_1\ \frac{w}{2}\ \coth\left(\frac{w}{2}(M+P_3)\right), \\ J_2 &= -\frac{1}{2\ \text{sh}(wM)}\ \{\text{sh}(wP_3), K_1\}_+ \pm S\ P_2\ \frac{w}{2}\ \coth\left(\frac{w}{2}(M+P_3)\right), \\ J_3 &= \frac{w}{\text{sh}(wM)}\ (K_1P_2 - K_2P_1) + S\ , \end{aligned} \tag{2.27}$$

$$C_1 = \frac{4}{w^2}\ \text{sh}^2(\frac{wM}{2}), \qquad C_2 = \frac{1}{w}\text{sh}(wM)S.$$

where the minus sign refers to $E(2,1)$. Obviously, for $w \to 0$ we recover the Lie algebra structures.

We shall conclude the section with a remark on the Hopf algebra aspects of the nonlinear realizations as our previous (2.27). The coproducts of the elements J_i as given in (2.27) are to be calculated starting from the coproducts of the generators (2.23). It turns out that ΔJ_i have a different form from the original expressions (2.16–17) and this difference is maintained also in the limit $w \to 0$, where the J_i thus obtained, while satisfying the correct Lie algebra commutation relations, are no more primitive elements: but of course, due to the nonlinearity of the realization, there is no contradiction with the Friedrichs theorem [Ps] and the possibility of having a group multiplication.

3. Contracted R-matrices

The universal R-matrix of a quantum group $\mathcal{U}$ is defined by its property of being an intertwining operator on the coproduct Δ, namely

$$R\Delta(u)R^{-1} = \sigma\Delta(u)\,, \qquad u \in \mathcal{U}\,, \tag{3.1}$$

where $\sigma(a \otimes b) = b \otimes a$. For $SU(2)_q$ the R-matrix has the well known expression

$$R = e^{zJ_3 \otimes J_3} \sum_{k\geq 0} \frac{(1-e^{-z})^k}{[k]!}\, e^{-zk(k-1)/4} \left(e^{zkJ_3/2}(J_+)^k \otimes e^{-zkJ_3/2}(J_-)^k\right)\,, \tag{3.2}$$

with $e^z = q$ [Dr].

3.1 The R-matrix of $H(1)_q$ [C2]. We shall determine the R-matrix of $H(1)_q$ by contracting (3.2) with the rules that produce $H(1)_q$ from $SU(2)_q$, i.e., according to (2.7), by defining

$$A = \varepsilon J_+,\ A^\dagger = \varepsilon J_-,\ N = \varepsilon^{-2}K - J_3,\ H = 2K,\ w = \varepsilon^{-1}z, \tag{3.3}$$

and then letting $\varepsilon \to 0$. Therefore, by a straightforward substitution and neglecting higher orders in ε, we find that

$$R \to e^{(w/4\varepsilon)H \otimes H} \sum_{k\geq 0} \frac{1}{k!} \left(w^{1/2} e^{wH/4}\, A\right)^k \otimes \left(w^{1/2} e^{-wH/4}\, A^\dagger\right)^k\,, \tag{3.4}$$

but the $\varepsilon \to 0$ limit of the first exponential is singular and requires an appropriate discussion.

Indeed, we see that the multiplication of R by any central factor leaves (3.1) invariant. When taking the limit of (3.2), we shall thus neglect any possibly emerging central factor, even a singular one: this is just the case of the term $\exp[(w/4\varepsilon)H \otimes H]$. The finite contribution of $\exp[zJ_3 \otimes J_3]$ in the limit is coming from the next order of the expansion in ε. Since $J_3 = H/(2\varepsilon^2) - N$, this finite contribution results in $\exp[-(w/2)(H \otimes N + N \otimes H)]$. We then have:

Proposition.

(i) *The non-singular part of the contraction of* (3.2) *leads to*

$$\begin{aligned} R &= e^{-(w/2)(H \otimes N + N \otimes H)} \sum_{k=0}^{\infty} \frac{1}{k!} \left(w^{1/2} e^{wH/4} A\right)^k \otimes \left(w^{1/2} e^{-wH/4} A^\dagger\right)^k \\ &= e^{-(w/2)\Omega}\, e^{\Lambda}\,, \end{aligned} \tag{3.5}$$

where $\Omega = H \otimes N + N \otimes H$ *and* $\Lambda = we^{w\Gamma/4}\,(A \otimes A^\dagger)$ *with* $\Gamma = H \otimes 1 - 1 \otimes H$.

(ii) *The relation* (3.1) *is satisfied in* $H(1)_q$.

(iii) *The R-matrix* (3.5) *is quasi-triangular and therefore it solves the QYBE*

$$R_{12}\, R_{13}\, R_{23} \;=\; R_{23}\, R_{13}\, R_{12}\,. \tag{3.6}$$

Proof. The point (i) has already been discussed. (ii) and (iii) are proved by direct calculations.

From the universal R-matrix, by standard techniques, we can readily deduce a new R-matrix depending on a parameter x. In fact, defining the operator T_x by its action on the generators

$$T_xA = xA,\ T_xA^\dagger = \frac{A^\dagger}{x},\ T_xH = H,$$

we can define

$$R(x) = (T_x \otimes \mathbf{1})R = e^{-(w/2)\Omega}\ e^{x\Lambda} \tag{3.7}$$

and again by a direct calculation we have:

Proposition. *The matrix $R(x)$ defined in* (3.7) *satisfies the QYBE*

$$R_{12}(x)\ R_{13}(xy)\ R_{23}(y) \ = \ R_{23}(y)\ R_{13}(xy)\ R_{12}(x)\ . \tag{3.8}$$

Corollary. *The classical r-matrix corresponding to* (3.5) *is*

$$r = a \otimes a^\dagger - \frac{1}{2}(h \otimes n + n \otimes h)\ , \tag{3.9}$$

where $a,\ a^\dagger,\ h,\ n$ are defined to be the classical limit of $A,\ A^\dagger,\ H,\ N$ and satisfy the commutation rules

$$[a, a^\dagger] = h,\ [n, a] = -a,\ [n, a^\dagger] = a^\dagger\ .$$

Moreover (3.9) *solves the CYBE*

$$[r_{12}, r_{13}] + [r_{12}, r_{23}] + [r_{13}, r_{23}] = 0\ . \tag{3.10}$$

Similarly, the classical limit of (3.7), *with $x = e^u$, is*

$$r(u) = e^u\ a \otimes a^\dagger - \frac{1}{2}(h \otimes n + n \otimes h) \tag{3.11}$$

and solves the parameter-dependent CYBE

$$[r_{12}(u), r_{13}(u+v)] + [r_{12}(u), r_{23}(v)] + [r_{13}(u+v), r_{23}(v)] = 0\ . \tag{3.12}$$

Proof. Recalling that the classical r-matrix is obtained by a first order expansion in w of the quantum R-matrix, i.e. $R = \mathbf{1} \otimes \mathbf{1} + w\ r$, it is immediate to verify that (3.9–12) are the classical counterparts of (3.5–8) respectively.

Remark. The contraction technique can again be used to produce trigonometric and rational solutions of (3.12). Indeed it is well known that for $SU(2)_q$ such a solution is of the form [J2]

$$r(u) = s - t + \frac{2t}{1 - e^u}\ , \tag{3.13}$$

where

$$s = J_+ \otimes J_- - J_- \otimes J_+, \; t = J_1 \otimes J_1 + J_2 \otimes J_2 + J_3 \otimes J_3 \, .$$

The same formula (3.13) gives then a trigonometric solution of the CYBE for $H(1)_q$, where now

$$s = \frac{1}{2}(a \otimes a^\dagger - a^\dagger \otimes a), \; t = \frac{1}{2}(a \otimes a^\dagger + a^\dagger \otimes a - h \otimes n - n \otimes h) \, .$$

As in the $SU(2)_q$ case, t/u is the rational solution of (3.12).

We conclude noticing that following the same scheme no other contracted quantum groups introduced in this section has a defined contracted R–matrix. The singularities that appear are not renormalizable and there is not a finite contribution solving the defining relation (3.1) and the Yang–Baxter equation.

3.2 The R-matrix of $E(3)_q$ [C4]. We want now to apply the contraction procedure in order to obtain the R-matrix of $E(3)_q$ from the expression of the R-matrix of $SO(4)_q$. The latter is of the form

$$R \;=\; R^1_a \; R^1_b \; R^2_a \; R^2_b \;=\; R^1_a \; R^2_a \; R^1_b \; R^2_b \, ,$$

where

$$\begin{aligned} R^1_a &= \exp\{z J^1_3 \otimes J^1_3\} \, , \; R^2_a = \exp\{-z J^2_3 \otimes J^2_3\} \, , \\ R^1_b &= \sum_{k\geq 0} \frac{\left(1-e^{-z}\right)^k}{[k]_q!} \left(e^{zJ^1_3/2} \, J^1_+\right)^k \otimes \left(e^{-zJ^1_3/2} \, J^1_-\right)^k \, , \\ R^2_b &= \sum_{k\geq 0} \frac{(1-e^{z})^k}{[k]_q!} \left(e^{-zJ^2_3/2} \, J^2_+\right)^k \otimes \left(e^{zJ^2_3/2} \, J^2_-\right)^k \end{aligned} \tag{3.14}$$

are the expression giving the R-matrix of $SU(2)_q$.

The contraction limit of the product $R^1_a \; R^2_a$ is easily calculated. Indeed, since

$$\begin{aligned} R^1_a &\simeq \exp\left\{\frac{w}{2}\left(\varepsilon^{-1} \; P_3 \otimes P_3 + (J_3 \otimes P_3 + P_3 \otimes J_3)\right)\right\} \, , \\ R^2_a &\simeq \exp\left\{\frac{w}{2}\left(-\varepsilon^{-1} \; P_3 \otimes P_3 + (J_3 \otimes P_3 + P_3 \otimes J_3)\right)\right\} \, , \end{aligned}$$

we see that the limit of the product gives the finite result

$$R^1_a \; R^2_a = \exp\Big\{ w \; (J_3 \otimes P_3 + P_3 \otimes J_3) \Big\} \, . \tag{3.15}$$

To calculate $R^1_b \; R^2_b$ we first give some definitions and we let

$$[k]_q! = [k]_q \; [k-1]_q \; \cdots \; [1]_q = q^{-k(k-1)/4} \; \Gamma_q(k+1) \, ,$$

where

$$\Gamma_q(k) = (1+q)\,(1+q+q^2) \; \cdots \; (1+q+ \; \cdots \; + q^{k-2}) \, .$$

We find it also useful to introduce the q-binomial coefficients

$$\begin{bmatrix} n \\ k \end{bmatrix}_q = \frac{[n]_q!}{[k]_q!\,[n-k]_q!} = q^{-k(n-k)/2}\,\frac{\Gamma_q(n+1)}{\Gamma_q(k+1)\,\Gamma_q(n-k+1)} .$$

We then see that we can write

$$R_b^1 = \sum_{k\geq 0} \frac{1}{[k]_q!}\,X^k ,$$

where

$$\begin{aligned} X &= \frac{1-e^{-2\varepsilon w}}{(2\varepsilon)^2}\{e^{w(P_3+\varepsilon J_3)/2}(P_+ + \varepsilon J_+)\} \otimes \{e^{-w(P_3+\varepsilon J_3)/2}(P_- + \varepsilon J_-)\} \\ &= D(\varepsilon^2)/\varepsilon \;+\; F(\varepsilon^2) , \end{aligned}$$

$D(\varepsilon^2)$ and $F(\varepsilon^2)$ being integer functions of ε^2. Analogously

$$R_b^2 = \sum_{h\geq 0} \frac{1}{[h]_q!}\,Y^h ,$$

with

$$\begin{aligned} Y &= \frac{1-e^{2\varepsilon w}}{(2\varepsilon)^2}\{e^{w(P_3-\varepsilon J_3)/2}(P_+ - \varepsilon J_+)\} \otimes \{e^{-w(P_3-\varepsilon J_3)/2}(P_- - \varepsilon J_-)\} \\ &= -D(\varepsilon^2)/\varepsilon \;+\; F(\varepsilon^2). \end{aligned} \tag{3.16}$$

Since $[X,Y]=0$ we have

$$R_b^1\, R_b^2 = \sum_{n\geq 0} \frac{1}{[n]_q!} \sum_{m=0}^{n} \begin{bmatrix} n \\ m \end{bmatrix}_q X^{n-m}\, Y^m . \tag{3.17}$$

and the explicit form of the R-matrix is got provided that we are able to calculate the limit for vanishing ε of the above sums. This calculation will be done in three steps.

STEP 1. - Representation of the sum $\sum_{m=0}^{n} \begin{bmatrix} n \\ m \end{bmatrix}_q X^{n-m}\, Y^m$.

This representation can be deduced from the following general result.

Lemma. *If α, β are commuting, then*

$$T(\alpha,\beta) \equiv \alpha\, q^{\beta\partial_\beta/2} + \beta\, q^{-\alpha\partial_\alpha/2}$$

is such that

$$T^n(\alpha,\beta)1 = \sum_{k=0}^{n} \begin{bmatrix} n \\ k \end{bmatrix}_q \alpha^{n-k}\beta^k . \tag{3.18}$$

Proof. By induction:

$$T(\alpha,\beta)1=\alpha+\beta\ .$$

Use then the relation [An]

$$\begin{bmatrix} n \\ k+1 \end{bmatrix}_q q^{(k+1)/2} + \begin{bmatrix} n \\ k \end{bmatrix}_q q^{-(n-k)/2} = \begin{bmatrix} n+1 \\ k+1 \end{bmatrix}_q ,$$

to prove that

$$T(\alpha,\beta)\sum_{k=0}^{n}\begin{bmatrix} n \\ k \end{bmatrix}_q \alpha^{n-k}\,\beta^k = \sum_{k=0}^{n+1}\begin{bmatrix} n+1 \\ k \end{bmatrix}_q \alpha^{n+1-k}\,\beta^k = T^{n+1}(\alpha,\beta)1\ .$$

STEP 2. - Calculation of $\lim_{\varepsilon\to 0} T^n(X,Y)1$.

Let $A=\lim_{\varepsilon\to 0} D(\varepsilon^2)$, $B=\lim_{\varepsilon\to 0} F(\varepsilon^2)$. Then

Proposition. *We have*

$$\lim_{\varepsilon\to 0} T^n(X,Y)1=\left((2B+w\,A\,(A\partial_A+B\partial_B))e^{-(w/2)A\partial_B}\right)^n 1\equiv \mathcal{P}_n(A,B)\ ,$$

where $\mathcal{P}_n(A,B)$ *is a homogeneous polynomial of degree* n *satisfying the recurrence relation*

$$(2B+nwA)\ \mathcal{P}_n(A,B-wA/2)=\mathcal{P}_{n+1}(A,B) \tag{3.19}$$

and where

$$\begin{aligned} A&=\frac{w}{2}\ Q_+\otimes Q_-\ ,\\ B&=\frac{w}{2}\ (L_+\otimes Q_-+Q_+\otimes L_-)\\ &\qquad -\frac{w^2}{2}\left(Q_+\otimes Q_- -\frac{1}{2}\ (J_3Q_+\otimes Q_- + Q_+\otimes J_3Q_-)\right), \end{aligned} \tag{3.20}$$

with

$$Q_\pm=e^{\pm wP_3/2}\ P_\pm\ ,\ L_\pm=e^{\pm wP_3/2}\ J_\pm\ . \tag{3.21}$$

Proof. Again by induction.

STEP 3. - The calculation of $R^1_b\ R^2_b$.

First observe that (3.19) can be solved yielding

$$\mathcal{P}_n(A,B)=(-2wA)^n\ \frac{\Gamma(-B/(wA)+(n+1)/2)}{\Gamma(-B/(wA)-(n-1)/2)}\ .$$

Then

$$R_b^1\ R_b^2 = \sum_{n\geq 0} \frac{1}{[n]_q!} \sum_{m=0}^{n} \begin{bmatrix} n \\ m \end{bmatrix}_q X^{n-m}\ Y^m = \sum_{n\geq 0} \frac{1}{[n]_q!} T^n(X,Y)1 .$$

In the limit $\varepsilon \to 0$ this gives

$$\sum_{n\geq 0} \frac{1}{n!} \mathcal{P}_n(A,B) = \sum_{n\geq 0} \frac{(-2wA)^n}{n!} \frac{\Gamma(-B/(wA)+(n+1)/2)}{\Gamma(-B/(wA)-(n-1)/2)} .$$

This sum can be expressed in terms of known functions. Indeed the sum over even values of n gives ${}_2F_1(a, 1-a; 1/2; -w^2A^2)$ while the sum over odd terms yields $wA(1-2a)\ {}_2F_1(a+1/2, 3/2-a; 3/2; -w^2A^2)$, with $a = -B/(wA)+1/2$. Finally, taking into account the special values of the parameters of the hypergeometric functions, we get the result

$$R_b^1\ R_b^2 = \exp\{2B\ \operatorname{arcsh}(wA)/(wA)\}\left(1+w^2A^2\right)^{-1/2} .$$

We finally have:

Proposition. *The expression*

$$R = \exp\{w(J_3 \otimes P_3 + P_3 \otimes J_3)\}\ \exp\{2B\ \operatorname{arcsh}(wA)/(wA)\}\left(1+w^2A^2\right)^{-1/2} \quad (3.22)$$

with A and B defined in (3.20–21), satisfies the relation $R\Delta R^{-1} = \sigma\Delta$ and is therefore the R-matrix of $E(3)_q$. Moreover R solves the Yang-Baxter equation

$$R_{12}\ R_{13}\ R_{23} = R_{23}\ R_{13}\ R_{12} .$$

Remark. From the classical limit of R we easily find the expression of the universal r-matrix, namely

$$r = J_+ \otimes P_- + P_+ \otimes J_- + P_3 \otimes J_3 + J_3 \otimes P_3$$

as well as the rational form of its λ-dependent counterpart:

$$r(\lambda) = \frac{1}{\lambda}\left(J_3 \otimes P_3 + P_3 \otimes J_3 + \frac{1}{2}(J_+ \otimes P_- + P_- \otimes J_+ + P_+ \otimes J_- + J_- \otimes P_+)\right) . \quad (3.23)$$

A related integrable dynamical system [Fd, FT] is then obtained by the Lax pair

$$U(\lambda) = \frac{1}{\lambda}\left(\mathbf{p}\cdot\mathbf{J} + \mathbf{j}\cdot\mathbf{P}\right),$$

$$V(\lambda) = i\kappa\,\frac{\mathbf{p}^2}{\lambda}\,U(\lambda) + $$

$$+\frac{\kappa}{\lambda}\left(-(\partial_x\mathbf{p}\times\mathbf{j})\cdot\mathbf{P}) + (\partial_x\mathbf{j}\times\mathbf{p})\cdot\mathbf{P}) + (\partial_x\mathbf{p}\times\mathbf{p})\cdot\mathbf{J})\right) .$$

The Poisson brackets and the equation of motion are straightforward and read

$$\{p_i(x), p_j(x')\} = 0\ ,\ \{j_i(x), p_j(x')\} = -i\epsilon_{ijk}\ p_k(x)\ \delta(x-x') ,$$

$$\{j_i(x), j_j(x')\} = -i\epsilon_{ijk}\ j_k(x)\ \delta(x-x') ,$$

$$\partial_t\mathbf{p} = \kappa\ \partial_{xx}^2\mathbf{p}\times\mathbf{p}\ ,\ \partial_t\mathbf{j} = \kappa\ \partial_{xx}^2(\mathbf{j}\times\mathbf{p}) .$$

3.3 The R-matrix of $G(2)_q$. We finally conclude this section by observing that, from (3.20–22), and the transformation (2.22) the R-matrix for $G(2)_q$ results in

$$R = \exp\{v(J \otimes M + M \otimes J)\}$$
$$\cdot \exp\left\{-iv\left(e^{vM/2} K_+ \otimes e^{-vM/2} P_- - e^{vM/2} P_+ \otimes e^{-vM/2} K_-\right)\right\},$$

that verifies the Yang–Baxter equation.

4. The matrix quantum groups

4.1 The matrix quantum group $H(1)_q$ [C2]. For those q–algebras whose R matrices are known it is straightforward to construct the C^*-algebra of the "functions" on the group by means of the defining relations [F1]. We present here the results for the two very important cases of $H(1)_q$ and $E(3)_q$.

Let us then consider a 3×3 matrix of the form:

$$T = \begin{pmatrix} 1 & \alpha & \beta \\ 0 & 1+\gamma & \delta \\ 0 & 0 & 1 \end{pmatrix} \tag{4.1}$$

where the matrix elements α, β, γ and δ generate a C^*-algebra $\mathcal{A}$ whose relations are to be determined. This will be done by first giving a 3×3 representation of the generators A, $A^\dagger$, H, N of $H(1)_q$, with commutation relations (2.8). As a consequence, the R-matrix (3.5) will be represented on the tensor product and the defining equation

$$R\, T_1\, T_2 = T_2\, T_1\, R\,, \tag{4.2}$$

with $T_1 = T \otimes \mathbf{1}$ and $T_2 = \mathbf{1} \otimes T$, will provide the required relations. A 3×3 representation of (2.8) is given by

$$\rho(A) = \begin{pmatrix} 0 & 1 & 0 \\ 0 & 0 & 0 \\ 0 & 0 & 0 \end{pmatrix} \qquad \rho(A^\dagger) = \begin{pmatrix} 0 & 0 & 0 \\ 0 & 0 & 1 \\ 0 & 0 & 0 \end{pmatrix}$$

$$\rho(H) = \begin{pmatrix} 0 & 0 & 1 \\ 0 & 0 & 0 \\ 0 & 0 & 0 \end{pmatrix} \qquad \rho(N) = \begin{pmatrix} 0 & 0 & 0 \\ 0 & 1 & 0 \\ 0 & 0 & 0 \end{pmatrix}$$

Correspondingly, (3.5) is represented by the 9×9 matrix

$$\rho(R) = \begin{pmatrix} I_3 & w\rho(A^\dagger) & -\frac{w}{2}\rho(N) \\ 0 & I_3 - \frac{w}{2}\rho(H) & 0 \\ 0 & 0 & I_3 \end{pmatrix},$$

I_3 being the 3×3 identity matrix.

In the following we shall suppress the explicit indication of the representation ρ and we shall simply denote the representatives of the generators by the generators themselves. Therefore the explicit form of (3.5) reads

$$\begin{pmatrix} T & \alpha T + wA^\dagger(1+\gamma)T & \beta T + wA^\dagger\delta T - \frac{w}{2}NT \\ 0 & (1+\gamma)T - \frac{w}{2}H(1+\gamma)T & \delta T - \frac{w}{2}H\delta T \\ 0 & 0 & T \end{pmatrix} =$$

$$\begin{pmatrix} T & wTA^\dagger + T\alpha - \frac{w}{2}T\alpha H & -\frac{w}{2}TN + T\beta \\ 0 & T(1+\gamma) - \frac{w}{2}T(1+\gamma)H & T\delta \\ 0 & 0 & T \end{pmatrix}.$$

The relations between the generators of $\mathcal{A}$ are then the following:

$$\alpha\beta - \beta\alpha = \frac{w}{2}\alpha, \; \alpha\delta - \delta\alpha = 0, \; \beta\delta - \delta\beta = -\frac{w}{2}\delta \,,$$

while γ is commuting with all the elements. We shall now use the tensor product and the inverse of representative elements of type (4.1) in order to define a Hopf algebra structure on $\mathcal{A}$, and we find

$$\begin{aligned} &\Delta(\alpha) = 1\otimes\alpha + \alpha\otimes 1\,, && \gamma(\alpha) = -\alpha\,, && \epsilon(\alpha) = 0\,, \\ &\Delta(\beta) = 1\otimes\beta + \beta\otimes 1 + \alpha\otimes\delta\,, && \gamma(\beta) = -\beta + \alpha\delta\,, && \epsilon(\beta) = 0\,, \\ &\Delta(\delta) = 1\otimes\delta + \delta\otimes 1\,, && \gamma(\delta) = -\delta\,, && \epsilon(\delta) = 0\,. \end{aligned} \tag{4.3}$$

4.2 The matrix quantum group $E(3)_q$ [C4]. We expose now the results concerning the 3–dim quantum Euclidean group. We consider the generators of $so(3)$ in the 3–dim representation

$$\Sigma_\pm = -i\begin{pmatrix} 0 & 0 & \mp i \\ 0 & 0 & 1 \\ \pm i & -1 & 0 \end{pmatrix}, \qquad \Sigma_3 = -i\begin{pmatrix} 0 & 1 & 0 \\ -1 & 0 & 0 \\ 0 & 0 & 0 \end{pmatrix}$$

from which the generators of $E(3)_q$ can be represented by

$$J_3 = \{\Sigma_3; (0,0,0)^T\}\,, \; J_\pm = \{\Sigma_\pm; (0,0,0)^T\}\,,$$
$$P_3 = \{0; (0,0,1)^T\}\,, \; P_\pm = \{0; (1,\pm i,0)^T\}\,.$$

where we have used the notation

$$\{\Sigma;\xi\} = \begin{pmatrix} \Sigma & \xi \\ 0 & 0 \end{pmatrix}.$$

Then, since $P_s P_{s'} = P_s J_{s'} = 0$ the 16×16 matrix representing R is

$$R = \mathbf{1} \otimes \mathbf{1} + w \left(J_3 \otimes P_3 + P_3 \otimes J_3 + J_+ \otimes P_- + P_+ \otimes J_- \right) = \mathbf{1} \otimes \mathbf{1} + w\, r$$

$$= \begin{pmatrix} I_4 & & & \\ & I_4 & & \\ & & I_4 & \\ & & & I_4 \end{pmatrix} + w \begin{pmatrix} & -iP_3 & -P_- & J_- \\ iP_3 & & -iP_- & iJ_- \\ P_- & iP_- & & J_3 \\ & & & \end{pmatrix}.$$

Take now

$$M = \begin{pmatrix} \exp(i\Sigma) & \mathbf{v} \\ 0 & 1 \end{pmatrix} \tag{4.4}$$

with $\Sigma \in so(3)$ and study the condition

$$R\, M_1\, M_2 \;=\; M_2\, M_1\, R\,, \tag{4.5}$$

where $\exp(i\Sigma)$ is expressed by the Euler angles θ, ϕ, χ [BL] and $\mathbf{v} \;=\; (x, y, z)^T$.

In order to determine the Hopf algebra of the representative functions we find it most evident to present the relations of the Hopf algebra in terms of the angles (θ, $\psi = \phi - \chi$, $\omega = \phi + \chi$) and of the coordinates ($\tilde{x} = x \cos\phi + y \sin\phi$, $\tilde{y} = -x \sin\phi + y \cos\phi$, z). Since the representative functions are entire functions of the Euler angles, we can present in a most compact form the (algebraic!) relations coming from (4.5), namely

$$[\theta, \tilde{x}] = w\, \sin\theta\, \operatorname{tg}(\theta/2)\,,\; [\theta, z] = w\, \sin\theta\,,\; [\omega, \tilde{y}] = -2w\, \operatorname{tg}(\theta/2)\,,$$
$$[z, \tilde{x}] = w\, z\,,\; [\tilde{y}, \tilde{x}] = w\, \tilde{y}\,,$$

all other commutators being vanishing.

The coproduct and the antipode of the representative functions are determined, as usual, by the multiplication and the inverse of the representative matrix,

$$\Delta M_{ij} = M_{ik} \otimes M_{kj}\,,\; \gamma(M_{ij}) = (M^{-1})_{ij}\,.$$

For instance, for M_{33},

$$\Delta M_{33} = \Delta \cos\theta = \cos(\theta \otimes \mathbf{1})\, \cos(\mathbf{1} \otimes \theta) - \sin(\theta \otimes \mathbf{1})\, \sin(\mathbf{1} \otimes \theta)\, \cos(\chi \otimes \mathbf{1} - \mathbf{1} \otimes \phi)\,.$$

Hence we can state that the functions given in (4.4) together with the relations specified above form the Hopf algebra of the representative functions of $E(3)_q$.

5. Concluding remarks

We shall conclude this presentation of the quantum group contractions by showing how some results obtained by means of this technique can be relevant to gain insight into the relation between classical and quantum R-matrices as well as they can add some new physical applications of the q-deformed Hopf algebras to the already existing ones (see, *e.g.*, [Ch, Kg, WZ]).

5.1 The quantization of the classical r-matrix. Recently the question has been raised [Op] whether the R-matrix obtained from a general quantum group admits an expansion uniquely in terms of its classical r counterpart. In the affirmative case the consequence would be the possibility of "quantizing" , at least in principle, any classical r-matrix. This is certainly the case of $SU(2)_q$ and of the triangular Hopf algebras [Mj] but it is not true in general, even for quasitriangular Hopf algebras. In fact we present here a simple calculation proving that for the Heisenberg quantum group $H(1)_q$ the quantum R-matrix cannot be expressed as a function of the classical r-matrix only. Let us recall that the R-matrix for the Heisenberg quantum group has been given in (3.5). Since

$$[\Omega,\Lambda]=\Gamma\Lambda\ ,\ [\Omega,\Gamma]=[\Lambda,\Gamma]=0\ ,$$

we see that

$$[\Omega[\Omega\ \dots\ [\Omega,\Lambda],\ \dots\ ]=\Gamma^n\Lambda\ ,$$
$$[\Omega,A\otimes A^\dagger]=\Gamma\ (A\otimes A^\dagger)\ .$$

The classical r-matrix is given by

$$r=A\otimes A^\dagger-\Omega/2$$

and we have

$$[A\otimes A^\dagger,r]=-(\Gamma/2)\ (A\otimes A^\dagger)\ .$$

Since all the commutators involving more than one factor Λ are vanishing, it is straightforward to cast the R-matrix in the form

$$R=e^{-(w/2)\Omega}\ e^{\Lambda}=\exp\left\{w\left(r+(\frac{w\Gamma/4}{\mathrm{sh}(w\Gamma/4)}-1)\ (A\otimes A^\dagger)\right)\right\}$$
$$=e^{wr}\ \exp\left\{w\left(e^{-w\Gamma/4}(1-\frac{w\Gamma/4}{\mathrm{sh}(w\Gamma/4)})\ (A\otimes A^\dagger)\right)\right\}\ ,$$

from which

$$R^{-1}\ r\ R\neq r\ ,$$

thus proving that R cannot be only function of r.

5.2 A physical implication of the q-deformation. The Euclidean and Poincaré groups are fundamental for the description of the physical spacetime. It is therefore conceivable that their quantum deformations deserve similar interest. We have been able to describe the 2-dim Galilei and the (2+1)-dim Poincaré quantum groups, which, presumably, give definite hints on their 4-dim analogues. The technical point playing the most important role in the Hopf algebra contraction is the rescaling of the parameter z, involved in the definition of semisimple quantum groups, by the contraction parameter ε: in fact, since ε becomes dimensional if some abelian generators acquire dimensions, the quantum parameter for the "non–semisimple" groups also gets a physical dimension and becomes a fundamental physical constant characterizing the model.

We shall conclude by illustrating an application of the q-deformed Poincaré kinematical symmetry. If we analyze the definition of the energy for $E(2,1)_q$ given in (2.25):

$$E(\mathbf{p},w)=\frac{2}{w}\ \text{arcsh}\left(\frac{w}{2}\left(\frac{4}{w^2}\ \text{sh}^2(\frac{wM}{2})+\mathbf{p}^2\right)^{1/2}\right),$$

one sees that w plays the role of a fundamental length. The energy at rest is $E(0,w)=M$, while, when expanding E in $\mathbf{p}^2$ we have:

$$E(\mathbf{p},w)\simeq M+\frac{\mathbf{p}^2}{2M'(w)}$$

where $M'(w)=\frac{1}{w}\,\text{sh}(wM)$ is the just obtained deformed Galilei mass. This was to be expected because of the general commutativity of contractions and q-deformations.

It is to be stressed that the inertial mass is controlled by the length w and is different from the energy at rest. This fact can be tested experimentally at relativistic regimes when the first correction in w of (2.25), namely

$$E^2-p^2-M^2\simeq-\frac{1}{12}\,w^2p^2(2M^2+p^2),$$

would appear not negligible.

Acknowledgements. We would like to thank the organizers of the Semester for their warm hospitality and acknowledge useful discussions with the participants and in particular with P.P. Kulish.

References

[Ab] Abe E., *Hopf algebras*, Cambridge Tracts in Math. no. 74 (1980), Cambridge Univ. Press.

[An] Andrews G.E., Reg. Conf. Series in Math. no. 66 (Providence, R.I., 1986, AMS).

[Bc] Bacry H., *Leçons sur la Théorie des Groupes et les Symétries des Particules Elémentaires*, (New York, N.Y., 1967, Gordon and Breach).

[Bd] Biedenharn L.C., J. Phys. A: Math. Gen. **22** (1989), L873; *A q-boson realization of the quantum group $SU_q(2)$ and the theory of q-tensor operators*, invited paper at Clausthal Summer Workshop on Mathematical Physics (1989).

[BL] Biedenharn L.C., Louck J.D., *The Angular Momentum in Quantum Physics*, Enc. of Math. and its Appl. **8**, (Reading, Mass., 1981, Addison-Wesley).

[BT] Biedenharn L.C. and Tarlini M., *On q-tensor operators for quantum groups*, Lett. Math. Phys. **20** (1990), 271.

[Bx] Baxter R.J., *Exactly solved Models in Statistical Mechanics*, Academic Press, New York, 1982.

[C1] Celeghini E., Giachetti R., Sorace E. and Tarlini M., J. Math. Phys. **31** (1990), 2548.

[C2] Celeghini E., Giachetti R., Sorace E. and Tarlini M., *The quantum Heisenberg group $H(1)_q$*, J. Math. Phys. (to appear).

[C3] Celeghini E., Palev T.D., Tarlini M., *The Quantum Superalgebra $B_q(0|1)$ and q-deformed Creation and Annihilation Operators*, Kyoto preprint YITP/K-865; Mod. Phys. Lett. B, (in press).

[C4] Celeghini E., Giachetti R., Sorace E. and Tarlini M., *The three dimensional Euclidean quantum group $E(3)_q$ and its R-matrix*, preprint Firenze (1990).

[CT] Celeghini E. and Tarlini M., Nuovo Cim. **B61** (1981), 265 **B65** (1981), 172 **B68** (1982), 133.

[Ch] Chaichian M., Ellinas D. and Kulish P.P., Phys. Rev. Lett. **65** (1990), 980.

[Dr] Drinfeld V.G., Proc. ICM-86, Berkeley **1** (1987), 798.

[F1] Faddeev L.D., Reshetikhin N.Yu. and Takhtajan L.A., "Braid Group, Knot Theory and Statistical Mechanics" eds. C.N. Yang and M.L. Ge, World Scientific, Singapore, 1989.

[Fd] Faddeev L.D., *Integrable models in* $(1+1)$*-dimensional quantum field theory*, Les Houches Lectures 1982, (Amsterdam, 1984, Elsevier).

[FT] Faddeev L.D., Takhtajan L.A., *Hamiltonian Methods in the Theory of Solitons*, Springer-Verlag, Berlin, 1987.

[Gc] Giachetti R., *Studio Algebrico della Dinamica di un Oscillatore Quantistico con Reazione*, Rend. Sem. Fac. Sci. di Cagliari (1991), in press.

[Gl] Gilmore R., *Lie groups, Lie algebras and some of their applications*, Wiley, New York, 1974.

[J1] Jimbo M., Int. J. of Mod. Phys. **A4** (1989), 3759.

[J2] Jimbo M., in [F1].

[K1] Kulish P.P., *Kontraktzia kvantovikh algebr i q-ostzilliatori*, LOMI preprint (1990); Teor. Mat. Fiz. **86** no. 1 (1991), 158.

[K2] Kulish P.P., private communication.

[Kg] Kagramanov E.D. Mir-Kasimov R.M. and Nagiyev Sh.M., J. Math. Phys. **31** (1990), 1733.

[KR1] Kulish P.P., Reshetikhin N.Yu., Zap. Nauchn. Sem. LOMI **101** (1981), 101; J. Sov. Math. **23** (1983), 2435.

[KR2] Kulish P.P., Reshetikhin N.Yu., Lett. Math. Phys. **18** (1989), 143.

[KS] Klimyk A.U. and Smirnov Yu.F., Kiev ITP preprint 90-36E (1990).

[Mf] Macfarlane A.J., J. Phys. A: Math. Gen. **22** (1989), 4581.

[Mj] Majid S., Int. J. of Mod. Phys. A **5** (1990), 1.

[Mn] Manin Yu.I., *Quantum groups and Non-Commutative Geometry*, (Montreal, 1988, Centre de Recherches Mathématiques).

[Ng] Ng Y.J., J. Phys. A: Math. Gen. **23** (1990), 1023.

[Op] October session on open problems in this conference.

[Ps] Postnikov M., *Groupes et Algèbres de Lie*, (Moscou, 1985, Editions Mir).

[VK] Vaksman L.L. and Korogodskii L.I., Soviet Math. Dokl. **39** (1989), 173.

[W1] Woronowicz S., Publ. RIMS (Kyoto University) **23** (1987), 117; Commun. Math. Phys. **111** (1987), 613 **122** (1989), 125.

[W2] Woronowicz S., private communication.

[WZ] Wess J., Zumino B., *Covariant differential calculus on the quantum hyperplane*, preprint CERN-TH- 5697/90.

[Ya] Yan H., J. Phys. A **23** (1990), L1155.

Dipartimento di Fisica dell' Università, Largo E. Fermi 2, 50125 Firenze, Italia

NEW SOLUTIONS OF YANG–BAXTER EQUATIONS AND QUANTUM GROUP STRUCTURES

MO-LIN GE

Theoretical Physics Department, Nankai Institute of Mathematics, Tianjin

ABSTRACT. New solutions of Yang-Baxter equations including those associated with the fundamental representations of B_n, C_n and D_n, $\check{R}^{j\frac{1}{2}}$ for $V^j \otimes V^{\frac{1}{2}}$ and colored R-matrix for $SL_q(2)$ with q a root of a unity are explicitly given. The related quantum group structures have been also set up. The Yang-Baxterization is performed to generate spectral parameter-depended solutions of YBE.

1. Introduction

It is well known that the Drinfeld-Jimbo approach is universal in the construction of quantum group [1-13]. The centralizer $\mathcal{R}$ is closely related to a solution of braid relation (without the spectral parameter x)

$$\check{R}_{12}\check{R}_{23}\check{R}_{12} = \check{R}_{23}\check{R}_{12}\check{R}_{23}. \tag{1.1}$$

It has been shown that there exist new types of solutions of YBE which are different from the "standard" ones given by the usual q-deformation theory for Lie algebras [14-20]. Indeed a lot of solutions of braid relation was found and with the help of Yang-Baxterization prescription [15] new solutions of YBE can be derived to satisfy

$$\check{R}_{12}(x)\check{R}_{23}(xy)\check{R}_{12}(y) = \check{R}_{23}(y)\check{R}_{12}(xy)\check{R}_{23}(x). \tag{1.2}$$

To my knowledge the known new solutions associated with simple Lie algebras are as follows

(1) For $G = SU(2)$ with any spin the new solutions were presented in [18].
(2) For the fundamental representations of A_n two types of new solutions have been found. One of them possesses the same eigenvalues as those of the standard ones [19], whereas the other does not. The later was discussed in many references.
(3) The braid group representations (BGR) associated with the fundamental representations of B_n, C_n and D_n were presented by Jimbo for the standard cases [7]. The non-standard cases as well as their Yang-Baxterizations can be found in [14,16,20]. It is shown that these new solutions still satisfy the Birman-Wenzl algebra [22].

(4) For $V^i \otimes V^j$ we find new solutions of $\check{R}_{j\frac{1}{2}}$ which cannot be obtained in terms of the usual fusion rule.

(5) The above new solutions are valid for $q^p \neq 1$. For $q^p = 1$ and taking color into account new solutions of $R^{j_1(\lambda)j_2(\mu)}$ have been given in [23] where λ and μ denote the color degrees.

In this talk we would like to give a brief description of new solutions $\check{R}$. Of course, a general construction of link polynomials is interesting, however there is not enough room to cover lengthy discussions in this work.

2. New solution of BGR for B_n, C_n and D_n

In this section we would like to present a general form of BGR associated with the fundamental representations of B_n, C_n and D_n that after Yang-Baxterization automatically generate the twisted ones, for example, $A_{2n}^{(2)}$ and $A_{2n-1}^{(2)}$ corresponding to B_n and D_n. This solution systems contain the standard family as a particular case, however there exists a new family of BGR. For completeness we first write the standard solutions namely the BGR denoting by T for $B_n^{(1)}, C_n^{(1)}$ and $D_n^{(1)}$, given by Jimbo [7]

$$T = q\sum_{k\neq 0} e_{kk}\otimes e_{kk} + w\sum_{\substack{k<m\\ k+m\neq 0}} e_{kk}\otimes e_{mm} + \sum_{\substack{k\neq m\\ k+m\neq 0}} e_{km}\otimes e_{mk}$$
$$+\sum_{k,m} a_{km}e_{k-m}\otimes e_{-km},\ (e_{km})_{ab} = \delta_{ka}\delta_{mb}, \tag{2.1}$$

where $w = q - q^{-1}$ and

$$a_{km} = \begin{cases} 1 & (k = m = 0) \\ q^{-1} & (k = m \neq 0) \\ w(\delta_{k-m} - \epsilon_k\epsilon_m q^{\bar{k}-\bar{m}}) & (k < m) \end{cases} \tag{2.2}$$

$\epsilon_k = 1 \quad (-(2N-1)/2 \leq k \leq -\frac{1}{2}), \qquad \epsilon_k = -1 \quad (\frac{1}{2} \leq k \leq (2N-1)/2)$ for $C_n^{(1)}$ and $\epsilon_k = 1$ for $B_n^{(1)}, D_n^{(1)}$.

$$\tilde{k} = \begin{cases} k+\frac{1}{2} & (-(N-1)/2 \leq k < 0) \\ k & k = 0 \\ k-\frac{1}{2} & (0 < k \leq (2n-1)/2) \end{cases} \tag{2.3}$$

for $B_n^{(1)}, D_n^{(1)}$.

$$\tilde{k} = \begin{cases} k-\frac{1}{2} & (-(2n-1)/2 \leq k \leq -\frac{1}{2}) \\ k+\frac{1}{2} & (\frac{1}{2} < k \leq (2n-1)/2) \end{cases} \tag{2.4}$$

for $C_n^{(1)}$.

The labeling set here is taken to be

$$L = [(N-1)/2, (N-1)/2-1, \ldots, -(N-1)/2] \tag{2.5}$$

which is slightly different from Jimbo [7]. $N = 2n+1$ for B_n and $N = 2n$ for D_n and C_n.

Now we extend the result to cover more general cases.

By the weight conservation and CP invariance of the BGR T has the general form

$$\begin{aligned} T = \sum_k u_k e_{kk} \otimes e_{kk} + \sum_{k<m} w^{(m)}_{k+m} e_{kk} \otimes e_{mm} \, , \\ \sum_{k\neq m} p^{(k,m)}_{k+m} e_{km} \otimes e_{mk} + \sum_{k,m} q^{(k,m)} e_{k-m} \otimes e_{-km} \, , \\ (k, m \in L, \qquad (e_{km})_{ab} = \delta_{ka}\delta_{mb}) \, , \end{aligned} \tag{2.6}$$

where

$$p^{(a,b)}_{a+b} = p^{(b,a)}_{a+b} \, , \tag{2.7}$$

$$q^{(a,c)} = q^{(c,a)} \tag{2.8}$$

and

$$q^{(a,c)}|_{a=\pm c} = q^{(a,c)}|_{\substack{|a|<|c| \\ c>0}} = q^{(a,0)}|_{a>0} = 0, \tag{2.9}$$

in which

$$a, b, c, d \in [(N-1)/2, (N-2)/2-1, \ldots, -(N-1)/2]. \tag{2.10}$$

The eq.(2.7) and eq.(2.8) come from the restrictions of the "6-vertex type" solutions. Following our previous works [14] and substituting eq.(2.6) into the braid relations after the tedious calculations including the use of the extended diagrammatic technique [14] we derive the following general solutions of the BGR:

$$\begin{aligned} T = q \sum_{k\neq 0} u_k e_{kk} \otimes e_{kk} + w \sum_{\substack{k<m \\ k+m\neq 0}} e_{kk} \otimes e_{mm} + \sum_{\substack{k\neq m \\ k+m\neq 0}} e_{km} \otimes e_{mk} \\ + \sum_{k,m} a_{km} e_{k-m} \otimes e_{-km} , \quad (e_{km})_{ab} = \delta_{ka}\delta_{mb} , \end{aligned} \tag{2.11}$$

where $u_k = q$ or $-q^{-1}$ for $k \neq 0$ and $u_{-k} = u_k$.

The a_{km} are given by

$$a_{km} = \begin{cases} 1 & (k=m=0) \\ u_k^{-1} & (k=m\neq 0) \\ w[1-u_m^{-1}(\prod_{j=1}^{m-1} u_j^{-2})] & (k=-m<0) \\ (-1)^{k+m+1} w u_{k+m}^{-\frac{1}{2}} \prod_{j=1}^{|k+m|-1} u_j^{-1} & (k=0<m, \text{or } k<m=0) \\ (-1)^{k+m+1} w u_m^{-\frac{1}{2}} u_k^{-\frac{1}{2}} \prod_{j=|k|+1}^{|m|-1} u_j^{-1} & (0<k<m, \text{or } k<m<0) \\ (-1)^{k+m+1} w u_m^{-\frac{1}{2}} u_k^{-\frac{1}{2}} (\prod_{j=|k|}^{m-1} u_j^{-1})(\prod_{i=1}^{|k|-1} u_i^{-2}) & (k<0, m>0, k+m <\neq 0) \end{cases} \tag{2.12}$$

for $B_n^{(1)}$.

$$a_{km} = \begin{cases} u_k^{-1} & (k = m) \\ w[1 - \epsilon u_m^{-1}(\prod_{j=1}^{m-\frac{1}{2}} u_{j-\frac{1}{2}}^{-2})u_{\frac{1}{2}}] & (k = -m < 0) \\ -wu_m^{-\frac{1}{2}} u_k^{-\frac{1}{2}}(\prod_{j=1}^{m-\frac{1}{2}} u_{j-\frac{1}{2}}^{-1})u_k^{-\frac{1}{2}} & (0 < k < m, \text{or } k < m < 0) \\ -\epsilon wu_m^{-\frac{1}{2}}(\prod_{j=|k|+\frac{1}{2}}^{m-\frac{1}{2}} u_{j-\frac{1}{2}}^{-1})(\prod_{i=1}^{|k|+\frac{1}{2}} u_{i-\frac{1}{2}}^{-2})u_k^{\frac{1}{2}} u_{\frac{1}{2}}^{\epsilon} & (k < 0, m > 0, k + m <\neq 0) \end{cases} \tag{2.13}$$

where $-\epsilon = 1$ for $C_n^{(1)}$ and $\epsilon = 1$ for $D_n^{(1)}$.

The distinct eigenvalues are given by

$$(T - \lambda_1)(T - \lambda_2)(T - \lambda_3) = 0 \tag{2.14}$$

where

$$\begin{array}{cccc} & \lambda_1 & \lambda_2 & \lambda_3 \\ B_n & q & -q^{-1} & (\prod_{j=1}^{n} u_j^{-2}) \\ C_n & -q^{-1} & q & -(\prod_{j=1}^{n} u_{j-\frac{1}{2}}^{-2})u_{\frac{1}{2}}^{-1} \\ D_n & q & -q^{-1} & (\prod_{j=1}^{n} u_{j-\frac{1}{2}}^{-2})u_{\frac{1}{2}} \end{array} . \tag{2.15}$$

when one of u_i's is not equal to q the eigenvalues are different from the standard ones, i.e. we meet new solutions. It is easy to understand that the solutions derived are natural generalization of the standard ones from the point of view of the block-diagonal matrix structures of BGR [14,19]. Based on the general discussion the BGR under the consideration possesses the form

$$T = \text{block diag}(T_1, T_2, \ldots, T_{n-1}, T_n, T_{n-1}, \ldots, T_2, T_1) \tag{2.16}$$

where T_n is $n \times n$ submatrix. For each odd-dimensional submatrix there is a "central element" denoting by u_{2m+1}. The parameters appearing in the sub-blocks will be determined by the substitution of one sub-block by one sub-block into the braid relations. The first one is simply q and the second one should be $\begin{bmatrix} 0 & q \\ q & w \end{bmatrix}$, however there are several possibility for the third one in solving the braid relations. It allows the central element to be either q or $-q^{-1}$ which gives rise to quite different forms of the other sub-blocks by braid relations. When all of the central elements are equal to q our solutions are the same as Jimbo's [7]. However, if one of the central elements equals to $-q^{-1}$ we will meet new solutions of BGR.

It is not difficult to check that the results in [20] are special cases of our general forms.

3. Birman-Wenzl Algebraic Properties

It has been known that the standard solutions obey the Birman-Wenzl algebra (BW) [10]. Now let us show that the new solutions eqs.(2.11)-(2.15) still satisfy the BW algebra. It is pointed out that a BGR T and the related $E = I - w^{-1}(T - T^{-1})$

$w = q - q^{-1}$) satisfy the BW algebra if they satisfy [22]

$$(1) \qquad E^{ab}_{cd} = r(a)r(c)\delta(a, b')\delta(c, d') \tag{3.1}$$

$$(2) \qquad r(a)r(a') = \pm 1 \qquad (a' = N + 1 - a) \tag{3.2}$$

$$(3) \qquad \sum_b T^{ab}_{ab} r^2(b) = \lambda_3^{-1} \tag{3.3}$$

and

$$T^{ab}_{cd} \neq 0 \qquad \text{only for} \qquad a + b = c + d \tag{3.4}$$

$$T^{ab}_{cd} = T^{b'a'}_{d'c'} \tag{3.6}$$

with $a' = N + 1 - a$, a belongs to eq.(2.5). $\delta(a, c) = 1$ for $a = c$ and $\delta(a, c) = 0$ for $a \neq c$. Under such a convention and the labelling set L (2.5) as well as the notations

$$(e_{ab})_{ij} = \delta(a, i) \cdot \delta(b, j) \tag{3.7}$$

the BGR T can be recast into the form

$$T = \sum_{i \neq i'} u_i e_{ii} \otimes e_{ii} + \sum_{\substack{i \neq j, j' \\ \text{or } i = j = j'}} e_{ij} \otimes e_{ji} + \sum_{i \neq i'} u_i^{-1} e_{ii'} \otimes e_{i'i}$$
$$+ w \sum_{i<j} e_{ii} \otimes e_{jj} - w \sum_{i<j} r(i) r(j') e_{j'i} \otimes e_{ji'} \tag{3.8}$$

where

$$u_{i'} = u_i, \qquad u_i = q \qquad \text{or} \quad -q^{-1} \tag{3.9}$$

$$u_0 = 1 \qquad (\text{only for } B_n) \tag{3.10}$$

and under (2.9)

$$r(i) = \begin{cases} (-1)^i u_i^{-\frac{1}{2}} \prod_{j=0}^{i} u_j & i \geq 0 \qquad \text{for } B_{n'}, \\ i u_{\frac{1}{2}}^{-\frac{1}{2}} u_i^{\frac{1}{2}} \prod_{j=\frac{1}{2}}^{i} u_i & i > \frac{1}{2} \qquad \text{for } C_{n'}, \\ u_{-\frac{1}{2}}^{\frac{1}{2}} u_i^{\frac{1}{2}} \prod_{j=\frac{1}{2}}^{i} u_i & i > \frac{1}{2} \qquad \text{for } D_{n'} \end{cases} \tag{3.11}$$

Noting that

$$r(i) = r^{-1}(-i) \qquad (i < 0) \tag{3.12}$$

in terms of eq.(3.11) it is not difficult to prove that the BGR T found in Sec.2 satisfies (3.1)-(3.6), namely obey the BW algebra. The operator E is defined by eq.(3.1) where $r(i)$ is given by eq.(3.11).

We emphasize that by interchanging λ_1 and λ_2 and leaving λ_3 unchanged we still obtain the BW algebra.

4. Yang-Baxterization

According to [21,15] the Birman-Wenzl algebra is easy to be Yang-Baxterized in two ways. One of them is given by

$$\check{R}_a(x) = \lambda_1 x(x-1)T^{-1} + (1+\lambda_1/\lambda_2+\lambda_2/\lambda_3+\lambda_1/\lambda_3)xI - \lambda_3^{-1}(x-1)T \quad (4.1)$$

and the other one denoted by $\check{R}_b(x)$ can be obtained by $\lambda_1 \leftrightarrow \lambda_2$ and $\lambda_3 \leftrightarrow \lambda_3$ in $\check{R}_a(x)$. For simplicity we only write the explicit form of the case (a) under (2.9)

$$\begin{aligned}\check{R}_a(x) &= \sum_{k\neq 0} u_k e_{kk}\otimes e_{kk} \\ &-(q^2-1)(x-\xi)\Big(\sum_{\substack{k<m\\ k+m\neq 0}} + x\sum_{\substack{k>m\\ k+m\neq 0}}\Big)\cdot e_{kk}\otimes e_{mm} \\ &+q(x-1)(x-\xi)\sum_{\substack{k\neq m\\ k+m\neq 0}} e_{km}\otimes e_{mk} + \sum_{k,m} a_{km} e_{k-m}\otimes e_{-km}\end{aligned} \quad (4.2)$$

where

$$u_k(x) = \begin{cases} (x-q^2)(x-\xi) & (u_k = q) \\ -q^2(x-q^{-2})(x-\xi) & (u_k = -q^{-1}) \end{cases} \quad (4.3)$$

and

$$\begin{aligned} a_{km}(x) &= q(x-1)(x\tilde{a}_{km} - \xi a_{km}) + (\xi-1)(q^2-1)x\delta_{k-m} \\ \xi &= \begin{cases} q^{-1}\lambda_3^{-1} & \text{for } B_n^{(1)} \text{ and } D_n^{(1)} \\ -q\lambda_3^{-1} & \text{for } C_n^{(1)} \end{cases} \end{aligned} \quad (4.4)$$

$$\tilde{a}_{km}(u_k) = a_{mk}(u_k^{-1}). \quad (4.5)$$

and $\check{R}_b(x)$ can be obtained by the formal interchange $q \leftrightarrow -q^{-1}$ and keeping λ_3 unchanged. The solutions thus derived correspond to the "twisted" ones. For instance, corresponding to $\check{R}'_a$ of $B_n^{(1)}$ and $D_n^{(1)}$ the case (b) gives the solutions relating with $A_{2n}^{(2)}$ and $A_{2n-1}^{(2)}$, respectively. It is easy to check by means of the standard solutions. We would like to note that the case (a) is the "normal" solution for C_n, however the case (a) is still a solution. We do not know about the Lie algebraic description of the case (b) for C_n yet. All of the "twisted" new solutions deserve to be understood.

It is worth to discuss the eigenvalues, since they are related to the diagonalization of BGR. The first two eigenvalues $\lambda_1 = u = q$ and $\lambda_2 = v = -q^{-1}$ appear only in the sub-blocks other than the largest one, whereas the third eigenvalue $\lambda = \lambda_3$ only appears in the largest sub-blocks. We thus focus on considering the eigenvalue equation of the largest sub-blocks. By calculation we find that

$$(\lambda-u)^{2n-1-n_1}(\lambda-v)^{n_1}[\lambda + (\prod_{j=1}^{n} u_{j-\frac{1}{2}}^{-2})u_{\frac{1}{2}}^{-1}] = 0$$

when

$$(\prod_{j=1}^{n} u_{j-\frac{1}{2}}^{-2})u_{\frac{1}{2}}^{-1} \neq \pm u \text{ or } \pm v \text{ for } C_n, (n_1 = n \text{ or } n-1);$$

and

$$(\lambda - u)^{n-1-n_1}(\lambda - v)^{n_1}[\lambda - (\prod_{j=1}^{n} u_{j-\frac{1}{2}}^{-2})u_{\frac{1}{2}}] = 0$$

when

$$(\prod_{j=1}^{n} u_{j-\frac{1}{2}}^{-2})u_{\frac{1}{2}} \neq \pm u \text{ or } \pm v \text{ for } D_n, (n_1 = n \text{ or } n-1).$$

There are three distinct eigenvalues for B_n, whereas when

$$(\prod_{j=1}^{n} u_{j-\frac{1}{2}}^{-2})u_{\frac{1}{2}}^{-1} \text{ for } C_n, (\text{or }, (\prod_{j=1}^{n} u_{j-\frac{1}{2}}^{-2})u_{\frac{1}{2}} \text{ for } D_n)$$

is equal to $\pm u$ or $\pm v$ the eigenvalue equations are reduced to

$$(\lambda - u)^n(\lambda - v)^n = 0,$$

namely, the number of the distinct eigenvalues is no longer the same as the decomposition dimensions three but two. For instance, for C_2 we have the minimum polynomial

$$(T-u)^2(T-v) = 0. \tag{4.6}$$

The multiplicity two in (4.6) is essential. This indicates that the corresponding BGR cannot be diagonalized. One can check this statement generally, for example, C_2 explicitly. Even in such a case it was proved that the solutions of BGR can still be Yang-Baxterized by either case (a) or case (b).

5. Quantum group structure

It is known that for a given BGR the corresponding quantum group can be constructed with the help of Faddeev-Reshetikhin-Takhtajan (FRT) approach [9]. However, for our problem the direct application of FRT is complicated and we, therefore, follow [11] to use an equivalent but simple technique to set up desired relations satisfied by the generators of quantum group (algebra, in fact). Suppose

$$\check{R}(x) = A(x)T + B(x)I + C(x)T^{-1} \tag{5.1}$$

satisfies the YBE

$$\check{R}_{12}(x)\check{R}_{23}(xy)\check{R}_{12}(y) = \check{R}_{23}(y)\check{R}_{12}(xy)\check{R}_{23}(x), \tag{5.2}$$

the Yang-Baxter operator can be defined by [11]

$$(t_{ab}(x))_{cd} = \check{R}^{ca}_{bd}(x) \tag{5.3}$$

in accordance with

$$\check{R}(xy^{-1})(t(x)\otimes t(y))=(t(y)\otimes t(x))\check{R}(xy^{-1}). \tag{5.4}$$

Following [11] the asymptotic behavior of $t(x)_{x\to 0}$ and $t(x)_{x^{-1}\to 0}$ gives the generators of quantum group by omitting the factors $A(x)_{x\to 0}$ and $C(x)_{x^{-1}\to 0}$ (even the limits tend to infinity).

The calculations are lengthy and we only give the results for B_n. The other cases can be treated in a similar way. We have

$$\begin{gathered}(t_{ii}(x))|_{x\to 0}=A(x)|_{x\to 0}\{u_i\delta_{ab}\delta_{ai}\\ +\delta_{ab}|_{\substack{a\neq i,i'\\ a=n+1}}+u_i^{-1}\delta_{ab}\delta_{i'a}\}=A(x)|_{x\to 0}(k_i)_{ab}\\ (t_{ii}(x))|_{x\to\infty}=C(x)|_{x\to\infty}\{u_i^{-1}\delta_{ab}\delta_{ai}\\ +\delta_{ab}|_{\substack{a\neq i,i'\\ a=n+1}}+u_i^{-1}\delta_{ab}\delta_{i'a}\}=C(x)|_{x\to\infty}(k_i^{-1})_{ab}\\ (t_{i+1\ i}(x))|_{x\to 0}=A(x)|_{x\to 0}\{w\delta_{ai}\delta_{i+1\ b}\\ -wr(i)r(N-i)\delta_{N-i\ a}\delta_{i'b}\}=A(x)|_{x\to 0}w(e_i)_{ab}\\ (t_{i\ i+1}(x))|_{x\to\infty}=C(x)|_{x\to\infty}\{-w\delta_{a\ i+1}\delta_{i\ b}\\ +wr(i+1)r(i')\delta_{i'a}\delta_{N-i\ b}\}=C(x)|_{x\to\infty}(-w)(f_i)_{ab}\end{gathered}$$

Introducing

$$\begin{gathered}K_i=k_ik_{i+1}^{-1},\qquad K_n=k_n^2\\ X_i^+=e_i,\qquad X_i^-=f_i\qquad i=1,2,\dots,n-1,\\ X_n^+=(u_n+u_n^{-1})^{\frac{1}{2}}e_n,\qquad X_n^-=(u_n+u_n^{-1})^{\frac{1}{2}}f_n,\end{gathered}$$

we obtain

$$\begin{gathered}[X_i^+,X_j^-]=\delta_{ij}(K_i-K_i^{-1})/w\qquad i,j=1,\dots,n,\\ K_iX_i^{\pm}K_i^{-1}=(u_iu_{i+1})^{\pm 1}X_i\qquad i=1,\dots,n-1,\\ K_{i-1}X_i^{\pm}K_{i-1}^{-1}=u_i^{\pm 1}X_i^{\pm}\qquad i=1,\dots,n,\\ K_{i+1}X_i^{\pm}K_{i+1}^{-1}=u_{i+1}^{\mp 1}X_i\qquad i=1,\dots,n-2,\\ K_nX_n^{\pm}K_n^{-1}=(u_n)^{\pm 2}X_n^{\pm}\qquad K_nX_{n-1}^{\pm}K_n^{-1}=(u_n)^{\mp 2}X_n,\\ K_jX_i^{\pm}K_j^{-1}=X_i\qquad |i-j|>1\qquad K_iK_j=K_jK_i.\end{gathered}$$

The Serre relations become under the representation

$$(X_i^{\pm})^2=0,\qquad i=1,\dots,n-1\qquad (X_n^{\pm})^3=0.$$

The coproducts read

$$\begin{gathered}\Delta(X_i^+)=k_{i+1}\otimes X_i^++X_i^+\otimes k_i\qquad i=1,\dots,n-1,\\ \Delta(X_n^+)=I\otimes X_n^++X_n^+\otimes k_n\\ \Delta(X_i^-)=k_i^{-1}\otimes X_i^-+X_i^-\otimes k_{i+1}^{-1}\\ \Delta(X_n^-)=k_n^{-1}\otimes X_n^-+X_n^-\otimes I,\qquad \Delta(K_i^{\pm})=K_i^{\pm}\otimes K_i^{\pm},\end{gathered}$$

where I stands for the unit matrix.

The antipode and the counit are given by

$$\gamma(k_i) = k_i^{-1}, \qquad \gamma(I) = I, \qquad \epsilon(X_i^{\pm}) = 0, \qquad \epsilon(k_i^{\pm 1}) = 1,$$
$$\gamma(X_i^+) = -k_{i+1}^{-1} X_i^+ k_i^{-1}, \qquad \gamma(X_i^-) = -k_i^{-1} X_i^- k_{i+1}^{-1}, \qquad i = 1, \ldots, n-1,$$
$$\gamma(X_n^+) = -X_n^+ k_n^{-1}, \qquad \gamma(X_n^-) = -k_n X_n^-.$$

When $u_i = q$ the results are exactly the standard quantum group structure.

6. New solution of $\check{R}^{j\frac{1}{2}}(x)$

Consider the following lattice and equations

$$\check{R}_{12}^{j\frac{1}{2}}(x)\check{R}_{23}^{j\frac{1}{2}}(xy)\check{R}_{12}^{\frac{1}{2}\frac{1}{2}}(y) = \check{R}_{23}^{\frac{1}{2}\frac{1}{2}}(y)\check{R}_{12}^{j\frac{1}{2}}(xy)\check{R}_{23}^{j\frac{1}{2}}(x)$$
$$\check{R}_{12}^{\frac{1}{2}j}(x)\check{R}_{23}^{\frac{1}{2}\frac{1}{2}}(xy)\check{R}_{12}^{j\frac{1}{2}}(y) = \check{R}_{23}^{j\frac{1}{2}}(y)\check{R}_{12}^{\frac{1}{2}\frac{1}{2}}(xy)\check{R}_{23}^{\frac{1}{2}j}(x)$$
$$\check{R}_{12}^{\frac{1}{2}\frac{1}{2}}(x)\check{R}_{23}^{\frac{1}{2}j}(xy)\check{R}_{12}^{\frac{1}{2}j}(y) = \check{R}_{23}^{\frac{1}{2}j}(y)\check{R}_{12}^{\frac{1}{2}j}(xy)\check{R}_{23}^{\frac{1}{2}\frac{1}{2}}(x).$$

This is "incomplete" YBE, but meaningful for the spin-$(j, \frac{1}{2})$ statistical model. The CP-invariance reads

$$(\check{R}^{j\frac{1}{2}})_{c(\frac{1}{2})\ d(j)}^{a(j)\ b(\frac{1}{2})} = (\check{R}^{\frac{1}{2}j})_{-d(j)\ -c(\frac{1}{2})}^{-b(\frac{1}{2})\ -a(j)}$$

Only the standard $\check{R}^{\frac{1}{2}\frac{1}{2}}$ is allowed due to the YBE and

$$a(j) = (j, j-1, \ldots, -j+1, -j), \ a(\frac{1}{2}) = (\frac{1}{2}, -\frac{1}{2}).$$

Tedious calculations give the solution

$$\check{R}^{j\frac{1}{2}} = \sum_{(a(j),b(\frac{1}{2}))} u^{(a(j),b(\frac{1}{2}))} E_{a(j)\ b(\frac{1}{2})} \otimes E_{b(\frac{1}{2})\ a(j)}$$
$$+ \sum_{(a(j),b(\frac{1}{2}))} v^{(a(j),b(\frac{1}{2}))} E_{a(j)\ -b(\frac{1}{2})} \otimes E_{b(\frac{1}{2})\ a(j)+2b(\frac{1}{2})},$$

where E is the unit matrix and the parameters u and v are given by

$$u^{(j,\frac{1}{2})} = q, \qquad u^{(-j,-\frac{1}{2})} = Q$$
$$u^{(a(j),\frac{1}{2})} = q^{a(j)+1-j}, \qquad u^{(a(j),-\frac{1}{2})} = q^{-a(j)-j}Q,$$
$$v^{(a(j),-\frac{1}{2})} = 0,$$
$$u^{(-a(j)-1)}u^{(a(j),\frac{1}{2})} = q^{-2j}QW^2[j-a(j)][j+a(j)+1] \qquad (a(j) \neq \pm j)$$
$$[n] = (q^n - q^{-n})/W, \qquad W = q - q^{-1}.$$

For instance

$$\check{R}^{1\frac{1}{2}} = (A_1, A_2, A_3, A_4)$$

where

$$A_1 = q,\ A_2 = \begin{bmatrix} 0 & q^{-2}Q \\ 1 & q_1 \end{bmatrix},\ A_3 = \begin{bmatrix} 0 & q^{-1}Q \\ q^{-1} & q_2 \end{bmatrix},\ A_4 = q,$$
$$q_1 q_2 = Q(q - q^{-1})(1 - q^{-4}).$$

The Yang-Baxterization can be made through

$$\check{R}^{j\frac{1}{2}}(x) = \alpha_1 x(\check{R}^{\frac{1}{2}j})^{-1} + \alpha_2 \check{R}^{j\frac{1}{2}}$$
$$\check{R}^{\frac{1}{2}j}(x) = \alpha_1 x(\check{R}^{j\frac{1}{2}})^{-1} + \alpha_2 \check{R}^{\frac{1}{2}j}$$

where the parameters are restricted by

$$\alpha_1^2 = Q^2 q^{-4} \alpha_2^2.$$

The simple choice is $\alpha_1 = q$ and $\alpha_2 = -q^{2j}Q^{-1}$. $\check{R}^{j\frac{1}{2}}(x)$ thus derived satisfies the "incomplete" YBE. It is easy to check that the standard solutions $\check{R}_s^{j\frac{1}{2}}$ given by [13] is a special case of our general solutions. In general our new solutions contain $j+2$ (for j =integer) or $j+3/2$ (for j = half integer) free parameters. Moreover our general solutions can be written in the form ($\rho^{[i]}$ stands for representation)

$$\mathcal{R}^{\frac{1}{2}j} = \mathcal{R}_S^{\frac{1}{2}j} + \mathcal{R}_1^{\frac{1}{2}j} \qquad (\mathcal{R}^{\frac{1}{2}j} = P^{\frac{1}{2}j}\check{R}^{\frac{1}{2}j})$$

where

$$\mathcal{R}_1^{\frac{1}{2}j} = \sum_{k=0}^{2j} A_k \rho^{[\frac{1}{2}]}(J_-)\rho^{[\frac{1}{2}]}(J_+) \otimes (\rho^{[\frac{1}{2}]}(J_-))^k (\rho^{[\frac{1}{2}]}(J_+))^k$$
$$+ \sum_{k=1}^{2j} B_k \rho^{[\frac{1}{2}]}(J_+) \otimes (\rho^{[\frac{1}{2}]}(J_+))^{k-1} (\rho^{[\frac{1}{2}]}(J_-))^k$$

and A_k and B_k are determined through

$$A_0 = q^{-j-1}(Q - q)$$
$$A_0 + [2j]A_1 = q^{-j}(Q - q)$$
$$A_0 + [2][2j-1]A_1 + [2j][2j-1]A_2 = q^{-j+1}(Q - q)$$
$$\cdots$$
$$A_0 + \{\sum_{k=1}^{2j}\sum_{r=1}^{k}[j+m+r][j-m-r+1]\}A_k = q^{-m-1}(Q - q)$$
$$(m \neq j)$$

whereas

$$([j+m][j-m+1])^{\frac{1}{2}}\{B_1 + \sum_{k=2}^{2j}(\sum_{r=2}^{k}[j+m-r+1][j-m+r])B_k\}$$
$$= v_{m-1} - q^{-\frac{1}{2}}W([j+m][j-m+1])^{\frac{1}{2}} \qquad (m \neq -j).$$

v_m are determine by the braid relations).

The corresponding quantum group structure can be derived in terms of F-R-T approach [9] with taking

$$\check{R}^{j\frac{1}{2}}(L_\pm^{\frac{1}{2}} \otimes L_\pm^j) = (L_\pm^j \otimes L_\pm^{\frac{1}{2}})\check{R}^{j\frac{1}{2}}$$
$$\check{R}^{\frac{1}{2}j}(L_\pm^j \otimes L_\pm^{\frac{1}{2}}) = (L_\pm^{\frac{1}{2}} \otimes L_\pm^j)\check{R}^{\frac{1}{2}j}$$
$$\check{R}^{j\frac{1}{2}}(L_+^{\frac{1}{2}} \otimes L_-^j) = (L_-^j \otimes L_+^{\frac{1}{2}})\check{R}^{j\frac{1}{2}}$$
$$\check{R}^{\frac{1}{2}j}(L_+^j \otimes L_-^{\frac{1}{2}}) = (L_-^{\frac{1}{2}} \otimes L_+^j)\check{R}^{\frac{1}{2}j}.$$

into account. By introducing

$$p^{(\mu,\frac{1}{2})} = q^{\mu+1-j}, \qquad p^{(\mu,-\frac{1}{2})} = q^{-\mu-j}Q,$$
$$q^{(-\mu-1,\frac{1}{2})}q^{(\mu,\frac{1}{2})} = q^{-2j}QW^2[j-\mu]_q[j+\mu+1]_q \qquad (\mu \neq \pm j),$$

after calculations we find that

$$(L_+)^\mu_\nu = \alpha^\mu_\nu K_1^{-2\nu}K_2^{2(\mu-j)}(X^+)^{\mu-\nu}$$
$$(L_-)^\mu_\nu = \beta^\mu_\nu K_1^{2\mu}K_2^{2(\nu-j-2\mu)}(X^-)^{\nu-\mu}$$
$$(L_+)^{\frac{1}{2}}_{\frac{1}{2}} = K_1^{-1}K_2, \qquad (L_+)^{-\frac{1}{2}}_{-\frac{1}{2}} = K_1K_2^{-1}, \qquad (L_+)^{\frac{1}{2}}_{-\frac{1}{2}} = K_1K_2X^+,$$
$$(L_-)^{\frac{1}{2}}_{\frac{1}{2}} = K_1K_2^{-1}, \qquad (L_-)^{-\frac{1}{2}}_{-\frac{1}{2}} = K_1^{-1}K_2, \qquad (L_-)^{-\frac{1}{2}}_{\frac{1}{2}} = K_1^{-1}K_2^3X^-,$$
$$(L_+)^{-\frac{1}{2}}_{\frac{1}{2}} = (L_-)^{\frac{1}{2}}_{-\frac{1}{2}} = 0,$$

where

$$\alpha^\mu_\nu = 0 \quad \text{for } \mu < \nu, \qquad \beta^\mu_\nu = 0 \quad \text{for } \mu > \nu, \qquad \alpha^\mu_\mu = \beta^\mu_\mu = 1$$
$$\alpha^{\nu+n}_\nu = \prod_{m=1}^{n}(q^{(-\nu-m,\frac{1}{2})}q^{\nu+j}(Qq^{-1})^{-\frac{m-1}{j}})/(q^m - q^{-m})$$
$$n = 1,2,\ldots,j-\nu,$$
$$\beta^{\nu-n}_\nu = \prod_{m=1}^{n}(q^{(\nu-m,\frac{1}{2})}q^{\nu+j+1}Q^{-1}q^{-2m+1}(Qq^{-1})^{\frac{m-1}{j}})/(q^m - q^{-m})$$
$$n = 1,2,\ldots,j+\nu,$$

and the operators satisfy the relations

$$K_2\tilde{X}^\pm K_2^{-1} = (Qq^{-1})^{\pm\frac{1}{2j}}\tilde{X}^\pm, \qquad [\mathcal{K}, K_2] = 0,$$
$$\mathcal{K}\tilde{X}^\pm\mathcal{K}^{-1} = q^{\pm 1}\tilde{X}^\pm,$$
$$[\tilde{X}^+, \tilde{X}^-] = q^{-1}W(\mathcal{K}^{-2} - \mathcal{K}^2)$$

with

$$\tilde{X}^{\pm} = X^{\pm}K_2^2, \qquad \mathcal{K} = K_1K_2^{-1}.$$

The co-products read

$$[K_1, K_2] = 0,$$
$$\Delta(X^+) = K_1^{-2} \otimes X^+ + X^+ \otimes K_2^{-2}, \qquad \Delta(K_1) = K_1 \otimes K_1,$$
$$\Delta(K_2) = K_2 \otimes K_2, \qquad \Delta(X^-) = X^- \otimes K_1^2K_2^{-4} + K_2^{-2} \otimes X^-$$
$$\epsilon(K_1) = \epsilon(K_2) = 1, \qquad \epsilon(X^{\pm}) = 0, \qquad \gamma(K_1) = K_1^{-1}, \qquad \gamma(K_2) = K_2^{-1}$$
$$\gamma(X^+) = -Q^{-1}qK_1^2K_2^2X^+, \qquad \gamma(X^-) = -Qq^{-3}K_1^{-2}K_2^6X^-.$$

Obviously besides a new central element K_2 the commutation relations are standard since we cannot go beyond the basic construction of quantum group. However in this example there are new parameters in the $\check{R}^{j\frac{1}{2}}$ matrices because the corresponding statistical model satisfies the "incomplete YBE".

7. New Colored Solutions

For the colored spin 1/2 case the YBE

$$\check{R}_{12}(\lambda,\mu)\check{R}_{23}(\lambda,\nu)\check{R}_{12}(\mu,\nu) = \check{R}_{23}(\mu,\nu)\check{R}_{12}(\lambda,\nu)\check{R}_{23}(\lambda,\mu). \tag{7.1}$$

admits the general solutions with colors λ and μ

$$\begin{aligned}\check{R}_{12}(\lambda,\mu) = &\sum_{a=+,-} u_a(\lambda,\mu)e_{aa} \otimes e_{aa} + w(\lambda,\mu)e_{--} \otimes e_{++} \\ &+ \sum_{\substack{a=b \\ a,b=+,-}} p^{(a,b)}(\lambda,\mu)e_{ab} \otimes e_{ba}\end{aligned} \tag{7.2}$$

where u_a, w and $p^{(a,b)}$ are to be determined by (7.1).

After calculations we find two types of solutions

(I)

$$\check{R}_{\rm I}(\lambda,\mu) = \begin{bmatrix} q & & & \\ & 0 & X(\lambda)\eta & \\ & \eta^{-1}X^{-1}(\mu) & \overline{w}(\lambda,\mu) & \\ & & & qX(\lambda)X^{-1}(\mu) \end{bmatrix} \tag{7.3}$$

where q and η are arbitrary parameters and $X(\lambda)$ is function of $\lambda, \overline{w}(\lambda,\mu) = (g(\lambda)/g(\mu))(q-q^{-1})$ with g being arbitrary function of the argument.

(II)

$$\check{R}_{\rm II}(\lambda,\mu) = \begin{bmatrix} q & & & \\ & 0 & X(\lambda)\eta & \\ & \eta^{-1}X^{-1}(\mu) & \overline{w}(\lambda,\mu) & \\ & & & -qX(\lambda)X^{-1}(\mu) \end{bmatrix}. \tag{7.4}$$

Obviously (7.4) is super-extension of (7.3). The "standard" colored solution (7.3) can also derived by rescaling the standard quantum group construction that contains new central element.

8. Bosonization and Solution of $\mathcal{R}^{j_1(\lambda)j_2(\mu)}$

It can be proved that the following bosonization satisfies the standard quantum group commutation rules [23]

$$J_+ = a^+\alpha(N), \qquad J_- = a\beta(N), \qquad J_3 = h(N)$$

if $\quad h(N) = 2N - \lambda \quad$ and $\quad \alpha(N-1)\cdot\beta(N) = [\lambda+1-N]$. $(\lambda \in \mathbb{C})$

On the q-deformed Fock space $(m = 1,2,\dots)$

$$J_+F(m) = \alpha(m)F(m+1), \; J_-F(m) = [m]\beta(m)F(m-1)$$
$$J_3F(m) = (2m-\lambda)F(m), \qquad \beta(m)\cdot\alpha(m-1) = [\lambda+1-m].$$

After finding invariant sub-spaces we give color dependent $\mathcal{R}$-matrix [23]

$$(\mathcal{R}^{j_1(\lambda)j_2(\mu)})^{m_1'\; m_2'}_{m_1\; m_2} = a^{\frac{1}{2}(2(j_1+m_1')-\lambda)(2(j_2+m_2')-\mu)}.$$

$$\{\delta^{m_1'}_{m_1}\delta^{m_2'}_{m_2} + \sum_{n=1}^{k}(\frac{(1-q^{-2})^n}{[n]!}\cdot q^{-\frac{1}{2}n(n-1)+n(j_1-j_2+m_1'-m_2'-\frac{1}{2}(\lambda-\mu))})$$

$$\cdot\prod_{r=1}^{n}\alpha_{j_1,m_1+r-1}(\lambda)\beta_{j_2,m_2-r+1}(\mu)[j_2+m_2-r+1]\delta^{m_1'}_{m_1+n}\delta^{m_2'}_{m_2-n}\}$$

$$k = Min(2j_1,2j_2) \qquad \text{and } \beta_{j,m}(\lambda)\alpha_{j,m-1}(\lambda) = [\lambda-j-m+1].$$

It is a general version of colored R-matrix with $q^p = 1$ and other continuous parameter for spin model.

It is a pleasure to thank Professors L.D.Faddeev, P.P.Kulish, A.Kirillov, E.K.Sklyanin for warm hospitality at Leningrad and for many enlightening discussions. I also acknowledge the cooperation with Dr. K.Xue and C.P.Sun.

References

[1] *Yang-Baxter Equation in Integrable Systems*, (M.Jimbo, ed.), World Scientific, Singapore, 1990.
Pierre Cartier, *Recent Development on The Applications of Braid Groups to Topology and Algebra*, Seminare BOURBAKI,no 716 (November,1989).
L.C.Biedenharn, *An Overview of Quantum Group,XVIIIth ICGTMP*, Moscow.

[2] C.N.Yang, Phys.Rev.Lett. **19** (1967), 1312; Phys.Rev. **168** (1968), 1920.

[3] *Braid Group, Knot Theory and Statistical Mechanics* (C.N.Yang and M.L.Ge, eds.), World Scientific, Singapore, 1989.

[4] T.Kohno, *in Braid Group, Knot Theory and Statistical Mechanics* (C.N.Yang and M.L.Ge, eds.), World Scientific, Singapore, 1989, p. 135.
L.D.Faddeev, Commun.Math.Phys **132** (1990), 131.

[5] V.G.Drinfeld, Sov.Math.Dokl. **32** (1985), 254; Yang-Baxter Equation in Integrable Systems, (M.Jimbo ed.), World Scientific, Singapore, 1990.

[6] M.Jimbo, Lett.Math.Phys. **10** (1985), 63.

[7] M.Jimbo, Commun.Math.Phys. **102** (1986), 537.

[8] N.Yu.Reshetikhin, *Preprint LOMI E-4-87*, 1988.

[9] L.D.Faddeev,N.Yu.Reshetikhin and L.A.Takhtajan, *Preprint LOMI E-14-87*.
L.A.Takhtajan, Nankai Lecture on Math.Phys."Introduction to Quantum Group and Integrable Massive Models of Quantum Field Theory" (M.L.Ge and B.H.Zhao, eds.), World Scientific, 1990, pp. 69–197.

[10] M.Wadati,T.Deguchi and Y.Akutsu, Phys.Reports. **180** (1989), 247.
Braid Group, Knot Theory and Statistical Mechanics (C.N.Yang and M.L.Ge, eds.), World Scientific, Singapore, 1989, pp. 151–200.

[11] H.J.de Vega, Adv.Studies in Pure Math. **19** (1989), 567; Inter.J.Mod.Phys. **B4** (1990), 735.

[12] A.A.Belavin and V.G.Drinfeld, Funct.Anal.Appl. **16** (1982), 159.

[13] A.N.Kirillov and N.Yu.Reshetikhin, Commun. Math. Phys. **134** (1990), 421.

[14] M.L.Ge,L.Y.Wang,Y.S.Wu and K.Xue, Inter.J.Mod.Phys. **4A** (1989), 3351.
M.L.Ge,Y.Q.Li and K.Xue, J.Phys. **23A** (1990), 605; 619.
M.L.Ge and K.Xue, Phys.Lett. **146A** (1990), 245 **152A** (1990), 266.

[15] M.L.Ge,Y.S.Wu and K.Hue, *Explicit trigonometric Yang-Baxterization*, ITP-SB-90-02; Inter.J.Mod.Phys. (to appear).

[16] M.L.Ge and K.Hue, *Proceedings of Workshop of Argonne National Laboratory* (T.Curtright, D.B.Fairlie and C.K.Zachos, eds.), World Scientific, Singapore, 1990.

[17] L.D.Faddeev,N.Yu.Reshetikhin and L.A.Takhtajan, Braid Group, Knot Theory and Statistical Mechanics (C.N.Yang and M.L.Ge, eds.), World Scientific, Singapore, 1989, p. 97.

[18] H.C.Lee,M.Couture and N.C.Schmeing, *CRNL-TP-1125R*, Canada.

[19] E.Cremmer and J.-L.Gervais, *Preprint LPTENS 89/19*, 1989.

[20] M.Couture,Y.Cheng,M.L.Ge and K.Xue, Inter.J.Mod.Phys. **6A no 4.** (1990).

[21] V.Jones, Commun.Math.Phys. **125** (1989), 459.

[22] Y.Cheng,M.L.Ge and K.Xue, Commun.Math.Phys. **136** (1990), 195; *ITP-SB-90-38*.

[23] Y.Cheng,M.L.Ge and K.Xue, *Construction of Non-standard Solutions for YBE in Terms of q-Deformed Boson Representations of $SL_q(2)$ with q a Root of Unity*, Nankai preprint, 1990.

THEORETICAL PHYSICS DEPARTMENT, NANKAI INSTITUTE OF MATHEMATICS, TIANJIN, 300071, CHINA

QUANTUM GROUP SYMMETRY OF 2D GRAVITY

JEAN-LOUP GERVAIS

Laboratoire de Physique Théorique
de l'École Normale Supérieure, Paris

ABSTRACT. Current progresses in understanding quantum gravity from the operator viewpoint are reviewed. They are based on the $U_q(sl(2))$-quantum-group structure recently put forward[1,2] for the chiral components of the metric in the conformal gauge.

1. Introduction

The algebraic approach to quantum gravity in the conformal gauge — that is to the Liouville field theory — is presently making substantial progress[1–6]. This comes about because there exist decompositions of inverse powers of the metric into operators that precisely transform under irreducible representations of the underlying quantum group [1,2] in the standard form. Their non-commutativity as quantum-field operators coincides with the non-commutativity that is induced by the "quantum" deformation of this group in the mathematical sense. Their braiding and fusion properties are known explicitly, since they are given by the universal R matrix and q-Clebsch-Gordan coefficients respectively[2,5]. Some related progress have also been made for the WZNW models[7].

The most recent advance [3,4,5] concerns the challenging problem of the strong-coupling regime. In our early works [8,9] Neveu and myself had put forward the special cases of central charges $C_{grav} = 7$, 13, and 19, for which we showed that a particular operator with a real Virasoro-weight is closed under braiding. We were thus led to the idea that the theory should make sense for these special cases. Making use of the properties of the operator-algebra derived from the quantum group structure just mentioned, it has been possible to prove a unitary truncation theorem [4,5] which shows that indeed 2D gravity is a consistent conformal theory at the above central charges. These notes review the main lines of the above-mentioned topics. Some background material scattered in the early papers [10–15] are included for completeness. The present discussion has been extended the Toda field theories, leading to new deformations for Lie algebras of rank higher than one [16]. We shall not discuss this topic here.

2. The Classical Structure

First recall some basic points about the weak coupling regime. In the conformal gauge, the classical dynamics is governed by the action:

$$S = \frac{1}{4\pi\gamma}\int d_2x\sqrt{\hat{g}}\Big\{\frac{1}{2}\hat{g}^{ab}\partial_a\Phi\partial_b\Phi + e^{2\Phi} - \frac{1}{4}R_0\Phi\Big\} \tag{2.1}$$

$\widehat{g}_{ab}$ is the fixed background metric. We work for fixed genus, and do not integrate over the moduli. As is well known, one can choose a local coordinate system such that $\widehat{g}_{ab} = \delta_{ab}$. Thus we are reduced to the action

$$S = \frac{1}{4\pi\gamma} \int d\sigma d\tau \left(\frac{1}{2} (\frac{\partial \Phi}{\partial \sigma})^2 + \frac{1}{2} (\frac{\partial \Phi}{\partial \tau})^2 + e^{2\Phi} \right) \tag{2.2}$$

where σ and τ are the local coordinates. The complex structure is assumed to be such that the curves with constant σ and τ are everywhere tangent to the local imaginary and real axis respectively.

This last action corresponds to a conformal theory such that $\exp(2\Phi)$ is conformal with weights (1,1). The chiral modes may be separated very simply using the fact that the field $\Phi(\sigma, \tau)$ satisfies the equation

$$\frac{\partial^2 \Phi}{\partial \sigma^2} + \frac{\partial^2 \Phi}{\partial \tau^2} = 2e^{2\Phi} \tag{2.3}$$

if and only if[1]

$$e^{-\Phi} = \frac{i}{\sqrt{2}} \sum_{j=1,2} f_j(x_+) g_j(x_-); \quad x_\pm = \sigma \mp i\tau \tag{2.4}$$

where f_j (resp. g_j), which are functions of a single variable, are solutions of the same Schrödinger equation

$$-f_j'' + T(x_+) f_j = 0, \quad (\text{ resp. } - g_j'' + \overline{T}(x_-) g_j). \tag{2.5}$$

The solutions are normalized such that their Wronskians $f_1' f_2 - f_1 f_2'$ and $g_1' g_2 - g_1 g_2'$ are equal to one. The proof goes as follows.

1) First check that (2.4) is indeed solution. Taking the Laplacian of the logarithm of the right-hand side gives

$$\frac{\partial^2 \Phi}{\partial \sigma^2} + \frac{\partial^2 \Phi}{\partial \tau^2} \equiv 4\partial_+ \partial_- \Phi = -4 \Big/ \Big(\sum_{i=1,2} f_i g_i \Big)^2$$

where $\partial_\pm = (\partial/\partial\sigma \pm i\partial/\partial\tau)/2$. The numerator has been simplified by means of the Wronskian condition. This is equivalent to (2.3).

2) Conversely check that any solution of (2.3) may be put under the form (2.4). If (2.3) holds one deduces

$$\partial_\mp T^{(\pm)} = 0; \quad \text{with } T^{(\pm)} := e^{\Phi} \partial_\mp^2 e^{-\Phi} \tag{2.6}$$

$T^{(\pm)}$ are thus functions of a single variable. Next the equation involving $T^{(+)}$ may be rewritten as

$$(-\partial_+^2 + T^{(+)}) e^{-\Phi} = 0 \tag{2.7}$$

[1] *The factor i means that these solutions should be considered in Minkowsky space-time*

with solution

$$e^{-\Phi} = \frac{i}{\sqrt{2}} \sum_{j=1,2} f_j(x_+) g_j(x_-); \quad \text{with } -f_j'' + T^{(+)} f_j = 0$$

where the g_j are arbitrary functions of x_-. Using the equation (2.5) that involves $T^{(-)}$, one finally derives the Schrödinger equation $-g_j'' + T^{(-)} g_j = 0$. Thus the theorem holds with $T = T^{(+)}$ and $\overline{T} = T^{(-)}$ □

One may deduce from (2.6) that the potentials of the two Schrödinger equations coincide with the two chiral components of the stress-energy tensor. **Thus these equations are the classical equivalent of the Ward identities that ensure the decoupling of Virasoro null vectors.** From the canonical Poisson brackets (P.B.) one finds two P.B. realizations of the Virasoro algebra such that the f_j's (resp. g_j's) are primary fields with weights $(-1/2, 0)$ (resp. $(0, -1/2)$). At the classical level it is trivial to compute powers of $e^{-\Phi}$:

$$e^{-N\Phi} = \left(\frac{i}{\sqrt{2}}\right)^N \sum_{p=0}^{N} \frac{N!}{p!(N-p)!} (f_1 g_1)^p (f_2 g_2)^{N-p} \tag{2.8}$$

which is primary with weight $(-N/2, -N/2)$. $e^{-N\Phi}$ is thus built up from powers of the solutions of the basic fields f_i and g_i. For positive N one has a finite number of terms but the weights are negative. Operators with positive weights have N negative so that (2.8) involves an infinite number of terms. Setting $N = -2$ gives weights (1,1) in agreement with the fact that the potential term of (2.2) is equal to $e^{2\Phi}$ which must be a marginal operator. An important point to keep in mind is that, a priori, any two pairs f_j and g_j of linearly independent solutions of (2.5) are suitable. In this connection one easily sees that (2.4) is left unchanged if f_j and g_j are replaced by $\sum_k M_{jk} f_k$ and $\sum_k (M^{-1})_{kj} g_k$, respectively, where M_{jk} is an arbitrary constant matrix. It is natural to choose the determinant of M to be equal to one in order to preserve the normalization of the Wronskians. Eq. (2.4) is a $sl(2,C)$-invariant with f_j transforming as a representation of spin 1/2. The higher powers (2.8) may be regarded in a similar way. The set of functions of x_+ which appear, that is $(f_1)^p (f_2)^{N-p}$, $0 \le p \le N$, transforms as a representation of spin $N/2$ under the above transformation of the f_j. This group structure will be replaced by a quantum group one when we turn to the quantum mechanical problem. This is natural, since the f's and the g's become non-commutative objects. At the classical level, the present viewpoint moreover shows that the definition of positive powers of the metric is connected with the problem of representation with negative spins (more about this below).

For the time being we shall concentrate on one of the two chiral components. Consider for instance the chiral components which are analytic functions of $z = \tau + i\sigma$. In a typical situation, σ and τ may be taken as coordinates of a cylinder obtained by conformal mapping from a particular handle of the Riemann surface considered. τ plays the role of imaginary time and σ is a space variable. One may work at $\tau = 0$ without loss of generality. The potential $T(\sigma)$ is periodic with period say 2π and we are working on the unit circle. Any two independent solutions of the Schrödinger equation is suitable.

It seems natural at first sight to diagonalize the monodromy matrix, that is to choose two solutions noted ψ_j, $j = 1, 2$, that are periodic up to a multiplicative constant[2]. It is convenient to introduce

$$\phi_j(\sigma) := \ln(\psi_j)/\sqrt{\gamma} - \ln d_j,$$

d_j are suitable normalization constants. The fields ϕ_j are periodic up to additive constants and have the expansion

$$\phi_j(\sigma) = q_0^{(j)} + p_0^{(j)}\sigma + i\sum_{n\neq 0} e^{-in\sigma}\, p_n^{(j)}/n, \quad j = 1, 2, \tag{2.9}$$

As shown in [11,12], the canonical P.B. structure of the action (2.2) is such that the chiral fields ϕ_j satisfy

$$\{\phi_1'(\sigma_1), \phi_1'(\sigma_2)\}_{\text{P.B.}} = \{\phi_2'(\sigma_1), \phi_2'(\sigma_2)\}_{\text{P.B.}} = 2\pi\,\delta'(\sigma_1 - \sigma_2), \tag{2.10}$$

$$\{q_0^{(j)}, p_0^{(j)}\}_{\text{P.B.}} = 1, \tag{2.11}$$

$$T/\gamma = (\phi_1')^2 + \phi_1''/\sqrt{\gamma} = (\phi_2')^2 + \phi_2''/\sqrt{\gamma}, \tag{2.12}$$

$$p_0^{(1)} = -p_0^{(2)}. \tag{2.13}$$

Equations (2.12) are trivial to derive from the Schrödinger equation (2.5). They are the associated Riccati equation. Eq. (2.13) follows from the fact that the product of the two eigenvalues of the monodromy matrix is equal to one, as a standard Wronskian argument shows.

At this point a parenthetical remark is in order. Equations (2.6) give

$$T^{(-)} = (\partial_-\Phi)^2 - \partial_-^2\Phi \tag{2.14}$$

that is very similar to (2.12). As a result, there is often a confusion, in the current literature, between ϕ and Φ. It should be stressed that the quadratic expressions (2.14) cannot be directly used to set up the canonical Hamiltonian formalism since it involves the second derivative of Φ with respect to τ, while a point in phase space is entirely determined by Φ and $\partial\Phi/\partial\tau$ at a given time. Indeed, in the canonical formalism, $\partial^2\Phi/\partial\tau^2$ is a dependent variable to be eliminated by using the field equations, before writing Hamilton's equations. When one does this, the potential term $\exp(2\Phi)$ reappears in (2.14) which is thus not trivially equivalent to a free-field expression. The equivalence is more involved and was just summarized. For the ϕ_j fields, the field equation is simply $\partial\phi_j/\partial\tau = i\partial\phi_j/\partial\sigma$ and (2.12) does remain quadratic when $\partial^2\phi_j/\partial\tau^2$ is eliminated.

In the language of field theory, the ϕ's are two equivalent free fields such that (2.12) takes the form of a U_1–Sugawara stress-tensor with a linear term. The latter is responsible for the classical Virasoro central charge $C_{class.} = 3/\gamma$. Clearly the two free fields play a symmetric role and one could as well build $e^{-\Phi}$ from different sets of Schrödinger

[2] *We assume that the monodromy matrix is diagonalizable.*

olutions. Such a possibility is at the origin of the quantum group action, as recalled bove. This is discussed in [2,4,5] and below.

The Poisson bracket of ϕ_1 with ϕ_2 is complicated and not very illuminating. We shall ome back to it below in a suitable form. It is simple to derive from (2.10) that

$$-\frac{1}{4\pi}\{\frac{T(\sigma)}{\gamma}, \psi_j(\sigma')\}_{\mathrm{P.B.}} = [\delta(\sigma-\sigma')\partial_{\sigma'} + \frac{1}{2}\delta'(\sigma-\sigma')]\, \psi_j(\sigma') \qquad (2.15)$$

which means that ψ_j is a primary field with weight $\Delta = -1/2$. More generally, a lassical primary field $\mathcal{A}_\Delta$ with weight Δ is such that

$$-\frac{1}{4\pi}\{\frac{T(\sigma)}{\gamma}, \mathcal{A}_\Delta(\sigma')\}_{\mathrm{P.B.}} = [\delta(\sigma-\sigma')\partial_{\sigma'} - \Delta\delta'(\sigma-\sigma')]\, \mathcal{A}_\Delta(\sigma'). \qquad (2.16)$$

This means that $\mathcal{A}_\Delta dz^\Delta$ is invariant by analytic mapping $z \to F(z)$. Moreover it also follows from (2.10) that

$$-\frac{1}{4\pi}\{\frac{T(\sigma)}{\gamma}, \frac{T(\sigma')}{\gamma}\}_{\mathrm{P.B.}} = [\delta(\sigma-\sigma')\partial_{\sigma'} - 2\delta'(\sigma-\sigma')]\frac{T(\sigma')}{\gamma} + \frac{\delta'''(\sigma-\sigma')}{\gamma} \qquad (2.17)$$

so that T satisfy a Poisson-bracket realization of the Virasoro algebra with central charge $C = 3/\gamma$ in standard notations.

The last formulae were recalled as a preparation for the following theorem. For $\sigma \neq \sigma'$, any primary field $\mathcal{A}_\Delta$ satisfies Poisson-bracket algebra of the form

$$\{\mathcal{A}_\Delta(\sigma), \psi_j(\sigma')\}_{\mathrm{P.B.}} = \sum_k C_k(\sigma,\epsilon)\, \psi_k(\sigma') \qquad (2.18)$$

where C_k depends upon the primary field $\mathcal{A}_\Delta$, ϵ is the sign of $\sigma - \sigma'$. The proof goes as follows. The Poisson bracket (2.16) vanishes for $\sigma \neq \sigma'$, and it follows from the Schrödinger equation (2.5) that

$$[-\partial^2_{\sigma'} + T(\sigma')]\,\{\mathcal{A}_\Delta(\sigma), \psi_j(\sigma')\}_{\mathrm{P.B.}} = 0 \qquad (2.19)$$

so that the left-member is a linear combination of ψ_1 and ψ_2 and (2.18) follows. Since the right-member of (2.17) is a sum of δ-functions at $\sigma = \sigma'$, the linear combination is different for $\sigma > \sigma'$ and $\sigma < \sigma'$.

Applying this result to the fields ψ themselves, one moreover sees, by reversing the role of σ and σ', that they satisfy Poisson bracket algebras of the form

$$\{\psi_j(\sigma), \psi_k(\sigma')\}_{\mathrm{P.B.}} = \sum_{lm} s_{jk}^{lm}(\epsilon)\, \psi_l(\sigma'), \psi_m(\sigma) \qquad (2.20)$$

where ϵ is the sign of $\sigma - \sigma'$. The relevance of this fact to our discussion is that this last relation is equivalent[3] to the Poisson-bracket relation which is the

[3] *After taking linear combinations that are the classical analogues of the ξ fields of [1,2] and will be discussed in the coming section.*

limit of the corresponding braiding property of $U_q(sl(2))$ for $q \to 0$. This discussion may be generalized to the higher powers $(\psi_1)^\mu(\psi_2)^\nu$ which we rewrite as $\psi_m^{(J)}$ with $2J = \mu + \nu$, and $2m = \nu - \mu$. One may verify[14] that they are primary fields with weight $\Delta = -J/2$, and that, for $-J \le m \le J$, they satisfy a differential equation of order $2J+1$. As a result their Poisson bracket algebra is closed in a way similar to (2.20) and again corresponds to the Poisson-bracket relations which are the limits of the $U_q(sl(2))$ quantum-group braiding-structure. In this limit, one recovers the standard $sl(2)$ algebra, which is precisely the one mentioned above that leaves the Φ field invariant. What is its geometrical meaning ? A well known fact is that the functions f_j and g_j may be written as $f_1 = A/\sqrt{A'}$, $f_2 = 1/\sqrt{A'}$, $g_1 = B/\sqrt{B'}$, $g_2 = 1/\sqrt{B'}$. Thus one has

$$e^{2\Phi} \propto \frac{A'B'}{(A-B)^2} \tag{2.21}$$

A (resp B) is a function of z (of z^*) and primes denote derivatives. Since Φ is real one may choose $B = A^*$. Perform the conformal mapping $Z = A(z)$ the metric becomes

$$e^{2\Phi} \propto \frac{1}{(Z+Z^*)^2} \tag{2.22}$$

Changing f_j into $\sum_k M_{jk} f_k$, with $M = \begin{pmatrix} a & b \\ c & d \end{pmatrix}$, corresponds to the Möbius transformation $Z \to (aZ+b)/(cZ+d)$. Thus the $sl(2)$ group describes translations, dilatations and inversions in the local coordinate-system where the metric becomes equivalent to the one of Poincaré-Lobatchevski. This group will be replaced by a quantum group when we turn to the quantum mechanical problem. This is natural, since we have seen that the canonical Poisson brackets are the classical limit of the associated braid relations. The above Z variable thus becomes non-commutative.

Before leaving the classical problem, it is worth pointing out another interpretation of the existence of the two free fields we just recalled. This is related with the Drinfeld-Sokolov[17] Hamiltonian reduction from the affine Kac-Moody algebra $sl(2)^A$ to the Virasoro algebra. In standard notations, one imposes the constraint $J_- = 1$. After reduction, the currents have different conformal weights, that is, 0 for J_-, 1 for J_0, and 2 for J_+. This type of operators may be found in the present scheme as follows: Eq. (2.12) trivially gives:

$$\frac{T}{\gamma} = \frac{1}{2}\left[(\phi_1')^2 + \frac{\phi_1''}{\sqrt{\gamma}} + (\phi_2')^2 + \frac{\phi_2''}{\sqrt{\gamma}}\right] \equiv \frac{1}{2}\left[(J_0)^2 + J_+J_- + \frac{J_0'}{\sqrt{\gamma}}\right], \tag{2.23}$$

where we have let

$$J_- = 1, \quad J_0 \equiv \phi_1' + \phi_2', \quad J_+ = -2\phi_1'\phi_2' \tag{2.24}$$

One arrives at a deformed $SU(2)$-Sugawara expression for T, and the relation between the ϕ_j fields and the current does give the correct spectrum of conformal weights, up to central terms. In this way, one should be able to regard the present discussion as coming from a WZW model by Hamiltonian reduction with a special choice of gauge.

3. The Quantum (Group) Structure

Let us now come to the quantum case. The basic point of the method[12,13,15] is to quantize the above classical structure in such a way that the conformal structure is maintained. In particular powers of the metric tensor must be primary fields. This is ensured by the following result of [12]:

On the unit circle, $z = e^{i\sigma}$, and for generic γ, there exist two equivalent free fields:

$$\phi_j(\sigma) = q_0^{(j)} + p_0^{(j)}\sigma + i\sum_{n\neq 0} e^{-in\sigma}\, p_n^{(j)}/n, \quad j=1,\,2, \tag{3.1}$$

such that

$$\left[\phi_1'(\sigma_1), \phi_1'(\sigma_2)\right] = \left[\phi_2'(\sigma_1), \phi_2'(\sigma_2)\right] = 2\pi i\, \delta'(\sigma_1-\sigma_2), \qquad p_0^{(1)} = -p_0^{(2)}, \tag{3.2}$$

$$N^{(1)}(\phi_1')^2 + \phi_1''/\sqrt{\gamma} = N^{(2)}(\phi_2')^2 + \phi_2''/\sqrt{\gamma}. \tag{3.3}$$

$N^{(1)}$ (resp. $N^{(2)}$) denote normal orderings with respect to the modes of ϕ_1 (resp. of ϕ_2). Eq.(3.3) defines the stress-energy tensor and the coupling constant γ of the quantum theory. The former generates a representation of the V with central charge $C_{Liou} = 3 + 1/\gamma$.

The chiral family is built up[2,8,13,14] from the following operators

$$\psi_j = d_j\, N^{(j)}\big(e^{\sqrt{h/2\pi}\,\phi_j}\big), \quad \widehat{\psi}_j = \widehat{d}_j\, N^{(j)}\big(e^{\sqrt{\widehat{h}/2\pi}\,\phi_j}\big), \quad j=1,\,2, \tag{3.4}$$

$$h = \frac{\pi}{12}\Big(C_{Liou} - 13 - \sqrt{(C_{Liou}-25)(C_{Liou}-1)}\Big),$$
$$\widehat{h} = \frac{\pi}{12}\Big(C_{Liou} - 13 + \sqrt{(C_{Liou}-25)(C_{Liou}-1)}\Big), \tag{3.5}$$

where d_j and $\widehat{d}_j$ are normalization constants. They are determined as solutions of the equations

$$-\frac{d^2\psi_j(\sigma)}{d\sigma^2} + \big(\frac{h}{\pi}\big)\Big(\sum_{n<0} L_n\, e^{-in\sigma} + \frac{L_0}{2} + \big(\frac{h}{16\pi} - \frac{C-1}{24}\big)\Big)\psi_j(\sigma) + \big(\frac{h}{\pi}\big)\psi_j(\sigma)\Big(\sum_{n>0} L_n\, e^{-in\sigma} + \frac{L_0}{2}\Big) = 0, \tag{3.6}$$

$$-\frac{d^2\widehat{\psi}_j(\sigma)}{d\sigma^2} + \big(\frac{\widehat{h}}{\pi}\big)\Big(\sum_{n<0} L_n\, e^{-in\sigma} + \frac{L_0}{2} + \big(\frac{h}{16\pi} - \frac{C-1}{24}\big)\Big)\widehat{\psi}_j(\sigma) + \big(\frac{\widehat{h}}{\pi}\big)\widehat{\psi}_j(\sigma)\Big(\sum_{n>0} L_n\, e^{-in\sigma} + \frac{L_0}{2}\Big) = 0. \tag{3.7}$$

These are operator Schrödinger equations equivalent to the decoupling of Virasoro null-vectors[8,13.14]. They are the quantum versions of Eq.(2.5). Since there are two possible quantum modifications h and $\widehat{h}$, there are four solutions. By operator product ψ_j, $j=1, 2$, and $\widehat{\psi}_j$, $j=1, 2$, generate two infinite families of chiral fields which are denoted $\psi_m^{(J)}$, $-J \le m \le J$, and $\widehat{\psi}_{\widehat{m}}^{(\widehat{J})}$, $-\widehat{J} \le \widehat{m} \le \widehat{J}$; respectively, with $\psi_{-1/2}^{(1/2)} = \psi_1$, $\psi_{1/2}^{(1/2)} = \psi_2$, and $\widehat{\psi}_{-1/2}^{(1/2)} = \widehat{\psi}_1$, $\widehat{\psi}_{1/2}^{(1/2)} = \widehat{\psi}_2$. An easy computation shows that the standard screening charges $-\alpha_\pm$ are such that

$$\alpha_- = \sqrt{\frac{2h}{\pi}}, \quad \alpha_+ = \sqrt{\frac{2\widehat{h}}{\pi}} \tag{3.8}$$

$\psi_m^{(J)}$, $\widehat{\psi}_{\widehat{m}}^{(\widehat{J})}$, are of the type $(1, 2J+1)$ and $(2\widehat{J}+1, 1)$, respectively, in the BPZ classification. For the zero-modes, it is simpler[2] to define the rescaled variables

$$\varpi = ip_0^{(1)}\sqrt{\frac{2\pi}{h}}, \quad \widehat{\varpi} = ip_0^{(1)}\sqrt{\frac{2\pi}{\widehat{h}}}, \quad \widehat{\varpi} = \varpi\,\frac{h}{\pi}, \quad \varpi = \widehat{\varpi}\,\frac{\widehat{h}}{\pi}. \tag{3.9}$$

At this point a pedagogical parenthesis may be in order: the hatted and unhatted ψ fields have the same chirality; if we go to $\tau \neq 0$ they are both functions of x_-; there are two counterparts $\overline{\psi}_m^{(J)}(x_+)$ and $\widehat{\overline{\psi}}_m^{(J)}(x_+)$ which may be discussed in essentially the same way. Returning to our main line we recall that the Hilbert space in which the operators ψ and $\widehat{\psi}$ live, is a direct sum[2,3,4,5,15] of Fock spaces $\mathcal{F}(\varpi)$ spanned by the harmonic excitations of highest-weight Virasoro states denoted $|\varpi, 0>$. They are eigenstates of the quasi momentum ϖ, and satisfy $L_n|\varpi,0>=0$, $n>0$; $(L_0 - \Delta(\varpi))|\varpi,0> = 0$. The corresponding highest weights $\Delta(\varpi)$ may be rewritten as

$$\Delta(\varpi) \equiv \frac{1}{8\gamma} + \frac{(p_0^{(1)})^2}{2} = \frac{h}{4\pi}(1+\frac{\pi}{h})^2 - \frac{h}{4\pi}\varpi^2. \tag{3.10}$$

The commutation relations (3.2) are to be supplemented by the zero mode ones:

$$\left[q_0^{(1)}, p_0^{(1)}\right] = \left[q_0^{(2)}, p_0^{(2)}\right] = i.$$

The fields ψ and $\widehat{\psi}$ shift the quasi momentum $p_0^{(1)} = -p_0^{(2)}$ by a fixed amount. For an arbitrary c-number function f one has

$$\psi_m^{(J)}\, f(\varpi) = f(\varpi + 2m)\,\psi_m^{(J)}, \quad \widehat{\psi}_{\widehat{m}}^{(\widehat{J})}\, f(\varpi) = f(\varpi + 2\widehat{m}\,\pi/h)\,\widehat{\psi}_{\widehat{m}}^{(\widehat{J})}. \tag{3.11}$$

The fields ψ and $\widehat{\psi}$ together with their products may be naturally restricted to discrete values of ϖ. They thus live in Hilbert spaces[4] of the form

$$\mathcal{H}(\varpi_0) \equiv \bigoplus_{n,\widehat{n}=-\infty}^{+\infty} \mathcal{F}(\varpi_0 + n + \widehat{n}\,\pi/h). \tag{3.12}$$

[4] *Mathematically they are not really Hilbert spaces since their metrics are not positive definite.*

ϖ_0 is a constant which is arbitrary so far. The $sl(2,C)$–invariant vacuum corresponds to $\varpi_0 = 1 + \pi/h$,[2] but other choices are also appropriate, as we shall see.

At the quantum level, one makes use of the above chiral conformal family, since the quantum field equation is likely to imply that the quantum Schrödinger equation holds for each chiral component. Associated with each quantum modification one finds a quantum version of (2.4). Since (3.4) involves h or $\widehat{h}$ instead of γ, it should be considered as defining different powers of the metric than in the classical case. In order to agree with standard notations, we write them as $\exp(-\alpha_\pm \Phi/2)$. For $\gamma \to 0$, $\alpha_- \sim 2\sqrt{\gamma}$, and $\exp(-\alpha_- \Phi/2) \sim \exp(-\sqrt{\gamma}\Phi)$. This does give back the classical metric field of section 2 since we have chosen a quantum definition that agrees with the conventional one, but corresponds to rescaling our classical field by $\sqrt{\gamma}$. By short-distance operator-product expansion, one generates the set of fields $\exp[-(J\alpha_- + \widehat{J}\alpha_-)\Phi]$. From the quantum-group viewpoint, J and $\widehat{J}$ are the spins of the representations. Thus one sees that the cosmological term, that is $\exp[\alpha_- \Phi]$ corresponds to $J = -1$! We shall come back to this below. The basic point of introducing the two normal orderings $N^{(i)}$, $i = 1, 2$, was to obtain a conformal regularization of the metric tensor operators $\exp(-\alpha_\pm \Phi/2)$. In terms of the Liouville field Φ, it is rather involved and field dependent. For γ going to zero, α_+ blows up. Thus if one wants to keep a smooth classical limit, only α_- should appear. This is possible with open boundary condition [6,13]. With closed boundary conditions, both α's should be kept in order to couple rational theories with gravity [15].

In any case, the quantum modifications are real only if $C_{Liou} > 25$ or $C_{Liou} < 1$ (The latter case describes 2D matter by analytic continuation of 2D gravity). Thus the construction of the metric tensor operator just recalled fails for $1 < C_{Liou} < 25$, which is the region of strongly coupled gravity. The chiral families may be continued, however, and this is taken to be the way to deal with 2D gravity in the strong coupling regime, if a consistent truncation may be found as shown in [4,5].

Next we display the quantum-group structure of the chiral fields. The operators ψ and $\widehat{\psi}$ are closed under O.P.E. and braiding. Each family obeys a quantum group symmetry of the $U_q(sl(2))$ type. However, the fusion coefficients and R–matrix elements depend upon ϖ and thus do not commute with the ψ's and $\widehat{\psi}$'s. Their explicit form is unusual, therefore. One may exhibit the standard $U_q(sl(2))$-quantum-group structure by changing basis to new families. Following my recent work,[2] let us introduce

$$\xi_M^{(J)}(\sigma) := \sum_{-J \le m \le J} |J,\varpi)_M^m \, \psi_m^{(J)}(\sigma), \quad -J \le M \le J; \tag{3.13}$$

$$|J,\varpi)_M^m = \sqrt{\binom{2J}{J+M}}\; e^{ihm/2} \times$$

$$\sum_{\left(\frac{J-M+m-t}{2}\right)\ \text{integer}} e^{iht(\varpi+m)} \binom{J-M}{(J-M+m-t)/2} \binom{J+M}{(J+M+m+t)/2};$$

$$\binom{P}{Q} \equiv \frac{\lfloor P \rfloor !}{\lfloor Q \rfloor ! \lfloor P-Q \rfloor !} \quad \lfloor n \rfloor ! \equiv \prod_{r=1}^{n} \lfloor r \rfloor \quad \lfloor r \rfloor \equiv \frac{\sin(hr)}{\sin h}. \tag{3.14}$$

The last equation introduces q–deformed factorials and binomial coefficients. The other fields $\widehat{\xi}^{(\widehat{J})}_{\widehat{M}}$ are defined in exactly the same way replacing h by $\widehat{h}$ everywhere. The symbols are the same with hats, e.g.

$$\lfloor \hat{n} \rfloor! \equiv \prod_{r=1}^{n} \lfloor \hat{r} \rfloor, \qquad \lfloor \hat{r} \rfloor \equiv \frac{\sin(\widehat{h}r)}{\sin \widehat{h}}, \qquad \text{and so on.} \tag{3.15}$$

The above transformation may be explicitly inverted [5]. One has

$$\psi_m^{(J)}(\sigma) = \sum_{M=-J}^{J} \xi_M^{(J)}(\sigma)\,(J,\varpi|_m^M, \tag{3.16}$$

$$(J,\varpi|_m^M := (-1)^{J+M}\, e^{ih(J+M)}\, |J,\varpi)_{-M}^{-m} \Big/ C_{-m}^{(J)}(\varpi), \tag{3.17}$$

$$C_m^{(J)}(\varpi) := (-1)^{J-m}\,(2i\sin h)^{2J} e^{ihJ} \frac{\binom{2J}{J-m}\lfloor \varpi - J + m \rfloor_{2J+1}}{\lfloor \varpi + 2m \rfloor}. \tag{3.18}$$

The quantum group structure of the exchange algebra is exhibited by introducing group theoretic states $|J,M>$, $-J \le M \le J$ and operators $J_\pm$, J_3 such that

$$J_\pm |J,M> = \sqrt{\lfloor J \mp M \rfloor \lfloor J \pm M + 1 \rfloor}\, |J, M \pm 1>, \quad J_3 |J,M> = M\, |J,M>. \tag{3.19}$$

These operators satisfy the $U_q(sl(2))$ - commutation relations

$$\Big[J_+, J_-\Big] = \lfloor 2J_3 \rfloor, \quad \Big[J_3, J_\pm\Big] = \pm J_\pm. \tag{3.20}$$

In [2,5] the operator algebra of the ξ fields was completely determined.
1) For $\pi > \sigma > \sigma' > 0$, these operators obey the exchange algebra

$$\xi_M^{(J)}(\sigma)\,\xi_{M'}^{(J')}(\sigma') = \sum_{-J \le N \le J;\, -J' \le N' \le J'} (J,J')_{M\,M'}^{N'\,N}\, \xi_{N'}^{(J')}(\sigma')\,\xi_N^{(J)}(\sigma), \tag{3.21}$$

$$(J,J')_{M\,M'}^{N'\,N} = \Big(< J,M| \otimes < J',M'|\Big)\, \mathbf{R}\, \Big(|J,N> \otimes |J',N'>\Big), \tag{3.22}$$

$$\mathbf{R} = e^{(-2ihJ_3 \otimes J_3)} \Big(1 + \sum_{n=1}^{\infty} \frac{(1-e^{2ih})^n\, e^{ihn(n-1)/2}}{\lfloor n \rfloor!} e^{-ihnJ_3}(J_+)^n \otimes e^{ihnJ_3}(J_-)^n\Big). \tag{3.23}$$

$\mathbf{R}$ coincides with the universal R matrix of $U_q(sl(2))$.
2) For $0 < \sigma < \sigma' < \pi$, the ξ fields obey the exchange algebra

$$\xi_M^{(J)}(\sigma)\,\xi_{M'}^{(J')}(\sigma') = \sum_{-J \le N \le J;\, -J' \le N' \le J'} \overline{(J,J')}_{M\,M'}^{N'\,N}\, \xi_{N'}^{(J')}(\sigma')\,\xi_N^{(J)}(\sigma), \tag{3.24}$$

$$\overline{(J,J')}_{M\,M'}^{N'\,N} = \Big(< J,M| \otimes < J',M'|\Big)\, \overline{\mathbf{R}}\, \Big(|J,N> \otimes |J',N'>\Big), \tag{3.25}$$

$$\overline{\mathbf{R}} = e^{(2ihJ_3\otimes J_3)}\Big(1+\sum_{n=1}^{\infty}\frac{(1-e^{-2ih})^n\, e^{-ihn(n-1)/2}}{\lfloor n\rfloor!}e^{-ihnJ_3}(J_-)^n\otimes e^{ihnJ_3}(J_+)^n\Big). \quad (3.26)$$

3) The two exchange formulae are related by the inverse relation

$$\sum_{-J\le N\le J;\,-J'\le N'\le J'} (J,J')_{M\,M'}^{N'\,N}\,\overline{(J',J)}_{N'\,N}^{P\,P'} = \delta_{M,P}\,\delta_{M',P'}. \quad (3.27)$$

4) The short-distance operator-product expansion of the ξ fields is of the form:

$$\xi_{M_1}^{(J_1)}(\sigma)\,\xi_{M_2}^{(J_2)}(\sigma') = \sum_{J=|J_1-J_2|}^{J_1+J_2}\Big\{\big(d(\sigma-\sigma')\big)^{\Delta(J)-\Delta(J_1)-\Delta(J_2)}$$

$$(J_1,M_1;J_2,M_2|J_1,J_2;J,M_1+M_2)\,\Big(\xi_{M_1+M_2}^{(J)}(\sigma)+\text{descendants}\Big)\Big\}, \quad (3.28)$$

where $d(\sigma-\sigma')\equiv 1-e^{-i(\sigma-\sigma')}$, $(J_1,M_1;J_2,M_2|J_1,J_2;J,M_1+M_2)$ denotes the Clebsch-Gordan coefficients of $U_q(sl(2))$, and $\Delta(J):=-hJ(J+1)/\pi-J$ is the Virasoro-weight of $\xi_M^{(J)}(\sigma)$.

5) Define the quantum group action on the ξ fields by

$$J_3\,\xi_M^{(J)} = M\xi_M^{(J)},\quad J_\pm\,\xi_M^{(J)} = \sqrt{\lfloor J\mp M\rfloor\lfloor J\pm M+1\rfloor}\,\xi_{M\pm1}^{(J)}. \quad (3.29)$$

Then the operator-product $\xi_{M_1}^{(J_1)}(\sigma)\xi_{M_2}^{(J_2)}(\sigma')$ gives a representation of the quantum group algebra (3.20) with the co-product generators

$$\mathbf{J}_\pm := J_\pm\otimes e^{ihJ_3}+e^{-ihJ_3}\otimes J_\pm,\quad \mathbf{J}_3 := J_3\otimes 1+1\otimes J_3, \quad (3.30)$$

where the tensor product is defined so that

$$(A\otimes B)\Big(\xi_{M_1}^{(J_1)}(\sigma)\,\xi_{M_2}^{(J_2)}(\sigma')\Big) := \big(A\xi_{M_1}^{(J_1)}(\sigma)\big)\,\big(B\xi_{M_2}^{(J_2)}(\sigma')\big), \quad (3.31)$$

and where each term in the expansion over J transforms according to a representation of spin J.

Similar formulae hold in the other half circle. Eq. (3.30) coincides with the standard co-product, and is thus non-symmetric, the two definitions being related by the universal R matrix. The exchange properties of the ξ fields (Eqs (3.21) – (3.26)) show that their quantum mechanical structure precisely matches this asymmetry, so that the transformation law (3.29) is fully consistent with the operator-algebra.

Obviously the same structure holds for the hatted fields. One replaces h by $\widehat{h}$ everywhere. Moreover the hatted and unhatted fields have simple braiding and fusions [2]. The most general $(2\widehat{J}+1,2J+1)$ field $\xi_{M\,\widehat{M}}^{(J\,\widehat{J})}\sim\xi_M^{(J)}\,\widehat{\xi}_{\widehat{M}}^{(\widehat{J})}$ has weight

$$\Delta_{Kac}(J;\widehat{J};C_{Liou}) = \frac{C_{Liou}-1}{24} - \frac{1}{24}\Big((J+\widehat{J}+1)\sqrt{C_{Liou}-1}-(J-\widehat{J})\sqrt{C_{Liou}-25}\Big)^2, \quad (3.32)$$

in agreement with Kac's formula.

4. The Case of Real Screening Charges

If $C_{Liou} > 25$ the screening charges $\alpha_\pm$ are real. This is the weak coupling regime which is connected with the classical limit ($\gamma \to 0$). In this region h and $\widehat{h}$ are real and the structure recalled above is directly handy. Let us briefly discuss how the powers of the metric are reconstructed. Consider for instance $\exp(-J\alpha_-\Phi)$. There are two types of cases one may distinguish.
1) One may consider, as is most usual, closed surfaces without boundary [24]. Then the natural region is the whole circle $0 \leq \sigma \leq 2\pi$. We are aiming at the quantum version of Eq.(2.8). It will involve the fields $\xi_M^{(J)}$, together with their counterparts $\overline{\xi}_M^{(J)}(x_+)$ whose exchange properties are similar. Concerning the latter one should remember that they are functions of z^*, that is, are anti-analytic functions. so that the orientation of the complex plane is reversed. This may be taken into account simply by replacing i by $-i$ in the above formulae for the ξ-fields, that is by taking the complex conjugate of all the c-numbers **without taking the Hermitian conjugate of the operators.** The appropriate definition of $\overline{\xi}_M^{(J)}(\sigma)$ is

$$\overline{\xi}_M^{(J)}(\sigma) := \sum_{-J\leq m\leq J} \left(|J,\varpi)_M^m\right)^* \overline{\psi}_m^{(J)}(\sigma), \quad -J \leq M \leq J; \tag{4.1}$$

where $\left(|J,\varpi)_M^m\right)^*$ is the complex conjugate of $|J,\varpi)_M^m$. In addition to taking the complex conjugate of $|J,\varpi)_M^m$, we have introduced an additional phase factor for later convenience. One may see that it does not change the braiding and fusion properties, up to overall normalizations. Concerning braiding, for instance, this is true because $(J_1, J_2)_{M_1,M_2}^{P_1,P_2}$ is nonzero only if $P_1 + P_2 = M_1 + M_2$. As is usual in conformally invariant field theory we assume that the right- and left- movers commute. Thus we take the ξ-fields to commute with the $\overline{\xi}$-fields.

There are two basic requirements that determine $\exp(-J\alpha_-\Phi)$. The first one is locality, that is, that it commutes with any other power of the metric at equal τ. The second one concerns the Hilbert space of states where the physical operator algebra is realized. The point is that, since we took the fields $\xi_M^{(J)}$ and $\overline{\xi}_M^{(J)}$ to commute, the quasi momenta ϖ and $\overline{\varpi}$ of the left- and right-movers are unrelated, while periodicity in σ requires that they be equal. This last condition is replaced by the requirement that $\exp(-J\alpha_-\Phi)$ leave the subspace of states with $\varpi = \overline{\varpi}$ invariant. The latter condition defines the physical Hilbert space $\mathcal{H}_{phys}$ where it must be possible to restrict the operator-algebra consistently. At $\tau = 0$, the appropriate definition is [24]:

$$e^{-J\alpha_-\Phi(\sigma)} = c_J \sum_{M=-J}^{J} (-1)^{J-M}\, e^{ih(J-M)}\, \xi_M^{(J)}(\sigma)\,\overline{\xi}_{-M}^{(J)}(\sigma), \tag{4.2}$$

where c_J is a normalization constant. It is invariant under the quantum group action (3.29) if the $\overline{\xi}$ fields transform in the same way as the ξ fields:

$$J_3\,\overline{\xi}_M^{(J)} = M\overline{\xi}_M^{(J)}, \quad J_\pm\,\overline{\xi}_M^{(J)} = \sqrt{\lfloor J\mp M\rfloor\lfloor J\pm M+1\rfloor}\,\overline{\xi}_{M\pm1}^{(J)}, \tag{4.3}$$

and if we assume that $\xi_M^{(J)}(\sigma)\overline{\xi}_{-M}^{(J)}(\sigma)$ transforms by the appropriate co-product. Locality is checked by making use of Eqs (3.21) - (3.23) . Choose $\pi > \sigma > \sigma' > 0$. In agreement with the above discussion we have

$$\overline{\xi}_M^{(J)}(\sigma)\overline{\xi}_{M'}^{(J')}(\sigma') = \sum_{-J\leq N\leq J;\, -J'\leq N'\leq J'} \left((J,J')_{M\,M'}^{N'\,N}\right)^* \overline{\xi}_{N'}^{(J')}(\sigma')\overline{\xi}_N^{(J)}(\sigma), \quad (4.4)$$

In checking locality, one encounters the product of two R matrices. It is handled by means of the identities

$$\left((J_1,J_2)_{-M_1\,-M_2}^{-N_2\,-N_1}\right)^* = \left((J_2,J_1)_{-N_2\,-N_1}^{-M_1\,-M_2}\right)^* = \overline{(J_2,J_1)_{N_2\,N_1}^{M_1\,M_2}} \quad (4.5)$$

that follow from the explicit expressions (3.23) , (3.26). In this way one deduces the equation

$$\sum_{M_1 M_2} (J_1,J_2)_{M_1\,M_2}^{P_2\,P_1} \left((J_1,J_2)_{-M_1\,-M_2}^{-N_2\,-N_1}\right)^* = \delta_{P_1,N_1}\,\delta_{P_2,N_2} \quad (4.6)$$

from the inverse relation (3.27) , and the desired locality relation follows:

$$e^{-J_1\alpha_-\Phi(\sigma_1)}\, e^{-J_2\alpha_-\Phi(\sigma_2)} = e^{-J_2\alpha_-\Phi(\sigma_2)}\, e^{-J_1\alpha_-\Phi(\sigma_1)}. \quad (4.7)$$

On the other hand, Eq.(4.2) is such that $\mathcal{H}_{phys}$ is left invariant, as we show next. This is seen by re-expressing (4.2) in terms of ψ fields. One gets, at first

$$e^{-J\alpha_-\Phi(\sigma,\tau)} = c_J \sum_{M=-J}^{J} (-1)^{J-M}\, e^{ih(J-M)}\, |J,\varpi)_M^m \left(|J,\overline{\varpi})_{-M}^p\right)^* \psi_m^{(J)}(\sigma)\overline{\psi}_p^{(J)}(\sigma). \quad (4.8)$$

Using Eq.(3.22) of [2] one writes

$$\sum_{M=-J}^{J} (-1)^{J-M}\, e^{ih(J-M)}\, |J,\varpi)_M^m \left(|J,\overline{\varpi})_{-M}^p\right)^* =$$

$$\sum_{M=-J}^{J} (-1)^{J-M}\, e^{ih(J-M)}\, |J,\varpi)_M^m\, |J,\overline{\varpi}+2p)_{-M}^{-p}. \quad (4.9)$$

If $\varpi = \overline{\varpi}$, this becomes, according to (3.16,3.17),

$$\sum_{M=-J}^{J} (-1)^{J-M}\, e^{ih(J-M)}\, |J,\varpi)_M^m\, |J,\varpi+2p)_{-M}^{-p} = \delta_{m,p}\, C_m^{(J)}(\varpi). \quad (4.10)$$

As a consequence, and when it is restricted to $\mathcal{H}_{phys}$, Eq.(4.2) is equivalent to

$$e^{-J\alpha_-\Phi(\sigma)} = c_J \sum_{m=-J}^{J} C_m^{(J)}(\varpi)\, \psi_m^{(J)}(\sigma)\overline{\psi}_m^{(J)}(\sigma) \quad (4.11)$$

and the condition $\varpi = \overline{\varpi}$ is indeed left invariant, according to (3.11). See [24] for details.

2) One may also consider gravity with boundary, following [6,10,11]. A typical situation is the half circle $0 \leq \sigma \leq \pi$. One may set up boundary conditions such that the system remains conformal, albeit with one type of Virasoro generators only. The left- and right-movers become related as in the case for open strings. The appropriate definition of the metric becomes [6]:

$$e^{-J\alpha_-\Phi(\sigma)} = c_J \sum_{M,N} A^{(J)}_{M,N}\, \xi^{(J)}_M(\sigma)\, \xi^{(J)}_N(2\pi - \sigma) \tag{4.12}$$

where

$$A^{(J)}_{M,N} =< J, M| \left\{ e^{-ihJ_3^2} \sum_{r,s=0}^{\infty} e^{ih(r+s)J_3} \frac{(J_+)^{r+s}}{\lfloor r \rfloor! \lfloor s \rfloor!} e^{-2iha(r-s)} e^{-ihrs} \right\} |J, N> \tag{4.13}$$

where a depends upon the boundary condition chosen. These operators are mutually local and closed by fusion [6]. A similar structure appears for factorizable scattering on the half line [18,19].

So far the present discussion assumes that q is not a root of unity, that is, deals with irrational theories. The problem of specializing q to a root of unity has, however, been essentially reduced to the equivalent limit in the representation theory of $U_q(sl(2))$ which is a much studied problem. Clearly, the present discussion applies whenever the screening charges are real, so that it also describes the $C < 1$ models, as a continuation of the Liouville theory.

There still remains the difficulty already pointed out in the classical case, that positive powers of the metric are difficult to handle. This is a major difference between the proper region of the Liouville theory ($C_{Liou} > 1$) and the region of statistical models ($C < 1$), since (3.32) gives negative or complex weights for positive J and $\widehat{J}$ in the former case, so that one must deal with negative J or $\widehat{J}$. The above discussion must be continued to negative spins [4,5]. One may show that equation (3.14) is equivalent to

$$|J, \varpi)^M_m = \sqrt{\binom{2J}{J+M}} e^{ih(m/2+(\varpi+m)(J-M+m))}\, F_q(a, b; c; e^{-2ih(\varpi+m)}), \tag{4.14}$$

where $a = M - J$ (resp. $a = -M - J$), $b = -m - J$ (resp. $b = m - J$), $c = 1 + M - m$ (resp. $c = 1 - M + m$) for $M > m$ (resp. $M < m$) and $F_q(a, b; c; z)$ is a q-deformed— so called basic— hypergeometric function. The continuation to negative J is a direct consequence of Rodgers identity [20]

$$F_q(a, b; c; e^{-2ihu}) = (2i \sin h)^{c-a-b}\, e^{ihu(a+b-c)} \frac{\Gamma_q(u - (a+b-c-1)/2)}{\Gamma_q(u + (a+b-c+1)/2)} \times$$
$$F_q(c-a, c-b; c; e^{-2ihu}). \tag{4.15}$$

where Γ_q denotes the q-deformed gamma-function. Equations (4.14,4.15) give

$$|J, \varpi)^M_m = (2i \sin(h))^{1+2J} \binom{2J}{J+M} \frac{\Gamma_q(\varpi + m + J + 1)}{\Gamma_q(\varpi + m - J)} | - J - 1, \varpi)^M_m. \tag{4.16}$$

This exhibits a symmetry between J and $-J-1$ which is also shared by the R matrices and Clebsch-Gordan coefficients [4,5]. It is the basis for defining operators with negative J. The crucial point of (4.16) is to show that $\psi^{(-J-1)}_m$ and $\xi^{(-J-1)}_M$ are to be considered for $-J \leq m \leq J$, and $-J \leq M \leq J$, respectively.

5. Solving The Reality Problem of Strongly Coupled Gravity: The Unitary Truncation Theorem

Next, we consider the region $1 < C_{Liou} < 25$, which is relevant to the strong coupling regime of 2D gravity. In this case, h and $\widehat{h}$ are complex and $\widehat{h} = h^*$. We choose the imaginary part of h to be negative for definiteness. In the weak coupling regime of gravity, the solution of the conformal bootstrap we just outlined arised in a natural way from the chiral decomposition of the 2D metric tensor in the conformal gauge, that is by solving Liouville's equation. It is thus legitimate to study the strong coupling regime by continuing this chiral structure below $C_{Liou} = 25$. Complex numbers appear all over the place. However—in a way that is reminiscent of the truncations that give the minimal unitary models— for $C_{Liou} = 7, 13, 19$; there is a consistent truncation of the above general family down to a unitary theory involving operators with real Virasoro conformal weights only. The main points of this theorem [4,5] are briefly summarized next. The truncated family is as follows:

a) The physical Hilbert space. It is given by [3,4,5,9]:

$$\mathcal{H}_{phys} \equiv \bigoplus_{r=0}^{1-s} \mathcal{H}_{-}(\varpi_0^r) \equiv \bigoplus_{r=0}^{1-s} \bigoplus_{n=-\infty}^{\infty} \mathcal{F}(\varpi_{r,n}), \tag{5.1}$$

$$\varpi_{r,n} \equiv \varpi_0^r + n(1-\frac{\pi}{h}) \equiv \left(\frac{r}{2-s}+n\right)\left(1-\frac{\pi}{h}\right). \tag{5.2}$$

The integer s is such that the special values correspond to

$$C_{Liou} = 1 + 6(s+2), \quad s = 0, \pm 1; \quad h + \widehat{h} = s\pi. \tag{5.3}$$

$\Delta(\varpi_{r,n})$ is positive and in $\mathcal{H}_{phys}$ the representation of the Virasoro algebra is unitary. The torus partition function corresponds to compactification on a circle with radius $R = \sqrt{2(2-s)}$ (see [3]).

b) The restricted set of conformal weights. The truncated family only involves operators of the type $(2J+1, 2J+1)$ noted $\chi_-^{(J)}$ and $(2(-J-1)+1, 2J+1)$ noted $\chi_+^{(J)}$. Their Virasoro conformal weights [3,4,5,15] which are respectively given by

$$\Delta^-(J) = -\frac{C_{Liou}-1}{6} J(J+1), \quad \Delta^+(J) = 1 + \frac{25 - C_{Liou}}{6} J(J+1), \tag{5.4}$$

are real. $\Delta^-(J)$ is negative for all J (except for $J = -1/2$ where it becomes equal to $\Delta^+(-1/2) = (s+2)/4$). $\Delta^+(J)$ is always positive, and is larger than one if $J \neq -1/2$.

c) The truncated families: $\mathcal{A}^{\pm}_{phys}$ is the set of operators $\chi_\pm^{(J)}$, $J \geq 0$ of the form [4,5]

$$\chi_-^{(J)} = \sum_{M=-J}^{J} \kappa^{J-M} (-1)^{s(J-M)(J-M-1)/2} \xi_{M,-M}^{(J,J)}, \tag{5.5a}$$

$$\chi_+^{(J)} = \sum_{M=-J}^{J} \kappa^{J-M} (-1)^{s(J-M)(J-M-1)/2} \xi_{M,-M}^{(-J-1,J)}. \tag{5.5b}$$

κ will be defined below.

THE UNITARY TRUNCATION THEOREM:

For $C_{Liou} = 1+6(s+2)$, $s = 0, \pm 1$, *and when it acts on* $\mathcal{H}_{phys}$; *the set* A^{+}_{phys} *(resp.* A^{-}_{phys}*) of operators* $\chi^{(J)}_{+}$ *(resp.* $\chi^{(J)}_{-}$*) is closed by fusion and braiding, and only gives states that belong to* $\mathcal{H}_{phys}$.

PROOF

Conditions (5.2) and (5.3) are instrumental since they allow us to relate hatted and unhatted quantities. In particular, for N integer, one has

$$\lfloor \widehat{N} \rfloor = e^{-i(N-1)s\pi} \lfloor N \rfloor, \qquad \widehat{h}\widehat{\varpi}_{r,n} - h\varpi_{r,n} = \pi(r + n(2-s)). \tag{5.6}$$

The truncation was originally observed [3,9] in terms of ψ and $\widehat{\psi}$ fields, for the braiding of $\chi^{(1/2)}_{-}$ with itself. The expression of $\chi^{(1/2)}_{-}$ using ξ fields was written in [3]. From this one may derive the general formula for $\chi^{(J)}_{-}$ recursively: one deduces $\chi^{(J+1/2)}_{-}$ from $\chi^{(J)}_{-}$ by fusion with $\chi^{(1/2)}_{-}$ to leading order in the singularity, using (3.28). The form of $\chi^{(1/2)}_{+}$ may be inferred from the symmetry between J and $-J-1$ recalled above. The proof of the unitary truncation theorem relies on the following three special properties:

Theorem (1). *If (5.3) holds, the hatted and unhatted Clebsch-Gordan coefficients are related by*

$$\widehat{(J_1, M_1; J_2, M_2|J_1, J_2; J, M)} = (J_1, -M_1; J_2, -M_2|J_1, J_2; J, -M)\times$$
$$(-1)^{s\{(J_1-M_1)(J_2+M_2)+(J_1+J_2-J)(J_1+M_1+J_2-M_2)\}}(-1)^{J_1+J_2-J}. \tag{5.7}$$

Thus they satisfy the orthogonality relation

$$\sum_{M_1, M_2} \widehat{(J_1, -M_1; J_2, -M_2|J_1, J_2; \widehat{J}, -M)}(J_1, M_1; J_2, M_2|J_1, J_2; J, M)\times$$
$$(-1)^{s\{(J_1+M_1)(J_2-M_2)+(J_1+J_2-J)(J_1-M_1+J_2+M_2)\}} = (-1)^{J_1+J_2-J}\delta_{J\widehat{J}}. \tag{5.8}$$

Theorem (2). *If (5.3) holds, the hatted and unhatted R-matrices are related by the relation*

$$\widehat{(J_1, J_2)}^{N_2 N_1}_{M_1 M_2} = \overline{(J_2, J_1)}^{-M_1-M_2}_{-N_2-N_1} e^{is\pi\{(J_1-M_1)(J_2+M_2)-J_1(J_2+N_2)+N_1(J_2-N_2)\}}. \tag{5.9}$$

Theorem (3). *Introduce the operator*

$$\kappa := -e^{i(\widehat{h}\widehat{\varpi}-h\varpi)}e^{i(h-\widehat{h})/2}. \tag{5.10}$$

In $\mathcal{H}_{phys}$ *one has*

$$\kappa^{J+M} \widehat{|J, \widehat{\varpi})}^{m}_{M} = (-1)^{s(J+M)(J+M-1)/2+J+M}\times$$
$$e^{ih(J+M-m)}e^{is\pi(-J^2+m/2)}e^{-im(\widehat{h}\widehat{\varpi}-h\varpi)}|J, \varpi-2m)^{m}_{M}. \tag{5.11}$$

Closure by fusion and braiding are consequences of Theorems 1 and 2 respectively. Theorem 3 combined with relation (4.10) shows that the χ fields may be rewritten in terms of the ψ fields with only terms with $m+\widehat{m}=0$ appearing, so that $\mathcal{H}_{phys}$ is indeed left invariant. Details are given in [5].

6. Physical Aspects of The Strongly Coupled Gravity Theories

1) String theories

First, taking D free fields as worldsheet matter, [9,15,21,22] one sees that one may construct consistent string emission vertices if $D = 26 - C_{grav} = 19, 13, 7$. The mass squared of the emitted string ground state is $m^2 = 2(\Delta - 1)$, where Δ is the conformal weight of the 2D-gravity dressing-operator. Since an infinite number of tachyons is unacceptable, this selects the $\mathcal{A}^+_{phys}$ family with positive weights Δ^+. Bilal and I have already unravelled striking properties of the associated Liouville strings [21,22]. Remarkably, one finds that the spectrum of conformal weights, which is selected by the truncation theorem, automatically gives modular partition functions on the torus so that the associated string theories are consistent, at least up to one loop. The extension of the present discussion to $N = 1$ super-Liouville theory is in progress [23]. The main features of the corresponding Liouville superstrings may be predicted [21,22]. The possible dimensions are $D = 3, 5, 7$. The first two models have striking features: First their total number of degree of freedom, that is 4 and 6 coincide with two choices of space-time dimensions where classical Green-Schwarz actions may be written, but could not be quantized consistently. These two theories may be regarded as the correct quantum theories associated with these classical actions where Lorentz invariance is broken from $D+1$ to D dimensions. Indeed, the physical number of degrees of freedom— that is $D-1$— are equal to 2 and 4 which coincide with the real dimensions of the division algebras of complex numbers and quaternions, so that light-cone Green-Schwarz formalism exist. The $O(8)$ triality of the ten-dimensional model is replaced by those of $U(1) \otimes U(1)$ and $SU(2) \otimes SU(2) \otimes SU(2)$ respectively. Both theories are space-time supersymmetric once the appropriate GSO projections are performed.

The existence of a three-dimensional model raises the hope of verifying the long-standing conjecture of Polyakov that the 3D Ising model is equivalent to a string theory. Indeed under plausible assumptions, it has been possible [22] to obtain a relation between critical exponents that is exactly satisfied by the numerical studies of the 3D Ising model.

2)Two-dimensional critical systems

Clearly, $\mathcal{A}^+_{phys}$ is also selected if we consider the associated conformal theories by themselves, in order to avoid correlation functions that grow at very large distance. One may play the game of fractal gravity, since Eq.(5.4) shows that $\Delta^-(J,C) + \Delta^+(J,26-C) = 1$, and since the set of values 7, 13, 19 is left invariant by $C \to 26 - C$. One will have two copies of the theories discussed above. One describes gravity, and, following what happens for non-critical strings, one would make use of the family $\mathcal{A}^+_{phys}$. Then the other copy, would only involve $\mathcal{A}^-_{phys}$ and correspond to a non-unitary matter theory. It is a challenge to derive these models from the matrix approach to 2D gravity.

Finally, the truncation theorem holds for any integer s so that it applies to $C_{Liou} = 1$ $(s = -2)$, and $C_{Liou} = 25$ $(s = 2)$, as well as for $C_{Liou} < 1$ $(s < -2)$ and $C_{Liou} > 25$ $(s > 2)$.

References

[1] O. Babelon, Phys. Lett. **B215** (1988), 523.
[2] J.-L. Gervais, Commun. Math. Phys. **130** (1990), 257.
[3] J.-L. Gervais and B. Rostand, Nucl. Phys. **B346** (1990), 473.

[4] J.-L. Gervais, Phys. Lett. **B243** (1990), 85.

[5] J.-L. Gervais, *Solving the strongly coupled 2D gravity: unitary truncation and quantum group structure*, LPTENS preprint 90/13; Commun. Math. Phys. (to appear).

[6] E. Cremmer, J.-L. Gervais, *The quantum strip: Liouville theory for open strings*, LPTENS preprint 90/32; Commun. Math. Phys. (to appear).

[7] see A. Alekseev, L. Faddeev, M. Semenov-Tian-Shansky and A. Volkov, *The unravelling of the quantum group structure in the WZNW theory*, CERN preprint TH-5981/91 and refs therein.

[8] J.-L. Gervais and A. Neveu, Nucl. Phys. **B257[FS14]** (1985), 59.

[9] J.-L. Gervais and A. Neveu, Phys. Lett. **151B** (1985), 271.

[10] J.-L. Gervais and A. Neveu, Nucl. Phys. **B199** (1982), 59.

[11] J.-L. Gervais and A. Neveu, Nucl. Phys. **B202** (1982), 125.

[12] J.-L. Gervais and A. Neveu, Nucl. Phys. **B224** (1983), 329.

[13] J.-L. Gervais and A. Neveu, Nucl. Phys. **B238** (1984), 125; Nucl. Phys. **B238** (1984), 396.

[14] J.-L. Gervais and A. Neveu, Nucl. Phys. **B264** (1986), 557.

[15] J.-L. Gervais, *Liouville Superstrings*, Perspectives in string the proceedings of the Niels Bohr, World Scientific, Nordita Meeting, 1987; *DST workshop on particle physics-Superstring theory , proceedings of the I.I.T. Kanpur meeting*, World Scientific, 1987.

J.-L. Gervais, *Systematic approach to conformal theories, Nucl. Phys. B (Proc. Supp.)* **5B** (1988), 119–136.

A. Bilal and J.-L. Gervais, *Conformal theories with non-linearly-extended Virasoro symmetries and Lie-algebra classification*, Conference Proceedings: "Infinite dimensional Lie algebras and Lie groups" (V. Kac, Marseille, eds.), World-Scientific, 1988.

[16] E. Cremmer, J.-L. Gervais, Commun. Math. Phys. **134** (1990), 619.

[17] see, e.g. B. Feigin and E. Frenkel, *Quantization of the Drinfeld-Sokolov reduction*, Harvard preprint (to appear)Phys. Lett. B..

[18] E. Sklyanin, J.Phys. **A 21** (1988), 2375.

[19] I. Cherednik, Theor.Math.Phys. **61** (1984), 977.

[20] G. Andrews, Conference board of the math. sciences, Regional conference series in math., vol. 66, A.M.S..

[21] A. Bilal and J.-L. Gervais, Nucl. Phys. **B284** (1987), 397; Phys. Lett. **B187** (1987), 39; *for reviews see [15]*, Nucl. Phys. **B293** (1987), 1.

[22] A. Bilal and J.-L. Gervais, Nucl. Phys. **B295[FS21]** (1988), 277.

[23] J.-L. Gervais and B. Rostand, *On two-dimensional supergravity and the super Möbius group*, preprint LPTENS 91/3,; Commun. Math. Phys. (to appear).

[24] J.-L. Gervais, *Gravity matter couplings from Liouville theory*, LPTENS preprint (to appear).

LABORATOIRE DE PHYSIQUE THÉORIQUE DE L'ÉCOLE NORMALE SUPÉRIEURE, 24, RUE LHOMOND, 75231, PARIS, CEDEX 05 - FRANCE.

EXTENDED CHIRAL CONFORMAL THEORIES WITH A QUANTUM SYMMETRY

L.K.HADJIIVANOV, R.R.PAUNOV AND I.T.TODOROV

Institute for Nuclear Research and Nuclear Energy, Sofia

ABSTRACT. A properly extended chiral part of rational conformal field theory (RCFT) possesses an internal quantum symmetry. There exist self dual chiral Green functions defined up to an overall phase factor which transform under a 1-dimensional (unitary) representation of the braid group B_n. A B_n-invariant inner product is constructed in the space of quantum group invariants which allows to reproduce 2-dimensional monodromy free euclidean Green functions by pairing the correlation functions of the left and right sectors.

1. Introduction

Beginning with the work of Belavin, Polyakov and Zamolodchikov (BPZ) [1] the 2-dimensional euclidean Green functions of a RCFT are written as finite sums of products of (multivalued) analytic and antianalytic conformal blocks (for recent reviews see [2,3]). In the case of "diagonal theories" (of zero "spin" fields, i.e. of primary fields of equal "left" and "right" conformal weights) we can write, e.g. a 2-dimensional 4-point function in either of the two dual forms

$$\langle\sigma_1(z_1,\bar{z}_1)\sigma_2(z_2,\bar{z}_2)\sigma_3(z_3,\bar{z}_3)\sigma_4(z_4,\bar{z}_4)\rangle = \sum_\nu |S_\nu\begin{pmatrix}\nu_1 & \nu_2 \\ z_1 & z_2\end{pmatrix};\begin{pmatrix}\nu_3 & \nu_4 \\ z_3 & z_4\end{pmatrix}|^2 \tag{1.1s}$$

$$= \sum_\mu |U_\mu\begin{pmatrix}\nu_1 \\ z_1\end{pmatrix};\begin{pmatrix}\nu_2 & \nu_3 \\ z_2 & z_3\end{pmatrix};\begin{pmatrix}\nu_4 \\ z_4\end{pmatrix}|^2. \tag{1.1u}$$

Here ν_i, ν, μ are labels for the representation of the underlying chiral algebra. (In the simplest case of minimal conformal models [1] studied here ν_i is a label of the conformal weight which takes a finite number of admissible values for a given Virasoro central charge $c < 1$.) The "s-channel blocks" S_ν are characterized by the following factorization properties for small $z_{34} = z_3 - z_4$ (or z_{12}):

$$\lim_{\substack{z_{34}\to 0 \\ (z_{12}\to 0)}} \frac{S_\nu\begin{pmatrix}\nu_1 & \nu_2 \\ z_1 & z_2\end{pmatrix};\begin{pmatrix}\nu_3 & \nu_4 \\ z_3 & z_4\end{pmatrix}}{\langle 0|\phi(z_1,\nu_1)\phi(z_2,\nu_2)|\nu\rangle\langle\nu|\phi(z_3,\nu_3)\phi(z_4,\nu_4)|0\rangle} = 1. \tag{1.2}$$

Here $\phi(z,\nu)$ is a short hand for a chiral *vertex operator* (CVO) [4,5,6,3] of weight ν which is specified, in general, by also giving the labels λ and μ of its target space:

$$\phi(z,\nu) \Leftrightarrow \begin{pmatrix} & \mu & \\ \nu & & \lambda \end{pmatrix}_z : \mathcal{H}_\lambda \to \mathcal{H}_\mu. \tag{1.3}$$

The short hand can only be used without ambiguity for 2- and 3-point functions, since $\phi(z,\nu)$ maps, by definition, the vacuum sector $\mathcal{H}_0$ into $\mathcal{H}_\nu$, so that we have

$$\langle 0|\phi(z_1,\nu_1)\phi(z_2,\nu_2)|\nu\rangle = \langle 0| \begin{pmatrix} & 0 & \\ \nu_1 & & \nu_1 \end{pmatrix}_{z_1} \begin{pmatrix} & \nu_1 & \\ \nu_2 & & \nu \end{pmatrix}_{z_2} |\nu\rangle. \tag{1.4a}$$

$|\nu\rangle$ being a lowest weight vector in $\mathcal{H}_\nu$:

$$|\nu\rangle = \phi(0,\nu)|0\rangle = \begin{pmatrix} & \nu & \\ \nu & & 0 \end{pmatrix}_\nu |0\rangle \tag{1.4b}$$

We are assuming that the CVOs are primary with respect to the chiral algebra [1,2], so that, in particular,

$$\left[L_n, \begin{pmatrix} & \mu & \\ \nu & & \lambda \end{pmatrix}_z\right] = z^n \left\{z\frac{d}{dz} + (n+1)\Delta_\nu\right\} \begin{pmatrix} & \mu & \\ \nu & & \lambda \end{pmatrix}_z \tag{1.5a}$$

where L_n are the generators of the Virasoro algebra Vir and Δ_ν is the conformal dimension (or weight) corresponding to the label ν and a lowest weight vector $|\nu\rangle$ is characterized by the properties

$$L_n|\nu\rangle = 0 \quad \text{for} \quad n = 1,2,\ldots, L_0|\nu\rangle = \Delta_\nu|\nu\rangle, \quad \langle\nu|\nu\rangle = 1. \tag{1.5b}$$

A similar factorization formula — valid for small z_{23} (or z_{14}) — characterizes the "u-channel blocks" U_μ. S_ν and U_μ span the same finite dimensional space singled out by the fusion rules of the theory. Typically, this space consists of the Möbius invariant solutions of some partial differential equations (that reflect the presence of null vectors in the corresponding Verma modules of the local chiral algebra). Thus S_ν and U_μ are related by a linear transformation which defines the *fusion matrix* $F = F\begin{bmatrix} \nu_2 & \nu_3 \\ \nu_1 & \nu_4 \end{bmatrix}$

$$U_\mu \begin{pmatrix} \nu_1 & ; & \nu_2 & \nu_3 & ; & \nu_4 \\ z_1 & & z_2 & z_3 & & z_4 \end{pmatrix} = \sum_\nu F_{\mu\nu} S_\nu \begin{pmatrix} \nu_1 & \nu_2 & ; & \nu_3 & \nu_4 \\ z_1 & z_2 & & z_3 & z_4 \end{pmatrix}. \tag{1.6}$$

Eq. (1.1) tells us that F is an unitary matrix. More generally, the finite dimensional space of 4-point conformal blocks carries a unitary representation of the braid group B_4 which leaves the euclidean Green function (1.1) invariant. (Such "monodromy representations" of the braid group B_n were first exhibited in [4] for the case of an $\widehat{su}(2)$-current algebra model, and later in [5,7] for the case of minimal models. Similar concepts appear earlier [8] in the context of a Liouville string theory.

The fact that instead of having a single chiral field $\phi(z,\nu)$ we are led to deal with a collection of CVOs (1.3) (labelled by λ and μ) suggest that the chiral building blocks of a 2-dimensional CFT might be endowed with internal degrees of freedom which are hidden in the 2-dimensional local fields. (One could think of an analogy with colour regarded as a (global) gauge symmetry of the first kind. Quark fields, the analogues of CVOs, are endowed with a colour index while "observable" hadron fields are colour singlets.) The presence of a braid group (rather than permutation group) statistics for CVOs puts severe restrictions on the possible type of such an internal symmetry. It cannot be described by a conventional group action but may be related to a deformed *quantum universal enveloping* (QUE) algebra (see [9,10,11]) or even to a more general structure like [12] (depending on the model under consideration [13] or on the a priori assumptions — like Wightman positivity — imposed on the unobservable chiral part of a RCFT, cf.[14]).

Moreover, the requirement that the statistics matrix (to be defined below) for the chiral RCFT with an internal quantum symmetry gives rise to a 1-dimensional representation of the braid group practically fixes the type of internal symmetry we can have. Various aspects of the relation between RCFTs and QUE algebras have been previously studied in [3,11,13,15–27].

The main result of our work can be formulated as follows. Given the quantum numbers and arguments $\left({}_{z}\,{}^{\nu}\,{}_{m}\right)$ of the underlying fields (where ν stands for chiral and internal weight, z is the "world sheet" argument restricted by the form (1.5) of the generators of Vir, and m is an internal index like "magnetic quantum number" which takes a finite number of values) there exists a chiral Green function

$$G_n = G\begin{pmatrix} & \nu_1 & & & \nu_2 & & & & \nu_n & \\ z_1 & & m_1 & , & z_2 & & m_2 & ,\dots, & z_n & & m_n \end{pmatrix}$$

that is determined up to an overall (non-zero) complex factor by some natural requirements which we proceed to formulate in an informal manner:

(i) Factorization properties; simplest example: if $\nu_{n-1} = \nu_n = \nu$

$$\begin{aligned} &\lim_{z_{n-1\,n}\to 0} z_{n-1\,n}^{2\Delta\nu} G_n \begin{pmatrix} & \nu_1 & & & & \nu & & & \nu & \\ z_1 & & m_1 & ,\dots, & z_{n-1} & & m & , & z_n & & m' \end{pmatrix} = \\ &= G_{n-2}\begin{pmatrix} & \nu_1 & & & & \nu_{n-2} & \\ z_1 & & m_1 & ,\dots, & z_{n-2} & & m_{n-2} \end{pmatrix} G_2\begin{pmatrix} \nu & & \nu \\ 1m & , & 0m' \end{pmatrix}. \end{aligned} \tag{1.7}$$

(ii) Braid group invariance; in the simplest case, for $\nu_1 = \nu_2 = \dots = \nu_n$ it says that G_n transforms under a 1-dimensional unitary representation of B_n.

(iii) There is a B_n-invariant sesquilinear form $(\bar{G}, G)$ on the space of QUE invariant tensors (defined, essentially, by a contraction over the indices $m_1,\dots,m_n$) which reproduces the monodromy free euclidean Green functions of the 2-dimensional RCFT.

The price for the first two properties is the necessity to include some non-unitary representations of the chiral algebra, which give vanishing contribution in the sesquilinear form of requirement (iii).

In this first announcement of our results we do not attempt to either formulate or prove them in full generality, but treat instead in detail a simple example: the 4-point

Green function involving a pair of step operators for a part of a minimal model [1] that is closed under operator product expansions (a p-model, see Sec.2). We study the conformal and quantum group properties of our models in Secs.2–3 and construct the braid covariant Green function of the combined theory in Sec.4.

2. Correlation functions and fusion relations in a chiral p-model

2A. The model. Fusion rules. 4-point blocks. We define a *chiral p-model* as a CFT whose superselection sectors (in the sense of the algebraic approach of Haag, Kastler, Doplicher and Roberts — see [28–30] and references therein) give rise to positive energy representations of Vir of central charge

$$c(p) = 1 - \frac{6}{p(p-1)}, \quad p = 4, 5, \ldots \tag{2.1}$$

and conformal weights

$$\Delta_\nu = \frac{\nu}{2}\lambda_\nu, \quad \lambda_\nu = \Delta_\nu + \Delta_1 - \Delta_{\nu-1} = \left(\frac{\nu}{2}+1\right) r - 1, \tag{2.2a}$$

$$r = 1 - \frac{1}{p}, \quad \nu = 0, 1, 2, \ldots \; . \tag{2.2b}$$

This set contains as a subset the unitary (irreducible) representations corresponding to $\nu \leq p-2$, a subset, that is closed under the BPZ fusion rule

$$\lambda \times \mu = \sum_{\substack{\nu=|\lambda-\mu| \\ \lambda+\mu-\nu \text{ even}}}^{\lambda_{p\mu}} \nu, \qquad \lambda_{p\mu} \equiv \min(\lambda+\mu, 2p-4-\lambda-\mu). \tag{2.3}$$

We shall need, however, for later purposes also the presence of some non-unitary lowest weight Vir modules whose minimal weights belong to the set (2.2).

If one of the factors in (2.3) is $\lambda = 1$ (the corresponding CVO is called a *step operator* [1]) then the sum in the fusion rule consists of at most two terms. Any conformal block involving a step operator satisfies a second order differential equation.

Consider the space of 4-point blocks spanned by

$$S_\mu \begin{pmatrix} 1 & \nu & ; & \nu & 1 \\ z_1 & z_2 & & z_3 & z_4 \end{pmatrix} \quad \left(\text{or} \quad U_\lambda \begin{pmatrix} 1 & ; & \nu & \nu & ; & 1 \\ z_1 & & z_2 & z_3 & & z_4 \end{pmatrix} \right).$$

Möbius invariance implies that each vector in this space can be written in the form

$$G\begin{pmatrix} 1 & \nu & \nu & 1 \\ z_1, & z_2, & z_3, & z_4 \end{pmatrix} = z_{14}^{-2\Delta_1} G_{11} \begin{pmatrix} \nu & \nu \\ z & w \end{pmatrix}, \qquad z = \frac{z_{13}z_{24}}{z_{14}}, \quad w = \frac{z_{12}z_{34}}{z_{14}}. \tag{2.4}$$

The level two null vector condition

$$L_n(L_{-1}^2 - rL_{-2})|\Delta_1\rangle = 0 \quad \text{for} \quad n = 1, 2, \ldots \tag{2.5}$$

implies the second order differential equation

$$\left\{(\partial_z + \partial_w)^2 + r(\frac{1}{z}\partial_z + \frac{1}{w}\partial_w - \frac{\Delta_\nu}{z^2} - \frac{\Delta_\nu}{w^2})\right\} G_{11}\begin{pmatrix} \nu & \nu \\ z & w \end{pmatrix} = 0. \tag{2.6}$$

Möbius invariance further implies that G_{11} is a homogeneous function of degree $-2\Delta_\nu$ of z and w. There are two natural bases in the 2-dimensional space of homogeneous solutions of (2.6): the s-channel basis

$$S_{\nu\pm1}\begin{pmatrix} \nu & \nu \\ z\,, & w \end{pmatrix} = (z-w)^{-2\Delta_\nu}\eta^{-\lambda_\nu} f_{\nu\pm1}(\eta,\nu), \quad \eta = \frac{w}{z} = \frac{z_{12}z_{34}}{z_{13}z_{24}} \tag{2.7}$$

where $f_{\nu\pm1}$ are specified by the initial conditions

$$f_{\nu-1}(0,\nu) = C^2_{1\nu\nu-1}, \quad f_{\nu+1}(\eta,\nu) \sim C^2_{1\,\nu\,\nu+1}\eta^{(\nu+1)r-1} \quad \text{for} \quad \eta\to 0, \tag{2.8}$$

the structure constants being related to 3-point functions

$$C_{\lambda\mu\nu} = \langle\lambda|\phi(1,\mu)|\nu\rangle \quad \text{for} \quad \langle\lambda|\lambda\rangle = 1 = z_{12}^{2\Delta_\mu}\langle\phi(z_1,\mu)\phi(z_2,\mu)\rangle; \tag{2.9}$$

the u-channel basis

$$U_\lambda\begin{pmatrix} \nu & \nu \\ z\,, & w \end{pmatrix} = (z-w)^{-2\Delta_\nu}\eta^{-\lambda_\nu} f_\lambda(1-\eta,\nu), \quad \lambda = 0,2, \tag{2.10}$$

$$\tilde{f}_0(0,\nu) = 1, \quad \tilde{f}_2(1-\eta,\nu) \approx C_{\nu\nu2}C_{121}(1-\eta)^{2r-1}, \quad \text{for} \quad \eta\to 1. \tag{2.11}$$

2B. Duality between s- and u-channel blocks. In order to write down the duality (or fusion) relation between the two bases we shall recall the integral representation for the f's coming from the so called Coulomb gas picture [31].

Eq.(2.6) implies the hypergeometric equation

$$\eta(1-\eta)f''_\mu + \{2-(\nu+1)r - [4-(\nu+3)r]\eta\}f'_\mu + 2\lambda_\nu(1-r)f_\mu = 0 \tag{2.12}$$

($2\lambda_\nu = (\nu+2)r - 2$ — see (2.2a)). The basis solutions of (2.12) have the form

$$\begin{aligned} \check{f}_{\nu-1}(\eta) \equiv C^{-2}_{1\nu\nu-1}f_{\nu-1}(\eta,\nu) &= F(-2\lambda_\nu, 1-r; 2-(\nu+1)r;\eta) = \\ &= \frac{\Gamma(2-(\nu+1)r)}{\Gamma(1-r)\Gamma(1-\nu r)} I_1(-r,-\nu r, 2\lambda_\nu;\eta). \end{aligned} \tag{2.13a}$$

$$\begin{aligned} \check{f}_{\nu+1}(\eta) \equiv C^{-2}_{1\,\nu\,\nu+1}f_{\nu+1}(\eta,\nu) &= \eta^{(\nu+1)r-1}F(\nu r, 1-r; 2\lambda_\nu;\eta) = \\ &= \frac{\Gamma((\nu+1)r)}{\Gamma(1-r)\Gamma((\nu+2)r-1)} I_2(-r,-\nu r, 2\lambda_\nu;\eta). \end{aligned} \tag{2.13b}$$

Similarly, for the u-channel basis we have

$$\begin{aligned} \check{\tilde{f}}_0 \equiv \tilde{f}_0(1-\eta,\nu) &= F(-2\lambda_\nu, 1-r; 2-2r; 1-\eta) = \\ &= \frac{\Gamma(2-2r)}{\Gamma^2(1-r)} I_1(-\nu r, -r, 2\lambda_\nu; 1-\eta), \end{aligned} \tag{2.14a}$$

$$\tilde{\tilde{f}}_2(1-\eta) = C^{-1}_{\nu\nu 2}C^{-1}_{121}\tilde{f}_2(1-\eta,\nu) = (1-\eta)^{2r-1}F(r,1-\nu r;2\lambda_\nu;1-\eta) = \\ = \frac{\Gamma(2r)}{\Gamma(1-\nu r)\Gamma((\nu+2)r-1)}I_2(-\nu r,-r,2\lambda_\nu;1-\eta). \tag{2.14b}$$

Here $I_{1,2}$ are the familiar integral representations for hypergeometric functions:

$$I_1(a,b,c;\eta) = \int_1^\infty u^a(u-1)^b(u-\eta)^c du = \\ = \int_0^1 t^{-a-b-c-2}(1-t)^b(1-\eta t)^c dt, \tag{2.15a}$$

$$I_2(a,b,c;\eta) = \int_0^\eta u^a(1-u)^b(\eta-u)^c du. \tag{2.15b}$$

The fusion matrix $F = F\begin{bmatrix}\nu & \nu\\ 1 & 1\end{bmatrix}$ (see (1.6)) is computed from the following simple crossing relations for the integrals (2.15):

$$\begin{pmatrix} I_1(b,a,c;1-\eta)\\ I_2(b,a,c;1-\eta)\end{pmatrix} = \frac{-1}{\sin\pi(a+c)}\begin{pmatrix} -\sin\pi b & \sin\pi c\\ \sin\pi(a+b+c) & \sin\pi a\end{pmatrix}\begin{pmatrix} I_1(a,b,c;\eta)\\ I_2(a,b,c;\eta)\end{pmatrix}. \tag{2.16}$$

(One way to check (2.16) is to note that both sides satisfy the same hypergeometric equation and then compare their values for $\eta = 0$ and $\eta = 1$ — cf. [32]. In verifying the involutivity of (2.16) we use the identity $\sin\pi a\sin\pi b + \sin\pi c\sin\pi(a+b+c) = \sin\pi(a+c)\sin\pi(b+c)$.

We can write the result in the form

$$F = \begin{pmatrix} x_\nu & 0\\ 0 & C_{\nu\nu 2}C_{121}y_\nu\end{pmatrix}\check{F}\begin{pmatrix} C^{-2}_{1\nu\nu-1} & 0\\ 0 & \frac{z_\nu}{x_\nu y_\nu}C^{-2}_{1\nu\nu+1}\end{pmatrix} \tag{2.17a}$$

$$x_\nu = \frac{\Gamma(r)\Gamma((\nu+1)r-1)}{\Gamma(2r-1)\Gamma(\nu r)}, \qquad y_\nu = \frac{\Gamma(2r)\Gamma((\nu+1)r-1)}{\Gamma(r)\Gamma((\nu+2)r-1)}, \\ z_\nu = \frac{2r-1}{(\nu+1)r-1}, \tag{2.17b}$$

$$\check{F} = \begin{pmatrix} \frac{1}{[2]} & \frac{[\nu+2]}{[2][\nu+1]}\\ 1 & -\frac{[\nu]}{[\nu+1]}\end{pmatrix} \tag{2.18}$$

Here we are using the "q-number" notation

$$[n] = \frac{\sin\pi n(1-r)}{\sin\pi(1-r)} = \frac{q^n-q^{-n}}{q-q^{-1}} \quad \text{for} \quad q = e^{\pm i\pi/p}. \tag{2.19}$$

The identity

$$[\nu]+[\nu+2] = [2][\nu+1] \tag{2.20}$$

implies $\det \check{F} = -1$. We restrict the choice of structure constants in such a way that this property remains valid for F:

$$\det F = -1 \Rightarrow C^2_{1\,\nu\,\nu-1}C^2_{1\,\nu\,\nu+1} = C_{121}C_{\nu\,\nu\,2}z_\nu. \tag{2.21}$$

Remark 2.1. We could further restrict and eventually determine the structure constants by imposing consecutively the conditions that

(i) F is involutive, i.e.

$$F^2 = 1 \Rightarrow \operatorname{tr} F = 0, (\det F)^2 = 1; \tag{2.22}$$

(ii) F is an orthogonal matrix, $F \in O(2)$.

An elementary analysis shows that (ii)⇒(i)⇒(2.21) (the latter implication being valid for non-negative structure constants). Eqs. (2.21) and (2.22) give

$$C^4_{1\nu\nu-1} = \frac{\Gamma(2-2r)\Gamma(1-\nu r)}{\Gamma(1-r)\Gamma(2-(\nu+1)r)}x_\nu = \frac{\Gamma(r)\Gamma(2-2r)\Gamma(1-\nu r)\Gamma((\nu+1)r-1)}{\Gamma(1-r)\Gamma(2r-1)\Gamma(\nu r)\Gamma(2-(\nu+1)r)}. \tag{2.23}$$

If we substitute $C_{\nu\nu 2}$ by another undetermined constant, ξ_ν, setting

$$\frac{C_{\nu\nu 2}C_{121}}{C^2_{1\nu\nu-1}} = \frac{\Gamma(r)\Gamma((\nu+2)r-1)\xi_\nu}{\sqrt{[2]}\Gamma(2r)\Gamma((\nu+1)r-1)} \left(= \frac{(\nu+1)r-1}{2r-1}C^2_{1\nu\nu+1}\right) \tag{2.24a}$$

we find a fusion matrix determined up to a similarity transformation:

$$F = \frac{1}{\sqrt{[2]}}\begin{pmatrix} \sqrt{\frac{[\nu]}{[\nu+1]}} & \frac{[\nu+2]}{[\nu+1]\xi_\nu} \\ \xi_\nu & -\sqrt{\frac{[\nu]}{[\nu+1]}} \end{pmatrix} \quad (\nu = 0, 1, \ldots, p-2). \tag{2.24b}$$

Requirement (ii) leads to the standard BPZ normalization (computed for a general minimal model in [7]) which guarantees the symmetry of 2-dimensional euclidian Green functions (and makes F a symmetric matrix):

$$\xi_\nu = ([\nu+2]/[\nu+1])^{1/2}. \tag{2.25}$$

We have, in particular, for the upper limit of the allowed range of ν in (2.24b)

$$\xi_{p-2} = 0 = C_{p-2\ p-2\ 2} = C_{1\ p-2\ p-1}. \tag{2.26}$$

thus reproducing the upper limit of the BPZ fusion rule (2.3). This normalization discards the second solution (2.13b) of the hypergeometric equation (2.12) for $\nu = p-2$ (while the first solution (2.13a) is a monodromy free polynomial of degree $p-3$ in η in that case). Anticipating the results of Secs.3 and 4 about an extended p-model with an internal quantum symmetry which require the presence of *all* solutions to the differential equations involved we shall not impose for the moment the orthogonality condition (or the symmetry of structure constants along with (2.21)) leading to (2.25). For $\xi_{p-2} \neq 0 \neq C_{\nu\nu 2}$, Eqs. (2.17), (2.18) or (2.24b) provide a "triangular" fusion relation which will be used in Sec.4. Only the product of conformal and QUE structure constants will be determined in our extended chiral theory.

2C. Statistics matrices. A Hecke algebra representation of the braid group. Each 4-point block is a well defined analytic function in a (complex) neighbourhood of the real open set

$$z_1 > z_2 > z_3 > z_4 > -z_3. \tag{2.27}$$

In order to define a permitted block we need to specify a path (or rather a homotopy class of paths) in C^4 with no coinciding arguments along which the conformal block is to be continued analytically. We find in this way the basic statistic matrices $R_i^{\nu_i \nu_{i+1}}$ which exchange the neighbouring arguments z_i and z_{i+1} in the positive direction amounting to the substitution

$$R_i \ : \ z_{i\,i+1} \to e^{i\pi} z_{i\,i+1}. \tag{2.28}$$

The operators R_1 and R_3 appear as diagonal matrices in the s-channel basis. They are determined by their action on the corresponding 3-point functions.

$$(R_1^{\nu_1 \nu_2})_{\mu\mu'} = e^{i\pi(\Delta_\mu - \Delta_{\nu_1} - \Delta_{\nu_2})} \delta_{\mu\mu'} = (R_3^{\nu_2 \nu_1})_{\mu\mu'} \tag{2.29}$$

The statistic matrices R_i satisfy the Yang-Baxter equation and are related to the fusion matrix as follows:

$$R_1^{\nu\nu} R_2^{1\nu} R_1^{1\nu} = e^{-2\pi i \Delta_\nu} F \begin{bmatrix} \nu & \nu \\ 1 & 1 \end{bmatrix} = R_2^{1\nu} R_1^{1\nu} R_2^{\nu\nu}. \tag{2.30}$$

For F given by (2.17) (2.18) and R_1 computed from (2.29) the first of these equations can be used to evaluate $R_2^{1\nu}$; then $R_2^{\nu\nu}$ can be found from the second one.

For $\nu = 1$ (the case of four equal weights) we thus obtain a Hecke algebra representation of B_4. Setting

$$b_i = e^{i\pi r/2} R_i^{11} = q - e_i, \quad q = e^{i\pi(1-r)} \tag{2.31a}$$

we find the following realization of the Temperley-Lieb projectors:

$$e_1 = \begin{pmatrix} 0 & 0 \\ 0 & [2] \end{pmatrix} = e_3, \; e_2 = \frac{1}{[2]} \begin{pmatrix} [3] & -\sqrt{[3]} \\ -\sqrt{[3]} & 1 \end{pmatrix}, \; [2] = q + q^{-1}, \; [3] = q^2 + 1 + q^{-2}, \tag{2.31b}$$

satisfying $e_i^2 = [2] e_i$, $e_i e_{i\pm1} e_i = e_i$ (cf. [5,30]).

2D. Conformal blocks involving indecomposable Vir-modules. Leaving the systematic study of conformal blocks involving no step operators to a subsequent publication, we shall indicate here, using the simplest example of $S_\mu \begin{pmatrix} 2 & \nu & ; & \nu & 2 \\ z_1 & z_2 & & z_3 & z_4 \end{pmatrix}$ $(\nu \geq 2)$ as a illustration, the new type of problems encountered in that case and a way to overcome the arising difficulties.

Writing a general (say, s-channel) 4-point block in the form

$$S_\mu \begin{pmatrix} \lambda & \nu & ; & \nu & \lambda \\ z_1 & z_2 & & z_3 & z_4 \end{pmatrix} = z_{14}^{-2\Delta_\lambda} z_{23}^{-2\Delta_\nu} \eta^{\Delta_{|\nu-\lambda|} - \Delta_\lambda - \Delta_\nu} f_\mu(\eta) \tag{2.32}$$

we find, for $\lambda = 2 \leq \nu$, as a consequence of a level 3 null-vector condition, a third order differential equation for f_μ. A basis of (unnormalized) solutions to this equation, singled out by initial conditions of the type

$$f_{\nu-2}(0) = C^2_{2\,\nu\,\nu-2}, \quad f_\nu(\eta) \sim \eta^{\nu r-1}, \quad f_{\nu+2} \sim \eta^{2(\nu+1)r-2} \quad \text{for } \eta \to 0, \tag{2.33}$$

is given, for $0 < \eta < 1$ and $\nu + 2 < p$ by the Coulomb gas integrals (cf. [31])

$$\check{f}_{\nu-2} = \gamma\alpha_\nu I_1, \; I_1(-2r, -\nu r, 2\lambda_\nu; 2r|\eta) = \int_1^\infty \frac{du_1}{u_1^{2r}} \int_1^{u_1} \frac{du_2(u_1-\eta)^{2\lambda_\nu}(u_2-\eta)^{2\lambda_\nu}}{u_2^{2r}(u_1-1)^{\nu r}(u_2-1)^{\nu r}} u_{12}^{2r}, \tag{2.34a}$$

$$\check{f}_\nu = \gamma\beta_\nu I_2, \; I_2(-2r, -\nu r, 2\lambda_\nu; 2r|\eta) = \int_1^\infty \frac{du_1}{u_1^{2r}} \int_1^{\eta} \frac{du_2(u_1-\eta)^{2\lambda_\nu}(\eta-u_2)^{2\lambda_\nu}}{u_2^{2r}(u_1-1)^{\nu r}(1-u_2)^{\nu r}} u_{12}^{2r}, \tag{2.34b}$$

$$\check{f}_{\nu+2} = \gamma I_3, \; I_3(-2r, -\nu r, 2\lambda_\nu; 2r|\eta) = \int_0^\eta \frac{du_1}{u_1^{2r}} \int_0^{u_1} \frac{du_2(\eta-u_1)^{2\lambda_\nu}(\eta-u_2)^{2\lambda_\nu}}{u_2^{2r}(1-u_1)^{\nu r}(1-u_2)^{\nu r}} u_{12}^{2r}. \tag{2.34c}$$

Making in I_2 the change of integration variables $u_1 - 1 = su_1$, $u_2 = \eta t$ we find

$$\begin{aligned} &\frac{1}{\Gamma^2(1-2r)} \lim_{\eta\to 0} \{\eta^{1-\nu r} I_2(-2r, -\nu r, 2\lambda_\nu; 2r|\eta)\} = \\ &\frac{\Gamma(1-\nu r)\Gamma((\nu+2)r-1)}{\Gamma(\nu r)\Gamma(2-(\nu+2)r)} \to 0 \quad \text{for } (\nu+2)r \to p-1. \end{aligned} \tag{2.35}$$

We find, in fact, that for $\nu = p - 2$ the solution I_2 becomes proportional to I_3. This is a reflection of the fact that the Verma module of lowest weight $\Delta_{p-2} = \frac{(p-2)(p-3)}{4}$ has a null vector at level $p - 2$ of weight $\Delta_p = \Delta_{p-2} + p - 2$. A linearly independent basis of solutions, applicable also in this limit, is given by I_1, I_3 and

$$\Gamma(2-(\nu+2)r)(I_2 - aI_3) \text{ for } a = \frac{I_2(-2r, 2r+1-p, p-3; 2r|\eta)}{I_3(-2r, 2r+1-p, p-3; 2r|\eta)} \to 1 \text{ for } r \to \frac{3}{4} (p = 4). \tag{2.36}$$

Defining the u-channel basis $\tilde{f}_{2n}(1-\eta, \nu)$, $n = 0, 1, 2$, by $\gamma a_\nu \tilde{I}_1$, $\gamma b_\nu \tilde{I}_2$, $\gamma \tilde{I}_3$ where

$$\tilde{I}_j = I_j(-\nu r, -2r, 2\lambda_\nu; 2r|1-\eta)$$

and specifying the normalization constants by

$$\alpha_\nu = \frac{[\nu-1]}{[\nu+1]}, \quad \beta_\nu = -\frac{[2][\nu]}{[\nu+2]}, \quad a_\nu = \frac{1}{[3]}, \quad b_\nu = (-1)^{\nu-1} \frac{[2][\nu]}{[4]}$$

we find (for a derivation — see [37], Eq (4.5) and Appendix B)

$$\check{F} = \left| \begin{matrix} \frac{1}{[3]} & -\frac{[\nu+2]}{[2][3][\nu]} & \frac{[\nu+3]}{[3][\nu+1]} \\ -\frac{[2]^2}{[4]} & [2]\frac{[\nu+2]-[2]}{[4][\nu]} & \frac{[2]^2[\nu][\nu+3]}{[4][\nu+1][\nu+2]} \\ 1 & \frac{[\nu-1]}{[\nu]} & \frac{[\nu-1][\nu]}{[\nu+1][\nu+2]} \end{matrix} \right| \tag{2.37}$$

For $\nu+2=p$ the elements of this fusion matrix become ill defined but modifying the basis as indicated in Eq. (2.36) we find a finite triangular matrix

$$F = A\check{F}B^{-1} = \begin{pmatrix} \frac{1}{[3]} & 0 & 0 \\ 0 & 1 & 0 \\ 1 & \frac{[3]}{[2]} & -[3] \end{pmatrix} \tag{2.38}$$

$$A = \begin{pmatrix} 1 & 0 & 0 \\ 0 & 1 & \frac{[2]^2}{[4]} \\ 0 & 0 & 1 \end{pmatrix}, \qquad B^{-1} = \begin{pmatrix} 1 & 0 & 0 \\ 0 & 1 & -\frac{[2][p-2]}{[p]} \\ 0 & 0 & 1 \end{pmatrix}. \tag{2.39}$$

3. U_q-blocks and fusion relations

It has been recognized [11,3,15–17] that the R-matrix of the QUE algebra $U_q = U_q(sl(2))$ (see Appendix) "mimics" the statistics matrices of p-models (the precise relation will be spelled out below) as well as of $su(2)$-current algebra models. The treatment of dual U_q-blocks below is based on concepts and techniques introduced in the Appendix.

3A. Dual 4-point blocks involving a pair of q-spin 1/2 operators. The basic ingredients in computing U_q-blocks are the 3j-symbols $V\begin{pmatrix} I_1 & I_2 & I \\ m_1 & m_2 & -m \end{pmatrix}$ which give rise to homogeneous polynomials in two undeterminates u_1 and u_2

$$V\begin{pmatrix} I_1 & I_2 & I \\ u_1 & u_2 & -m \end{pmatrix} = \sum_{\substack{m_1=-I_1 \\ m_1+m_2=m}}^{I_1} \sum_{m_2=-I_2}^{I_2} V\begin{pmatrix} I_1 & I_2 & I \\ m_1 & m_2 & -m \end{pmatrix} \begin{bmatrix} 2I_1 \\ I_1+m_1 \end{bmatrix}^{1/2} \begin{bmatrix} 2I_2 \\ I_2+m_2 \end{bmatrix}^{1/2} u_1^{I_1+m_1} u_2^{I_2+m_2} \tag{3.1}$$

of degree I_1+I_2+m. (The q-deformed binomial coefficients appearing in (3.1) are defined by (A.38)). This polynomial is evaluated (up to an overall normalization) as the expansion coefficient of a homogeneous polynomial in 3 variables (see (A.35)):

$$\begin{aligned} V^n\begin{pmatrix} I_1 & I_2 & I \\ u_1 & u_2 & u \end{pmatrix} &= N^{-1}_{I_1 I_2 I} V\begin{pmatrix} I_1 & I_2 & I \\ u_1 & u_2 & u \end{pmatrix} = \\ &= \left[q^{-\frac{I+1}{2}} u_1 (-) q^{\frac{I+1}{2}} u_2\right]^{I_1+I_2-I} \\ &\quad \left[q^{\frac{I_2-1}{2}} u_1 (-) q^{\frac{1-I_2}{2}} u\right]^{I_1+I-I_2} \left[q^{-\frac{1+I_1}{2}} u_2 (-) q^{\frac{1+I_1}{2}} u\right]^{I+I_2-I_1} = \\ &= \sum_{m=-I}^{I} \begin{bmatrix} 2I \\ I+m \end{bmatrix}^{1/2} V^n\begin{pmatrix} I_1 & I_2 & I \\ u_1 & u_2 & -m \end{pmatrix} u^{I-m}. \end{aligned} \tag{3.2}$$

Here $[x(-)y]^n$ is defined by (A.34) and can be characterized by the recursive formula

$$[q^{-1/2}u(-)q^{1/2}v]^n = (q^{-n/2}u - q^{n/2}v)[u(-)v]^{n-1}. \tag{3.3}$$

The subspace of invariant tensors in the 4-fold tensor product of irreducible U_q-modules

$$\langle I_1 I_2 I_2 I_1 \rangle \equiv \mathrm{inv}\{\mathcal{V}_{I_1} \otimes \mathcal{V}_{I_2} \otimes \mathcal{V}_{I_2} \otimes \mathcal{V}_{I_1}\}$$

is $I_1 + I_2 - |I_1 - I_2| + 1 (= 2I_1 + 1 \quad \text{for } I_1 \leq I_2)$-dimensional.

We shall be dealing with two dual bases in this subspace. The s-channel basis

$$\mathcal{S}_I \begin{pmatrix} I_1 & I_2 & ; & I_2 & I_1 \\ u_1 & u_2 & ; & u_3 & u_4 \end{pmatrix} = \sum_{m=-I}^{I} (-1)^{I-m} q^m V \begin{pmatrix} I_1 & I_2 & I \\ u_1 & u_2 & -m \end{pmatrix} V \begin{pmatrix} I & I_2 & I_1 \\ m & u_3 & u_4 \end{pmatrix} \tag{3.4}$$

and the u-channel basis

$$\mathcal{U}_{\tilde{I}} \begin{pmatrix} I_1 & ; & I_2 & I_2 & ; & I_1 \\ u_1 & ; & u_2 & u_3 & ; & u_4 \end{pmatrix} = \sum_{m=-\tilde{I}}^{\tilde{I}} (-1)^{\tilde{I}-m} q^m V \begin{pmatrix} I_2 & I_2 & \tilde{I} \\ u_2 & u_3 & -m \end{pmatrix} V \begin{pmatrix} I_1 & \tilde{I} & I_1 \\ u_1 & m & u_4 \end{pmatrix} \tag{3.5}$$

(These expressions can viewed as q-deformations of familiar $SU(2)$ invariants. Their invariance follows from the U_q-transformation properties displayed in Appendix A.)

For $I_1 = 1/2, I_2 = \nu/2 (\nu \geq 1)$ the space of 4-point invariant tensors is 2-dimensional and we can write the two dual bases explicitly. Defining the non-normalized blocks $\check{\mathcal{S}}_I$ and $\check{\mathcal{U}}_{\tilde{I}}$ by

$$\mathcal{S}_I = N^2_{I_1 I_2 I} \check{\mathcal{S}}_I, \qquad \mathcal{U}_{\tilde{I}} = N_{I_2 I_2 \tilde{I}} N_{I_1 \tilde{I} I_1} \check{\mathcal{U}}_{\tilde{I}} \tag{3.6}$$

we can write

$$\begin{aligned} &\check{\mathcal{S}}_{\frac{\nu-1}{2}} \begin{pmatrix} 1/2 & \nu/2 & ; & \nu/2 & 1/2 \\ u_1 & u_2 & ; & u_3 & u_4 \end{pmatrix} = \\ &= \left(q^{-\frac{\nu+1}{4}} u_1 - q^{\frac{\nu+1}{4}} u_2\right) \left(q^{-\frac{\nu+1}{4}} u_3 - q^{\frac{\nu+1}{4}} u_4\right) [q^{-1} u_2 (-) q u_3]^{\nu-1}, \end{aligned} \tag{3.7a}$$

$$\begin{aligned} \mathcal{S}_{\frac{\nu+1}{2}} = &\frac{1}{[\nu+1]} \{ (q^{\nu-\frac{1}{2}} u_1 - q^{\frac{1}{2}-\nu} u_4)(q^{-\frac{\nu}{2}} u_2 - q^{\frac{\nu}{2}} u_3) [u_2(-)u_3]^{\nu-1} + \\ &+ [\nu] (q^{\frac{\nu-5}{2}} u_1 u_2 + q^{\frac{5-\nu}{2}} u_3 u_4 - u_1 u_4 - u_2 u_3) [q^{-1} u_2 (-) q u_3]^{\nu-1} \}, \end{aligned} \tag{3.7b}$$

$$\begin{aligned} \mathcal{U}_0 = \check{\mathcal{U}}_0 &= (q^{-1/2} u_1 - q^{1/2} u_4) [q^{-1/2} u_2 (-) q^{1/2} u_3)]^{\nu} = \\ &= (q^{-1/2} u_1 - q^{1/2} u_4)(q^{-\nu/2} u_2 - q^{\nu/2} u_3) [u_2(-)u_3]^{\nu-1}, \end{aligned} \tag{3.8a}$$

$$\check{\mathcal{U}}_1 = [q^{-1} u_2 (-) q u_3)]^{\nu-1} \{ q^{\nu/2} u_2 + q^{-\nu/2} u_3) \frac{q^{-3/2} u_1 + q^{3/2} u_4}{[2]} - u_1 u_4 - u_2 u_3 \}. \tag{3.8b}$$

The U_q fusion matrix $\mathcal{F}$, defined by

$$\mathcal{U}_\lambda = \sum_{\mu=0}^{1} \mathcal{F}_{\lambda\mu} \mathcal{S}_{\frac{\nu-1}{2}+\mu}, \tag{3.9}$$

has the form

$$\mathcal{F} = \begin{pmatrix} 1 & 0 \\ 0 & N_{\nu/2\ \nu/2\ 1} N_{1/2\ 1\ 1/2} \end{pmatrix} \check{\mathcal{F}} \begin{pmatrix} N^{-2}_{\frac{1}{2}\frac{\nu}{2}\frac{\nu-1}{2}} & 0 \\ 0 & N^{-2}_{\frac{1}{2}\frac{\nu}{2}\frac{\nu+1}{2}} \end{pmatrix}, \tag{3.10a}$$

$$\check{\mathcal{F}} = \begin{pmatrix} -\frac{[\nu]}{[\nu+1]} & 1 \\ \frac{[\nu+2]}{[\nu+1][2]} & \frac{1}{[2]} \end{pmatrix}. \tag{3.10b}$$

(In verifying (3.10b) we use the identity

$$[q^{-1/2}u_2(-)q^{1/2}u_3]^\nu = (q^{\frac{\nu-2}{2}}u_2 - q^{\frac{2-\nu}{2}}u_3)[q^{-1}u_2(-)qu_3]^{\nu-1} \tag{3.11}$$

which can be derived from (3.3).) The identity (2.20) implies the condition

$$\det \check{\mathcal{F}} = -1. \tag{3.12a}$$

We postulate that this condition is respected by the normalized fusion matrix $\mathcal{F}$:

$$\det \mathcal{F} = -1, \tag{3.12b}$$

yielding the relation

$$N_{\nu/2\,\nu/2\,1} N_{1/2\,1\,1/2} = N^2_{\frac{1}{2}\frac{\nu}{2}\frac{\nu-1}{2}} N^2_{\frac{1}{2}\frac{\nu}{2}\frac{\nu+1}{2}} \tag{3.13}$$

In analogy with the case of a p-model (Remark 2.1) we can fix $N^4_{\frac{1}{2}\frac{\nu}{2}\frac{\nu-1}{2}}$ or all U_q-structure constants by assuming that

(i) $\mathcal{F}$ is involutive (and satisfies (3.12b)), so that

$$\operatorname{tr}\mathcal{F} = 0 \Rightarrow N^4_{\frac{1}{2}\frac{\nu}{2}\frac{\nu-1}{2}} = \frac{[2][\nu]}{[\nu+1]}, \tag{3.14}$$

(ii) $\mathcal{F}$ is orthogonal (and hence symmetric).

If, exploiting (3.14), we set

$$N^2_{\frac{1}{2}\frac{\nu}{2}\frac{\nu-1}{2}} = \varepsilon_\nu \sqrt{\frac{[2][\nu]}{[\nu+1]}}, \quad N^2_{\frac{1}{2}\frac{\nu}{2}\frac{\nu+1}{2}} = \frac{\varepsilon_{\nu+1}}{\rho_\nu}\sqrt{[2]} \quad (\varepsilon_\mu = \pm 1, \rho_\nu > 0) \tag{3.15}$$

we shall have

$$\mathcal{F} = \frac{1}{\sqrt{[2]}} \begin{pmatrix} -\varepsilon_\nu \sqrt{\frac{[\nu]}{[\nu+1]}} & \varepsilon_{\nu+1}\rho_\nu \\ \frac{\varepsilon_{\nu+1}}{\rho_\nu}\frac{[\nu+2]}{[\nu+1]} & \varepsilon_\nu \sqrt{\frac{[\nu]}{[\nu+1]}} \end{pmatrix}. \tag{3.16}$$

The orthogonality condition (ii) leads to

$$\rho_\nu = ([\nu+2]/[\nu+1])^{1/2} \Rightarrow \rho_{p-2} = 0 \tag{3.17}$$

so that $N_{\frac{1}{2}\frac{p-2}{2}\frac{p-1}{2}}$ and hence $S_{\frac{p-1}{2}}$ are ill defined for this choice of normalization. (We note that adopting (ii) leads to the coincidence of the fusion matrix with the $6j$-symbols of [33].)

The following observation plays a crucial role for the applications (Sec.4).

Proposition 3.1. *The product of the transposed to $\mathcal{F}$ with the p-model fusion matrix F is a multiple of the unit matrix iff the two sign factors in (3.15) are opposite to each other and $\rho_\nu = \xi_\nu$:*

$$ {}^t\mathcal{F}F = \varepsilon 1 \Leftrightarrow \varepsilon_\nu \varepsilon_{\nu+1} = -1, \quad \rho_\nu = \xi_\nu \quad (\varepsilon = -\varepsilon_\nu). \tag{3.18}$$

If the same condition is applied to the more general F-matrices (2.17), (2.18) and (3.9) it gives

$$N_{\nu/2\ \nu/2\ 1} N_{1/2\ 1\ 1/2} C_{\nu\ \nu\ 2} C_{1\ 2\ 1} = -\frac{x_\nu}{y_\nu} = -\frac{\Gamma^2(r)\Gamma((\nu+2)r-1)}{(2r-1)\Gamma^2(2r-1)\Gamma(\nu r)}, \tag{3.19a}$$

$$N^2_{\frac{1}{2}\,\frac{\nu}{2}\,\frac{\nu-1}{2}} C^2_{1\ \nu\ \nu-1} = \varepsilon_\nu x_\nu\,, \tag{3.19b}$$

$$N^2_{\frac{1}{2}\,\frac{\nu}{2}\,\frac{\nu+1}{2}} C^2_{1\ \nu\ \nu+1} = -\varepsilon_\nu \frac{z_\nu}{y_\nu} = -\varepsilon_\nu x_{\nu+1}. \tag{3.19c}$$

The proof is straightforward (and uses (2.20)).

We see, in particular, that the relevant products (3.19a) and (3.19c) neither diverge nor vanish for $\nu = p-2$.

3B. Semidefinite braid-invariant inner product in the space of U_q-blocks. Following the pattern of Sec.2C we can define U_q-statistic matrices (yielding a representation of B_4 on U_q-blocks) in terms of the fusion matrix (3.9) (or(3.16)) and a pair of diagonal matrices (in the s-channel basis) whose eigenvalues are determined from the analysis of 3-point functions.

Using a result of [33] (and correcting on the way the expression for the universal $\mathcal{R}$-matrix — cf. [15]) we find

$$\mathcal{R}_1^{I_1 I_2} \mathcal{V}\begin{pmatrix} I_1 & I_2 & I \\ u_1 & u_2 & u \end{pmatrix} = P_{12} \check{\mathcal{R}} \begin{bmatrix} I_1 & I_2 \\ 0 & I \end{bmatrix} \mathcal{V}\begin{pmatrix} I_1 & I_2 & I \\ u_1 & u_2 & u \end{pmatrix} \tag{3.20}$$

where P_{12} is a permutation of the first two arguments,

$$P_{12}\mathcal{V}\begin{pmatrix} I_1 & I_2 & I \\ u_1 & u_2 & u \end{pmatrix} = \mathcal{V}\begin{pmatrix} I_2 & I_1 & I \\ u_2 & u_1 & u \end{pmatrix} \tag{3.21}$$

while $\check{\mathcal{R}} \begin{bmatrix} I_1 & I_2 \\ I_i & I_f \end{bmatrix}$ is the matrix corresponding to the braiding of I_1 and I_2 between states of q-spins I_i and I_f. If either I_i or I_f is 0 (the "U_q-vacuum") then $\check{\mathcal{R}}$ is just a complex number (for $|q| = 1$ it is a phase factor); in particular

$$\check{\mathcal{R}} \begin{bmatrix} I_1 & I_2 \\ 0 & I \end{bmatrix} = (-1)^{I_1+I_2-I} q^{I(I+1)-I_1(I_1+1)-I_2(I_2+1)} = \check{\mathcal{R}} \begin{bmatrix} I_2 & I_1 \\ I & 0 \end{bmatrix} \tag{3.22}$$

Defining the diagonal matrix $\check{\mathcal{R}}_1^{I_1 I_2}$ by setting

$$\left(\check{\mathcal{R}}_1^{I_1 I_2}\right)_{II'} = \check{\mathcal{R}} \begin{bmatrix} I_1 & I_2 \\ 0 & I \end{bmatrix} \delta_{II'} \tag{3.23}$$

we can compute $\check{\mathcal{R}}_2$-matrices corresponding to the braiding of the arguments (2,3) in a 4-point block from the following analogue of Eq. (2.30):

$$\check{\mathcal{R}}_1^{I_2 I_2} \check{\mathcal{R}}_2^{I_1 I_2} \check{\mathcal{R}}_1^{I_1 I_2} = (-1)^{2I_2} q^{-2I_2(I_2+1)} \mathcal{F} = \check{\mathcal{R}}_2^{I_1 I_2} \check{\mathcal{R}}_1^{I_1 I_2} \check{\mathcal{R}}_2^{I_2 I_2} \tag{3.24}$$

In the special case of $I_1 = I_2 = 1/2$ we obtain the following U_q-counterpart of the Hecke algebra representation (2.31) of B_4:

$$q^{1/2} \mathcal{R}_i^{1/2\,1/2} = q - \tilde{e}_i, \qquad \tilde{e}_1 = \begin{pmatrix} [2] & 0 \\ 0 & 0 \end{pmatrix} = \tilde{e}_3, \qquad \tilde{e}_2 = \frac{1}{[2]} \begin{pmatrix} 1 & \sqrt{[3]} \\ \sqrt{[3]} & [3] \end{pmatrix} . \tag{3.25}$$

In writing down $\tilde{e}_2$ we have used the orthogonal normalization (3.17) of $\mathcal{F}$ (3.16) and we have set (in accord with (3.18))

$$\varepsilon_\nu = (-1)^{\nu - 1} . \tag{3.26}$$

We thus have — at least for $2I_1 + 2I_1 \leq p - 2$ — unitary (or unitarizable) statistics and fusion matrices. This implies the existence of a braid invariant Hermittian inner product in $\langle I_1 I_2 I_2 I_1 \rangle$ that is positive for sufficiently small $I_1 + I_2$. This inner product is unique up to an overall normalization and we shall construct it starting with a complex valued sesquiliner form on tensor products of irreducible U_q-modules.

We define it as a factorizable form: $v_1 \otimes v_2$ and $v_1' \otimes v_2' \in \mathcal{V}_1 \otimes \mathcal{V}_2$ than

$$(\overline{v_1 \otimes v_2}, v_1' \otimes v_2') = (\bar{v}_1, v_1')_{\nu_1} (\bar{v}_2, v_2')_{\nu_2} . \tag{3.27}$$

Then it is fixed by its values on irreducible U_q-modules which are defined in the polynomial basis by

$$(u^{-I-\bar{m}}, u^{I+m})_I = e^{i\pi I} \begin{bmatrix} 2I \\ I+m \end{bmatrix}^{-1} \delta_{\bar{m}m} . \tag{3.28}$$

The phase factor is chosen in such a way that the norm square of the invariant 2-point function (A.34) is equal to the quantum dimension

$$\begin{aligned} &\left(\overline{[q^{1/2}u_1(-)q^{1/2}u_2]^{2I}}, [q^{1/2}u_1(-)q^{1/2}u_2]^{2I} \right) = \\ &= \sum_{m,\bar{m}=-I}^{I} \begin{bmatrix} 2I \\ I-\bar{m} \end{bmatrix} \begin{bmatrix} 2I \\ I+m \end{bmatrix} (-1)^{2I+\bar{m}-m} q^{-m-\bar{m}} (\bar{u}_1^{I-\bar{m}} \bar{u}_2^{I+\bar{m}}, u_1^{I+m} u_2^{I-m}) = \\ &= (-1)^{2I} e^{2\pi i I} \sum_{m=-I}^{I} q^{-2m} = [2I+1] . \end{aligned} \tag{3.29}$$

(and hence is positive for $2I \leq p - 2$).

The space $\langle I_1 I_2 I_3 \rangle$ of 3-point invariant tensors is non-empty (for $I_1 + I_2 + I_3 \leq p - 1$ if $I_i + I_j - I_k$ is a non-negative integer for any permutation (i, j, k) of (1,2,3); then it is again one dimension and is given by multiples of V'' (3.2) (or (A.35)). A rather lengthy computation (which will be reproduced elsewhere) yields the following expression for its norm square:

$$\|\check{V}\begin{pmatrix} I_1 & I_2 & I_3 \\ u_1 & u_2 & u_3 \end{pmatrix}\|^2 \equiv \left(\overline{\check{V}\begin{pmatrix} I_1 & I_2 & I_3 \\ u_1 & u_2 & u_3 \end{pmatrix}}, \check{V}\begin{pmatrix} I_1 & I_2 & I_3 \\ u_1 & u_2 & u_3 \end{pmatrix}\right) = \frac{[I_1+I_2-I_3]!\,[I_2+I_3-I_1]!\,[I_1+I_3-I_2]!\,[I_1+I_2+I_3+1]!}{[2I_1]!\,[2I_2]!\,[2I_3]!}. \tag{3.30}$$

Remark 3.1. The normalization that renders the fusion matrix (= $6j$ symbol) orthogonal (and is restricted by (A.36)),

$$N^4_{I_1 I_2 I_3} = \frac{[2I_1]!\,[2I_1+1]!\,[2I_2]!\,[2I_2+1]\,!\,[2I_3]!\,[2I_3+1]!}{([I_1+I_2-I_3]!\,[I_1+I_3-I_2]!\,[I_2+I_3-I_1]!\,[I_1+I_2+I_3+1]!)^2}, \tag{3.31}$$

corresponds to the following simple norm square of V:

$$\|V\begin{pmatrix} I_1 & I_2 & I_3 \\ u_1 & u_2 & u_3 \end{pmatrix}\|^2 = \sqrt{[2I_1+1][2I_2+1][2I_3+1]}. \tag{3.32}$$

However, the structure constants (3.21) diverge for $I_1+I_2+I_3 \geq p-1$ and are, hence, inconvenient for studying the non-physical contribution to the chiral Green function.

The 4-point block provide the first example of a multidimensional space of U_q-invariants:

$$\dim\langle I_1 I_2 I_2 I_1\rangle = 2I_{1+1} \quad \text{for } I_1 \leq I_2 \tag{3.33}$$

Their inner products can be computed as follows. The unitarity of statistic matrices (including the orthogonality of the fusion matrix) imply that $\{S_I\}$ and $\{\mathcal{U}_{\bar{I}}\}$ form two equally normalized orthogonal bases. One of the u-channel basis vectors, $\mathcal{U}_0$, is factorized into a product of 2-point functions, whose norm squares are given, according to (3.29) by the corresponding quantum dimensions. We, therefore, have

$$\left(\overline{\mathcal{U}_{\bar{I}}\begin{pmatrix} I_1 & ; & I_2 & I_2 & ; & I_1 \\ u_1 & ; & u_2 & u_3 & ; & u_4 \end{pmatrix}}, \mathcal{U}_{\bar{I}}\begin{pmatrix} I_1 & ; & I_2 & I_2 & ; & I_1 \\ u_1 & ; & u_2 & u_3 & ; & u_4 \end{pmatrix}\right) = \tag{3.34a}$$
$$= \delta_{\bar{I}\bar{I}'}\,(\overline{\mathcal{U}_0}, \mathcal{U}_0) = [2I_1+1][2I_2+1]\delta_{\bar{I}\bar{I}'};$$

it then also follows that

$$(\overline{S_I}, S_{I'}) = [2I_1+1][2I_2+1]\delta_{II'}. \tag{3.34b}$$

The knowledge of (3.34) and of the structure constants $N_{I_1 I_2 I}$ (3.31) allows to compute the norm square of the unnormalized blocks:

$$\|\check{S}_I\begin{pmatrix} I_1 & I_2 & ; & I_2 & I_1 \\ u_1 & u_2 & ; & u_3 & u_4 \end{pmatrix}\|^2 = \frac{[2I_1+1][2I_2+1]}{N^4_{I_1 I_2 I}} \tag{3.35}$$

Eq. (3.35) has an important consequence: the norm square of $\check{S}_I$ vanishes whenever $N_{I_1 I_2 I}$ diverges (which happens whenever the triple I_1, I_2, I violates the BPZ fusion rules); similarly, $\|\check{\mathcal{U}}_{\bar{I}}\|^2$ vanishes if $N^2_{I-2I_2\bar{I}}\; N^2_{I-1\bar{I}I_1}$ diverges.

The above expressions exhibit (for $I_3 = I_2$, $I_4 = I_1$) the following simple factorization property

$$\|[q^{-1/2}u(-)q^{1/2}u']^{2I}\|^2 \, \|\check{S}_I \begin{pmatrix} I_1 & I_2 \\ u_1 & u_2 \end{pmatrix}; \begin{pmatrix} I_3 & I_4 \\ u_3 & u_4 \end{pmatrix} \|^2 = \\ \|\check{\mathcal{V}} \begin{pmatrix} I_1 & I_2 & I \\ u_1 & u_2 & u \end{pmatrix} \|^2 \, \|\check{\mathcal{V}} \begin{pmatrix} I & I_3 & I_4 \\ u & u_3 & u_4 \end{pmatrix} \|^2. \tag{3.36}$$

We conjecture that such a property remains valid for higher point correlation functions so that there are simple rules for computing the norm square of an U_q-block corresponding to an arbitrary tree diagram.

3C. Remark about U_q-blocks involving indecomposable representations. The basis (3.4) (or (3.5)) is not appropriate for $2I \geq p$ (a value that appears for $I_1 + I_2 \geq p/2$). We shall illustrate the problem and the way out for the simplest example of $p = 4, I_1 = I_2 = 1$. In this case

$$\check{S}_2 = q^{-2}u_1^2u_2^2 + q^2u_3^2u_4^2 - \frac{[2]^2}{[4]}(q^{-1}u_1u_2 + qu_3u_4)(qu_1 + q^{-1}u_2)(qu_3 + q^{-1}u_4) + \\ + \begin{bmatrix} 4 \\ 2 \end{bmatrix}^{-1} \{(qu_1 + q^{-1}u_2)^2 + ([3]-1)u_1u_2\}\{qu_3 + q^{-1}u_4)^2 + ([3]-1)u_3u_4\} \tag{3.37}$$

is ill defined since $[4] = 0$ for $q^4 = -1$. On the other hand, $[4]\check{S}_2$ has a limit for $q^4 \to -1$ that is proportional to $\check{S}_1$:

$$\frac{[4]}{[2]^2}\check{S}_2 \underset{\text{for } q^4=-1}{=} \check{S}_1 = \\ = (q^{-1}u_1 - qu_2)(q^{-1}u_3 - qu_4)\{q^{-1}u_1u_2 + qu_3u_4 - \frac{1}{[2]}(qu_1 + q^{-1}u_2)(qu_3 + q^{-1}u_4)\}. \tag{3.38}$$

The space $\langle 1111 \rangle$ or 4-point U_q-blocks is, however, still 3-dimensional. It is spanned by $\mathcal{S}_0 (= \check{S}_0)$, $\check{S}_1$ and

$$\mathcal{S}_2 \equiv \check{S}_2 - \frac{[2]^2}{[4]}\check{S}_1 = q^{-2}u_1^2u_2^2 + q^2u_3^2u_4^2 - [2](u_1u_3 + u_2u_4)(q^{-1}u_1u_2 + qu_3u_4) + \\ + \frac{1}{[3]}\{[2](q^{-1}u_1u_3 + qu_2u_4) - u_1u_4 - u_2u_3\}(qu_1 + q^{-1}u_2)(qu_3 + q^{-1}u_4). \tag{3.39}$$

The invariants $\check{S}_1$ and (3.39) corresponding to an (8-dimensional) indecomposable intermediate state space [27] generate an isotropic 2-space with respect to the above inner product; in particular,

$$\|\check{S}_2 - \frac{[2]^2}{[4]}\check{S}_1\|^2 = [5] + [3] \ (= 0 \text{ for } [4] = 0). \tag{3.40}$$

We note that the fusion matrix corresponding to the regular basis $\mathcal{S}_0$, $\check{S}_1$, $\mathcal{S}_2$ (3.39) (and to the corresponding u-channel basis) coincides — for $[4] = 0$ — with the (inverse) transposed of F (2.38):

$$\mathcal{F} = {}^tF = \begin{pmatrix} 1 & 0 & 1 \\ 0 & 1 & 1/[2] \\ 0 & 0 & -1 \end{pmatrix}, \qquad \mathcal{F}^2 = 1 \ (= F^2 \text{ for } [3] = 1). \tag{3.41}$$

4. Combined (p,q)-model with an U_q chiral symmetry

4A. 4-point (p,q)-Green function involving a pair of step operators. We define the combined (p,q)-theory of a chiral Green functions depending on both the z- and u- variables. (An underlying operator theory won't be unitary and requires more work to be formulated with full precision.) In the simplest case of a diagonal (p,q)-theory a 4-point Green function involving pair of step operators can be written in the form of a, say, s-channel expansion

$$G_4 = G\begin{pmatrix} 1 & \nu & \nu & 1 \\ z_1 u_1 & z_2 u_2 & z_3 u_3 & z_4 u_4 \end{pmatrix} = \\ = \sum_{\mu=0}^{1} B_{\nu-1+\mu} \check{S}_{\nu-1+2\mu} \begin{pmatrix} 1 & \nu & ; & \nu & 1 \\ z_1 & z_2 & ; & z_3 & z_4 \end{pmatrix} \check{\mathcal{S}}_{\frac{\nu-1}{2}+\mu} \begin{pmatrix} 1/2 & \nu/2 & ; & \nu/2 & 1/2 \\ u_1 & u_2 & ; & u_3 & u_4 \end{pmatrix}. \tag{4.1}$$

Here $\check{\mathcal{S}}_{\frac{\nu\pm1}{2}}$ are defined by (3.6), $\check{S}_{\nu\pm1} = C^{-2}_{1\,\nu\,\nu\pm1} S_{\nu\pm1}$ (see (2.7)) and B_λ substitute the product of structure constants in the two theories

$$B_\nu = C^2_{1\,\nu\,\nu+1} N^2_{\frac{1}{2}\frac{\nu}{2}\frac{\nu+1}{2}} \quad (B_0 = 1). \tag{4.2}$$

The symmetric extrapolation of (2.23),

$$C^4_{1\,\nu\,\nu+1} = \frac{\Gamma(r)\Gamma(2-2r)\Gamma(1-(\nu+1)r)\Gamma((\nu+2)r-1)}{\Gamma(1-r)\Gamma(2r-1)\Gamma((\nu+1)r)\Gamma(2-(\nu+2)r)}, \tag{4.3}$$

tends to zero for $\nu = p-2$ and $r \to 1-1/p$, while $N_{\frac{1}{2}\frac{\nu}{2}\frac{\nu+1}{2}}$ (given by (3.31)) diverges at that point so that their product (4.2) remains finite. We shall determine directly the B_ν's and their u-channel counterpart β_ν, defined by

$$\tilde{G}_4 = U_0 \mathcal{U}_0 + \beta_\nu \check{U}_2 \check{\mathcal{U}}_1 \tag{4.4}$$

where U_λ are given by (2.10) (2.11) while $\check{\mathcal{U}}_{\lambda/2}$ are the U_q-invariants (3.7), by demanding

$$\tilde{G}_4 = (-1)^\nu G_4. \tag{4.5}$$

Indeed, using (2.17) (2.18) and (3.9) we can write

$$\frac{\Gamma(2-2r)}{\Gamma(1-r)}\left\{\frac{\Gamma((\nu+1)r-1)}{\Gamma(\nu r)}\check{S}_{\nu-1} + \frac{\Gamma(1-(\nu+1)r)}{\Gamma(2-(\nu+2)r)}\check{S}_{\nu+1}\right\}\left(\frac{-[\nu]}{[\nu+1]}\check{\mathcal{S}}_{\frac{\nu-1}{2}} + \check{\mathcal{S}}_{\frac{\nu+1}{2}}\right) + \\ + \frac{\Gamma(2r)}{[2]\Gamma(r)}\beta_\nu\left\{\frac{\Gamma((\nu+1)r-1)}{\Gamma((\nu+2)\rho-1)}\check{S}_{\nu-1} + \frac{\Gamma(1-(\nu+1)r)}{\Gamma(1-\nu r)}\check{S}_{\nu+1}\right\}\left(\frac{[\nu+2]}{[\nu+1]}\check{\mathcal{S}}_{\frac{\nu-1}{2}} + \check{\mathcal{S}}_{\frac{\nu+1}{2}}\right) = \\ = (-1)^\nu\left(B_{\nu-1}\check{S}_{\nu-1}\check{\mathcal{S}}_{\frac{\nu-1}{2}} + B_\nu \check{S}_{\nu+1}\check{\mathcal{S}}_{\frac{\nu+1}{2}}\right) \tag{4.6}$$

The absence of mixed terms ($\check{S}_{\nu-1}\check{\mathcal{S}}_{\frac{\nu+1}{2}}$ and $\check{S}_{\nu+1}\check{\mathcal{S}}_{\frac{\nu-1}{2}}$) gives two compatible equation for $\beta\nu$ each yielding

$$\frac{\Gamma(2r)}{[2]\Gamma(r)}\beta_\nu = -\frac{\Gamma(2-2r)\Gamma((\nu+2)r-1)}{\Gamma(1-r)\Gamma(\nu r)}. \tag{4.7}$$

Inserting this value back in Eq. (4.6) gives

$$(-1)^{\nu-1}B_{\nu-1} = \frac{\Gamma(r)\Gamma((\nu+1)r-1)}{\Gamma(2r-1)\Gamma(\nu r)} \tag{4.8}$$

(where we have used the identity $[\nu]+[\nu+2]=[2][\nu+1]$). In verifying the fact that B_ν in (4.6) is obtained from (4.8) by substituting ν by $\nu+1$ (as a consequence of the symmetry of structure constants) we use the identity

$$\frac{2\Gamma(2-2r)\Gamma(1-(\nu+1)r)\sin\pi(\nu+1)r\cos\pi r}{\Gamma(1-r)\Gamma(2-(\nu+2)r)\sin\pi(\nu+2)r} = \frac{\Gamma(r)\Gamma((\nu+2)r-1)}{\Gamma(2r-1)\Gamma((\nu+1)r)}\ (=(-1)^\nu B_\nu).$$

For equal spins, i.e. for $\nu=1$ the Green function of four (p,q)-step operators

$$G \equiv G\begin{pmatrix} 1 & 1 & ; & 1 & 1 \\ z_1u_1 & z_2u_2 & & z_3u_3 & z_4u_4 \end{pmatrix} =$$

$$= (z_{14}z_{23}\eta)^{1-\frac{3}{2}r}\{F(2-3r,1-r;2-2r;\eta)(q^{-1/2}u_1-q^{1/2}u_2)\times(q^{-1/2}u_3-q^{1/2}u_4)-$$
$$-\frac{\Gamma(r)\Gamma(3r-1)}{[2]\Gamma(2r-1)\Gamma(2r)}\eta^{2r-1}F(r,1-r;3r-2;\eta)((u_1-u_4)(q^{-1/2}u_2-q^{1/2}u_4)+$$
$$+q^{-2}u_1u_2+q^2u_3u_4-u_1u_4-u_2u_3)\},\quad q=e^{i\pi(1-r)} \tag{4.9}$$

gives rise to the following q-independent, 1-dim. representation of the braid group B_4

$$R_1G = iG = R_2G = R_3G, \qquad FG = -q^{3/2}e^{2\pi i\Delta_1}R_1R_2R_1G = -G. \tag{4.10}$$

In deriving the first relation (for R_1G) we have combined (2.29) and (3.20) (3.22); in writing down the expression for the fusion matrix F we have used (2.30) and (3.24).

We stress that although in some sense trivial this 1-dimensional representation of B does not reduce to a representation of the permutation group S_4 but is homomorphi to its double covering, since $R_i^2=-1\,(\neq 1)$ for $i=1,2,3$, a representation that als appears at the level 1 $\widehat{su}(2)$-current algebra theory [3]. (Its appearance can, in fact, b understood through the coset construction of a minimal model.)

Using the inner product on the space of U_q-invariants (introduced and studied in Sec. 3B) we can reproduce — for arbitrary $\nu(=1,2,\ldots,p-2)$ — the (euclidean 2-dimensional 4-point function:

$$(\overline{G_4},G_4) = [2][\nu+1]\sum_{\mu=\nu\pm1} C^4_{1\,\nu\,\mu}|S_\mu\begin{pmatrix} 1 & \nu & ; & \nu & 1 \\ z_1 & z_2 & & z_3 & z_4\end{pmatrix}|^2. \tag{4.11}$$

For $\nu=p-2$ the structure constant $C_{1\,p-2\,p-1}$ (4.3) vanishes so that the non-unitary rep resentation of Vir of weight Δ_{p-1} does not contribute to the U_q-invariant 2-dimensiona correlation function.

4B. Concluding remarks. The present paper presents a first step in the proof o the following statement concerning correlation functions in a chiral (p,q) model and it 2-dimensional RCFT counterpart.

Given a set of U_q- (and conformal) weights $\nu_1 = 2I_1, \dots, \nu_n = 2I_n$ such that a non-vanishing conformal block corresponding to a tree graph with such external legs' labels is consistent with the BPZ fusion rules, there exists a braid covariant Green function

$$G_n = G\begin{pmatrix} \nu_1 & \nu_2 & \dots & \nu_n \\ z_1 u_1 & z_2 u_2 & \dots & z_n u_n \end{pmatrix} \tag{4.12}$$

with the properties (i–iii) listed in the introduction. It splits into sums of products of conformal and U_q-blocks corresponding to a given tree diagram (including internal line labels forbidden by the BPZ fusion rules) changing when passing from one tree to another by at most a phase factor (in effect, a power of i). The corresponding 2-dimensional euclidean Green function is given by the norm square of G_n defined according to (3.28). (Its normalization is fixed by the norm (3.29) of the 2-point function and would hence differ from the BPZ normalization [1] by an overall factor $([\nu_1+1][\nu_2+1]\dots[\nu_n+1])^{1/2}$.) The non-unitary terms in G_n do not contribute to $\|G_n\|^2$.

We have verified this statement for the 4-point function involving a pair of step operators (for $\nu_1 = \nu_4 = 1$) and for $\nu_2 = \nu_3 (= \nu)$:

$$G_4 = G\begin{pmatrix} 1 & \nu & \nu & 1 \\ z_1 u_1 & z_2 u_2 & z_3 u_3 & z_4 u_4 \end{pmatrix} =$$

$$= \frac{\Gamma(r)}{\Gamma(2r-1)}\left\{(-1)^{\nu-1}\frac{\Gamma((\nu+1)r-1)}{\Gamma(\nu r)}\check{S}_{\nu-1}\check{\mathcal{S}}_{\frac{\nu-1}{2}} + (-1)^{\nu}\frac{\Gamma((\nu+2)r-1)}{\Gamma((\nu+1)r)}\check{S}_{\nu+1}\check{\mathcal{S}}_{\frac{\nu+1}{2}}\right\} \tag{4.13a}$$

$$= (-1)^{\nu}\left\{U_0\mathcal{U}_0 - \frac{\Gamma^2(r)\Gamma((\nu+2)r-1)}{\Gamma(2r)\Gamma(2r-1)\Gamma(\nu r)}\check{U}_2\check{\mathcal{U}}_1\right\} \tag{4.13b}$$

where the s- and u- channel blocks are computed in terms of hypergeometric functions of the cross-ratio η and homogeneous polynomials in the u_i's. For $\nu = p-2$ we encounter the first example of a chiral Green function with a non-unitary contribution (the second term in either (4.13a) or (4.13b)) that disappears in the 2-dimensional correlation function $\|G_4\|^2$.

The next step in proving our statement should involve a study of indecomposable representations of both Vir and U_q. The arising difficulties and the way to overcome them are briefly discussed in Secs.2D and 3C. The problem already appears in the Ising model ($p = 4$) for 4 q-spin-1 fields. The details of its solution will be published elsewhere.

Similar results for the $s\hat{u}(2)$-current algebra can be obtained with the same case. (A more conceptual approach to that case is being developed by K.Gawedzki [34] who relates it to new results of G.Felder.) Here one also observes that models with 2-dimensional fields of non-zero spin can be treated by considering non-diagonal pairings between the current algebra and the U_q-blocks. Thus the (extended) chiral counterpart of a 2-dimensional RCFT appears not just as an ingredient but as a faithful image of the full theory in which the antianalytic pieces are substituted by U_q-blocks. This picture can be completed by introducing a chiral partition function [35] which reflects the operator content of the theory and allows to reproduce the ADE classification of $s\hat{u}(2)$-current algebra models (see the work of Cappelli, Itzykson and Zuber in CIASM [1] and further references cited there).

Acknowledgements. The present paper was preceded and stimulated by a joint effort of D.Buchholz, I.Frenkel, G.Mack and one of the authors (I.T.) to understand the role of quantum groups in RCFT.

It a pleasure to thank J.Fröhlich, K.Gawedzki and G.Mack for acquainting us with their related current work and for helpful discussions.

R.P. and I.T. thank la Division de Physique Théorique, Institut de Physique Nucléaire, Orsay for hospitality; L.H. and R.P. acknowledge financial support from D.F.G. and the hospitality of the II. Institut für Theoretische Physik, Univesität Hamburg where parts of this work was done. The results of the paper were reported at the 4th Annecy Meeting on Theoretical Physics (March, 1990) at the 50ième Rencontre entre Physiciens Théoriciens et Mathématiciens à Strasbourg (April, 1990) and at the Workshop on Quantum Groups, Symmetries of Dynamical Systems and Conformal Field Theory at the Euler International Mathematical Institute in Leningrad (November, 1990). I.T. thanks the organizers for invitation. The work was supported in part by the Ministry of Science, Culture and Education of Bulgaria under Contract N403.

Appendix A

The U_q algebra for $q+q^{-1}=2\cos\pi/p$
Admissible representations and tensor operators

We start by summarizing the relevant facts about the (quasitriangular) QUE algebra [9,10]

$$U_q \equiv U_q(sl(2)) \quad \text{for } q+q^{-1}=2\cos\pi/p \quad (q^p=-1), \quad p=3,4,\ldots \tag{A.1}$$

fixing on the way our conventions (which are essentially those of [11]).

U_q is defined as a Hopf algebra generated by four elements, $q^{\pm J_3}$ and $J_\pm$, subject to the relations specified below. It has, first of all, the structure of an associative algebra with a unit whose generators satisfy

$$q^{J_3}q^{-J_3}=1\,(=q^{-J_3}q^{J_3}),\ q^{J_3}J_\pm=J_\pm q^{J_3\pm1}, \tag{A.2a}$$

$$[J_+,J_-]\equiv J_+J_- - J_-J_+ = [2J_3], \tag{A.2b}$$

where

$$[X]\,(\equiv[X]_q)=\frac{q^X-q^{-X}}{q-q^{-1}} \tag{A.3}$$

($X=aJ_3+b,\quad a,b\in\frac{1}{2}\mathbf{Z}$). It is also equipped with a coproduct $\Delta: U_q\to U_q\otimes U_q$, a deformation of the tensor product of U_q with itself, that is an algebra homomorphism satisfying

$$\Delta(q^{\pm J_3})=q^{\pm J_3}\otimes q^{\pm J_3},\quad \Delta(J_\pm)=q^{J_3}\otimes J_\pm + J_\pm\otimes q^{-J_3}. \tag{A.4}$$

If σ is the permutation in $U_q\otimes U_q$,

$$\sigma(X\otimes Y)=Y\otimes X, \tag{A.5}$$

then we can define a second coproduct Δ' setting

$$\Delta'(X)=\sigma\cdot\Delta(X). \tag{A.6}$$

There exists an algebra homomorphism $\varepsilon : U_q \to \mathbb{C}$, called the *counit*, satisfying

$$\varepsilon(q^{\pm J_3}) = 1, \quad \varepsilon(J_\pm) = 0, \tag{A.7}$$

and an antihomomorphism $\gamma : U_q \to U_q$, the *antipode*, such that

$$\gamma(q^{\pm J_3}) = q^{\mp J_3}, \quad \gamma(J_\pm) = -q^{\mp 1} J_\pm. \tag{A.8}$$

The associative multiplication m in U_q, the identity map $id : U_q \to U_q$ and the above (anti)homomorphisms satisfy the following *defining relations for a Hopf algebra*:

$$(id \otimes \Delta)\Delta(X) = (\Delta \otimes id)\Delta(X) \quad \text{for } X \in U_q, \tag{A.9a}$$

$$m(id \otimes \gamma)\Delta(X) = m(\gamma \otimes id)\Delta(X) = \varepsilon(X)1, \tag{A.9b}$$

$$(\varepsilon \otimes id)\Delta(X) = (id \otimes \varepsilon)\Delta(X) = X. \tag{A.9c}$$

Drinfeld [6] has demonstrated that every QUE algebra is almost cocommutative in the sense that there exists an invertible element

$$R \in U_q \otimes U_q, \quad R = \sum_n X_n \otimes Y_n, \tag{A.10a}$$

the *universal R-matrix* such that

$$\Delta'(X) = R\Delta(X)R^{-1}. \tag{A.10b}$$

Moreover, U_q is a quasitriangular algebra, i.e. it has the property which we proceed to define. Let $R_{ij}, i \neq j = 1,2,3$ be elements of $U_q \otimes U_q \otimes U_q$ given by

$$R_{12} = R \otimes 1, \quad R_{23} = 1 \otimes R, \quad R_{13} = \sum_n X_n \otimes 1 \otimes Y_n. \tag{A.11}$$

The condition that (U_q, R) is a quasitriangular Hopf algebra reads

$$(\Delta \otimes id)(R) = R_{12} \cdot R_{23}, \quad (id \otimes \Delta)(R) = R_{13} R_{12}. \tag{A.12}$$

For quasitriangular algebras one proves [6] the Yang-Baxter equation

$$R_{12} R_{13} R_{23} = R_{23} R_{13} R_{12} \tag{A.13}$$

and the relations

$$(\gamma \otimes id)(R) = R^{-1} = (id \otimes \gamma)R \tag{A.14a}$$

$$(\varepsilon \otimes id)(R) = 1 = (id \otimes \varepsilon)R \tag{A.14b}$$

For $q^p = -1$ — see (A.1) — each of the elements $J_\pm^p$ of U_q generates an ideal, and we shall set, following [11]

$$J_+^p = 0 = J_-^p. \tag{A.15}$$

(The relations (A.15) are thus a part of our definition of U_q.)

U_q has finite dimensional representations which can be labeled by their q-spin $2I \in \mathbf{Z}_+$. The representation of q-spin I is $2I+1$ dimensional and admits a canonical basis $|I\,m\rangle$ $m = -I, 1-I, \ldots, I$ such that

$$(q^{\pm J_3} - q^{\pm m})|I\,m\rangle = 0, \tag{A.16a}$$

$$J_\pm |I\,m\rangle = \sqrt{[I \mp m][I \pm m + 1]}\ |\,I\,m \pm 1\rangle \tag{A.16b}$$

(in particular, $J_+|I\,I\rangle = 0 = J_-|I\ -I\rangle$). We can also define a q-deformation of the familiar $su(2)$ Casimir operator

$$C_2 = J_-J_+ + [J_3 + 1/2]^2 - [1/2]^2 = J_-J_+ + [J_3][J_3+1] \tag{A.17}$$

satisfying

$$(C_2 - [I][I+1])|I\,m\rangle = 0. \tag{A.18}$$

For $2I \le p-1$ the representations so constructed are irreducible. For $p \le 2I \le 2p-2$ they are fully reducible and split into a direct sum of three irreducible U_q modules spanned by the vectors $|I\,m\rangle$ with

$$-I \le m \le I-p, \quad I+1-p \le m \le p-1-I, \quad p-I \le m \le I, \tag{A.19}$$

respectively. (The U_q invariance of these subspaces is a direct consequence of (A.16) noting that $[p] = 0$. The representations of q-spins I and $p-1-I$ are characterized by the same eigenvalue of C_2 (A.18), since

$$[I][I+1] + [1/2]^2 = [I+1/2]^2 = [p-I-1/2]^2. \tag{A.20}$$

In fact, for $p \le 2I \le 2p-2$ the subrepresentation with J_3 projection satisfying the middle inequalities (A.19) is equivalent to the irreducible representation of q-spin $p-1-I$. An useful additional characteristics (besides the value of C_2) of the finite dimensional representations D of U_q is their q-dimension

$$d_q(D) = tr_D q^{2J_3}. \tag{A.21}$$

The q-dimension of the representation D_I of q-spin I is

$$(d_q(D_I) \equiv) d_q(2I) = [2I+1] = q^{-2I} + q^{2-2I} + \cdots + q^{2I} \tag{A.22a}$$

satisfying

$$\begin{gathered} d_q(2I) = -d_q(2p-2-2I) \quad 0 \le 2I \le 2p-2, \qquad \text{(A.22b)} \\ 0 \le d_q(2I) = d_q(p-2-2I) = -d_q(2I+p) \quad -1 \le 2I \le p-1. \qquad \text{(A.22c)} \end{gathered}$$

Noting the relation

$$\begin{aligned} q^{2I-2p} + \cdots + q^{-2I} &= -(q^{2I-p} + \cdots + q^{p-2I}) = -d_q(2I-p) = \\ &= -d_q(2p-2-2I) = q^{2p-2I} + \cdots + q^{2I} \end{aligned} \tag{A.23}$$

we see that the q dimensions of all three pieces of the decomposition (A.19) have the same absolute value, but for the middle one d_q is positive while for the "upper" and "lower" parts it is negative.

We shall call a finite dimensional irreducible representation of U_q *admissible* if its q-dimension is non-negative

Consider the irreducible U_q tensor operators

$$A^I = \{A^I_m, \quad m = -I_p, 1 - I_p, \ldots, I_p\} \quad 2I = 0, 1, \ldots, 2p-2, \tag{A.24a}$$

$$I_p = \min(I, p-1-I) \tag{A.24b}$$

which map a U_q-invariant "vacuum" vector $|0\rangle$, such that

$$X|0\rangle = \varepsilon(X)|0\rangle \quad \text{for every } X \in U_q \tag{A.25}$$

into an irreducible, $(2I_p+1)$-dimensional U_q module $V_I(p)$ so that

$$A^I_m|0\rangle = |I\, m\rangle \quad (-I_p \le m \le I_p). \tag{A.26}$$

Postulating the commutation relations

$$q^{J_3} A^I_m = q^m A^I_m q^{J_3} \tag{A.27a}$$

$$J_\pm A^I_m = \sqrt{[I \mp m][I \pm m + 1]}A^I_{m\pm 1} q^{-J_3} + q^m A^I_m J_\pm \tag{A.27b}$$

we reproduce the U_q-transformation law (A.16) consistent with the additional (U_q-invariant) condition

$$J_\pm^{2I_p+1} A^I_m|0\rangle = 0. \tag{A.28}$$

In verifying the equivalence of the resulting representation of U_q with the one with q-spin I_p (for $2I \ge p$) we use the invariance of $[I \mp m][I \pm m + 1]$ with respect to the substitution $I \to p - 1 - I$.

The map $A^I : \{|0\rangle\} \to V_I$ can be extended to a U_q-vertex operator (qVO)

$$\begin{gathered} {}^{I_2}A^I_{I_1} : V_{I_1} \to V_{I_2} \\ \text{for } I_2 = |I - I_1|, \quad |I - I_1| + 1, \ldots, \min(I_1 + I, 2p - 2 - I_1 - I) \end{gathered} \tag{A.29}$$

whose components still satisfy (A.27). (In the above special case when $I_1 = 0$ range of I_2 consists of a single point $I_2 = I$ and we use the simplified notation of Eq.(A.26) rather than writing ${}^I(A^I_m)_{0\cdot}$.) By an abuse of notation we shall denote by A^I both the collection of all qVOs (A.29) and its vacuum component. The only non-vanishing (tensor) product of two qVOs is ${}^{I_1}A^I_J \otimes {}^J A^{I'}_{I_2}$. We shall be particularly interested in U_q-invariant vacuum U_q-blocks

$${}^0A^{I_1}_{I_1} \otimes {}^{I_1}A^{I_2}_{I_2'} \otimes \cdots \otimes {}^{I_n}A^{I_n}_0 = \langle 0|A^{I_1} \otimes A^{I_2}_{I_2'} \otimes {}^{I_2'}A^{I_3}_{I_3'} \otimes \cdots \otimes A^{I_n}|0\rangle \tag{A.30}$$

which are determined by their U_q invariance up to multiplicative factors. In particular, the 2-points block is proportional to

$$\langle 0|A^{I_1}_{m_1} A^{I_2}_{m_2}|0\rangle = \delta_{I_1 I_2} q^{-m_1}(-1)^{I_1 - m_1}\delta_{m_1+m_2} \quad (\delta_I \equiv \delta_{I_0}). \tag{A.31}$$

Following [16] we shall exchange the set of tensor components $\{A^I_m\}$ for $2I \le p-1$ by a polynomial in a single variable u of degree $2I$:

$$A^I(u) = \sum_{m=-I}^{I} \begin{bmatrix} 2I \\ I+m \end{bmatrix}^{1/2} A^I_m u^{I+m} \quad (2I = 2I_p \le p-1), \tag{A.32}$$

where the bracket stands for the q-deformed binomial coefficients

$$\begin{bmatrix} 2I \\ I+m \end{bmatrix} = \frac{[2I]!}{[I+m]![I-m]!}, \quad [n]! = [n][n-1]\dots 1. \tag{A.33}$$

(The undeformed limits of such operators and associated states are known as $su(2)$ coherent states — see [36].)

This allows us to write the normalized 2-point function (A.31) in the form

$$\begin{aligned} \langle A^I(u_1)A^I(u_2)\rangle &= \left[q^{-1/2}u_1(-)q^{1/2}u_2\right]^{2I} \equiv \\ &\equiv \sum_{m=-I}^{I} \begin{bmatrix} 2I \\ I+m \end{bmatrix} (-1)^{I-m} q^{-m} u_1^{I+m} u_2^{I-m}. \end{aligned} \tag{A.34}$$

The general U_q-invariant 3-point function can then be written as a product of three propagators:

$$\langle A_{I_1}(u_1)A_{I_2}(u_2)A_{I_3}(u_3)\rangle \equiv V\begin{pmatrix} I_1 & I_2 & I_3 \\ u_1 & u_2 & u_3 \end{pmatrix} N_{I_1I_2I_3} \check{V}\begin{pmatrix} I_1 & I_2 & I_3 \\ u_1 & u_2 & u_3 \end{pmatrix}, \tag{A.35a}$$

$$\begin{aligned} \check{V}\begin{pmatrix} I_1 & I_2 & I_3 \\ u_1 & u_2 & u_3 \end{pmatrix} &= \left[q^{-\frac{1}{2}(I_3+1)}u_1(-)q^{\frac{1}{2}(I_3+1)}u_2\right]^{I_1+I_2+I_3} \times \\ \times \left[q^{\frac{1}{2}(I_2-1)}u_1(-)q^{\frac{1}{2}(1-I_2)}u_3\right]^{I_1+I_3-I_2} &\left[q^{-\frac{1}{2}(I_1+1)}u_2(-)q^{\frac{1}{2}(I_2+1)}u_3\right]^{I_2+I_3-I_1}. \end{aligned} \tag{A.35b}$$

We restrict the choice of the structure constants $N_{I_1I_2I_3}$ in Sec.3A by demanding involutivity type properties of a suitable defined fusion matrix and by using the symmetric normalization condition

$$N_{II0} = N_{0II} = N_{I0I} = 1 \tag{A.36}$$

(related to the choice (A.34) for the normalized 2-point function).

To prove (A.34) and (A.35) one uses the following substitute for the covariance relation (A.27):

$$q^{J_3}A^I(u)q^{-J_3} = q^{-I}A^I(qu) \tag{A.37a}$$

$$J_+A^I(u) = q^{-I}A^I(qu)J_+ + \mathcal{D}A^I(u)q^{-J_3} \tag{A.37b}$$

where the "q-derivative" $\mathcal{D}(=\mathcal{D}_u)$ is the finite difference operator

$$\mathcal{D}f(u) = \frac{f(qu)-f(q^{-1}u)}{u(q-q^{-1})} \quad (\Rightarrow \mathcal{D}u^n = [n]u^{n-1}). \tag{A.38}$$

Eq. (A.37a) implies the homogeneity condition

$$\langle A^{I_1}(qu_1)\dots A^{I_n}(qu_n)\rangle = q^{I_1+\dots+I_n}\langle A^{I_1}(u_1)\dots A^{I_n}(u_n)\rangle. \tag{A.39}$$

The fact that the A's are the polynomials then tells us that $I_1+\dots+I_n$ should be a positive integer. J_+ invariance gives, on the other hand (for $\mathcal{D}_i \equiv \mathcal{D}_{u_i}$),

$$\begin{aligned}&\mathcal{D}_1\langle A^{I_1}(u_1)A^{I_2}_{I_2'}(q^{-1}u_2)\dots A^{I_n}(q^{-1}u_n)\rangle q^{I_2+\dots+I_n}+\\ &+\mathcal{D}_2\langle A^{I_1}(qu_1)A^{I_2}_{I_2'}(u_2)\dots A^{I_n}(q^{-1}u_n)\rangle q^{I_3+\dots+I_n-I_1}+\dots+\\ &+\mathcal{D}_n\langle A^{I_1}(qu_1)A^{I_2}_{I_2'}(q\,u_2)\dots A^{I_n}(u_n)\rangle q^{-I_1+\dots+I_{n-1}}=0.\end{aligned} \tag{A.40}$$

We shall continue the argument for the 2-point function $\langle A^{I_1}(u_1)A^{I_2}(u_2)\rangle$. Setting $I = \min(I_1, I_2)$ and using (A.39) we can write

$$\langle A^{I_1}(u_1)A^{I_2}(u_2)\rangle = \sum_{m=-I}^{I} \alpha_m u_1^{I_1+m}u_2^{I_2-m}.$$

Eq. (A.40) then gives $\alpha_m[I_2-m]+\alpha_{m+1}[I_1+m+1]=0$ for $-I\le m\le I-1$, and $\alpha_m[I_1+m]+\alpha_{m-1}[I_2-m+1]=0$ for $1-I\le m\le I$, while $\alpha_I[I_2-I]=0=\alpha_{-I}[I_1-I]$. Thus a non-trivial U_q-invariant 2-point function only exists for $[\,|I_1-I_2|\,]=0$ or (for $q=e\left(\frac{\pm1}{2p}\right)$) $|I_1-I_2|\in\mathbf{Z}_p$. The assumption $0\le I_i\le p-1$ then implies $I_1=I_2=I$. The above recurrence relation for α_m then leads to a 2-point function proportional to (A.34). The proof of (A.35) is similar (though longer).

References

[1] A.A.Belavin, A.M.Polyakov, A.B.Zamolodchikov, Nucl. Phys. **B241** (1984), 333, this and other basis original papers are collected in: *Conformal Invariance and Applications to Statistical Mechanics* (CLASM) Eds. C.Itzykson, H.Saleur, J-B.Zuber (World Scientific, Singapore 1988).

[2] P.Furlan, G.M.Sotkov, I.T.Todorov, Riv. Nuovo. Cim. **12** no. 6 (1989), 1.

[3] G.Moore, N.Seiberg, *Lectures on RCFT*, Rutgers and Yale Univ. preprint RU-89-32 and YCTP-P13-89; see also Commun. Math. Phys. **123** (1989), 177.

[4] A.Tsuchiya, Y.Kanie, Conformal Field Theory and Solvable Lattice Models, Advanced Studies in Pure Mathematics **16** (1988), 297–372; Lett. Math. Phys. **13** (1987), 303.

[5] K.H.Rehren, Commun. Math. Phys. **116** (1988), 675; Nucl. Phys. **B312** (1989), 715.

[6] J.Fröhlich, *Statistics of fields, the Yang-Baxter equation and the theory of knots and links, 1987 Cargèse lectures*, Non-Perturbative Quantum Field Theory, Eds. G.'t Hooft et al., New York, Plenum Press (to appear).

[7] G.Felder, J.Fröhlich, G.Keller, Commun. Math. Phys. **124** (1989), 647.

[8] J.-L.Gervais, A.Neveu, Nucl. Phys. **B238** (1984), 125.

[9] V.G.Drinfeld, *Quantum groups*, Proc. ICM-86, Vol.1, Berkeley, CA, 1987, pp. 798–820; Algebra and Anal. **1** no. 2 (1989), 30. (Russian)

[10] L.D.Faddeev, N.Yu.Reshetikhin, L.A.Takhtajan, Alg. Anal. **1** no. 1 (1989), 178. (Russian)

[11] V.Pasquier, H.Saleur, Nucl. Phys. **B330** (1990), 523.

[12] V.G.Drinfeld, Alg. Anal. **1** no. 6 (1989), 114; **2** no. 4 (1990), 149.

[13] R.Dijkgraaf, V.Pasquier, R.Roche, *Quasi-quantum groups related to orbifold models*, Talk at Inter. Colloq. Modern Quantum Field Theory, Tata Inst. Fund. Research, Jan. 1990.

[14] B.Schroer, *Algebraic QFT as a framework for classification and model building. A heretic view of the New Kinematics*, Annecy-le-Vieux preprint LAPP-TH-280/90, (and references therein).

[15] L.Alvarez-Gaumé, C.Gomes, G.Sierra, Nucl. Phys. Lett. **B319** (1989), 155; Phys. Lett. **B220** (1989), 142; Nucl. Phys. **B330** (1990), 347.
[16] A.Ch.Canchev, V.B.Petkova, Phys. Lett. **B233** (1989), 374.
[17] G.Moore, N.Reshetikhin, Nucl. Phys. **B328** (1989), 557.
[18] P.Furlan, A.Ch.Ganchev, V.B.Petkova, *Quantum groups and fusion rules multiplicities*, Trieste preprint INFN/AE-89/15.
[19] E.Witten, Nucl. Phys. **B330** (1990), 285.
[20] E.Guadagnini, M.Martellini, M.Mintchev, Phys. Lett. **B235** (1990), 275.
[21] D.J.Smit, Commun. Math. Phys. **128** (1990), 1.
[22] N.Reshetikhin, F.Smirnov, *Hidden quantum group symmetry and integrable perturbations of conformal field theories*, LOMI-Harvard preprint (October 1989).
[23] O.Babelon, Phys. Lett. **B215** (1988), 523.
A.Alexeev, S.Shatashvili, Nucl. Phys. **B323** (1989), 719.
L.D.Faddeev, Commun. Math. Phys. **132** (1990), 131.
[24] J.-L.Gervais, Commun. Math. Phys. **130** (1990), 257.
E.Cremmer, J.-L.Gervais, Commun. Math. Phys. **134** (1990), 619.
J.-L.Gervais, *Critical dimensions for non critical strings*, Paris preprint LPTENS 90/4.
[25] J.Fröhlich, C.King, Int. J. Mod. Phys. **A4** (1989), 5321.
[26] C.Gomez, G.Sierra, Phys. Lett. **B240** (1990), 149; *The quantum symmetry of rational conformal field theories*, Genève preprint UGVA-DPT 1990/04-669.
[27] J.Fröhlich, T.Kerler, *On the role of quantum groups in low dimensional local quantum field theory*, ETH, Zürich preprint (1990).
[28] *Algebraic Theory of Superselection Sectors: Introduction and Recent Results*, Proc. of the Convegno Internazionale di Presentazione dell' Instituto Scientifico Internazionale G.B. Guccia "Algebraic Theory of Superselection Sectors and Field Theory" held in Palermo (Italy) 23–30 November 1989, Ed. D.Kastler, Word Scientific, Singapore, 1990, (See, in particular, the introductory article by D.Kastler, M.Mebkhout and K.H.Rehren, as well as the contributions by D.Buchholz, G.Mack, I.T.Todorov; G.Mack, V.Schomerus; R.H.Rehren).
[29] D.Buchholz, G.Mack, I.T.Todorov, Conformal Field Theory and Related Topics, Eds. P.Binetruy, P.Sorba; Nucl. Phys. B (proc. Supl.) **5B** (1988), 20.
[30] K.Fredenhagen, R.H.Rehren, B.Schroer, Commun. Math. Phys. **125** (1989), 201.
[31] R.Longo, Commun. Math. Phys. **126** (1989), 217, and to be published.
L.S.Dotsenko, V.A.Fateev, Nucl. Phys. **B240** **[FS12]** (1984), 312; CLASM [1], 214–250.
[32] I.T.Todorov, Lect. Notes Phys. **370** (1990), 231–277.
[33] A.N.Kirillov, N.Yu.Reshetikhin, *Representations of the algebra* $U_q(sl(2))$, *q-orthogonal polynomials and invariant of links*, LOMI (Leningrad) preprint E-9-88.
[34] K.Gawedzki, *Geometry of Wess-Zumino-Witten models of conformal field theory*, Proceedings of the 4th Annecy Meeting in Theoretical Physics, Spring 1990, (in honour of Raymond Stora).
[35] I.-G.Koh, S.Ouvry, I.T.Todorov, *Quantum dimensions and modular forms in chiral conformal theory*, Orsay preprint IPNO/TH 90-17, Phys. Lett. **B**.
[36] J.Kurchan, P.Lebeuf, M.Saraceno, Phys. Rev. **A40** (1989), 6800.
[37] L.K.Hadjiivanov, R.R.Paunov, I.T.Todorov, *Quantum group extended chiral p-models* (to appear).

INSTITUTE FOR NUCLEAR RESEARCH AND NUCLEAR ENERGY, BG-1784 SOFIA, BULGARIA

FUSION RSOS MODELS AND RATIONAL COSET MODELS

ATSUO KUNIBA[(1)] AND TOMOKI NAKANISHI[(2)]

[(1)] School of Mathematical Sciences, IAS, Australian National University
[(2)] Department of Mathematics, Nagoya University

ABSTRACT. **A series of conjectures is presented for a family of fusion $\widehat{sl}(n)_\ell$ RSOS models labeled by rank n, level ℓ, degree N of the fusion and primitive roots of unity $e^{\frac{2\pi it}{n+\ell}}$. Based on it, we introduce rational coset models which are the universality classes of their critical behaviors.**

1. Introduction

The RSOS model in [1–3] is a 2-dimensional statistical model which by construction is related to the affine Lie algebra $\widehat{sl}(n)_\ell$ and the quantum group $U_q(sl(n))$ at $q = e^{\frac{2\pi i}{n+\ell}}$. In spite of being defined on a lattice, it realizes several important structures in rational conformal field theory (RCFT) e.g., fusion rules, monodromy matrices, level-rank dualities and so forth. Moreover, it is exactly solvable away from criticality and its 1-point functions (in regime III) are determined by the character identities in the coset construction $\widehat{sl}(n)_{\ell-1} \oplus \widehat{sl}(n)_1 \oplus \widehat{sl}(n)_\ell$. Thus the model realizes the critical behavior described by the minimal model of the W_n algebra at the critical point.

Besides extensions of such results to other affine Lie algebras [4–6], there are at least two directions to generalize the RSOS model. The first one is to consider the non-positive region [7–11] of the model parametrized by a discrete parameter $t(1 \leq t \leq s = n - \ell)$ coprime with s, where the deformation parameter in $U_q(sl(n))$ now becomes the root of unity $q = e^{\frac{2\pi it}{n+\ell}}$. The choice $t = 1$ reduces to the original case. For $1 < t < s$, the analysis of [11] using the level-rank duality [2,12] shows that the non-unitary minimal models of the W_n or W_ℓ algebras appear as the universality classes. The second direction is to consider the *fusion* models [13–15] parametrized by the finite dimensional irreducible representations Y of $sl(n)$. The original model [2] is included as the simplest case $Y =$ the simplest representation. The relevant coset pair describing the 1-point functions has been conjectured as [16]

$$\widehat{sl}(n)_{\ell_1} \oplus \cdots \oplus \widehat{sl}(n)_{\ell_k} \supset \widehat{sl}(n)_\ell, \tag{1}$$

where the integer k and the levels $\{\ell_i\}$ s.t. $\sum_{i=1}^k \ell_i = \ell$ are uniquely specified from the Y by a certain rule.

The purpose of this note is to study the fusion RSOS models corresponding to

$$\begin{aligned} &Y = \text{symmetric tensor representation of degree} N, \\ &1 \leq t \leq \ell/N, \qquad t \text{ and } s \text{ are coprime.} \end{aligned} \tag{2}$$

We shell present conjectures on H-function (6), the structure of the ground states (9) and the 1-point functions (14) in a regime III-like region. These are supported by a computer analysis of the combinatorial q-series (1-dimensional configuration sums) arising from the corner transfer matrix method [17].

We remark that for $t = 1$, these 1d sums have also been given Lie algebraic characterizations [18,19]. However, the 1-point functions themselves have been only analytically derived so far for $(n,t,N) = (2, 1, \text{general}$ [13], $(n,t,N) = (\text{general}, 1, 1)$ [2] and $(n,t,N) = (2\text{or}3, 1 \sim \ell, 1)$ [7,10]. Our conjecture generalizes the earlier one in [15] for $(n,t,N) = (\text{general}, 1, \text{general}$. Note that due to the level-rank duality in the fusion models [12], our analysis on the case (2) also yields informations in a regime II-like region of the $\widehat{sl}(n)_\ell$ RSOS model undergone the fusion of type $Y =$ antisymmetric tensor of degree N.

2. Fusion RSOS Models

Fix the integers $n \geq 2, \ell \geq 2, 1 \leq N \leq \ell$. Let us recall the level ℓ fusion RSOS model corresponding to the degree N symmetric tensor representation of $sl(n)$ [14,15]. Let $\Lambda_0, \ldots, \Lambda_{n-1}$ be the fundamental weights of $\widehat{sl}(n)$. We extend the suffix to all integers by $\Lambda_\mu = \Lambda_{\mu+n}, \forall\mu$ and set $\hat{\mu} = \Lambda_{\mu+1} - \Lambda_\mu$. Let $P_+(n,\ell)$ denote the set of level ℓ dominant integral weights (DIWs) of $\widehat{sl}(n)$, i.e., $P_+(n,\ell) = \{a^0\Lambda_0 + \cdots + a^{n-1}\Lambda_{n-1} | \forall a^\mu \in \mathbf{Z}_{\geq 0}, \sum a^\mu = \ell\}$. For each $a \in P_+(n,\ell)$ as above, we assign the coordinates $a_{\mu\nu}(0 \leq \mu,\nu \leq n-1)$ by $a_{\mu\nu} = -a_{\nu\mu} = a^{\mu+1} + a^{\mu+2} + \cdots + a^\nu + \nu - \mu (\mu \leq \nu)$. We shall call an ordered pair (a,b) of the DIWs *admissible* if and only if the following hold.

i) $b - a = \sum_{i=1}^N \hat{\mu}_i$ for some $0 \leq \mu_1, \ldots, \mu_N \leq n-1$.
ii) $a + \sum_{i=1}^k \hat{\mu}_{\sigma(i)} \in P_+(n,\ell)$ for all $1 \leq k \leq N$ and all permutations σ on N numbers $1, \ldots, N$.

In terms of the fusion rule [20,21] of the WZNW model, these conditions are simply stated as $N_{a[N]}^b = 1$. See [22] for this point.

Consider a planar square lattice with a fluctuation variable $a^{(i)} \in P_+(n,\ell)$ assigned to each lattice site i. For every adjacent sites i and j (j: either right or lower neighbor of i), we impose a constraint that $(a^{(i)}, a^{(j)})$ must be admissible. Elementary interaction takes place among the four DIWs a,b,c,d (ordered clockwise from the NW corner) round a face and is specified by the Boltzmann weights $W_N\begin{pmatrix} a & b \\ d & c \end{pmatrix}\!\!\Big| u\Big)$ that depend on the *spectral parameter* u and solve the Yang-Baxter equation (cf. [17]). For $N = 1$ they are given as follows [2]:

$$W_1\begin{pmatrix} a & a+\hat{\mu} \\ a+\hat{\mu} & a+2\hat{\mu} \end{pmatrix}\Big|u = \frac{[\lambda+u]}{[\lambda]},$$
$$W_1\begin{pmatrix} a & a+\hat{\mu} \\ a+\hat{\mu} & a+\hat{\mu}+\hat{\nu} \end{pmatrix}\Big|u = \frac{[\lambda a_{\mu\nu}-u]}{[\lambda a_{\mu\nu}]} \quad (\mu\neq\nu), \tag{3}$$
$$W_1\begin{pmatrix} a & a+\hat{\nu} \\ a+\hat{\mu} & a+2\hat{\mu} \end{pmatrix}\Big|u = \frac{[u]}{[\lambda]}\frac{[\lambda(a_{\mu\nu}+1)]}{[\lambda a_{\mu\nu}]} \quad (\mu\neq\nu).$$

Here the symbol $[u]$ is defined by

$$[u] = 2|p|^{1/8}\sin\pi u\prod_{j=1}^{\infty}(1-2p^j\cos 2\pi u+p^{2j})(1-p^j), \tag{4}$$

where $p(|p|<1)$ is the elliptic nome and plays a role of temperature ($p=0$: critical, $p=\pm1$: zero temperature). The λ is a coupling constant chosen to be the rational values:

$$\lambda = t/s,\quad 1\leq t\leq s-1,\quad t \text{ and } s \text{ are coprime}, \tag{5}$$

where $s=n+\ell$ and t may be regarded as non-positivity parameter as in Sec. 1. The Boltzmann weights $W_N\begin{pmatrix} a & b \\ c & d \end{pmatrix}\Big|u$ for general N are built from those for $N=1$ by the prescription in [14,15] if one replaces L and u therein with λ^{-1} and $y\lambda^{-1}$, respectively. Here we omit the formula but not that the resulting weights can be made symmetric so that $W_N\begin{pmatrix} a & b \\ c & d \end{pmatrix}\Big|u = W_N\begin{pmatrix} a & c \\ b & d \end{pmatrix}\Big|u$ by an appropriate gauge transformation. From now on we shall mean symmetrized weight by $W_N\begin{pmatrix} a & b \\ c & d \end{pmatrix}\Big|u$.

3. H-function and Local State Probability

Local state probability (LSP) is by definition the probability that the central-site variable takes a given state under a fixed boundary conditions. It can be calculated by the corner transfer matrix (CTM) method [17]. Let us present a series of conjectures on the LSPs for the fusion models in the range of eq. (2). We have checked them by computer calculations in several cases.

H-function. The first step of the CTM method is to determine the behavior of the Boltzmann weights in the ground state limit. Introducing the conjugate modulus parameters x and w by $p=e^{-\epsilon}, x=e^{-4\pi^2/\epsilon}, w=x^u, (0<x<1, w>1)$, we conjecture the following behavior.

$$\lim_{x\to0,w:\text{fix}} W_N\begin{pmatrix} a & b \\ c & d \end{pmatrix}\Big|u\; F_w^{-f_a+f_d+f_b-f_c} = \delta_{bd}w^{-H_N(\{\mu_i\},\{\nu_i\};a)},$$
$$-H_N(\{\mu_i\},\{\nu_i\};a) = \max_{\sigma'\in S_N}\left[\min_{\sigma\in S_N}\sum_{i=1}^{N}H_1(\mu_{\sigma'(i)},\nu_{\sigma(i)};a+\sum_{k=1}^{i-1}\hat{\mu}_{\sigma'(k)})\right], \tag{6}$$
$$H_1(\mu,nu;a) = [-ta_{\mu\nu}/s]_I+1.$$

Here $F = x^{-Nu(u+1+\lambda(\frac{2N}{n}+1-N))/2}$, $f_a = \frac{t}{2ns}\sum_{\mu<\nu} a_{\mu\nu}^2$ and the symbol $[y]_I$ denotes the largest integer not exceeding y. The sets $\{\mu_i\}$ and $\{\nu_i\}$ are determined by

$$b = a + \sum_{i=1}^{N} \hat{\mu}_i, \quad c = a + \sum_{i=1}^{N} (\hat{\mu}_i + \hat{\nu}_i). \tag{7}$$

The function H_N governs the ground state structure and LSPs. Eq. (6) can be verified directly for $N = 1$. It extends the conjecture for the case $t = 1$ and N general in [15] where one can set $\sigma' = id$ dropping the outer maximization. The resulting H-function has also been observed to occur in the $q \to 0$ limit of the fusion $sl(n)$ R matrix [23].

Ground states. The ground states are identified with those one dimensional sequences $b^{(j)}$ of the DIWs minimizing the function $\sum_j H_N(\{\mu_i^{(j)}\}, \{\mu_i^{(j+1)}\}; b^{(j)})$, where $(b^{(j)}, b^{(j+1)})$ is admissible and $b^{(j+1)} = b^{(j)} + \hat{\mu}_1^{(j)} + \cdots + \hat{\mu}_N^{(j)}$. For general N and $t = 1$, the ground states under the assumption of eq. (6) are given as follows: 1) The ground states are degenerated and parametrized by a doublet $(\xi, \eta) \in P_+(n, \ell - N) \times P_+(n, N)$. 2) The state $b^{(j)}$ at the jth site of the corresponding one-dimensional configuration along horizontal line is given by

$$b^{(j)} = \xi + \sigma^j(\eta), \tag{8}$$

where $\sigma : \Lambda_\mu \mapsto \Lambda_{\mu+1}$ is the generator of the Dynkin automorphism of $\widehat{sl}(n)_N$. On the other hand, for $N = 1$ and $1 \le t \le \ell$, it is conjectured that the ground states are parametrized by doublets $(\xi, \eta) \in P_+(n, \ell - t) \times P_+(n, 1)$ and its parameterization is given in ref. [10].

To present the generalized conjecture for general N and $1 \le t \le \ell/N$, we need some more definition. We divide region $P_+ = \{\lambda = \sum_{i=1}^{n-1} \lambda_i \bar{\Lambda}_i\}$ of the (classical) dominant integral weights of $sl(n)$ into an infinite number of domains by the hyperplanes $\lambda'_{\mu\nu} \equiv \lambda^{\mu+1} + \cdots + \lambda^{\nu} = N\, k_{\mu\nu}$ ($\mu < \nu, k_{\mu\nu}$: positive integer), where each domain D surrounded by these hyperplanes includes its boundaries. See fig.1 for example in the case $n = 3, N = 2$. For each domain D in $P_+(n, N) \to D$ recursively defined as follows For $D = \{\lambda \in P_+ | 0 \le \lambda'_{0n-1} \le N\}$, $\bar{e}_D$ is simply the projection i_N from the affine weights to their classical parts. For two adjacent domains D, D' along the hyperplanes $\lambda'_{\mu\nu} = const.$, let $B_{DD'}$ denote their common boundary. For any $\eta \in P_+(n, N)$, its image $\bar{e}_D(\eta)$ can be written as $p\hat{\mu} + q\hat{\nu} + c$ with $c \in B_{DD'}$ and $p, q \in \mathbf{Z}$. Then we set $\bar{e}_{D'}(\eta) = q\hat{\mu} + p\hat{\nu} + c$. Let i_{tN}^{-1} be the inverse of the classical projection from P_+ to the set of (not necessarily dominant) integral weight of level tN. We write the composition $i_{tN}^{-1} \circ \bar{e}_D$ as e_D.

We conjecture that the ground states are parametrized by doublets $(\xi, \eta) \in P_+(n, \ell - Nt) \times P_+(n, N)$ and given by

$$b^{(j)} = \xi + e_{D(\xi)}(\sigma^j(\eta)), \tag{9}$$

where $D(\xi)$ is the domain in P_+ determined by ξ as

$$\begin{aligned} N\, k_{\mu\nu} &\le \lambda'_{\mu\nu} \le N(k_{\mu\nu} + 1) \\ k_{\mu\nu} &= [t\xi_{\mu\nu}/(s - Nt)]_I, \quad (\mu < \nu). \end{aligned} \tag{10}$$

We remark that $e_{D(xi)}(\sigma^j(\eta))$ is always DIW of level Nt.

On the other hand, like the case $N = 1$, there is another characterization of the ground states in terms of the 'region' in $P_+(n,\ell)$ [10,11], though we omit its description here. The examples for $(n,\ell,N) = (3,8,2)$ are exhibit in fig.2.

Local State Probabilities. The LSP $P(\xi,\eta,a;x)$ is associated to a triplet $(\xi,\eta,a) \in \mathcal{A}$, where $\mathcal{A}$ is the set of the elements of $P_+(n,\ell-Nt) \times P_+(n,N) \times P_+(n,\ell)$ satisfying $\bar{a} \equiv \bar{\xi} + \bar{\eta}$ modulo the root lattice Q of $sl(n)$. In $P(\xi,\eta,a;x)$, a represents the state at the central site and the pair (ξ,η) represents the boundary condition determined from the ground state specified by (ξ,η) described as above.

Let us introduce the functions Θ, χ for a weight μ and a number m,

$$\begin{aligned}\Theta_{\mu,m}(u,q) &= \sum_{\alpha\in Q} q^{|m\alpha+\mu|^2/2m} e^{2\pi i(m\alpha+\mu,u)},\\ \chi_{\mu,m}(u,q) &= \frac{\sum_{w\in S_n} \det w\cdot\Theta_{w(\mu+\rho),m+n}(u,q)}{\sum_{w\in S_n}\det w\cdot\Theta_{w(\rho),n}(u,q)},\end{aligned} \tag{11}$$

Let $r = s - Nt$. It is convenient to introduce the quantities

$$S = \frac{s}{t},\quad R = \frac{r}{t},\quad S - R + N. \tag{12}$$

Let ξ,η,a be dominant integral weights of level $r-n, N, s-n$, respectively. Then we shall define the branching coefficient $b_{\xi\eta a}(q)$ by the decomposition

$$\chi_{\xi,R-n}(u,q)ch_{\eta,N}(u,q) = \sum b_{\xi\eta a}(q)\chi_{a,S-n}(u,q), \tag{13}$$

where the summation runs over $a \in P_+(n,s-n)$ such that $(\xi,\eta,a) \in \mathcal{A}$ and $ch_{\eta,N}(u,q)$ is the character of the h.w.r. associated to $\eta \in P_+(n,N)$ (See (19)).

Our conjecture for the LSP is

$$P(\xi,\eta,a;x) = \frac{b_{\xi\eta a}(x^n)\chi_{a,S-n}(x)}{ch_{\eta,N}(x)\chi_{\xi,R-n}(x)}, \tag{14}$$

where $\chi(x)$ and $ch(x)$ are the values of $\chi(u,q)$ and $ch(u,q)$ evaluated at the principal specialization $e^{2\pi i(\alpha_j,u)} = x$ (for all the simple roots α_j, $q = x^n$.

To write a more explicit form of $b_{\xi\eta a}(q)$, we recall that the character $ch_{\eta,N}(u,q)$ has the expression using the string function $c^\eta_\lambda(q)$ of $\widehat{sl}(n)_N$

$$ch_{\eta,N}(u,q) = \sum_{\substack{\lambda\in P/NQ\\ \lambda\equiv\eta \bmod Q}} c^\eta_\lambda(q)\Theta_{\lambda,N}(u,q). \tag{15}$$

There is a theta function identity [24] including the integer parameter t (Eq.(C.2) of [2]):

$$\begin{aligned}\Theta_{\mu_1,m_1}(u,q)\Theta_{\mu_2,m_2}(u,q) &= \sum_{\mu\in Q/t(m_1+m_2)Q} d_{\mu\mu_1\mu_2}(q)\Theta_{m_2\mu+\mu_1+\mu_2,m_1+m_2}(u,q),\\ d_{\mu\mu_1\mu_2}(q) &= \Theta_{t(m_1m_2\mu-m_2\mu_1+m_1\mu_2),t^2m_1m_2(m_1+m_2)}(0,q).\end{aligned} \tag{16}$$

After some manipulations about the summation indices, we get

$$b_{\xi\eta a}(q) = \sum_{\substack{\lambda \in P/N \cdot Q, \\ \bar{\lambda} \equiv \eta \bmod Q}} \sum_{\mu \in Q} \sum_{w \in S_n} \delta^{(N)}_{\lambda \xi w a \mu} \\ c^{\eta}_{\lambda}(q) \det w \cdot q^{|rs\mu + rw(a+\rho) - s(\xi+\rho)|^2/2Nrs}, \tag{17}$$

where we have introduced the delta function

$$\delta^{(N)}_{\lambda \xi w a \mu} = \begin{cases} 1 & \text{if } \lambda + \xi + \rho \equiv s\mu + w(a+\rho) \bmod NQ \\ 0 & \text{otherwise} \end{cases} \tag{18}$$

4. Rational Coset Model

By the modular transformation property of the theta functions (11) and the scaling relation, the leading powers of $b_{\xi\eta a}(q)$ provide the anomalous dimensions of the fields describing the scaling limit of the order parameters. Moreover, as we shall see below, the branching coefficient $b_{\xi\eta a}(q)$ is the character of the certain coset model including the rational levels. To see this, let us recall that the character of the admissible representation μ of the rational level m is given by [25]

$$ch_{\mu,m}(u,q) = \frac{\sum_{\alpha \in tQ} \sum_{w \in S_n} \det w \cdot q^{|m\alpha+\mu|^2/2m} e^{2\pi i (w(m\alpha+\mu),u)}}{\sum_{w \in S_n} \det w \cdot \Theta_{w(\rho),n}(u,q)}. \tag{19}$$

Notice that though two functions $ch_{\mu,m}(u,q)$ and $\chi_{\mu,m}(u,q)$ are very similar, the range of the summation of α is different (for $t = 1$ they coincide with each other). However it is important to notice that there is another decomposition formula:

$$ch_{\xi',r-n}(u,q) ch_{\eta,N}(u,q) = \sum b_{\xi\eta a}(q) ch_{a',s-n}(u,q), \tag{20}$$

where $\bar{\xi}' = \bar{\xi}$ and $\bar{a}' = \bar{a}$ and the region for a is the same as in eq. (13). Thus, the branching coefficient $b_{\xi\eta a}(q)$ is the character of the primary field of the coset model associated to the embedding

$$\widehat{sl}(n)_{R-n} \oplus \widehat{sl}(n)_N \supset \widehat{sl}(n)_{S-n}, \tag{21}$$

where the central charge and the conformal weight are

$$c = \frac{N(n^2-1)}{n+N} \left[1 - \frac{n(n+N)(s-r)^2}{rsN} \right],$$
$$h = \frac{|-s(\xi+\rho) + r(a+\rho)|^2 - |(s-r)\rho|^2}{2Nrs} + \frac{(\eta, \eta+2\rho)}{2(n+N)} - \frac{|\eta|^2}{2N}. \tag{22}$$

This is a generalization of the W_n minimal model and it is natural to call them (one of) the *rational coset models*. In the unitary case $t = 1$, the character formula (17) was written in [26–28] for $n = 2$ and was conjectured in [29] for general n.

It is worth mentioning that the Feigin-Fuchs construction of unitary coset models with the W_n scalar fields and the parafermion system can be easily extended to our models. The splitting of the central charge are

$$c = c_W + c_p,$$
$$c_W = (n-1)\left[1 - \frac{n(n+1)(s-r)^2}{rsN}\right], \quad c_p = \frac{N(n^2-1)}{n+N} - (n-1). \tag{23}$$

Here c_W is the central charge of the $n-1$ component scalar fields of the W_n algebra with the screening charge $\alpha_0^2 = (s-r)^2/2rsN$, and c_p is the one of the parafermionic algebra $\widehat{sl}(n)_N$.

To conclude, we must solve many very complicated (though interesting) combinatorial problems if we try to prove conjectures in Sec.3 in a straightforward way. These difficulties might be avoided if we extend the crystal base approach in [19]. Also it is an interesting open problem what kinds of rational coset models will appear outside the region (2).

We would like to thank M.Wakimoto for useful discussions. T.N. would like to thank V.V.Bazhanov, P.P.Kulish and the members of Institute for High Energy Physics, Protvino and Leningrad Branch of Steklov Mathematical Institute for their great hospitality.

References

[1] G.E.Andrews, R.J.Baxter and P.J.Forrester, J. Stat. Phys. **35** (1984), 193.

[2] M.Jimbo, T.Miwa and M.Okado, Nucl. Phys. **B300** **[FS22]** (1988), 74.

[3] V.Pasquier, Nucl. Phys **B295** **[FS21]** (1988), 491.
V.V.Bazhanov and N.Yu.Reshetikhin, Int. J. Mod. Phys. **A4** (1989), 115.

[4] M.Jimbo, T.Miwa and M.Okado, Commun. Math. Phys. **116** (1988), 507.

[5] E.Date, M.Jimbo, A.Kuniba, T.Miwa and M.Okado, Lett. Math. Phys. **17** (1989), 69.

[6] A.Kuniba, *Exact Solution of the Solid-on-Solid Models for Twisted Affine Lie Algebras $A_{2n}^{(2)}$ and $A_{2n-1}^{(2)}$*, Nucl. Phys. **B** (to appear).

[7] P.J.Forrester and R.J.Baxter, J. Stat. Phys. **38** (1985), 435.

[8] V.Pasquier, Nucl. Phys. **B285** **[FS21]** (1987), 162.
P.di Francesco, H.Saleur and J.-B.Zuber, Nucl. Phys. **B300** **[FS22]** (1988), 393.

[9] H.Riggs, Nucl. Phys. **B326** (1989), 673.

[10] T.Nakanishi, Nucl. Phys. **B334** (1990), 745.

[11] A.Kuniba, T.Nakanishi and I.Suzuki, *Ferro and Antiferro Magnetizations in RSOS Models*, Nucl. Phys. **B** (to appear).

[12] A.Kuniba, T.Nakanishi, Proc. of International Colloquium on Modern Quantum Field Theory, Bombay, World Scientific, Singapore, 1990, in press.

[13] E.Date, M.Jimbo, A.Kuniba, T.Miwa and M.Okado, Nucl. Phys. **B290** **[FS20]** (1987), 231; Adv. Stud. in Pure Math. **16** (1988), 17.

[14] M.Jimbo, T.Miwa and M.Okado, Mod. Phys. Lett. **B1** (1987), 73.

[15] M.Jimbo, A.Kuniba, T.Miwa and M.Okado, Commun. Math. Phys. **119** (1988), 543.

[16] E.Date, M.Jimbo, A.Kuniba, T.Miwa and M.Okado, *Talk at Taniguchi Conference, Kyoto*, unpublished, See ref. 12.

[17] R.J.Baxter, *Exactly Solved Models in Statistical Mechanics*, Academic Press, London, 1982.

[18] E.Date, M.Jimbo, A.Kuniba, T.Miwa and M.Okado, Lett. Math. Phys. **17** (1989), 51 (1989), 108, Infinite Dimensional Lie Algebras and Groups, ed. V.G.Kac, World Scientific, Singapore; Adv. Stud. in Pure Math. **19** (1989), 149.

[19] M.Jimbo, K.C.Misra, T.Miwa and M.Okado, *Combinatorics of Representations of* $U_q(\widehat{sl}(n))$ *at* $q = 0$, Kyoto preprint (1990).

[20] E.Verlinde, Nucl. Phys. **B300** (1988), 360.

[21] A.Tsuchiya, K.Ueno and Y.Yamada, Adv. Stud. in Pure Math. **19** (1989), 459.

[22] F.M.Goodman and T.Nakanishi, *Fusion Algebras in Integrable Systems in Two Dimensions*, Iowa Preprint (1990).

[23] E.Date, M.Jimbo and T.Miwa, Physics and Mathematics of Strings, eds. L.Brink, D.Friedan and A.M.Polyakov, World Scientific, 1990, pp. 185.

[24] V.G.Kac and D.H.Peterson, Adv. Math. **53** (1984), 125.

[25] V.G.Kac and M.Wakimoto, Infinite Dimensional Lie Algebras and Groups, ed. V.G.Kac, World Scientific, Singapore, 1989, p. 138.

[26] J.Bagger, D.Nameschansky and S.Yankielowicz, Phys. Rev. Lett. **60** (1988), 389.

[27] D.Kastor, E.Martinec and Z.Qiu, Phys. Lett. **B200** (1988), 434.

[28] F.Ravanini, Mod. Phys. Lett. **A3** (1988), 271.

[29] K.Hamada, Nucl.Phys. **B334** (1990).

Figure Captions

Fig.1. The Decomposition of P_+ for $(n, N) = (3, 2)$. We label each domain for later convenience.

Fig.2. The parameterizations of the ground states for $(n, \ell, N) = (3, 8, 2)$. For each $\xi \in P_+(3, 8, -2t)$, the label of the domain $D(\xi)$ in Fig.1 is attached by the rule (10). We put the barycentre of the corresponding ground state motion (9) in the space of the local state $P_+(3, 8)$. The open (filled) circle represents that the ground state motion move around the barycentre clockwise (counterclockwise), respectively.

I

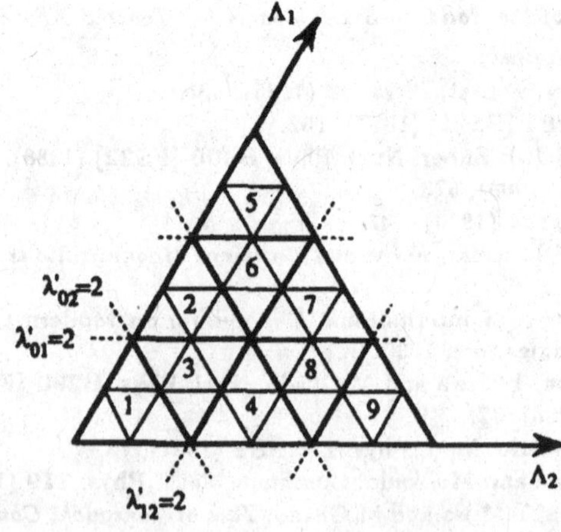

Fig.1

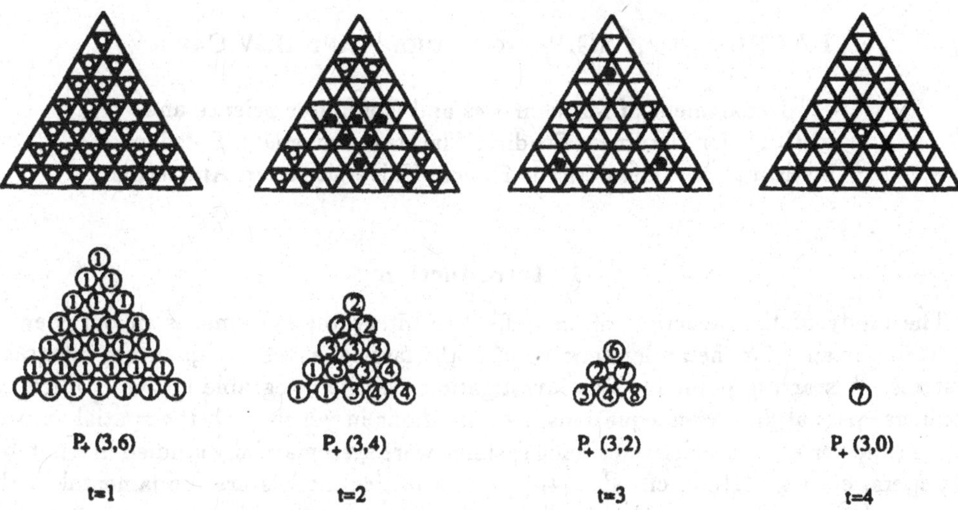

Fig.2

(2) DEPARTMENT OF MATHEMATICS, NAGOYA UNIVERSITY, CHIKUSA-KU, NAGOYA 464, JAPAN

INTEGRABLE TIME–DISCRETE SYSTEMS: LATTICES AND MAPPINGS

F.W.Nijhoff[1)], V.G.Papageorgiou[1)] and H.W.Capel[2)]

[1)] Department of Mathematics and Computer Science and
Institute for Nonlinear Studies, Clarkson University, Potsdam
[2)] Instituut voor Theoretische Fysica, Universiteit van Amsterdam

1. Introduction

The study of the structure of time-discrete integrable systems, is in our view of great importance for the understanding of both classical as well as quantum integrable systems. A starting point for this investigation are the integrable lattices, which are nonlinear partial *difference* equations, i.e. equations in which both the spatial- as well as the time-variables are discrete. Such systems were systematically studied in a number of papers, cf. e.g. [1]-[3], cf. also [4]-[6]. The lattice models are fundamental in the sense that their various continuous limits give rise to entire hierarchies of integrable PDE's [7]. Furthermore, they have the advantage that space- and time variables appear on the same footing, and typically there is on the lattice level a covariance between the various lattice directions, which is generally absent in the continuous situation. In addition, they are computationally simple allowing for numerical approaches for their integration. Finally, mappings in general are used to model dynamical systems which are important for concrete physical applications, and as such perturbations of *integrable* lattices and mappings could be useful for the study of non-integrable situations as well.

In our opinion classical *time-discrete* systems are the 'nearest step' to quantum systems, and as such (in the integrable case) to quantum groups. In fact, one may ask how far one may go to view the discrete time-step as a parameter for a quantum deformation of a classical (time-continuous) integrable system. In particular, the investigation of the analogue of a Hamiltonian structure for a discrete-time system, and the r-matrix formulation, could open up some new points of view on classical and quantum integrability in general.

In this talk we are dealing with two separate aspects of the integrable lattices that we consider. The two subjects can be viewed as two different classes of solutions of the lattice equations. In the first part we treat periodic initial data on infinite chains in the lattice corresponding to solutions of finite-gap type. This class of solutions gives rise to integrable *mappings*, which are time-discrete dynamical systems with finite degrees of freedom, cf. e.g. [8]-[16]. In the second part we consider similarity type of solutions that arise from localized configurations of initial data, and which propagate on the lattice analogue of the characteristics, i.e. they are subject to some well-defined constraints arising from an isomonodromic problem on the lattice. We shall illustrate our ideas by means of a special example, namely lattices of KdV type, i.e. lattice equations that under some well-defined continuum limits reduce to the (potential) KdV equation. It is needless to say that our methods can be applied to many other integrable lattices, and indeed at the end we will give a few preliminary facts about the generalization to lattices associated with the Gel'fand-Dikii hierarchy.

2. The KdV Lattice

In a previous note [13] we showed how the consideration of 'local' initial value problems on the lattice for lattice equations of KdV-type gives rise to integrable finite-dimensional mappings. For example it was shown in [13] that the (potential) KdV-lattice

$$(p-q+u_{n,m+1}-u_{n+1,m})(p+q-u_{n+1,m+1}+u_{n,m}) = p^2-q^2 , \qquad (1)$$

in which $p,q \in \mathbb{C}$ are the lattice parameters, $u=u_{n,m}$ the dynamical (field) variable at site (n,m), $n,m \in \mathbb{Z}$) gives rise to a family of $2P$-dimensional mappings. An action for the KdV lattice (1) is, [16],

$$\mathcal{S} = \sum_{n,m\in\mathbb{Z}} \left[u_{n,m+1}(u_{n+1,m+1}-u_{n,m}) + \epsilon\delta \ln(\epsilon + u_{n,m} - u_{n+1,m+1}) \right] . \qquad (2)$$

The Euler-Lagrange equations for (2), which are obtained by variation of $\mathcal{S}$ with respect to the variables $u_{n,m}$, i.e.

$$\frac{\delta \mathcal{S}}{\delta u_{n,m}} = 0 ,$$

leads to the equation

$$\begin{aligned} & u_{n+1,m} - u_{n,m+1} + u_{n-1,m} - u_{n,m-1} + \\ & + \frac{\epsilon\delta}{\epsilon + u_{n,m} - u_{n+1,m+1}} - \frac{\epsilon\delta}{\epsilon + u_{n-1,m-1} - u_{n,m}} = 0 , \end{aligned} \qquad (3)$$

which is an integrable lattice equation leading in a well-defined continuum limit to the (non-potential) KdV equation. Note that eq. (3) is a consequence of eq. (1).

The lattice equation (1) arises as the compatibility condition of a Zakharov-Shabat pair defining the shifts (translations) of an eigenfunction ϕ_k in the n- and m-directions,

$$(p-k)\phi_k(n+1,m) = L_k \cdot \phi_k(n,m) , \ (q-k)\phi_k(n,m+1) = M_k \cdot \phi_k(n,m) , \qquad (4)$$

where L_k is given by

$$L_k = \begin{pmatrix} p-u_{n+1,m} & 1 \\ k^2-p^2+(p+u_{n,m})(p-u_{n+1,m}) & p+u_{n,m} \end{pmatrix} , \qquad (5)$$

and where M_k is given by a similar matrix obtained from (5) by making the replacements $p \to q$ and $(n+1,m) \to (n,m+1)$. The parameter k is the spectral parameter.

Eq. (1) reduces to the potential KdV equation after two successive continuum limits [1,2], see also section 7 below), whereas eq. (3) leads to the KdV equation. Below, we shall investigate the initial value problem for these lattice equations in two situations: periodic initial data on 'staircases' on the lattice (which we describe in the next section), and localized initial data in the presence of an isomonodromic constraint (described in section 6). The first situation leads to finite-gap type of solutions, whereas the second situation will yield similarity type of solutions leading in particular to a lattice analogue of Painlevé II equation.

3. Mappings of KdV Type

The (potential) KdV lattice (1) has the property that the variables appearing in the equation are organized along simple plaquettes in the lattice. Because of this feature, a natural way of posing a *local* initial value problem on the lattice is not the one where one chooses initial data on the horizontal (or vertical) lines, but one in which we choose such data on 'broken lines', i.e *staircases*, cf. [13]. (The initial value problem of the first possibility has been investigated in [7], and leads to a nonlocal problem, in which one has to impose adequate decay conditions at infinity). Choosing such initial data for the variables along a 'standard' staircase on the lattice, namely

$$u_{j,j} =: a_{2j} \ , \quad u_{j+1,j} =: a_{2j+1} \ , \quad (j \in \mathbf{Z}),$$

we perform iterations by updating the lattice variables u along a vertical shift in the m-direction, i.e.

$$u_{j,j+1} = a'_{2j} \ , \quad u_{j+1,j+1} = a'_{2j+1}$$

using the lattice KdV (1). In this way we obtained the following mapping

$$a'_{2j} = a_{2j+1} - \delta + \frac{\epsilon\delta}{\epsilon - a_{2j+2} + a_{2j}} \ , \quad a'_{2j+1} = a_{2j+2}, \tag{6}$$

where $\delta = p-q$, $\epsilon = p+q$, which by imposing periodic initial conditions on the staircase, i.e.

$$a_{2(j+P)} = a_{2j} \ , \quad a_{2(j+P)+1} = a_{2j+1}$$

reduces to a finite-dimensional mapping of dimension $2P$. Introducing the variables for the differences on odd and even sites of the staircase , i.e.

$$x_j \equiv \epsilon + a_{2j-1} - a_{2j+1} \ , \quad y_j \equiv \epsilon + a_{2j} - a_{2j+2} \ , \tag{7}$$

the mapping can be even further reduced to a $2P-2$-dimensional one which reads

$$\begin{cases} x'_j = y_j \\ y'_j = x_{j+1} - \dfrac{\epsilon\delta}{y_{j+1}} + \dfrac{\epsilon\delta}{y_j} \end{cases} ,(j = 1, \cdots, P), \tag{8}$$

where we impose the periodicity condition

$$\sum_{j=1}^{P} x_j = C_1 \ , \quad \sum_{j=0}^{P-1} y_j = C_2 \ ,$$

C_1 and C_2 being constant.

The mapping (8) is a multidimensional generalization of the McMillan mapping [8], and it arises as the compatibility condition of a linear (Zakharov-Shabat type of) problem, which is obtained using a special property of the matrices L_k and M_k of (4,5) and performing at each site of the staircase a gauge transformation, cf. [16]. The compatibility equation for the mapping (8) in terms of the *reduced variables* x_j and y_j is

$$L'_j(k)\cdot M_j(k) \;=\; M_{j+1}(k)\cdot L_j(k)\,, \tag{9}$$

where

$$L_j(k)=\left[(k^2-q^2)F+\begin{pmatrix}0 & 1\\ * & x_{j+1}-\beta_{j+1}\end{pmatrix}\right]\cdot\left[(k^2-p^2)F+\begin{pmatrix}\beta_j & 1\\ * & y_j\end{pmatrix}\right] \tag{10}$$

$$M_j(k)=\left[(k^2-q^2)F+\begin{pmatrix}\beta_j-\frac{\epsilon\delta}{y_j} & 1\\ * & y_j-\beta'_j\end{pmatrix}\right]\,. \tag{11}$$

The variable β_j in (11) is an arbitrary function and can be chosen at our convenience (for instance by taking $\beta_j=x_j$ or $\beta_j=0$), this freedom of choice corresponding to a gauge freedom, and $*$ stands in any matrix for the product of its diagonal entries. In (11) and also later we use the following notation

$$E\;=\;\begin{pmatrix}0 & 1\\ 0 & 0\end{pmatrix}\quad,\quad S_-\;=\;\begin{pmatrix}0 & 0\\ 0 & 1\end{pmatrix}\quad,\quad F\;=\;\begin{pmatrix}0 & 0\\ 1 & 0\end{pmatrix}.$$

Having obtained the linear system in Zakharov-Shabat form, one can then construct the *monodromy matrix* T_k by gluing the elementary translation matrices L_j along the staircase over one period P, leading to

$$\mathcal{T}(k)\;\equiv\;\prod_{j=0}^{\overset{\frown}{P-1}} L_j(k)\,. \tag{12}$$

The trace of the monodromy matrix leads to the following expansion (again expressible only in terms of x_j and y_j), namely

$$Tr(\mathcal{T}(k))\;=\;\left(\prod_{j=0}^{2P-1} w_j\right)\left[1+\sum_{\substack{0\le J_1<\cdots<J_N\le 2P-1\\ J_{\nu+1}-J_\nu\ge 2,\,J_1-J_N+2P\ge 2}}\;\prod_{\nu=1}^{N}\frac{\lambda_{J_\nu}}{w_{J_\nu}w_{J_\nu+1}}\right]\,, \tag{13}$$

in which $w_{2j}=x_j$ and $w_{2j+1}=y_j$, and $\lambda_{2j}=k^2-p^2$, $\lambda_{2j+1}=k^2-q^2$. The trace (13) is invariant under the mapping. Furthermore, by expanding the trace in powers of k^2 we obtain P independent invariants.

Alternatively, probably more useful for the finite-gap integration of the mapping, we mention another linear problem (a Lax-Max pair) in terms of $2P\times 2P$ matrices

$$\mathbf{L}\cdot\varphi\;=\;(k^2-q^2)\varphi\quad,\quad \mathbf{L}\;=\;\begin{pmatrix}\epsilon\delta & -y_0 & 1 & & 0\\ 0 & 0 & -x_1 & 1 & 0\\ \vdots & & \epsilon\delta & -y_1 & \ddots & \vdots\\ \vdots & & & \ddots & \ddots & 1\\ h & 0 & & \cdots & \epsilon\delta & -y_{P-1}\\ -hx_0 & h & \cdots & \cdots & 0 & 0\end{pmatrix}\,, \tag{14}$$

and

$$\varphi' = \mathbf{M}\cdot\varphi \;,\; \mathbf{M} = \begin{pmatrix} -\frac{\epsilon\delta}{y_0} & 1 & & & & \\ 0 & 0 & 1 & & & \\ \vdots & 0 & -\frac{\epsilon\delta}{y_1} & 1 & & \\ \vdots & & \ddots & \ddots & \ddots & \\ 0 & & \cdots & 0 & -\frac{\epsilon\delta}{y_{P-1}} & 1 \\ h & 0 & \cdots & & 0 & 0 \end{pmatrix} , \tag{15}$$

and their compatibility

$$\mathbf{L}' = \mathbf{M}\cdot\mathbf{L}\cdot\mathbf{M}^{-1} \;, \tag{16}$$

yielding the mapping (8). The parameter h is a factor coming in from the quasi-periodicity of the eigenfunctions ϕ_k of (4).

The trace of the monodromy matrix (13) is expressible in terms of the determinant of the Lax matrix (14) via the relation

$$det\left(\mathbf{L} - (k^2 - q^2)1\right) = (p^2 - k^2)^P(q^2 - k^2)^P + h^2 - h\,tr\left(\mathcal{T}(k)\right) . \tag{17}$$

We note that the corresponding characteristic equation associated with (17) determines a hyperelliptic curve in the variables h and the spectral parameter $k^2 - q^2$.

We point out that the mapping (8) is integrable in the sense that there exists an underlying Lax representation, which in principle allows us to apply standard inverse techniques (finite-gap integration) to integrate the discrete-time dynamics, and in particular to generate a sufficient number of integrals of the motion, as we have demonstrated in [13].

4. Canonical Structure

Next we want to establish a canonical structure for the KdV mappings (symplectic, Poisson structure) in terms of which we can show the involutivity of the integrals, in order to establish integrability in the sense of Arnol'd-Darboux-Liouville. Clearly, in the time-discrete case such a symplectic structure does not imply the existence of a Hamiltonian structure in the usual sense, because we do not have any longer a continuous time-flow parameter present in the system. However, instead of a Hamiltonian one can introduce a generating function of the canonical transformation which is the mapping.

Starting from the action (2) for the KdV lattice, we can now choose the variables along the staircase and write the action in terms of the a_{2j} and a_{2j+1}, and then perform the Legendre transformation in order to obtain the canonically conjugate variables. It turns out that one choice of a (discrete) Lagrangian contains only the variables on the even sites of the staircase and their updatings

$$\mathcal{S} = \sum_{m\in\mathbf{Z}} \mathcal{L}[a_{2j}(m), a_{2j}(m+1)] \;,$$

where now the variable m in the argument is to indicate the mth iterate of the mapping.

i.e. $a_{2j}(m+1) \equiv a'_{2j}(m)$, and where (omitting the argument m)[1]

$$\mathcal{L}[a_{2j},a'_{2j}] = \sum_{j=0}^{P-1} \Big[\frac{1}{2}(a'_{2j} + a_{2j+2} - a_{2j})^2 - \frac{1}{2}a_{2j}^2 - \frac{1}{2}(a_{2j+2} - a_{2j})^2 + \epsilon\delta \ln(\epsilon + a_{2j} - a_{2j+2}) \Big] . \tag{18}$$

The Euler-Lagrange equations obtained from this action give us precisely the mapping (8) in terms of the reduced variables x_j, y_j. The conjugate momenta are found to be

$$p'_{2j} = \frac{\partial \mathcal{L}}{\partial a'_{2j}} = a'_{2j} + a_{2j+2} - a_{2j} ,$$

which by using (6) can be identified with

$$p_{2j} = a_{2j} + a_{2j+1} - a_{2j-1} .$$

The generating functional of the mapping (Hamiltonian) $\mathcal{H}$ is calculated by performing a Legendre transformation

$$\begin{aligned} \mathcal{H}[a_{2j},p'_{2j}] &= \sum_{j=0}^{P-1} p'_{2j}(a'_{2j} - a_{2j}) - \mathcal{L}[a_{2j},a'_{2j}] \\ &= \sum_{j=0}^{P-1} \left[\frac{1}{2}(p'_{2j} - a_{2j+2})^2 + \frac{1}{2}(a_{2j+2} - a_{2j})^2 - \epsilon\delta \ln(\epsilon + a_{2j} - a_{2j+2}) \right] . \end{aligned} \tag{19}$$

The Hamiltonian acts as the generating functional for the mapping, i.e. one has the discrete-time Hamilton equations

$$a'_{2j} - a_{2j} = \frac{\partial \mathcal{H}}{\partial p'_{2j}} , \quad p'_{2j} - p_{2j} = -\frac{\partial \mathcal{H}}{\partial a_{2j}} , \tag{20}$$

leading to the equation for the mapping under consideration. In terms of the conjugate variables a_{2j}, p_{2j} we have the standard symplectic structure

$$\Omega = \sum_{j=0}^{P-1} dp_{2j} \wedge da_{2j}$$

leading to standard Poisson brackets, and the mapping is in fact a canonical transformation (i.e. $\Omega' = \Omega$, which can be seen directly from the Hamilton equations). We mention that in order to perform the discrete Legendre transformation one could as well start from the action (2) and write it in terms of the *odd* variables only. In that

[1]Throughout this paper when we deal with periodic boundary conditions on the staircases, it will from now on be understood that the label j, $j = 0, \cdots, P-1$, is interpreted modulo P, i.e. $j \in \mathbf{Z}_P$. In particular all summations over variables bearing such a label are subject to the same periodicity convention without further explicit mentioning.

case, we have to choose a slightly different Lagrangian associated with the backward iteration of the mapping rather than with the forward iteration as we consider here. In both cases one finds canonical brackets between combinations of variables a_n ($n = 2j$ resp. $n = 2j + 1$) that can be summarized by the formula

$$\{a_n \, , \, a_{n'+1} - a_{n'-1}\} = \delta_{n,n'} \, , \tag{21}$$

i.e. the difference of two a-variables on neighbouring sites on the same diagonal on the staircase is canonically conjugate to the variable on the site in between. In terms of the reduced variables we thus have

$$\{x_j \, , y_{j'}\} = \delta_{j,j'} - \delta_{j,j'+1} \, , \quad \{x_j \, , x_{j'}\} = \{y_j \, , y_{j'}\} = 0 \, , \tag{22}$$

which are the brackets we need in order to calculate the Poisson brackets between the invariants. Note that the Poisson structure we obtain by performing the discrete Legendre transformation corresponds to the discrete Gardner bracket that was considered also in the time continuous case in ref. [17], cf. also [7].

5. Classical r-Matrix Structure

Having established the Poisson brackets and Hamiltonian structure for the mapping (8), we can now use these to prove involutivity of the invariants. In contrast to the method employed in [16], we use here the fact, that the system under consideration carries a r-matrix structure, from which then the involutivity of the integrals is manifest. What is particularly interesting is that the r-matrix structure we find is of ***non-ultralocal type***, cf. [18,19]. Non-ultralocal r-matrix structures have been studied in a number of papers, cf. e.g. [20]-[23]. For involutivity of the invariants, it is clearly sufficient to prove that the Poisson bracket between the trace of the monodromy matrix (13) for different values of k, k and k' say, vanishes.

Using the Poisson brackets (22) we can now calculate the (fundamental) Poisson relations between the Lax matrices $L_j(k)$ of (11) in the case $\beta_j = 0$, leading to

$$\begin{aligned}\{L_j(k) \otimes L_{j'}(k')\} = &- \delta_{j,j'+1} \, (L_j(k) \otimes 1) \; s^+ \; (1 \otimes L_{j'}(k')) \\ &+ \delta_{j+1,j'} \, (1 \otimes L_{j'}(k')) \; s^- \; (L_j(k) \otimes 1) \\ &+ \delta_{j,j'} \left[r^+ \; (L_j(k) \otimes L_{j'}(k')) - (L_j(k) \otimes L_{j'}(k')) \; r^- \right]\end{aligned} \tag{23}$$

in which

$$\begin{aligned} s^+(k,k') &= \frac{E \otimes F}{k^2 - p^2} \, , \quad s^-(k,k') = \frac{F \otimes E}{k'^2 - p^2} \\ r^+(k,k') &= \frac{\mathbf{P}}{k^2 - k'^2} \, , \quad r^-(k,k') = r^+(k,k') - s^+(k,k') + s^-(k,k') \, , \end{aligned} \tag{24}$$

where $\mathbf{P}$ denotes the permutation matrix acting on the matrix tensor product. In the (gauge-equivalent) choice $\beta_j = x_j$ we have a similar structure with the roles of s^+ and s^- and of r^+ and r^- interchanged and p replaced by q.

The (fundamental) Poisson bracket relations (23) are similar to the structure of the discrete non-ultralocal brackets that were introduced in ref. [24], cf. also [25]. Note that here, however, the matrices $s^{\pm}$ are not symmetric. To ensure the skew-symmetry of the bracket (23), we have, however, the relation

$$s^{-}(k,k') = \mathbf{P}s^{+}(k',k)\mathbf{P}\,, \tag{25}$$

whereas both $r^{\pm}$ are anti-symmetric

$$r^{\pm}(k,k') = -\mathbf{P}r^{\pm}(k',k)\mathbf{P}\,. \tag{26}$$

Calculating the Poisson brackets between the monodromy matrices (12) leads to the result

$$\begin{aligned}\{\mathcal{T}(k)\otimes\mathcal{T}(k')\} = {} & \left[r^{+}(k,k'),\,\mathcal{T}(k)\otimes\mathcal{T}(k')\right] \\ & + \left[s^{+}(k,k'),\,\mathcal{T}(k)\otimes\mathbf{1}\right](\mathbf{1}\otimes\mathcal{T}(k')) \\ & - \left[s^{-}(k,k'),\,\mathbf{1}\otimes\mathcal{T}(k')\right](\mathcal{T}(k)\otimes\mathbf{1})\,.\end{aligned} \tag{27}$$

The Jacobi identities for the Poisson bracket (23) lead to a set of equations for $r^{\pm}$ and $s^{\pm}$. For $r^{\pm}$ we obtain the usual classical Yang-Baxter equations, i.e.

$$\left[r^{\pm}_{12},\,r^{\pm}_{13}\right] + \left[r^{\pm}_{12},\,r^{\pm}_{23}\right] + \left[r^{\pm}_{13},\,r^{\pm}_{23}\right] = 0\,, \tag{28}$$

and in addition there are the following relations between $r^{\pm}$ and $s^{\pm}$

$$\left[s^{\pm}_{12},\,s^{\pm}_{13}\right] = \left[r^{\pm}_{23},\,s^{\pm}_{12}\right] + \left[r^{\pm}_{23},\,s^{\pm}_{13}\right]\,. \tag{29}$$

Eqs. (28, 29) differ from the non-ultralocal Yang-Baxter equations found in e.g. [20], cf. also [22,23], in that there the r- and s-matrices combine into one single r-matrix (which is neither symmetric nor antisymmetric). It is easy to show that the matrices $r^{\pm}$ and $s^{\pm}$ of (24) obey the equations (28) and (29). It is, furthermore, interesting to note that r^{+} is the usual rational solution of the classical YB equation arising in connection with the Yangian, and that the addition of a term $s^{-} - s^{+}$ does not disturb the fact that it remains a solution of the CYBE.

As a consequence of (27), taking the trace over both components of the tensor product, the right-hand side of (27) vanishes identically, and we find

$$\{\,Tr(\mathcal{T}(k))\,,\,Tr(\mathcal{T}(k'))\,\} = 0\,,$$

implying that the integrals of the mapping (8) are in involution. Thus, we establish integrability in the sense of Arnol'd-Darboux-Liouville, applying the general result of [15].

In [14] similar results for a different type of systems (discrete-time Toda-lattices in central difference scheme) have been obtained, using an r-matrix formalism. Furthermore, we point out that the systems we consider here and in [13] have a quite different behaviors in the kinetic terms, explaining why the underlying r-matrix structure is not of standard type. Having obtained the classical r-matrix structure the next obvious step is to look for quantization of the mappings, which could possibly lead to alternative quantization schemes for the KdV system, cf. e.g. [26] or alternative lattice generalizations of Virasoro algebra, cf. e.g. [27]-[30].

6. Similarity Reductions of Integrable Lattices

The similarity solutions of the lattice equations, [31], provide the other class of solutions that we want to consider in this report. As is well-known the similarity solutions of integrable nonlinear PDE's give rise to the various Painlevé transcendents. For instance the Painlevé II equation

$$y'' = 2y^3 + \xi y + \mu , \tag{30}$$

is related to the KdV equation as well as to the MKdV equation by similarity reduction [32]-[36]. Furthermore, the Painlevé equations are found in the calculation of correlation functions of quantum exactly solvable (spin) models, cf. e.g. [37,38], and more recently in connection with 2D quantum gravity and random matrix models they give rise to discrete analogues of Painlevé transcendents [39]-[42].

The similarity reduction in the lattice case exhibits some new features with respect to the continuum case. Let us illustrate our idea by the analogy to that case. For example the reduction to similarity solutions for the potential KdV can be formulated in terms of a system consisting of the equation itself and an integrable constraint, namely

$$0 = u + x u_x + 3t u_t, \tag{31}$$

from which one derives the similarity variable $\xi \equiv x/3(t^{\frac{1}{3}})$, i.e. $u(x,t) = t^{-\frac{1}{3}}\bar{u}(\xi)$. In the lattice case we supplement the original lattice equation, which is the lattice (potential) KdV equation (1), with an integrable constraint

$$0 = u + pn \frac{u_{n+1,m} - u_{n-1,m}}{2p - u_{n+1,m} + u_{n-1,m}} + qm \frac{u_{n,m+1} - u_{n,m-1}}{2q - u_{n,m+1} + u_{n,m-1}} , \tag{32}$$

which in contrast to the continuous case is nonlinear. Although it might not be possible to solve explicitly for a similarity variable, nevertheless it provides us with a *system* of difference equations, which carries an associated *isomonodromy problem*, which reads

$$\begin{aligned} k\frac{d}{dk}\phi_k &= S_- \cdot \phi_k + pn\left[-F + \frac{1}{R_p}\left(\frac{V_p}{p-k} + \frac{V_p'}{p+k}\right)\right]\cdot\phi_k \\ &\quad + qm\left[-F + \frac{1}{R_q}\left(\frac{V_q}{q-k} + \frac{V_q'}{q+k}\right)\right]\cdot\phi_k , \end{aligned} \tag{33}$$

using the following notation: $R_p = 2p - u_{n+1,m} + u_{n-1,m}$, $R_q = 2q - u_{n,m+1} + u_{n,m-1}$

$$V_p = \begin{pmatrix} p - u_{n+1,m} & 1 \\ * & p + u_{n-1,m} \end{pmatrix}, \quad V_p' = \begin{pmatrix} -(p + u_{n-1,m}) & 1 \\ * & -(p - u_{n+1,m}) \end{pmatrix}, \tag{34}$$

and V_q resp. V_q' denote the matrices obtained from (34) by interchanging $p \leftrightarrow q$ and $(n \pm 1, m) \leftrightarrow (n, m \pm 1)$. $*$ denotes in (34) as before the product of the diagonal entries.

Clearly, by imposing the constraint (32) arising from the compatibility of the linear system (4) of the lattice KdV and the isomonodromy problem (33), we have an integrable reduction corresponding to the lattice analogue of a similarity reduction. In order to give an interpretation of this reduction from the point of view of initial value problem let us note that global solutions can be obtained by imposing initial data on localize configurations of sites ('embryons') on the lattice, cf. [31]. This contrasts the situation of the mappings treated in the previous sections, where we have a periodic initial value problem on an infinite path in the two-dimensional lattice. The similarity reduction is completely analogous to the continuous case, in which the PDE allows for similarity solutions that are obtained from an ODE that requires initial data at a single point in space-time.

As we shall indicate in the following section, well-defined continuum limits will bring us back to the continuum situation, which reduces to PII. As such it is justified to refer to the system of partial difference equation plus discrete similarity constraint, as being a lattice or discrete analogue of Painlevé II equation.

7. Continuum Limits

We shall now discuss the continuum limits that enables one to recover the Painlevé type of equations from the lattice KdV (or lattice MKdV) plus the similarity constraint. The limit that leads from the lattice equation most straightforwardly to the (potential) KdV is performed as follows [1,2]. We first do the limit with respect to the m-variable after a change of variables $u_{n,m} =: u_{n'}(m)$, where $n' = n + m$ remains fixed, namely by taking

$$\delta \equiv p - q \mapsto 0 \quad , \quad m \mapsto \infty \quad , \quad \delta m \mapsto \tau \; .$$

This limit is motivated from the behaviors of discrete plane-wave factors, namely

$$\left(\frac{p+k}{p-k}\right)^n \left(\frac{q+k}{q-k}\right)^m \mapsto \left(\frac{p+k}{p-k}\right)^{n'} e^{\frac{2k\tau}{p^2-k^2}} \; ,$$

cf. [1,2,31]. By this limit, the lattice KdV (1) goes over into the following differential-difference equation (we omit the prime of the n' variable)

$$1 + \partial_\tau u_n \;=\; \frac{2p}{2p - u_{n+1} + u_{n-1}} \; , \tag{35}$$

which is directly related to the Kac-van Moerbeke equation, cf. [1]. The similarity constraint goes in this limit over into

$$0 = u_n + \frac{(pn-\tau)(u_{n+1} - u_{n-1}) + p\tau\,\partial_\tau(u_{n+1} + u_{n-1})}{2p - u_{n+1} + u_{n-1}} + \frac{2p\tau\,(u_{n+1} - u_{n-1})}{(2p - u_{n+1} + u_{n-1})^2} \; . \tag{36}$$

Next, the second continuum limit is performed by taking

$$p \to \infty \; , \quad n \to \infty \; , \quad \tau \to \infty \; ,$$

such that

$$2\frac{n}{p} + 2\frac{\tau}{p^2} \mapsto x \quad , \quad \frac{2}{3}\frac{n}{p^3} + 2\frac{\tau}{p^4} \mapsto t \; ,$$

in which case (35) goes over into the potential KdV equation, and the similarity constraint goes over into eq. (31). This, as we mentioned before leads immediately to the proper similarity reduction to obtain Painlevé II (30) for the special value $\mu = 0$. It is straightforward to check that the Lax pair of the lattice and the isomonodromy problem go over to the proper ones for the continuum case by performing the same limits as given above.

We mention that there is an alternative way to obtain Painlevé II more directly starting from a lattice version of the MKdV equation rather than the KdV equation, cf. [31]. In particular, on the differential-difference level, by eliminating the derivatives with respect to the variable τ, one can obtain an *ordinary difference equation*, namely

$$z_{n+1} + z_{n-1} + \frac{2p}{\tau}\frac{n\, z_n}{1-z_n^2} = 0\,, \tag{37}$$

which under the above limit leads also to Painlevé II. Eq. (37) can be viewed as a Schlesinger transformation for the corresponding ODE in terms of the variable τ which is an equation of Painlevé type, but which we will not give here, cf. [31]. This resembles the situation of ref. [42] for Painlevé I.

8. The Lattice Gel'fand-Dikii Hierarchy

We would like to finish with an indication how to extend the results obtained so far to more general lattice models, taking as an example the lattice analogue of the Gel'fand-Dikii (GD) hierarchy. This example is particularly interesting, because in contrast to the lattice KdV (1), and many other lattice systems of similar type that have the property that the variables are situated around a simple plaquette on the rectangular lattice, the higher-order lattice equations in the GD hierarchy involve variables which are more than one lattice spacing apart. In [43] we introduce a general class of linear problems associated with this hierarchy. The lattice equations that come out of this type of problems, i.e. discrete equations that in appropriate continuum limits reduce to PDE's in the Gel'fand-Dikii hierarchy, [44], are naturally enumerated by an integer N coming from a corresponding Nth root of unity $\omega \equiv \exp(2\pi i/N)$, $(N \in \mathbf{Z})$.

Here, we mention only the first member after the KdV lattice (which is the case $N=2$), leading to a lattice analogue of the Boussinesq (BSQ) equation ($N=3$). The linear system in this case is given by

$$(p+\omega k)\phi_k(n+1,m) = L_k\cdot\phi_k(n,m)\,, \quad (q+\omega k)\phi_k(n,m+1) = M_k\cdot\phi_k(n,m)\,, \tag{38}$$

in which

$$L_k = \begin{pmatrix} p-u_{n+1,m} & 1 & 0 \\ -v_{n+1,m} & p & 1 \\ k^3+p^3+* & w_{n,m} & p+u_{n,m} \end{pmatrix}\,, \tag{39}$$

and again M_k is obtained by replacing $p \to q$ and $(n+1,m) \to (n,m+1)$. In (39) the v and w are auxiliary fields, and the term $*$ in the left-lower corner of the matrix L_k is determined by the condition that the determinant $det(L_k) = p^3+k^3$.

The compatibility relations of (39) lead to the relations

$$\begin{aligned} v_{n,m+1} - v_{n+1,m} &= u_{n+1,m+1}(p-q+u_{n,m+1}-u_{n+1,m}) + q\,u_{n+1,m} - p\,u_{n,m+1}, \\ w_{n,m+1} - w_{n+1,m} &= -u_{n,m}(p-q+u_{n,m+1}-u_{n+1,m}) + p\,u_{n+1,m} - q\,u_{n,m+1}, \\ v_{n+1,m+1} - w_{n,m} &= pq - (p+q+u_{n,m})(p+q-u_{n+1,m+1}) + \\ &\quad + \frac{p^3-q^3}{p-q+u_{n,m+1}-u_{n+1,m}}, \end{aligned} \tag{40}$$

from which one obtains the lattice BSQ equation

$$\begin{aligned}
&\frac{p^3-q^3}{p-q+u_{n+1,m+1}-u_{n+2,m}} - \frac{p^3-q^3}{p-q+u_{n,m+2}-u_{n+1,m+1}} \\
&-u_{n,m+1}u_{n+1,m+2}+u_{n+1,m}u_{n+2,m+1} \\
&+u_{n+2,m+2}\left(p-q+u_{n+1,m+2}-u_{n+2,m+1}\right)+u_{n,m}\left(p-q+u_{n,m+1}-u_{n+1,m}\right) \\
=&(2p+q)\left(u_{n+1,m}+u_{n+1,m+2}\right)-(p+2q)\left(u_{n,m+1}+u_{n+2,m+1}\right) .
\end{aligned} \tag{41}$$

From (41) by appropriate continuum limits we recover the continuum BSQ equation. An intermediate continuum limit of the lattice BSQ yields the following differential-difference equation

$$3p^2\partial_t\left(\frac{1}{1+\dot{u}_n}\right) = (1+\dot{u}_{n+1})(u_{n+2}-u_{n-1}-3p) - (1+\dot{u}_{n-1})(u_{n+1}-u_{n-2}-3p) . \tag{42}$$

The BSQ lattice can now in principle be investigated along similar lines as the KdV lattice, i.e. to study the integrable mapping reductions, their invariants and their canonical structure. In fact, we have also an action for the BSQ lattice, given by (here: $\epsilon \equiv p^2+pq+q^2$, δ as before)

$$\begin{aligned}
\mathcal{S} = \sum_{n,m\in\mathbf{Z}} [& \epsilon\delta \ln(\delta+u_{n,m+1}-u_{n+1,m}) \\
& -(p+q+u_{n,m})(p+q-u_{n+1,m+1})(\delta+u_{n,m+1}-u_{n+1,m}) \\
& +q\, u_{n,m}\, u_{n,m+1} - p\, u_{n,m}\, u_{n+1,m}] ,
\end{aligned} \tag{43}$$

which allows us to apply our techniques to obtain the proper symplectic structure. On the basis of these results we can, finally, study the quantization of the systems in the lattice GD hierarchy. This is particularly interesting in connection with the study of lattice analogues of the W-algebras, [30], cf. also [28], for which we conjecture that the lattices and mappings coming from the GD hierarchy are the proper time-discrete counterparts.

9. Concluding Remarks

We have shown explicitly how (discrete-time) integrable lattices cf. e.g. [1,2] exhibit many features of integrable systems; a) their periodic problem gives rise to integrable finite dimensional mappings admitting Lax pairs for a finite-gap integration, and Zakharov-Shabat systems for an r-matrix formulation of their Poisson structure and involutivity of their integrals of motion and b) they posses similarity type of solutions that provide discrete analogues of the Painlevé transcendents associated to an isomonodromic problem. Other aspects such as alternative boundary conditions, bi-hamiltonian structure, transformation to action-angle variables and quantization are under current investigation. In particular, quantizing these mappings would then possibly lead to explicit models of time-discrete integrable systems for the study of the corresponding quantum group structures and, in the context of dynamical systems, such quantum mappings could serve as a starting point for the investigation of perturbation techniques on the quantum level and of the transition to 'quantum chaos'.

Acknowledgement

VGP and FWN were partially supported by AFOSR Grant No. 86-05100.

References

[1] F.W.Nijhoff, G.R.W.Quispel and H.W.Capel, Phys. Lett. **97A** (1983), 125.
[2] G.R.W.Quispel, F.W.Nijhoff, H.W.Capel and J. van der Linden, Physica **125A** (1984), 344.
[3 F.W.Nijhoff, H.W.Capel, G.L.Wiersma and G.R.W.Quispel] Phys. Lett. **105A** (1984), 267.
H.W.Capel, G.L.Wiersma and F.W.Nijhoff, Physica **138A** (1986), 76.
[4] M.J.Ablowitz and F.J.Ladik, Stud. Appl. Math. **55** (1976), 213 **57** (1977), 1.
[5] R.Hirota, J. Phys. Soc. Japan **43** (1977), 1424, 2074, 2079; ibid. **50** (1981), 3785.
[6] E.Date, M.Jimbo and T.Miwa, J. Phys. Soc. Japan **51** (1982), 4125 **52** (1983), 388,766.
[7] G.L.Wiersma and H.W.Capel, Physica **142A** (1987), 199; ibid. **149A** (1988), 49,75; Phys. Lett. **124A** (1987), 124.
[8] E.M.McMillan,, Topics in Physics, eds. W.E.Brittin and H.Odabasi,, Colorado Associated Univ. Press, Boulder, 1971, p. 219.
[9] G.R.W.Quispel, J.A.G.Roberts and C.J.Thompson, Phys. Lett. **A126** (1988), 419; Physica **D34** (1989), 183.
[10] A.P.Veselov, Funct. Anal. Appl. **22** (1988), 83; Theor. Math. Phys. **71** (1987), 446.
[11] P.A.Deift and L.C.Li, Commun. Pure Appl. Math. **42** (1989), 963.
[12] J.Moser and A.P.Veselov, Preprint ETH (Zürich) (1989).
[13] V.G.Papageorgiou, F.W.Nijhoff and H.W.Capel, Phys. Lett. **147A** (1990), 106.
[14] Yu.B.Suris, Phys. Lett. **145A** (1990), 113; Algebra i Anal. **2** (1990), 141. (Russian)
[15] M.Bruschi, O.Ragnisco, P.M.Santini and G.-Z.Tu, *Integrable Symplectic Maps*, Preprint Università di Roma I (June 1990).
[16] H.W.Capel, F.W.Nijhoff and V.G.Papageorgiou, *Complete Integrability of Lagrangian Lattices and Mappings of KdV Type*, Preprint INS # 165/90.
[17] S.V. Manakov, Sov. Phys. JETP.
[18] L.D.Faddeev, Développements Récents en Théorie des Champs et Mécanique Statistique, eds. J.-B. Zuber and R. Stora, North-Holland Publ. Co., 1984, p. 561.
[19] L.D.Faddeev and L.A.Takhtajan, *Hamiltonian Methods in the Theory of Solitons*, Springer Verlag, Berlin, 1981.
[20] J.M.Maillet, Phys. Lett. **162B** (1985), 137; Nucl. Phys. **B269** (1986), 54.
[21] A.G.Reyman and M.A.Semenov-Tian-Shanskii, Phys. Lett. **130A** (1988), 456.
[22] O.Babelon and C.Viallet, Phys. Lett. **237B** (1990), 411.
[23] J.Avan and M.Talon, Preprint PAR-LPTHE 90-30.
[24] L.-C.Li and S.Parmentier, C.R. Acad. Sci. Paris **307** no. Série I (1988), 279; Commun. Math. Phys **125** (1989), 545.
[25] O.Babelon and L.Bonora, *Quantum Toda Theory*, Preprint SISSA/ISAS-141/90/EP.
[26] S.L.Luk'yanov, Funct. Anal. Appl. **22** (1989), 255.
[27] J.L.Gervais, Phys. Lett. **160B** (1985), 277, 279.
[28] L.D.Faddeev and L.A.Takhtajan, Lect. Notes Phys. **246** (1986), 166.
[29] A.Yu.Volkov, Theor. Math. Phys. **74** (1988), 96.
[30] O.Babelon, Phys. Lett. **215B** (1988), 523; ibid. **238B** (1990), 234.
[31] F.W.Nijhoff and V.G.Papageorgiou, *Similarity Reductions of Integrable Lattices and Discrete Analogues of Painlevé II Equation*, Preprint INS # 155/90.
[32] M.J.Ablowitz and H.Segur, Phys. Rev. Lett. **38** (1977), 1103.
M.J.Ablowitz, A.Ramani and H.Segur, J. Math. Phys. **21** (1980), 715, 1006.
[33] H.Flaschka and A.C.Newell, Commun. Math. Phys. **76** (1980), 67.
[34] M.Jimbo, T.Miwa and K.Ueno, Physica **2D** (1981), 306.
M.Jimbo and T.Miwa, Physica **2D** (1981), 407 **4D** (1981), 47.
[35] A.S.Fokas and M.J.Ablowitz, Phys. Rev. Lett. **47** (1981), 1096.
[36] A.R.Its and V.Y.Novokshenov, *The Isomonodromic Deformation Theory in the Theory of Painlevé Equations*, Lect. Notes Math. **1191** (1986), Springer Verlag, Berlin.
[37] E.Barouch, B.M.McCoy and T.T.Wu, Phys. Rev. Lett. **31** (1973), 1409.

T.T.Wu, B.M.McCoy, C.A.Tracy and E.Barouch, Phys. Rev. **B13** (1976), 316.

[38] M.Jimbo, T.Miwa, Y.Mori and M.Sato, Physica **1D** (1980), 80.

[39] E.Brézin and V.A.Kazakov, Phys. Lett. **236B** (1990), 144.

[40] D.J.Gross and A.A.Migdal, *Nonperturbative Two-Dimensional Quantum Gravity*, Preprint PUTP-1148 (Oct. 1989).

[41] M.R.Douglas and S.H.Shenker, *Strings in less than one dimension*, Preprint Rutgers University, RU-89-34.

[42] A.R.Its, A.V.Kitaev and A.S.Fokas, *Isomonodromic Approach in the Theory of 2D Quantum Gravity*, Usp. Math. Nauk (to appear).

[43] F.W.Nijhoff, V.G.Papageorgiou, H.W.Capel and G.R.W.Quispel, *The Lattice Gel'fand-Dikii Hierarchy*, Preprint INS # 154/90.

[44] I.M.Gel'fand and L.A.Dikii, Funct. Anal. Appl. **10** (1976), 18; ibid. **13** (1979), 8; Russ. Math. Surv. **30** (1975), 77.

Department of Mathematics and Computer Science and Institute for Nonlinear Studies, Clarkson University, Potsdam NY 13699-5815, USA

ON RELATIONS BETWEEN POISSON GROUPS AND QUANTUM GROUPS

S. ZAKRZEWSKI

Department of Mathematical Methods in Physics,
Faculty of Physics, University of Warsaw

Introduction

In contrast to the compact case [14,15,16], a satisfactory definition of a *locally* compact quantum group is still under investigation. Clearly, any definition should provide us at least with the following two objects:

- a locally compact quantum space Λ (represented by a C^* -algebra), shortly: a C^* – *space*,

- an associative morphism from $\Lambda \times \Lambda$ to Λ(represented by a suitable morphism of C^*-algebras – the *comultiplication*.)

A suitable 'invertibility axiom' should be, additionally, assumed, in order to obtain a C^* – *group*(we use this short term to denote a 'C^*-space equipped with a *group* structure' or a 'locally compact quantum group'). Recently, several examples of (candidates for) non-compact C^*-groups have been constructed ([8,17,19,20,2]). The construction follows a general scheme, similar to that one used in the compact case. The starting point is a Hopf $*$-algebra deformation of a group of matrices. The underlying algebra is given in terms of generators and relations. Usually, there is a natural choice of a C^*-algebra associated with these relations (in the non-compact case, the generators are only affiliated [17] with the C^*-algebra). Then one has to construct the comultiplication using a formal expression for its values on generators. It turns out that in some cases such a comultiplication does not exist [17], i.e. some Hopf $*$-algebra deformations of a matrix groups do not give rise to C^*-groups.

Recently, symplectic models of C^* - groups (called S^* - ***groups*** or ***symplectic pseudogroups***) have been introduced [23] in order to develop a theory applicable to classical physical systems in the same way as the theory of C^*-groups is applicable (in principle) to quantum physical systems. At the same time, it has been realized that S^*-groups can be very useful for studying the relations between quantum groups and quasiclassical limit. The purpose of this article is to make an emphasis of the following two observations that can be made immediately after having a look at the structure of S^*-groups:

1. S^*-groups provide geometrical data which seem to be sufficient for constructing C^*-groups,
2. S^*-groups turn out to be integrated versions of Poisson (Lie) groups. Not all Poisson groups can be integrated to S^*-groups. We conjecture that a Hopf

$*$-algebra deformation of a Lie groups gives rise to a C^*-group if and only if the corresponding Poisson group can be (essentially) integrated to a S^*-group.

In the rest of the paper we discuss these observations and give some examples. Basic notions are introduced in the text two sections.

1. Symplectic noncommutative geometry

Consider the following (apparently incomplete) 'definition' of a category:

Objects – quadruples (X, m, e, s) such that

$$\begin{aligned} m: X \otimes X \to X \quad &\text{and} \quad m(m \otimes \mathrm{id}) = m(\mathrm{id} \otimes m) \\ e: \mathcal{K} \to X \quad &\text{and} \quad m(e \otimes \mathrm{id}) = \mathrm{id} = m(\mathrm{id} \otimes e) \\ s: X \xrightarrow{\frac{1}{2}} X \quad &\text{and} \quad s\, s = \mathrm{id},\ sm = m(s \otimes s)t \end{aligned} \tag{1}$$

Morphisms (from (X, m, e, s) to (X', m', e', s')) – such $h: X \to X'$ that

$$\begin{aligned} hm &= m'(h \otimes h) \\ he &= e' \\ hs &= s'h \end{aligned} \tag{2}$$

One possibility to complete the above definition is to make the following choice:

arrows = linear maps of complex vector spaces (anti-linear maps in the $\frac{1}{2}$-case), $\otimes$ = tensor product, $\mathcal{K} = \mathbb{C}$, t = permutation in the product.

This choice leads to the category of *complex $*$-algebras with unit*. The following (more familiar) notation for the *multiplication*, the *unit* and the *star* can be used:

$$ab = m(a \otimes b), \qquad I = e(1), \qquad a^* = s(a), \tag{3}$$

for $a, b \in X$. If an object (X, m, e, s) of this category is isomorphic to a $*$-subalgebra of bounded operators acting in a Hilbert space (this is the case for C^*-algebras), then it satisfies the following 'positivity condition':

$$a^*a \neq 0 \qquad \text{for each non-zero } a \in X.$$

Now we consider another choice. Let us assume that in (1) and (2)

arrows = symplectic relations between symplectic manifolds (anti-symplectic relations in the $\frac{1}{2}$-case),

$\otimes$ = the cartesian product (which plays the role of the tensor product when dealing with relations, cf. [22]),

$\mathcal{K}$ = $\{1\}$ is a distiguished one-point set,

t = permutation in the product.

This choice leads to symplectic analogues of $*$-algebras. They have been considered some time ago ([11]) but only on a heuristic level, due to troubles with composing the arrows (symplectic relations do not form a category). Only recently, a set of axioms has been found which circumvents those difficulties (see [23], see also [12,5]). Following [23], along with the symplectic choice we make the following two **assumptions**:

1. *transversality* of relations being composed in (1), (2) (see [23] for details),
2. we consider only such objects (X, m, e, s) which satisfy the *strong positivity*:

$$E \supset x^* x \neq 0 \qquad \text{for } x \in X,$$

where $E = e(1)$ (notation as in (3)).

With these assumptions, (1) and (2) define a category which is said to be the *category of S^*-algebras.*

Let (X, m, e, s) be a S^*-algebra. It follows from the first two lines in (1) that there exist unique two mapping $e_L, e_R : X \to E$ (said to be the *left* and the *right projection*,respectively) such that for $x \in X$ we have $e_L(x)x \neq 0$ and $xe_R(x) \neq 0$. A more detailed inspection shows that (X, m, e, s) is a symplectic groupoid [12,1], in which m is the *partially defined multiplication*, E is the *set of units*, e_L is the *target* map, e_R is the *source* map and s is the *inverse*. In particular, the projections are smooth submersions and they induce on E a pair of Poisson structures, π_L and π_R, which differ only by the sign. It is shown in [23] that each morphism $h : X \to X'$ from a S^*-algebra (X, m, e, s) to a S^*-algebra (X', m', e', s') induces (by projection, or, equivalently, by restriction) a map $f_0 : E' \to E$ (said to be the *base map of* h, which is smooth and Poisson, i.e. $f_0 * \pi'_L = \pi_L$. It is also *complete* in the sense that the pullback of any complete function on E is a complete function on E' (by definition, a function is *complete* if its hamiltonian vector field is complete). Summarizing, assignments $(X, m, e, s) \mapsto (E, \pi_L), h \mapsto f_0$, define a contravariant functor from the category of S^*-algebras to the category of Poisson manifolds with complete Poisson maps. Moreover after restricting to *integrable* Poisson manifolds (i.e. such Poisson manifolds which are the sets of units of S^*-algebras, cf. [13]) and to S^*-algebras with connected and simply connected fibers, both categories are anti-equivalent, cf. the following result.

Theorem 1 ([23]). *Let $(X, m, e, s), (X', m', e', s')$ be two S^*-algebras and let fibers of e_L be connected and simply connected. Then any complete Poisson map $f_0; E' \to E$ is a base map of a (unique) morphism $h : X \to X'$.*

It is clear that Poisson manifolds can provide examples of S^*-algebras. In particular, any symplectic manifold Z is integrable as a Poisson manifold and is isomorphic to the set of units of End$Z = (X, m, e, s)$, where $X = Z \otimes \overline{Z}$($\overline{Z}$ denotes Z with the opposite symplectic structure), $e(1) = E$ is the diagonal in $Z \otimes \overline{Z}$, $s(a, b) = (b, a)$ and

$$m((a,b),(c,d)) = \begin{cases} 0 & b \neq c \\ (a,d) & b = c \end{cases}$$

To decide whether a general Poisson manifold is integrable or not is not so easy. A construction of the S^*-algebra corresponding to an integrable Poisson manifold is also not simple. Similar problems arise with constructing a C^*-algebra from a set of relations between some unbounded generators!)

An effective, functorial constructions of S^*-algebras is provided by the phase functor applied to regular D^*-algebras [23]. *D^*-algebras* are defined similarly as S^*-algebras, but using only differential manifolds and differential relations. A D^*-algebra is *regular* if its projections are submersions. Regular D^*-algebra coincide with differentiable groupoids and they form a category (morphisms are differentiable relations satisfying (2) and transversality). If $A = (X, m, e, s)$ is a regular D^*-algebra, then $(\mathrm{P}X, \mathrm{P}m, \mathrm{P}e, -\mathrm{P}s)$ is a S^*-algebra, called the *phase lift of A* (the minus sign denotes the reflection in the cotangent bundle).

S^*-algebras play the role analogical to C^*-algebras. There are two main applications of these structures, reflecting the duality between Algebra and Geometry:

- *Algebra of the group multiplication.*

  A regular D^*-algebra (G, m, e, s) such that m is a map is a Lie group. The phase lift of this D^*-algebra is said to be the *group S^*-algebra* of G. Its left and right projection coincides with the right and left translation, respectively, of covectors to $T_e^*G \cong \mathfrak{g}^*$.

  Group C^*-algebras are used to study unitary group representations. Group S^*-algebras have a similar application [21,23].
- *Geometry of noncommutative spaces.*

  For any binary relation r we denote by r^T its transpose. A regular D^*-algebra (X, m, e, s) such that m^T is a map, can be identified with the manifold X itself ($m^T = d$ is then the diagonal map, $E = X, s =$id). The phase lift of this D^*-algebra is said to be the *cotangent bundle S^*-algebra of X*. Both projections coincide in this case. This construction describes an (contravariant!) embedding of the category of differential manifolds in the category of S^*-algebras. We can treat now general S^*-algebras as 'twisted cotangent bundles', or 'cotangent bundles to symplectic-noncommutative manifolds'.This point of view constitutes the *symplectic noncommutative geometry.*

In the S^*-case (unlike in C^*-case), it is easy to pass from the contravariant to a covariant embedding of usual manifolds in a noncommutative world. All we need is to transpose binary relations.

Definition ([23]). A *S^* − space* is a quadruple $\Lambda = (X, d, c, r)$ such that $\Lambda^T = (X, d^T, c^T, r^T)$ is a S^*-algebra.

(We define also D^*-spaces by transposing regular D^*-algebras.)

Any manifold X can be treated as a D^*-space (X, d, c, r), where d is the diagonal map, $c : X \rightarrow \{1\}$ is the constant map and $r =$id. The covariant embedding of usual manifolds in S^*-spaces is given by the phase functor.

The *product* (denoted by $\otimes$) of S^*-spaces (as well as S^*-algebras) is naturally defined.

2. Symplectic pseudogroups

Definition ([23]). A *S^* − group* is a pair (Λ, m), where $\Lambda = (X, d, c, r)$ is a S^*-space, m is a morphism from $\Lambda \otimes \Lambda$ to a Λ such that $m(m \otimes \mathrm{id}) = m(\mathrm{id} \otimes m)$ and

1. there exists a morphism $e : \mathcal{K} \rightarrow \Lambda$ satisfying $m(e \otimes \mathrm{id}) = \mathrm{id} = m(\mathrm{id} \otimes e)$
2. there exists a diffeomorphism $k : X \rightarrow X$ satisfying

$$m(k \otimes \mathrm{id})d = ec = m(\mathrm{id} \otimes k)d \tag{4}$$

(together with the transversality of m and $(k \otimes \mathrm{id})d$).

(We define also D^* − *groups* by replacing S^*-spaces by D^*-spaces. Phase functor applied to D^*-groups produces S^*-groups.)

Up to the end of this section we assume that (Λ, m) is a S^*-group as in the above definition. It can be shown ([23]) that $V = (X, m, e, kr)$ is a S^*-algebra. Moreover, (V^T, d^T) is a S^*-group which is said to be the *dual* of (Λ, m).

Example 1. Each Lie group G can be considered as a D^*-group. Its phase lift is a S^*-group. (This defines a embedding of the category of Lie groups into the category of S^*-groups.) The cotangent bundle T^*G carries the structure of a S^*-space whose both projections (on G) coincide and the structure of a S^*-algebra with projections (on $\mathfrak{g}^*$) which are different in general. This qualitative picture of a Lie group treated as a S^*-group shows that the dual S^*-group is in general not of this type because its space is noncommutative in general (the two projections can be different).

In general we have four different projections: c_L, c_R, projecting on $C = c^T(1)$ and e_L, e_R, projecting on $E = e(1)$. Note that $G = (C, m_0)$, where m_0 is the base map of m, is a Lie group. Also E is a Lie group (cf. Lemma 3.4 in [22]) under the structure induced by d^T. One may suspect that there is a group structure on X such that the four projections are related to four possible quotients of X by E and C.

Theorem 2 ([22]). *There is exactly one group structure on X such that the groups C and E are subgroups of X and for each $x \in X$,*

$$x = c_L(x)e_R(x) = e_L(x)c_R(x).$$

It is shown in [23] that $(X; C, E)$ is in fact a *double Lie Group*[6] equipped with a non degenerate invariant scalar product n which vanishes when restricted to C and E. Such a structure is called a *Manin group* by analogy with Manin triples (Manin algebras). The scalar product n and the symplectic form ω on X are related by a simple local algebraic operation, namely

$$n(u, v) = \omega(Ru, v)$$

where u, v are vectors tangent to X (at the same point) and $R = \frac{1}{2}(R_\lambda + R_\rho)$ is the 'arithmetic mean' of two reflections R_λ and R_ρ associated with the decomposition of the tangent space given by the left and right cosets, respectively. It is shown also that the correspondence between S^*-groups and Manin groups is in fact bijective. (The same is true for D^*-groups and double Lie groups.) This is very important for finding examples of S^*-groups.

Example 2. $(sl(N, \mathbb{C}); su(N), sb(N))$ with the scalar product given by

$$n(u, v) = \epsilon^{-1} \operatorname{Im} \operatorname{tr} uv \qquad \text{for } u, v \in sl(N, \mathbb{C}), \tag{5}$$

where ϵ is non-zero real number (deformation parameter), is a Manin triple and the corresponding triple of Lie groups, $(SL(N, \mathbb{C}); SU(N), SB(N))$, is a Manin group (see [6]).

Manin triples not always give rise to Manin groups (a splitting of a Lie algebra does not necessarily lead to a global splitting of the corresponding Lie group). Another explanation of the troubles can be given in terms of Poisson groups as follows. If (Λ, m) is a S^*-group then the group $G = (C, m_0)$ equipped with, say, the left Poisson structure π_L (coming from the left projection c_L in Λ) is a Poisson group and the group multiplication, m_0, is complete. Poisson groups with complete multiplication are said to be *complete*. Non-complete Poisson groups (and the corresponding Manin triples) cannot be used for constructing S^*-groups.

In [7] it was shown that each connected Poisson groups is integrable, hence it is a set of units of a S^*-space Λ with connected and simply connected fibers. By Theorem 1, the neutral element of the group can be lifted to a morphism e from $\mathcal{K}$ to Λ and the group inverse can be lifted to an anti-isomorphism from Λ to Λ. If the group is complete, then the group multiplication can be lifted to a morphism m from $\Lambda \otimes \Lambda$ to Λ. We are therefore close to prove that (Λ, m) is a S^*-group. Unfortunately, at present we are not able to prove (4)(see however Section 4, particularly lines after formulas (6)-(8)).

3. Geometric quantization

Geometric Quantization is a belief that there exist geometric data for constructing 'quantum' objects from 'classical' ones (eg. [9]). It has been in fact justified in situations intimately related to Lie groups (cf. the *orbit method*). The natural geometrical procedure consists in replacing (generating functions of) symplectic relations by (kernels of) linear operators ([11]). In the case of S^*-groups we have a set of symplectic relations 'waiting to be quantized'. The geometric procedure will be 'good' if the resulting operators satisfy the same algebraic equalities, so that we obtain a Hopf $*$-algebra (hopefully, close to a C^*-group).

Dealing with generating functions of symplectic relations requires a choice of a polarization (distinguished by the problem). It is not known how to choose appropriate polarizations in general. Here we describe briefly what has been done in [10] in the case of the (2n+1)-dimensional Heisenberg group H. On this group there is a (2n+1)-parameter family of Poisson structures for which one can construct very effectively corresponding S^*-groups (Λ, m) (they exist!). It is also proved that for each Poisson structure on H, the exponential map from $\mathfrak{h}$(the Heisenberg Lie algebra) to H is Poisson. Here the Poisson structure on $\mathfrak{h}$ is the linearized one and equal to the standard Poisson structure on $\mathfrak{g}^* \cong \mathfrak{h}$, where $\mathfrak{g}$ is the Lie algebra of the dual Poisson group, denoted by G. By Theorem 1, the exponential map lifts to an isomorphism from the transposition of the symplectic group algebra of G to $\Lambda = (X, d, c, r)$. In particular, this identifies X with a cotangent bundle. It turns out that also (X, m, e, s) can be standardized by a similar method. This enabled us in [10] to carry out the quantization successfully.

The above case can be considered as a vast generalization of the example of Kac and Paljutkin [4], perhaps the first natural example of a quantum group.

It would be interesting to study geometric quantization in the case of remaining Poisson brackets on the Heisenberg group and the other nilpotent groups, then also solvable ones. However, real challenge for the geometric methods seems to be provided by simple groups, like $SU(N)$ (Example 2).

4. Non-complete Poisson groups

When we try to construct a Manin group from a Manin triple, we often end up with a structure described in the following definition.

Definition. A Manin *splitting* consists of two objects:

1. a triple $(X; C, E)$ such that X is a Lie group, C and E are its closed subgroups, $C \cap E = \{0\}$ (the neutral element of X) and X is the smallest Lie group containing C and E,
2. a non-degenerate invariant scalar product on X vanishing when restricted to C and E.

According to [22,23], the set $P = CE \cap EC$ of *decomposable* elements of X carries the structure of a S^*-space $\Lambda = (X, d, c, r)$ with C as the set of units, and the structure of a S^*-algebra (P, m, e, s) with E as the set of units, such that e is a morphism from $\mathcal{K}$ to Λ, $cm = c \otimes c$, $rm = m(r \otimes r)$ and

$$dm \supset (m \otimes m)(\mathrm{id} \otimes t \otimes \mathrm{id})(d \otimes d), \tag{6}$$

$$ec \supset m(sr \otimes \mathrm{id})d, \tag{7}$$

$$ec \supset m(\mathrm{id} \otimes sr)d, \tag{8}$$

It is also shown that the equality in the one of the above inclusions implies the equality in remaining ones, and is equivalent to the statement $CE = EC = X$.

It is clear that the set $CE \setminus EC$ measures how (Λ, m) is far from being a S^*-group we distinguish the following three cases:

1. $CE \setminus EC = \emptyset$. The splitting is said to be *perfect*(Manin group case),
2. $CE \setminus EC$ is of a measure zero. The splitting is said to be ***admissible***,
3. $CE \setminus EC$ is not of a measure zero. The splitting is said to be ***not admissible***.

The group C is a Poisson group, said to be ***associated*** with the Manin splitting A Poisson group is said to be *perfect(admissible)* if there exists a perfect(admissible) Manin splitting such that Λ has connected fibers and the associated Poisson group is isomorphic to the given one.

One can try to a quantize the relations d, c, r, m, e and s given by a Manin splitting but, in the non-admissible case we should not expect the resulting operators to satisfy (6),(7) and (8) with inclusions replaced by equalities. (For the admissible case, let us remark that the quantization of classical flows which are not complete only on a set of measure zero, is possible.)

Conjecture. Quantization of Poisson groups which are not admissible should meet serious problems.

Example 3. *Poisson* $SU(1,1)$. The triple $(SL(2,\mathbf{C}); SU(1,1), SB(2))$ with the scalar product given by (5) is a Manin splitting. It is easy to see that $CE \setminus EC$ contains an open set.

Example 4. *Poisson* $SL(2,\mathbf{R})$. We set $G = SL(2,\mathbf{R})$. Consider the Manin splitting $(X; C, E)$, where $X = G \times G$, $C = \{(g,g) : g \in G\}$,

$$E = \left\{ \left(\begin{bmatrix} a & b \\ 0 & a^{-1} \end{bmatrix}, \begin{bmatrix} a^{-1} & 0 \\ c & a \end{bmatrix} \right) : a \neq 0, b, c \in \mathbf{R} \right\}$$

and the scalar product is given by

$$n((u,v),(u,v)) = \epsilon^{-1}(K(u,u) - K(v,v)) \qquad \text{for } u, v \in sl(2,\mathbf{R}),$$

where K is the Killing form on $sl(2,\mathbf{R})$. One can check that the splitting is not admissible.

Example 5. Let G be as in the preceding example. Consider the triple $(\mathrm{T}^*G; G, E)$, where E is the subgroup of $\mathrm{T}^*G \cong TG \cong G \ltimes \mathfrak{g}$ corresponding to the subalgebra of $\mathfrak{g}\otimes\mathfrak{g}$ spanned by the following three elements:

$$\left(\epsilon\begin{bmatrix} 1 & 0 \\ 0 & -1 \end{bmatrix}, \begin{bmatrix} 0 & 0 \\ 1 & 0 \end{bmatrix}\right), \left(0, \begin{bmatrix} 0 & 1 \\ 0 & 0 \end{bmatrix}\right), \left(-\epsilon\begin{bmatrix} 0 & 1 \\ 0 & 0 \end{bmatrix}, \frac{1}{2}\begin{bmatrix} 1 & 0 \\ 0 & -1 \end{bmatrix}\right).$$

This describes a Manin splitting which is not admissible.

The above three examples exhaust all types of Poisson structures on $G = SL(2,\mathbf{R}) = SU(1,1)$. All of them are not admissible. On the other hand, each of this type corresponds to a Hopf$*$-algebra deformation of G (for the third case see [24]; no other deformations are known!). Some attempts have been already made to formulate those deformations on the C^*-level. It has been shown in [17] that "something is essentially wrong" for the $SU(1,1)$ case. An experience with the "$ax+b$" group (cf. Examples 7,8 below) shows that there are also serious difficulties in the remaining two cases.

Example 6. Consider the Manin splitting $(SL(2,\mathbf{C}); E(2), SB(2))$ with the scalar product given by (5), where $E(2)$ is the (double covering of the) group of motions of Euclidean plane. This splitting is admissible. One can check it directly or to refer to the Gauss decomposition. The corresponding C^*-group exists and has been constructed in [17].

Example 7. The Manin splitting $(X; C, E)$, where $X = GL(2,\mathbf{R})$, $n(u,u) = \det u$ for $u \in gl(2,\mathbf{R})$ and

$$C = \left\{\begin{bmatrix} a & b \\ 0 & 1 \end{bmatrix} : a \neq 0\right\}, \qquad E = \left\{\begin{bmatrix} 1 & 0 \\ c & d \end{bmatrix} : d \neq 0\right\},$$

is admissible, but the Poisson group C is not admissible.

Example 8. Let X be the connected component of the unit I in $GL(2,\mathbf{R})$ and

$$C = \left\{\begin{bmatrix} a & b \\ 0 & 1 \end{bmatrix} : a > 0\right\}, \qquad E = \left\{\begin{bmatrix} 1 & 0 \\ c & d \end{bmatrix} : d > 0\right\}.$$

Then $(X; C, E)$ is not admissible. Set $X_0 = X/\Gamma$, $C_0 = p(C)$, $E_0 = p(E)$, $\Gamma = \{I, -I\}$ and $p: X \to X_0$ is the canonical projection. Then $(X_0; C_0, E_0)$ is admissible, but the Poisson group C_0 is not admissible.

The last two examples 'explain' the difficulties ([18,3]) with constructing the quantum "$ax+b$" group on the C^*-algebra level. They show also that it may be possible to construct a C^*-deformation of some (discrete) extensions of this group (the admissible Manin splitting leads to a S^*-space with **non connected** fibers in this case).

We summarize this discussion by the following diagram:

S^*–groups $\xrightarrow{\text{geometric quantization}}$ C^* – groups

↑ ↑

← integration problems (non-completeness) →

| |

Poisson groups $\xleftarrow[\text{quasiclassical limit}]{}$ Hopf – $*$-algebras

References

[1] Coste A., Dazord P. and Weinstein A., *Groupoïdes symplectiques*, Publ. du Dep. de Math. de l'Universitè de Lyon 1.

[2] Van Daele A., *Quantum deformation of the Heisenberg group*, Proceedings of Nara conference (1990) (to appear).

[3] Van Daele A., private communication.

[4] Kac G.I. and Palyutkin V.G., *An example of a ring group generated by Lie Groups*, Ukrain. Math. J. **16** (1964), 99–105.

[5] Karasev M.V., *Analogues of objects of Lie groups theory for nonlinear Poisson brackets*, Math. USSR Izvestiya **28** (1987), 497–527.

[6] Lu J.-H. and Weinstein A., *Poisson Lie groups, dressing transformations and Bruhat decompositions*, J. Diff. Geometry **31** (1990), 501–526.

[7] Lu J.-H. and Weinstein A., *Groupoïdes symplectiques doubles des groups de Lie-Poisson*, C. R. Acad. Sci. Paris **309** (1989), 951–954.

[8] Podleś and Woronowicz S.L., *Quantum Deformation of Lorentz group*, Commun. Math. Phys. **130** (1990), 381–431.

[9] Souriau J.-M., *Structure des systèmes dynamiques*, Dunod, Paris, 1970.

[10] Szymczak I. and Zakrzewski S., *Quantum Deformation of the Heisenberg group obtained by geometric quantization*, J. Geom. and Phys. (to appear).

[11] Weinstein A., *The symplectic category*, Lect. Notes Math. **905** (1982), 45–50.

[12] Weinstein A., *Symplectic groupoids and Poisson manifolds*, Bull Amer. Soc. **16** (1987), 101–104.

[13] Weinstein A., *Noncommutative geometry and geometric quantization*, preprint (1990).

[14] Woronowicz S. L., *Compact matrix pseudogroups*, Commun. Math. Phys. **111** (1987), 613–665.

[15] Woronowicz S. L., *A remark on compact matrix quantum groups*, Lett. Math. Phys. **21** (1991), 35–39.

[16] Woronowicz S. L., *A work in preparation.*

[17] Woronowicz S. L., *Unbounded elements affiliated with C^*-algebras and non-compact quantum groups*, Commun. Math. Phys. **136** (1991), 399–432.

[18] Woronowicz S. L., *private communication.*

[19] Woronowicz S.L. and Zakrzewski S., *Quantum deformation of Lorentz group related to Gauss decomposition*, in preparation.

[20] Zakrzewski S., *Matrix pseudogroups associated with anti-commutative plane*, Lett. Math. Phys. 2 (1991), 309–321.

[21] Zakrzewski S., *Hamiltonian group representations and phase actions*, Rep. Math. Phys. **28** (1989), 189–196.

[22] Zakrzewski S., *Quantum and classical pseudogroups, Part I Union pseudogroups and their quantization*, Commun. Math. Phys. **134** (1990), 347–370.

[23] Zakrzewski S., *Quantum and classical pseudogroups, Part II Differential and symplectic pseudogroups*, Commun. Math. Phys. **134** (1990), 371–395.

[24] Zakrzewski S., *Hopf $*$-algebra of polynomials on the quantum $SL(2,\mathbb{R})$ for a "unitary" R-matrix*, Lett. Math.Phys. **21** (1991).

DEPARTMENT OF MATHEMATICAL METHODS IN PHYSICS, FACULTY OF PHYSICS, UNIVERSITY OF WARSAW, HOZA, 74, 00-682 WARSAWA, POLAND

CHARACTERS OF HECKE AND BIRMAN–WENZL ALGEBRAS

S.V. Kerov

Leningrad Electrical Engeeneering Institute of Communications

It is well known now that the algebras in the title afford Markov traces which can be used to define new invariants of links: the original polynomial of Jones, its two-variable generalization and the Kauffman invariant. On the other hand, these algebras generate the commutants to tensor representations of (semisimple) quantum groups. In the appropriate limit one gets the Brauer algebra, responsible for the decomposition of tensor representations of groups $O(n)$ and $Sp(2n)$. In the first part of this paper I am going to describe the representations and characters of the Hecke algebra $H(q)$. These results were used in [VK] for the construction of factor-representations of the limiting algebra $H_\infty(q)$ and for a new characterization of Markov traces. In the second part I shall discuss the geometric construction of representations of Birman-Wenzl algebras BW_n. The characters of algebras BW_n and BW_∞ will be also considered. We are dealing with all traces of algebras H_∞, BW_∞, not only Markov traces.

1. Hecke algebra. Let $g_1, \ldots, g_{n-1}, \ldots$ be the generators of braid group B_∞, B_n being the subgroup, generated by the first $n-1$ generators. Schematically,

$$g_k = \text{[diagram]}$$

$k \; k+1$

Let $H_n(q)$ be the quotient algebra of $\mathbb{C}B_n$ by the relations

$$(g_k - q)(g_k + 1) = 0,$$

then $dim H_n(q) = n!$. Generically (that is, if $q^k \neq 1$ for $k \leq n$) $H_n(q)$ is semisimple and isomorphic to the group algebra of the symmetric group $\mathbb{C}\mathfrak{S}_n = H_n(1)$. If $x = \sigma_{i_1} \cdots \sigma_{i_m}$ is the reduced decomposition of a permutation $x \in \mathfrak{S}_n$, then the elements $gx = g_{i_1} \cdots g_{i_m}$ form a linear basis in $H_n(q)$.

There are two cases in which $H_n(q)$ has a natural involution: $g_x^* = g_{(x)}^{-1}$ if $q > 0$ and $g_x = (g_x)^{-1}$ if $|q| = 1$.

2. Characters of $H_n(q)$. For $q = 1$ the characters of $H_n(1) \cong \mathbb{C}\mathfrak{S}_n$ can be described by classical Frobenius formula

$$p_\rho(x) = \sum \chi_\rho^\lambda s_\lambda(x)$$

where $x = (x_1, x_2, \dots)$ and $\lambda = (\lambda_1, \lambda_2, \dots)$, $\rho = (\rho_1, \rho_2, \dots)$ are partitions of n. By $p_\rho(x) = p_{\rho_1}(x)p_{\rho_2}(x)\dots, p_{\rho_k}(x) = \sum x_i^k$ I denote power sum symmetric functions and by $s_\lambda(x)$ - Schur functions. Here ρ parametrizes the conjugacy classes of $\mathfrak{S}_n$. For $H_n(q)$ this is not the case: g_X and g_Y are not conjugate in $H_n(q)$, if $l(x) \neq l(y)$ where $l(x)$ is the length of the reduced decomposition of x. For example, there are four conjugacy classes in $H_3(q)$: $\{1\}, \{g_{(12)}, g_{(23)}\}, \{g_{(123)}, g_{(132)}\}, \{g_{(13)}\}$ .An element $g_x \in H_n(q)$ is said to be *square-free* , if all the transpositions in the reduced decomposition of x are distinct. For such g_x the indexes x are exactly the elements of least length in their conjugacy classes in $\mathfrak{S}_n$. Let c_ρ be the class in $H_n(q)$ of square-free elements g_x with $x \in \mathfrak{S}_n$ having the cycle structure ρ. We shall use symmetric functions

$$t_k(x;q) = (q-1)^{-1} \sum p_\mu(x)/z_\mu(q)$$

where $z_\mu(q) = \prod k^{\mu_k} \mu_k!(q-1)^{-\mu_k}$ for $\mu = (1^{\mu_1}, 2^{\mu_2}, \dots)$. These functions are closely related to Hall-Littlewood symmetric polynomials (see [Mac]): $t_k(x;q) = q^{k-1}P_k(x;1/q)$. The products $t_\rho = t_{\rho_1}, t_{\rho_2}, \dots$ with $\rho = (\rho_1, \rho_2, \dots)$ form a linear basis in the $\mathbb{C}$-algebra $\Lambda_\mathbb{C}$ of symmetric polynomials.

Theorem [VK]. *Let $\chi_\rho(q)$ be the value of the character χ^λ of $H_n(q)$ on the class c_ρ. Then*

$$t_k(x;q) = \sum \chi_\rho^\lambda(q) s_\lambda(x).$$

3. Characters of $H_\infty(q), q < 0$. Recall, that for a $*$-algebra the word "character" means an *indecomposable* positive central trace on it (i.e. trace of a factor-representation). For finite groups only the irreducible characters are characters in that sense.

Let us remind the description of characters of $\mathbb{C}\mathfrak{S}_\infty$ by E. Thoma. They are parametrized by a pair of sequences $\alpha = (\alpha_1 \geq \alpha_2 \geq \dots)$, $\beta = (\beta_1 \geq \beta_2 \geq \dots)$ with $\gamma = 1 - \sum \alpha_i - \sum \beta_i \geq 0$. It was shown in [VK2] that these alphas (betas) describe the growth rate of rows (columns) of a Young diagram λ while χ^λ is approximating $\chi^{\alpha,\beta}$:

$$\lim \frac{r_k(\lambda_n)}{|\lambda_n|} = \alpha_k, \ \lim \frac{c_k(\lambda_n)}{|\lambda_n|} = \beta_k \Leftrightarrow \lim \chi^\lambda(\sigma) = \chi^{\alpha\beta} \text{ for all } \sigma.$$

Since $H_\infty(q) \cong \mathbb{C}\mathfrak{S}$ for $q > 0$, the characters of $H_\infty(q)$ are the same. Let us consider their values on the classes $c \in H_\infty(q)$.

Theorem [VK].

1) $\chi^{(\alpha\beta}(c_{(n)}) = (q-1)^{-1} \sum p_\mu(\alpha,\beta)/z_\mu(q)$, where $p_k(\alpha,\beta) = \sum \alpha_i^k + (-1)^{k+1} \sum \beta_i^k$ and $p_\mu = p_{\mu_1} p_{\mu_2} \dots$;
2) $\chi^{(\alpha\beta)}(c_\rho) = \prod \chi^{(\alpha\beta)}(c_{(k)})^{\rho_k}$ *for square-free classes $c_\rho \in H_\infty(q)$.*

4. Markov traces. Consider the generating function

$$T(z) = e^{\gamma(q-1)z} \prod \frac{1 - q\alpha_i z}{1 - \alpha_i z} \prod \frac{1 - \beta_i z}{1 - q\beta_i z}.$$

The trace $\chi : H_\infty(q) \to \mathbb{C}$ is called Markov if $\chi(g_n^{\pm 1} h) = w_\pm \chi(h)$ for all $h \in H_n(q)$. For the Markov trace we have $t_n = \chi(c_{(n)}) = w_+^{n-1}, n \geq 2$ and so the function $T(z)$ is fractionary-linear. It follows that α and β are geometric progressions with ratio $1/q$. As usual, one can associate with these Markov traces link invariants, depending on two parameters: q and the balance $t = (\sum \alpha_i)/(\sum \beta_i)$ between α and β.

Remark. There is a nice Monte-Karlo procedure, generating random Young tableaux associated with Markov traces [K].

5. *R*-matrices and factor-representations. Consider the following two-side infinite matrix R with indices $i, j = \pm 1, \pm 2, \ldots$:

$$R_{ii}^{ii} = q \text{ if } i > 0, R_{ii}^{ii} = -1 \text{ if } i < 0$$

$$R_{ij}^{ij} = q - 1 \text{ if } i < j, \quad R_{ji}^{ij} = -\sqrt{q}$$

and $R_{ij}^{kl} = 0$ otherwise. Then R is a solution of Yang-Baxter equation. Denote by $R^{(k)}$ the matrix R, acting on the factors k and $k+1$ in the infinite tensor product $\overset{\infty}{\underset{1}{\otimes}} \mathbf{C}^{\infty}$. Then $g_k \mapsto R^{(k)}, k = 1, 2, \ldots$ is the representation of $H_\infty(q)$; let $M = diag(\alpha, \beta)$.

Theorem [VK]. *Suppose $q > 0$. Then $\phi = \phi^{\alpha,\beta}$ is the factor-representation of $H_\infty(q)$ with character $\chi^{\alpha,\beta}(g) = tr(\phi(g) M^{\otimes n}), g \in H_n(q)$.*

6. Representations of $H_\infty(q)$ with $q = 1$. There is a generalization of so called Young orthogonal form for representations of $H_n(q)$. The matrices of generators were independently described by H. Wenzl [W] and myself [K2]. We can take the matrix $S^{(k)}$, representing the generator g_k in the representation, associated with a partition λ, in the form

$$S_{s,t}(\tau) = (W_{r_k}(\tau) W_{-r_k}(\tau))^{1/2}.$$

In this formula s and t are Young tableaux of the shape $\lambda, r_k = (i_{k+1} - j_{k+1}) - (i_k - j_k) + 1$ is the axial distance between boxes k and $k+1$ of s and

$$W_{-r}(\tau) = 1 - \frac{\tau}{1 - \frac{\tau}{1 - \ldots}} = \frac{P_r(\tau)}{P_{r-1}(\tau)}; \quad W_r(\tau) = \frac{\tau}{1 - \frac{\tau}{1 - \ldots}} = \tau \frac{P_{r-2}(\tau)}{P_{r-1}(\tau)}$$

for $r \geq 1$. We write $P_r(1/x^2) = U_n(x)/(2x)^n$, where $U_n(\cos\phi) = \sin(n+1)\phi/\sin\phi$ are Tchebyschev polynomials of the second kind and $t = q/(q+1)^2$.

If $q = 1$ for $m \leq n$, the algebras $H_n(q)$ are not semisimple. Their representation theory, developed in [J], is very similar to that of modular representations of the symmetric group $\mathfrak{S}_n$ in non-zero characteristic m. For example, irreducible representations are parametrized by m-regular Young diagrams (with each row-length repeating less than m times).

Not all of these representations are $*$-representations. The exact condition is $\lambda_1 - \lambda_k + k \leq m$. In the definition of the matrices $S^{(k)}$ only the tableaux s, t satisfying that condition must be considered. The branching diagram for the sequence

$$\tilde{H}_1(q) \subset \tilde{H}_2(q) \subset \cdots \subset \tilde{H}_n(q) \subset \ldots$$

of semisimple quotients of Hecke algebras is the complete subgraph of the corresponding graph for $q = 1$ (Young graph).

Theorem [K2]. *For $q^m = 1$, $q \neq 1$ there are finitely many factor-representations of $H_\infty(q)$. Their characters have the form*

$$\chi^{(k)}(c_\rho) = t_\rho(\alpha; q), \ 1 \leq k \leq (m-1)/2$$

where $\alpha = \alpha^{(k)} = (1, q, \ldots, q^{k-1})/sum$, $\beta = (0, 0, \ldots)$ and $sum = (1 - q^k)/(1 - q)$.

7. Birman - Wenzl algebra $BW_\infty(q,r)$. This algebra can be defined as a quotient algebra of $\mathbf{C}B_\infty$ by the relations

$$(g_k - q)(g_k + q)(g_k - r) = 0,$$

$$f_k g_{k\pm 1} f_k = -(q + q^{-1}),\ f_k f_{k\pm 1} f_k = r^2(q + q^{-1})^2 f_k$$

where $f_k = (g_k - q)(g_k + q^{-1})$, $k = 1, 2, \dots$. By $BW_n(q,r)$ we denote subalgebra o BW_n, generated by $g_1, \dots, g_{n-1}$.

Proposition. *All the relations of the original definition in* [BW] *are the consequences of these.*

It is known that $\dim BW_n = (2n-1)!!$. We shall describe the linear basis in BW_n To this end, let us write that the product

$g_j g_{j-1} \cdots g_{i+1} g_i$	$(i \le j)$	is a g – block
$g_i g_{i+1} \cdots g_{j-1} f_j$	$(i \le j)$	is an f – block

We refer to g_j and f_j as leading terms of corresponding blocks with index j. Conside all the products of g-blocks and f-blocks, satisfying the following restrictions:

1) the indexes of all blocks are distinct
2) the indexes of f-blocks decrease and the indexes of g-blocks increase
3) all f-blocks are to the left of any g-block.

Theorem. *The products above form a linear basis in BW_∞.*

8. Geometric construction of BW_n. The threads of a braid $b \in B_n$ intersect each intermediate horizontal plane in exactly n points. We shall say that a tangle with n in/out vertices is a *chip*, if the intersection number does not exceed n. The *chip semigroup* C_n is generated by the chips

$g_k =$ ... and $h_k =$ ...

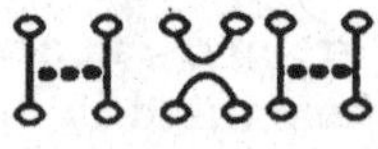

Let $L = L(a,b)$ denote the number of free loops, arising after the composition of chips a, b. Also, let $W = W(a,b)$ be the total twist (writhe) of the composition ab, provided the threads of both a and b were replaced by untwisted bands. Then $\rho(a,b) = u^L v^W$ where $u, v \in \mathbf{C}^*$ is a cocycle, that is $\rho(a,bc)\rho(b,c) = \rho(a,b)\rho(ab,c)$ for all $a,b,c \in C_n$ Equivalently, the modified multiplication $a \times b = \rho(a,b)ab$ is associative. Denote by $\mathbf{C}[C_n, \rho]$ the semigroup algebra for latter multiplication.

Theorem. *The quotient algebra of $\mathbf{C}[C_n,\rho]$ by the relations*

$$g_k - g_{k+1}^{-1} = (q + q^{-1})(1 - h_k)$$

is naturally isomorphic to $BW_n(q,r)$. Here $q - q^{-1} = (u - u^{-1})/(v-1)$ and $r = u$.

9. Representations of BW_n. Generically, the irreducible representation $(\lambda)_k$ of $BW_n(q,r)$ is parametrized by a partition λ of $m = |\lambda|$ and a nonnegative integer k with $|\lambda| + 2k = n$. Consider the space $M(n,m)$ of generalized chips with n in-vertices and m out-vertices, factorized as above. This is bimodule for $BW_n(q,r)$ and the Hecke algebra $H_m(q^2)$.

Theorem (cf.[K4]). *The actions of BW_n and H_m in $M(n,m)$ generate commutants of each other.*

This fact can be used for the explicit construction of all the irreducibles of $BW_n(q,r)$.

10. Factor-representations of BW_∞. Let us denote by J the ideal in BW_∞, generated by the elements $h_k, k = 1,2,\ldots$. Then $BW_\infty(q,r)/J \cong H_\infty(q^2)$ and factor-representations of H_∞ can be lifted to BW_∞. Strangely enough, *all* (finite type) factor-representations arise in such a way.

Theorem. *In the generic case any factor-representation of $BW_\infty(q,r)$ factors through Hecke algebra H_∞.*

11. Characters of $BW_n(q,r)$. For a partition ρ of $m = |\rho|$ and an integer $l = (n-m)/2 \geq 0$ consider the product $\alpha_\rho^{(l)} = f_1 \ldots f_{2l} g_x \in BW_n(q,r)$ where g_x is a square-free monomial in generators $g_{2l+1}, \ldots, g_{n-1}$ with $x \in \mathfrak{S}_n$ having cycle structure ρ. Let $\Lambda[z^2]$ be the algebra of symmetric functions with coefficients in $\mathbb{C}[z^2]$, z being a variable. Following [KT] (see also [K3]) we introduce the Weil functions $w_\lambda \in \Lambda[z^2]$ as

$$w_\lambda = \begin{vmatrix} h_{\lambda_1} & h_{\lambda_1+1} + z^2 h_{\lambda_1-1} & h_{\lambda_1+2} + z^4 h_{\lambda_1-2} & \cdots \\ h_{\lambda_2-1} & h_{\lambda_2} + z^2 h_{\lambda_2-2} & h_{\lambda_2+1} + z^4 h_{\lambda_2-3} & \cdots \\ \cdots & \cdots & \cdots & \cdots \end{vmatrix}$$

The elements $\phi_{(\lambda)_k} = z^{2k} w_\lambda$ where $0 \leq k \leq n/2$ and λ is a partition of $m = n - 2k$ form a linear basis in $\Lambda[z^2]$. Using symmetric polynomials t_k of section 2, consider the following elements of $\Lambda[z^2]$:

$$\varkappa_n = t_n + (1 - q^{-2}) \sum_{l=1}^{n/2} z^{2l} t_{n-2l} + \begin{cases} 0 & \text{if } n \text{ is odd} \\ (r + q^{-1})w & \text{if } n \text{ is even} \end{cases}$$

and let $\varkappa_\rho = \varkappa_{\rho_1} \varkappa_{\rho_2} \cdots$ for $\rho = (\rho_1, \rho_2, \ldots)$.

Theorem. *Suppose $|\rho| + 2l = n = |\lambda| + 2k$; then*

$$z^{2l} \varkappa_\rho = \sum_{\lambda,k} \chi^{(\lambda)_k}(\alpha_\rho^{(l)}) \cdot \phi_{(\lambda)_k}.$$

This is an analog for BW_n of the Frobenius theorem for $\mathfrak{S}_n$ and $H_n(q)$.

References

[DJ] R.Dipper, G.James, Proc. London Math. Soc. **52-3** (1986), 20–52.
[K1] S.Kerov, *Random Young tableaux and q-analog of hook formula*, Funk. Anal. Pril. (1991). (Russian)
[K2] S.Kerov, Journ. Sov. Math. **46-5** (1989), 2148–58.

[K3] S.Kerov, Zapiski Nauch. Sem LOMI **172** (1989), 68–77. (Russian)
[K4] S.Kerov, J. Sov. Math. Soc. **47-2** (1989), 2503–07.
[Mac] I.Macdonald, *Symmetric functions and Hall polynomials*, Oxford, 1979.
[T] E.Thoma, Math. Zap **85** (1964), 40–61.
[VK1] A.Vershik, S.Kerov, Sov. Math. Dokl. **38-1** (1989), 134–137.
[VK2] A.Vershik, S.Kerov, Funk. Anal. Pril. **15-4** (1981), 15–27. (Russian)
[VK3] A.Vershik, S.Kerov, Adv. Stud. Contemp. Math. **7** (1990).
[W] H.Wenal, Thesis.

LENINGRAD ELECTRICAL ENGEERING INSTITUTE OF COMMUNICATIONS, LENINGRAD, MOIKA 61 USSR

INVARIANTS OF 3–MANIFOLDS BASED ON CONFORMAL FIELD THEORY AND HEEGAARD SPLITTING

TOSHITAKE KOHNO

Department of Mathematics, Kyushu University, Fukuoka

1. Introduction

The purpose of this note is to give a brief description on the construction of topological invariants of 3-manifolds by means of projectively linear representations of the mapping class group of a closed orientable surface appearing in conformal field theory. First, we give a combinatorial description of the holonomy of $SU(2)$-Wess-Zumino-Witten model. More precisely, we derive the fusing matrices, conformal dimensions and switching operators by analyzing the monodromy representation of the Knizhnik-Zamolodchikov equation, and using the fact that these data give solutions to Moore and Seiberg polynomial equations [12], we construct projectively linear representations of the mapping class group on a vector space called the space of conformal blocks. Based on these representations, we define topological invariants of 3-manifolds using a Heegaard splitting. A more detailed description of this part is given in [10]. Shortly after the discovery of new 3-manifold invariants due to Witten [19], Reshetikhin and Turaev [16] gave a Dehn surgery formula using representations of the quantized universal enveloping algebra $U_q(sl(2,\mathbf{C}))$ with q a root of unity. Our approach described in this note is different from theirs.

Our principle to define 3-manifold invariants can be applied to other class of solutions to the Moore and Seiberg polynomial equations. It should be noted that a similar program was also proposed by Crane [2] (see also [3]). In this note we focus in particular on the invariants derived from cyclic group fusion rules. It turns out that these invariants are closely related to Gocho's geometric construction based on $U(1)$ gauge theory ([4]). We also give a Dehn surgery formula for these invariants. We are planning to give a more detailed account on this subject elsewhere.

Acknowledgement. I would like to thank the members of the Euler International Mathematical Institute for their hospitality and stimulating discussions.

2. $SU(2)$-Wess-Zumino-Witten model

Let Σ_g be a closed orientable surface of genus g. We denote by $\mathcal{M}_g$ the mapping class group $\pi_0 Diff^+(\Sigma_g)$. We are going to associate to Σ_g a finite dimensional complex vector space $Z_K(\Sigma_g)$, which is called the space of conformal blocks in $SU(2)$-Wess-Zumino-Witten model at level K, and then we define the action of the mapping class group on this vector space.

A marking μ of the closed orientable surface Σ_g is by definition a maximal collection of disjoint, non-contractible, pairwise non-isotopic smooth circles on Σ_g. We associate to μ a dual trivalent graph $\gamma(\mu)$ as shown in Fig. 1. We fix a positive integer K called a level. Now the vector space $Z_K(\gamma(\mu))$ is by definition a complex vector space with basis $\{e_f\}$, which is in one-to-one correspondence with a function $f : edge(\gamma(\mu)) \to \{0, 1/2, 1, \cdots, K/2\}$ satisfying

$$\begin{aligned} |f(c_1) - f(c_2)| \leq f(c_3) \leq f(c_1) + f(c_2) \\ f(c_1) + f(c_2) + f(c_3) \in \mathbf{Z} \\ f(c_1) + f(c_2) + f(c_3) \leq K \end{aligned} \tag{2.1}$$

for the edges c_1, c_2 and c_3 meeting at each vertex. Let us note that the first two conditions are so-called the Clebsch-Gordan condition for $sl(2, \mathbf{C})$.

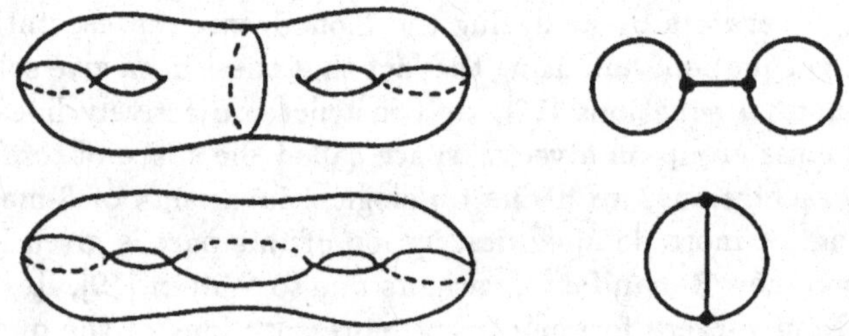

Fig. 1

The basic ingredients to define the action of the mapping class group $\mathcal{M}_g$ on this vector space are the fusing matrices, conformal dimensions and switching operators. These data are obtained in a natural way by analyzing the monodromy representations of the following Knizhnik-Zamolodchikov differential equation ([9]). For a half integer j, we denote by V_j the spin j representation of $sl(2, \mathbf{C})$, which is an irreducible representation of dimension $2j+1$. Let $j_p, 1 \leq p \leq K/2$ be half integers. We put

$$\Omega = \sum_{\mu} I_\mu \otimes I_\mu$$

where $\{I_\mu\}$ is an orthonormal basis of $sl(2, \mathbf{C})$ with respect to the Cartan-Killing form. We define the matrices $\Omega_{ij}, 1 \leq i, j \leq n$ by

$$\Omega_{ij} = \sum_{\mu} \pi_i(I_\mu)\pi_j(I_\mu) \in End(V_{j_1} \otimes \cdots \otimes V_{j_n})$$

where π_i and π_j stand for the operation on the i-th and j-th components respectively. The Knizhnik-Zamolodchikov equation is by definition

$$\frac{\partial \Phi}{\partial z_i} = \frac{1}{K+2} \sum_{j \neq i} \frac{\Omega_{ij}}{z_i - z_j} \Phi, \quad 1 \leq i \leq n. \tag{2.2}$$

Now we define the fusing matrix, which will be used to identify the space of conformal blocks associated with the two different "pants" decompositions as shown in Fig. 2.

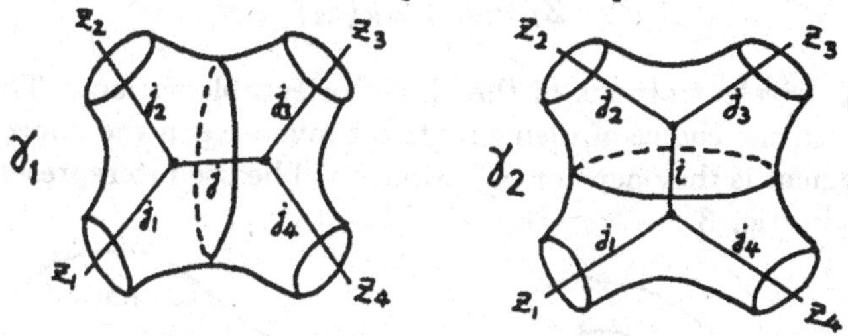

Fig. 2

Let us consider the Knizhnik-Zamolodchikov equation of four variables with values in

$$Hom_{sl(2,\mathbf{C})}(V_{j_1} \otimes V_{j_2} \otimes V_{j_3}, V_{j_4}).$$

Let us denote by

$$C_j^{j_1 j_2} : V_{j_1} \otimes V_{j_2} \to V_j$$

the $sl(2,\mathbf{C})$ homomorphism given by the Wigner's 3j-symbols (see [8]). To each weighted graph depicted in Fig. 2 we associate a solution of the Knizhnik-Zamolodchikov equation defined in the region $|z_1| \leq |z_2| \leq |z_3| \leq |z_4| = \infty$ in the following way. Let us suppose that the weights in the graphs in Fig. 2 satisfy the admissibility condition 2.1 at each vertex. For the weighted graph γ_1, we consider the solution normalized around $z_1 = z_2$ as

$$\Phi_{\gamma_1,j} = (z_2 - z_1)^{\Delta_j - \Delta_{j_1} - \Delta_{j_2}}(C_{j_4}^{j j_3} \cdot C_j^{j_1 j_2} + \text{higher order holomorphic terms}).$$

Here $\Delta_j = \frac{j(j+1)}{K+2}$, which is called the conformal dimension. In a similar way, we have

$$\Phi_{\gamma_2,i} = (z_3 - z_2)^{\Delta_i - \Delta_{j_2} - \Delta_{j_3}}(C_{j_4}^{j_1 i} \cdot C_i^{j_2 j_3} + \text{higher order holomorphic terms})$$

normalized around $z_2 = z_3$ associated with the weighted graph γ_2. Using an analytic continuation, it follows from a work of Tsuchiya and Kanie [17] that we have a constant matrix connecting these two solutions. We write it as

$$\Phi_{\gamma_1,j} = \sum_i F_{ij} \begin{bmatrix} j_2 & j_3 \\ j_1 & j_4 \end{bmatrix} \Phi_{\gamma_2,i}. \tag{2.3}$$

The above matrix is called a fusing matrix. In a similar way, we introduce the following braiding matrix which represents the action of the half monodromy on the solution $\Phi_{\gamma_1,i}$ interchanging z_2 and z_3.

$$B \begin{bmatrix} j_2 & j_3 \\ j_1 & j_4 \end{bmatrix}$$
$$= F \begin{bmatrix} j_3 & j_2 \\ j_1 & j_4 \end{bmatrix}^{-1} \cdot diag_i\left((-1)^{j_2+j_3-i} \exp \pi\sqrt{-1}(\Delta_i - \Delta_{j_2} - \Delta_{j_3})\right) \cdot F \begin{bmatrix} j_2 & j_3 \\ j_1 & j_4 \end{bmatrix}$$

Using a composition of fusing matrices we have an isomorphism

$$Z_K(\gamma_1) \cong Z_K(\gamma_2) \tag{2.4}$$

for any two dual trivalent graphs of the closed orientable surface. This isomorphism does not depend on the choice of fusing matrices involving in the above process.

Our last ingredient is the operator $S(j)$ which will be used to represent the switching operation shown in Fig. 3.

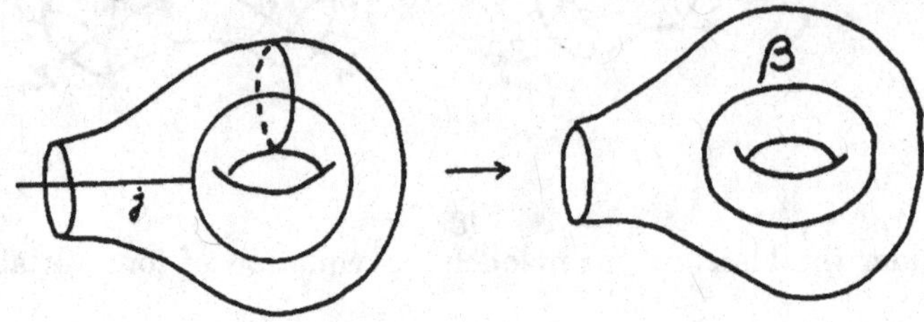

Fig. 3

The operator $S(0)$ is a $k \times k$ matrix given by

$$S(0)_{ij} = \left(\frac{2}{K+2}\right)^{\frac{1}{2}} \sin\frac{(2i+1)(2j+1)\pi}{K+2}, 0 \leq i, j \leq K/2 \tag{2.5}$$

which appeared in the work of Kac-Peterson [6] to describe the modular property of the characters of the integrable highest modules of level K of the affine Lie algebra of type $A_1^{(1)}$. The formula for $S(j)$ was obtained by Li and Yu [11]:

$$S(j)_{pq} = \sum_k \exp 2\pi\sqrt{-1}(\Delta_k - \Delta_p - \Delta_q) \cdot S(0)_{0k} B_{qp} \begin{bmatrix} j & k \\ p & q \end{bmatrix}, j/2 \leq p, q \leq (K-j)/2 \tag{2.6}$$

We put

$$T(j) = diag_{j/2 \leq i \leq (K-j)/2}\left(\exp 2\pi\sqrt{-1}(\Delta_i - \frac{c}{24})\right) \tag{2.7}$$

with $c = \frac{3K}{K+2}$. Then we have the following modular relations:

$$\begin{aligned} S(j)^2 &= (-1)^j \exp(-\pi\sqrt{-1}\Delta_j) \cdot id \\ (S(j)T(j))^3 &= S(j)^2 \end{aligned} \tag{2.8}$$

Thus we have defined the fusing matrices F, conformal dimensions Δ_j and the switching operators $S(j)$ based on the structure of the holonomy of the Knizhnik-Zamolodchikov equation. These provide solutions to the Moore and Seiberg polynomial equations. Let us now define the action of the mapping class groups. We start with a trivalent graph γ associated with a marking of the closed orientable surface Σ_g. Let V be a regular neighbourhood of the graph γ in $\mathbf{R}^3$ considered as a handlebody of genus g, and we realize Σ_g as its boundary. For an edge a of the graph γ, we take a disk Δ in V meeting transversely with a with one point and satisfying $\partial\Delta \subset \partial V$. Let α denote the Dehn twist about the circle $\partial\Delta$. We define the action of α on the vector space $Z(\gamma)$ by

$$\alpha \cdot e_f = \exp(-2\pi\sqrt{-1}\Delta_{f(a)})e_f . \tag{2.9}$$

Let us recall that according to Humphries [5] the mapping class group $\mathcal{M}_g$ is generated by the Dehn twists $\alpha_1, \cdots, \alpha_g, \beta_1, \cdots, \beta_g, \delta$ shown in Fig. 4.

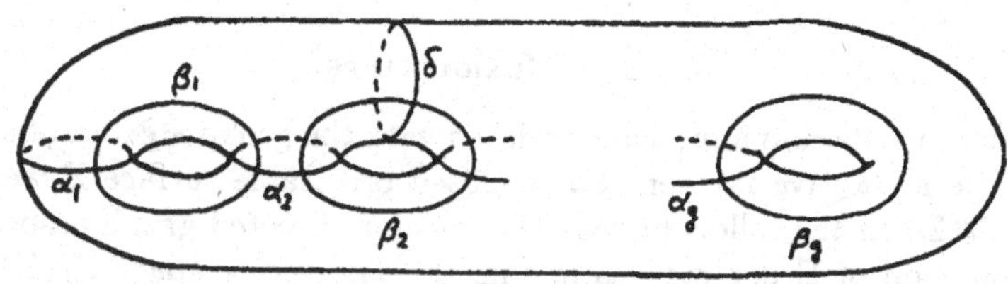

Fig. 4

Considering various trivalent graphs and by identifying the associated vector spaces by fusing matrices, we can define the action of the Dehn twists $\alpha_1, \cdots, \alpha_g, \delta$. The action of the Dehn twists β_i is defined in the following way. Let us go back to Fig. 3 and consider the Dehn twist β. We define the action of β by $T(j)S(j)T(j)$. Combining with fusing matrices we can define the action of the Humphries generators. More precisely, we have

Proposition 2.10. *Let Ω_K denote the cyclic group generated by $\exp \frac{\pi\sqrt{-1}}{4}c \cdot id$. Then, the above construction defines a well-defined homomorphism*

$$\rho_K : \mathcal{M}_g \longrightarrow GL(Z_K(\gamma))/\Omega_K$$

Remark. After a suitable normalization of solutions of the Knizhnik-Zamolodchikov equation, it is known that the fusing matrices are expressed as the $q-6j$–symbols at $q = \exp \frac{2\pi\sqrt{-1}}{K+2}$ (see [1,8,22]). Hence we can also start from these $q-6j$–symbol in a purely algebraic way and we might avoid the above analytic construction related to the holonomy of the Knizhnik-Zamolodchikov equation. In [18], Turaev and Viro gave a different construction of invariants using $q-6j$–symbols and a triangulation.

Now we are in position to define our 3-manifold invariants. Let M be a closed oriented 3-manifold. It is known that M admits a Heegaard splitting. Namely, there exists a handlebody V_1 and its second copy V_2 such that M is obtained from V_1 and V_2 by attaching their boundaries by some $h \in \mathcal{M}_g$. Let us denote by e_0 a member of the basis of $Z_K(\gamma)$ corresponding to the weight f such that $f(a) = 0$ for any edge a. We define the $(0,0)$-entry $\rho_K(h)_{00}$ by

$$\rho_K(h)e_0 = \rho_K(h)_{00}e_0 + \sum_{f \neq 0} \rho_K(h)_{f,0}e_f . \tag{2.11}$$

We put

$$\phi_K(M) = \left(\left(\frac{2}{K+2}\right)^{1/2} \sin\frac{\pi}{K+2}\right)^{-g} \rho_K(h)_{00} . \tag{2.12}$$

We have the following theorem.

Theorem 2.13 [10]. *Let M_1 and M_2 be closed oriented 3-manifolds. If there exists a orientation preserving homeomorphism $M_1 \cong M_2$, then we have*

$$\phi_K(M_1) = \phi_K(M_2)$$

in $\mathbf{C}^*/\Omega_K \cup \{0\}$.

3. Z/kZ fusion rules

In this section, we discuss a model associated with the group algebra of a finite cycli group. Let k be a positive integer. For a closed orientable surface Σ we construct vector space $Z_k(\Sigma)$ in the following way. Let γ be a directed graph associated with pants decomposition of Σ depicted as in Fig. 1. The vector space $Z_k(\Sigma)$ has a basi which is in one-to-one correspondence with weights $f : edge(\gamma) \to \mathbf{Z}/k\mathbf{Z}$ such that fo each vertex the sum of weights corresponding to the "ingoing" edges is congruent t the sum of weights corresponding to the "outgoing" edges modulo k. Here we use th convention that an edge with weight x is identified with the edge having the opposit direction with the weight $-x$. We see that the vector space $Z_k(\Sigma)$ is naturally isomorphi to the tensor product V^g with a k dimensional complex vector space, where g denote the genus of Σ.

Now we describe the action of the mapping class group $\mathcal{M}_g$ on $Z_k(\Sigma)$ using a solutio of the polynomial equations due to Moore and Seiberg ([12] Appendix E) associated wit $\mathbf{Z}/k\mathbf{Z}$ fusion rules. Let m be a positive integer such that m and k are relatively prim and we suppose that m is even if k is odd. We put

$$\Delta_x = \frac{mx^2}{2k}, \quad x \in \mathbf{Z}/k\mathbf{Z} \tag{3-1}$$

$$S_{xy} = \frac{1}{\sqrt{k}} \exp 2\pi\sqrt{-1}(-\Delta_{x+y} + \Delta_x + \Delta_y), \quad x, y \in \mathbf{Z}/k\mathbf{Z} \tag{3-2}$$

Let T be a diagonal matrix defined by

$$T = diag_{0 \le x \le k-1}(\exp 2\pi\sqrt{-1}\Delta_x) \tag{3-3}$$

One can check that the above matrices S and T satisfy the modular relations

$$\begin{aligned} S^2 &= C \\ (ST)^3 &= \xi(m,k)S^2 \end{aligned} \qquad (3\text{-}$$

where C is the duality matrix defined by $C_{xy} = \delta_{x,-y}$ and $\xi(m,k)$ is the Gauss sum

$$\xi(m,k) = \frac{1}{\sqrt{k}} \sum_{0 \le x \le k-1} \exp \frac{\pi\sqrt{-1}mx^2}{k} \qquad (3\text{-}4$$

which is known to be an eighth root of unity. We introduce the fusing matrices as

$$F_{g_2+g_3, g_1+g_2}\begin{bmatrix} g_2 & g_3 \\ g_1 & g_1+g_2+g_3 \end{bmatrix} = 1$$

for any g_1, g_2 and g_3 in $\mathbf{Z}/k\mathbf{Z}$.

By means of the above data, one can construct an action of the mapping class group $\mathcal{M}_g$ on $Z_k(\Sigma)$ as in the previous section. More precisely, we obtain a projectively linear representation

$$\varphi_{mk} : \mathcal{M}_g \longrightarrow GL(V^{\otimes g})/ < \xi(m,k) >$$

where the image of Dehn twists is described in the following way. We put $U = STS, W = T^{-1} \otimes T^{-1}$ and we adapt the notation for Dehn twists in the previous section. We set

$$\varphi_{mk}(\alpha_1) = T_1^{-1} = T^{-1} \otimes \cdots \otimes 1$$
$$\varphi_{mk}(\alpha_2) = W_{12}, \cdots, \varphi_{mk}(\alpha_g) = W_{g-1,g}$$
$$\varphi_{mk}(\delta_2) = T_2^{-1} = 1 \otimes T^{-1} \otimes \cdots \otimes 1$$
$$\varphi_{mk}(\beta_1) = U_1, \cdots, \varphi_{mk}(\beta_g) = U_g$$

Here the symbol $W_{k,k+1}, 1 \leq k \leq g-1$, stands for the operation of W on the k-th and $(k+1)$-st components of the tensor product $V^{\otimes g}$. One can show that the above representation factors through $Sp(2g, \mathbf{Z})$.

Let M be a closed oriented 3-manifold obtained as a Heegaard decomposition $V_1 \cup_h V_2$, where V_1 and V_2 are handlebodies of genus g. As in the previous section we consider the (0,0)-component $\varphi_{mk}(h)_{00}$. We have the following theorem

Theorem 3.5. *We put*

$$I_{mk}(M) = \sqrt{k^{-g}}\varphi_{mk}(h)_{00}$$

Then, $I_{mk}(M)$ is a topological invariant of M.

In the case $m = 1$ and k is even, the above invariant was discovered by Gocho [4] from a geometric viewpoint. In fact he constructed a vector bundle over the Siegel upper half plane with a projectively flat connection whose holonomy gives the above representation of $Sp(2g, \mathbf{Z})$.

Now we describe the Dehn surgery formula of the invariant I_{mk}. Let us suppose that the closed oriented 3-manifold M is obtained from the Dehn surgery on a framed link L with n components in S^3. Let A be the linking matrix whose diagonal entries are given by the framing. We denote by σ the signature of the linking matrix A. Using the above notations, we have the following theorem.

Theorem 3.6. *We put*

$$J_{mk}(M) = \sqrt{k^{-n}}\xi(m,k)^{-\sigma} \sum_{h \in (\mathbf{Z}/k\mathbf{Z})^n} \exp\left(\frac{\pi\sqrt{-1}m}{k}{}^t hAh\right)$$

Then, J_{mk} is a topological invariant of M. Moreover, the invariant I_{mk} computes this invariant up to some power of the Gauss sum $\xi(m,k)$.

Remark. The invariant J_{mk} can be written as the state sum

$$\xi(m,k)^{-\sigma} \sum_{\lambda} S_{0,\lambda(1)} \cdots S_{0,\lambda(n)} F(L,\lambda)$$

for any $\lambda : \{1, \cdots, n\} \to \mathbf{Z}/k\mathbf{Z}$, where $F(L, f)$ denotes the product for all crossing points in the link diagram given by

$$\prod \exp \pi\sqrt{-1}(\Delta_{\lambda(i)+\lambda(j)} - \Delta_{\lambda(i)} - \Delta_{\lambda(j)})$$

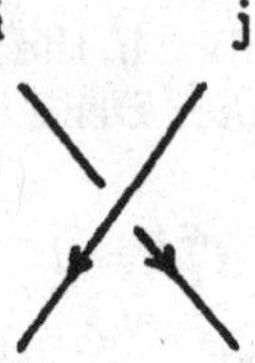

Fig. 5

Here to each crossing point of i-th and j-th components we associate the weight as shown in Fig. 5 and we take the product for all crossing points. The Dehn surgery formula corresponding to the case of Gocho's invariant was discussed by Ohtsuki [15]. We observe that the case $k = 2$ coincides with the Reshetikhin-Turaev invariant for $r = 3$. Generalizing the investigation due to Kirby and Melvin [7], Ohtsuki showed that the absolute value of J_{1k} is equal to the square root of the number of elements in $H^1(M; \mathbf{Z}/k\mathbf{Z})$ if we do not have $\alpha \in H^1(M; \mathbf{Z}/k\mathbf{Z})$ such that $\alpha \smile \alpha \smile \alpha \neq 0$ and is equal to 0 otherwise. In the case k is even, we have a slightly different representation of $Sp(2g, \mathbf{Z})$ by putting

$$\Delta_x = \frac{mx^2}{2k} + \frac{x}{2}$$

and by replacing the above Gauss sum by

$$\delta(m, k) = \frac{1}{\sqrt{k}} \sum_{0 \leq x \leq k-1} (-1)^x \exp \frac{\pi\sqrt{-1}m}{k} x^2$$

We have a similar construction and the resulting Dehn surgery formula

$$\sqrt{k^{-n}}\delta(m, k)^{-\sigma} \sum_{h \in (\mathbf{Z}/k\mathbf{Z})^n} (-1)^{<diag A, h>} \exp\left(\frac{\pi\sqrt{-1}m}{k} {}^t hAh\right)$$

was introduced by Murakami and Okada [14] related to the cyclotomic invariants for links discovered in [13] based on the IRF model due to Kashiwara and Miwa. Here $< diag A, h >$ stands for $\sum_i A_{ii} h_i$.

References

[1] L. Alvarez-Gaumé, G. Sierra and C. Gomez, *Topics in conformal field theory*, Physics and Mathematics of Strings, World Scientific (1990), 16–111.

[2] L. Crane, *Topology of 3-manifolds and conformal field theories*, preprint, Yale University (1989).

[3] S. E. Capell, R. Lee and E. Y. Miller, *Invariants of 3-manifolds from conformal field theory*, preprint (1990).

[4] T. Gocho, *The topological invariant of three-manifolds based on the $U(1)$ gauge theory*, preprint, University of Tokyo (1990).

[5] S. Humphries, *Generators for the mapping class group*, LNM, Springer (1979), 44–47.

[6] V. G. Kac and D. H. Peterson, *Infinite dimensional Lie algebras, theta functions and modular forms*, Advances in Math. **53** (1984), 125–264.

[7] R. Kirby and P. Melvin, *Evaluations of the 3-manifold invariants of Witten and Reshetikhin-Turaev*, London Math. Soc. Lect. Notes Series **151** (1990), 101–114.

[8] A. N. Kirillov and N. Y. Reshetikhin, *Representation of the algebra $U_q(sl(2,\mathbf{C}))$, q-orthogonal polynomials and invariants of links*, Infinite dimensional Lie algebras and groups, World Scientific (1988), 285–342.

[9] V. G. Knizhnik and A. B. Zamolodchikov, *Current algebra and Wess-Zumino models in two dimensions*, Nucl. Phys. **B247** (1984), 83–103.

[10] T. Kohno, *Topological invariants for 3-manifolds using representations of mapping class groups*, to appear in Topology.

[11] M. Li and M. Yu, *Braiding matrices, modular transformations and topological field theories in 2+1 dimensions*, Commun. Math. Phys. **127** (1990), 195–224.

[12] G. Moore and N. Seiberg, *Classical and quantum conformal field theory*, Comm. Math. Phys. **123** (1989), 177–254.

[13] T. Kobayashi, H. Murakami and J. Murakami, *Cyclotomic invariants for links*, Proc. Japan Acad. **64** (1988), 235–238.

[14] H. Murakami and M. Okada, *talk at Waseda University, Dec. 1990.*

[15] T. Ohtsuki, *private communication.*

[16] N. Y. Reshetikhin and V. G. Turaev, *Invariants of 3-manifolds via link polynomials and quantum groups*, Invent. Math **103** (1991), 547.

[17] A. Tsuchiya and Y. Kanie, *Vertex operators in conformal field theory on $\mathbf{P}^1$ and monodromy representations of braid groups*, Advanced Studies in Pure Math. **16** (1988), 297–372.

[18] V. G. Turaev and O. Y. Viro, *State sum invariants of 3-manifolds and quantum 6j-symbols*, preprint, LOMI (1990).

[19] E. Witten, *Quantum field theory and the Jones polynomial*, Commun. Math. Phys. **121** (1989), 351–399.

DEPARTMENT OF MATHEMATICS, KYUSHU UNIVERSITY, FUKUOKA 812 JAPAN

THE MULTI–VARIABLE ALEXANDER POLYNOMIAL AND A ONE—PARAMETER FAMILY OF REPRESENTATIONS OF $\mathcal{U}_q(\mathfrak{sl}(2, \mathbf{C}))$ AT $q^2 = -1$

JUN MURAKAMI

Department of Mathematics, Osaka University

Introduction

The purpose of this paper is to clarify connection between an R–matrix related to the multi–variable Alexander polynomial introduced in [6] and a family of 2–dimensional representations of $\mathcal{U}_q(\mathfrak{sl}(2, \mathbf{C}))$ at $q^2 = -1$. An R–matrix related to the reduced (one-variable) Alexander–Conway polynomial is given by several authors. In [2] and [4], such R–matrix is introduced by using the super version of $\mathcal{U}_q(\mathfrak{sl}(2, \mathbf{C}))$. In [5], such R–matrix is introduced by using a twisted version of $\mathcal{U}_q(\mathfrak{sl}(2, \mathbf{C}))$ at $q^2 = -1$. In [6], the author generalized the above R–matrix so that the related link invariant is the multi–variable Alexander polynomial ∇. The multi–variable Alexander polynomial $\nabla(L)$ of a link L has variables associated to connected components of L. A definition of ∇ is given in [1]

The R–matrix for ∇ is defined for each crossing point of a link diagram. It depends on variables associated to the strings of the under path and the over path of the crossing point. The R–matrix associated to the positive crossing is denoted by $R^{(s,t)}$, where s and t are variables corresponding to the components of the under path and the over path respectively. The R–matrix associated to the negative crossing point is denoted by $\bar{R}^{(s,t)}$, where s and t are variables corresponding to the components of the over path and the under path respectively.

The quantum group $\mathcal{U}_q(\mathfrak{sl}(2, \mathbf{C}))$ at $q^2 = -1$ has a one–parameter family of 2–dimensional representations, which is described in Sec.2. Let $\pi^{(s)}$ and $\pi^{(t)}$ be two representations of this family. The main result of this paper is to show that the R–matrices $R^{(s,t)}$ and $\bar{R}^{(s,t)}$ intertwine two representations $\pi^{(s)} \otimes \pi^{(t)}$ and $\pi^{(t)} \otimes \pi^{(s)}$ of $\mathcal{U}_q(\mathfrak{sl}(2, \mathbf{C}))$ at $q^2 = -1$.

1. R–matrix for the multi–variable Alexander polynomial

Let $\{\pi^{(t)} \mid t \in \mathbf{C} \setminus \{0\}\}$ be a one–parameter family of 2–dimensional complex vector spaces. The parameter t runs through non–zero complex numbers. We fix a basis $\{e_1^{(t)}, e_2^{(t)}\}$ of $V^{(t)}$. Let $R^{(s,t)}$ be a linear mapping from $V^{(s)} \otimes V^{(t)}$ to $V^{(t)} \otimes V^{(s)}$ defined by the following matrix

$$R^{(s,t)} = \begin{pmatrix} t & 0 & 0 & 0 \\ 0 & s - 1/s & 1 & 0 \\ 0 & t/s & 0 & 0 \\ 0 & 0 & 0 & -1/s \end{pmatrix} \tag{1.1}$$

with respect to the basis $\{e_1^{(s)} \otimes e_1^{(t)}, e_1^{(s)} \otimes e_2^{(t)}, e_2^{(s)} \otimes e_1^{(t)}, e_2^{(s)} \otimes e_2^{(t)}\}$ of $V^{(s)} \otimes V^{(t)}$ and $\{e_1^{(t)} \otimes e_1^{(s)}, e_1^{(t)} \otimes e_2^{(s)}, e_2^{(t)} \otimes e_1^{(s)}, e_2^{(t)} \otimes e_2^{(s)}\}$ of $V^{(t)} \otimes V^{(s)}$. $R^{(s,t)}$ satisfies the braid relation:

$$(R^{(t_2,t_3)} \otimes id)(id \otimes R^{(t_1,t_3)})(R^{(t_1,t_2)} \otimes id) = (id \otimes R^{(t_1,t_2)})(R^{(t_1,t_3)} \otimes id)(id \otimes R^{(t_2,t_3)}) \quad (1.2)$$

as linear mappings from $V^{(t_1)} \otimes V^{(t_2)} \otimes V^{(t_3)}$ to $V^{(t_3)} \otimes V^{(t_2)} \otimes V^{(t_1)}$. Let $\bar{R}^{(s,t)} = (R^{(s,t)})^{-1}$ and $R^{(s,t)}(x) = R^{(s,t)}x - \bar{R}^{(s,t)}x^{-1}$. Then $R^{(s,t)}(x)$ satisfies the Yang–Baxter equation:

$$(R^{(t_2,t_3)}(x) \otimes id)(id \otimes R^{(t_1,t_3)}(xy))(R^{(t_1,t_2)}(y) \otimes id) = (id \otimes R^{(t_1,t_2)}(y))(R^{(t_1,t_3)}(xy) \otimes id)(id \otimes R^{(t_2,t_3)}(x)) \quad (1.3)$$

In the case $s = t$, this solution is known as the six vertex model of free–fermion type. (See, for example, [7])

Let L be a link. Then, by the Alexander's Theorem, there is a positive integer n and an n–string braid b such that the closure $\hat{b}$ of b is isotopic to the link L. Let $c(i)$ denote the component of $\hat{b}$ containing the i-th point at the top of b and let t_c denote the parameter corresponding to the component c. We define an element $\rho^{(t_{c(1)},t_{c(2)},\cdots t_{c(n)})}(b) \in \mathrm{End}(V^{(t_{c(1)})} \otimes V^{(t_{c(2)})} \otimes \cdots \otimes V^{(t_{c(n)})})$. Let $b = \sigma_{i_1}^{\varepsilon_1}\sigma_{i_2}^{\varepsilon_2}\cdots\sigma_{i_r}^{\varepsilon_r}$ $(1 \le i_j \le n-1,\ \varepsilon_k = \pm 1)$. Then

$$\rho^{(t_{c(1)},t_{c(2)},\cdots t_{c(n)})}(b) = R_{i_1}^{\varepsilon_1} R_{i_2}^{\varepsilon_2} \cdots R_{i_r}^{\varepsilon_r} \quad (1.4)$$

where

$$R_{i_k} = id^{\otimes(k-1)} \otimes R^{(s_k,t_k)} \otimes id^{\otimes(n-k-1)}$$

and

$$R_{i_k}^{-1} = id^{\otimes(k-1)} \otimes \bar{R}^{(t_k,s_k)} \otimes id^{\otimes(n-k-1)}.$$

The parameters s_k and t_k correspond to the components of the under path and the over path of σ_{i_k}. Let

$$S^{(t_{c(1)},t_{c(2)},\cdots t_{c(n)})}(b) = \frac{1}{t_{c(1)} - t_{c(1)}^{-1}} \sum_{i_2,i_3,\cdots,i_n=1}^{2} \left(\prod_{k=2}^{n}(-1)^{i_k-1}t_{c(k)}\right) \rho^{(t_{c(1)},t_{c(2)},\cdots t_{c(n)})}(b)_{1i_2i_3\cdots i_n}^{1i_2i_3\cdots i_n}, \quad (1.5)$$

where $\rho^{(t_{c(1)},t_{c(2)},\cdots t_{c(n)})}(b)_{q_1q_2q_3\cdots q_n}^{p_1p_2p_3\cdots p_n}$ is a matrix element of $\rho^{(t_{c(1)},t_{c(2)},\cdots t_{c(n)})}(b)$ with respect to the basis $\{e_{k_1}^{t_{c(1)}} \otimes \cdots \otimes e_{k_n}^{t_{c(n)}}\}$ of $V^{(t_{c(1)})} \otimes \cdots \otimes V^{(t_{c(n)})}$. Then, as it is shown in [6], $S(b)$ is invariant under the Markov's moves and is essentially equal to the multi-variable Alexander polynomial $\nabla(\hat{b})$ of the link $\hat{b}$.

2. A family of 2–dimensional representations of $\mathcal{U}_q(\mathfrak{sl}(2,\mathbf{C}))$ at $q^2=-1$

The Hopf algebra $\mathcal{U}_q(\mathfrak{sl}(2,\mathbf{C}))$ is generated by K, K^{-1}, X^+, X^- and they satisfy the following relations.

$$\begin{aligned} K\,K^{-1} &= K^{-1}\,K = 1, \\ K\,X^{\pm}\,K^{-1} &= q^{\pm 2}\,X^{\pm}, \\ [X^+,X^-] &= \frac{K-K^{-1}}{q-q^{-1}}. \end{aligned} \tag{2.1}$$

The Hopf algebra structure of $\mathcal{U}_q(\mathfrak{sl}(2,\mathbf{C}))$ is given by the following:

$$\begin{aligned} &\Delta(K^{\pm1}) = K^{\pm1}\otimes K^{\pm1}, \\ &\Delta(X^+) = X^+\otimes 1 + K\otimes X^+, \quad \Delta(X^-) = X^-\otimes K^{-1} + 1\otimes X^-, \\ &\varepsilon(K^{\pm1}) = 1, \quad \varepsilon(X^{\pm}) = 0, \\ &S(K^{\pm1}) = K^{\mp1}, \quad S(X^+) = -K^{-1}X^+, \quad S(X^-) = -X^-K. \end{aligned} \tag{2.2}$$

The vector representation π of $\mathcal{U}_q(\mathfrak{sl}(2,\mathbf{C}))$ is the 2–dimensional representation given by the following:

$$\pi(X^+) = \begin{pmatrix} 0 & 1 \\ 0 & 0 \end{pmatrix}, \quad \pi(X^-) = \begin{pmatrix} 0 & 0 \\ 1 & 0 \end{pmatrix}, \quad \pi(K) = \begin{pmatrix} q & 0 \\ 0 & q^{-1} \end{pmatrix}. \tag{2.3}$$

In Sec.4 of [3], all the finite–dimensional representations of $\mathcal{U}_q(\mathfrak{sl}(2,\mathbf{C}))$ are classified for all q. Especially, in the case $q=\sqrt{-1}$, there is a one–parameter family $\pi^{(t)}$ of 2–dimensional representations including the vector representation. This family is given by the following:

$$\pi^{(t)}(X^+) = \begin{pmatrix} 0 & 1 \\ 0 & 0 \end{pmatrix}, \quad \pi^{(t)}(X^-) = \begin{pmatrix} 0 & 0 \\ -\frac{t-t^{-1}}{2}\sqrt{-1} & 0 \end{pmatrix}, \quad \pi^{(t)}(K) = \begin{pmatrix} t & 0 \\ 0 & -t \end{pmatrix}. \tag{2.4}$$

If $t=\sqrt{-1}$, then this representation is equal to the vector representation. The above family of representations and the family of R–matrices given in Sec.1 satisfies the following

Theorem. *For every $x\in\mathcal{U}_q(\mathfrak{sl}(2,\mathbf{C}))$, we have*

$$\begin{aligned} R^{(t_1,t_2)}\,(\pi^{(t_1)}\otimes\pi^{(t_2)})(\Delta(x)) &= (\pi^{(t_2)}\otimes\pi^{(t_1)})(\Delta(x))\,R^{(t_1,t_2)}, \\ \bar{R}^{(t_1,t_2)}\,(\pi^{(t_1)}\otimes\pi^{(t_2)})(\Delta(x)) &= (\pi^{(t_2)}\otimes\pi^{(t_1)})(\Delta(x))\,\bar{R}^{(t_1,t_2)}. \end{aligned} \tag{2.5}$$

Proof. From the definitions (2.4) of the representation $\pi^{(t)}$ and (2.2) of the coproduct Δ, we have

$$(\pi^{(t_1)}\otimes\pi^{(t_2)})(\Delta(X^+)) = \begin{pmatrix} 0 & t_1 & 1 & 0 \\ 0 & 0 & 0 & 1 \\ 0 & 0 & 0 & -t_1 \\ 0 & 0 & 0 & 0 \end{pmatrix},$$

$$(\pi^{(t_2)} \otimes \pi^{(t_1)})(\Delta(X^+)) = \begin{pmatrix} 0 & t_2 & 1 & 0 \\ 0 & 0 & 0 & 1 \\ 0 & 0 & 0 & -t_2 \\ 0 & 0 & 0 & 0 \end{pmatrix},$$

$$(\pi^{(t_1)} \otimes \pi^{(t_2)})(\Delta(X^-)) = \begin{pmatrix} 0 & 0 & 0 & 0 \\ -\frac{t_2-t_2^{-1}}{2}\sqrt{-1} & 0 & 0 & 0 \\ -\frac{t_1-t_1^{-1}}{2t_2}\sqrt{-1} & 0 & 0 & 0 \\ 0 & \frac{t_1-t_1^{-1}}{2t_2}\sqrt{-1} & -\frac{t_2-t_2^{-1}}{2}\sqrt{-1} & 0 \end{pmatrix},$$

$$(\pi^{(t_2)} \otimes \pi^{(t_1)})(\Delta(X^-)) = \begin{pmatrix} 0 & 0 & 0 & 0 \\ -\frac{t_1-t_1^{-1}}{2}\sqrt{-1} & 0 & 0 & 0 \\ -\frac{t_2-t_2^{-1}}{2t_1}\sqrt{-1} & 0 & 0 & 0 \\ 0 & \frac{t_2-t_2^{-1}}{2t_1}\sqrt{-1} & -\frac{t_1-t_1^{-1}}{2}\sqrt{-1} & 0 \end{pmatrix},$$

$$(\pi^{(t_1)} \otimes \pi^{(t_2)})(\Delta(K)) = \begin{pmatrix} t_1 t_2 & 0 & 0 & 0 \\ 0 & -t_1 t_2 & 0 & 0 \\ 0 & 0 & -t_1 t_2 & 0 \\ 0 & 0 & 0 & t_1 t_2 \end{pmatrix},$$

$$(\pi^{(t_2)} \otimes \pi^{(t_1)})(\Delta(K)) = \begin{pmatrix} t_1 t_2 & 0 & 0 & 0 \\ 0 & -t_1 t_2 & 0 & 0 \\ 0 & 0 & -t_1 t_2 & 0 \\ 0 & 0 & 0 & t_1 t_2 \end{pmatrix}.$$

The relation (2.5) in the theorem comes from a direct computation.

References

[1] Birman, J. S., *Braids, Links, and Mapping Class Groups*, Princeton: Princeton University Press.

[2] Deguchi, T., Akutsu, Y., *Graded solutions of the Yang–Baxter relations and link polynomials*, J. Phys. A: Math. Gen. **23** (1990), 1861-1875.

[3] Izumi, Y., *The q-analogue of the universal enveloping algebras of simple Lie algebras of type A*, Master thesis (in Japanese), February, 1988, Osaka University.

[4] Kauffman, L. H., Saleur, H., *Free Fermions and the Alexander–Conway polynomial*, preprint EFI 90-42, July, 1990, Enrico Fermi Institute.

[5] Lee, H. C., *Twisted quantum groups of A_n and the Alexander–Conway link polynomial*, preprint TP-90-0220, 1990, Chalk River Nuclear Laboratories.

[6] Murakami, J., *A state model for the multi-variable Alexander polynomial*, preprint, August, 1990, Osaka University.

[7] Sogo, K., Uchinami, M., Akutsu, Y., Wadati, M. , *Classification of exactly solvable two-component models*, Prog. Theo. Phys. **68** (1982), 508–526.

DEPARTMENT OF MATHEMATICS, OSAKA UNIVERSITY, TOYONAKA, OSAKA, 560, JAPAN

PREPARATION THEOREMS FOR ISOTOPY INVARIANTS OF LINKS IN 3–MANIFOLDS

A. B. SOSSINSKY

Moscow Institute of Electronic Machinebuilding

By a *preparation theorem* I mean a geometric fact that this sets the stage for introducing a well-defined algebraic invariant. For example, in the context of knots and links in $\mathbf{R}^3$ (or $\mathbf{S}^3$), the classical property of Redemeister moves is the preparation theorem for the Kauffman polynomial [Ka], while the Alexander and Markov (see [Al], [Bi]) theorems set the stage for the Jones polynomial [Jo].

This paper describes the first steps of an attempt to carry over Vaugman Jones' strategy [Jo] from the sphere S^3 to arbitrary 3-manifold M^3. Here we obtain an Alexander theorem for closed 3-manifolds and handlebodies and a Markov theorem for handlebodies. Our approach differs from that of R.Skora [Sk], who proves Alexander and Markov theorems in the case of an arbitrary 3-manifold M^3; in particular, our definition of closed braids differs from the one in [Sk]. My perseverance in continuing work in the direction set out in [Sol] is due to the hope that the expected invariant will turn out to be easier to calculate than those outlined in the remarkable work of E.Witten [Wi], V.Turaev, N.Reshetikhin, O.Viro [RT], [TV].

This paper is organized as follows. In Sec.1 we observe that all ambient isotopy invariants of (k+g)-component links in $\mathbb{R}^3$ (e.g. the Jones polynomial) provide invariants for k-component links in handlebodies H_g of genus $g \geq 1$ via what I call *associated links*. In Sec.2 Alexander and Markov theorems are presented for $g = 1$ (the solid torus case, considered earlier [Sol]). In Sec.3 we define (geometrically and algebraically) the group B_n^g of what I call $g-$*braids* (generalizing the 1-braids of [Sol]) which plays the role of the classical Artin group B_n in our case. The corresponding Alexander and Markov theorems are obtained in Sec.4. Finally, in Sec.5 we speculate about eventual further steps in our approach.

The author is grateful to I. Cherednik for stimulating conversations, to Vaughan Jones for a useful discussion (and for sending the preprints [Sk] and [Xi]) and to V. P. Maslov for creating working conditions that have made this research possible.

1. Associated links

Let H_g be a handlebody of genus $g \geq 1$ and $L \subset H_g$ a k-component link (knot if $k = 1$) in H_g, i.e. a one-dimensional oriented compact closed PL-submanifold of Int H_g with k connectivity components. Call two such links $L_1, L_2 \subset H_g$ *equivalent* if L_1 can be taken to L_2 by a finite sequence of *triangular moves*

$$L' \leftrightarrow L'' = (L' - [a,b]) \cup [a,c] \cup [c,b]$$

where the triangle [a,b,c] intersects L' only along the edge $[a,b] \subset L'$, intersects L'' only along the two edges [a,c],[c,b], and lies inside H_g. The corresponding equivalence classes will be called g − *links*. It can be shown that these equivalence classes are classes of links in H_g isotopic via isotopies of H_g that restrict to the identity on the boundary ∂H_g.

We now fix a standard embedding $H_g \subset \mathbb{R}^3$ and call *g-link diagram* L of the g-link $\{L\}$ the projection on a fixed horizontal plane $\mathbb{R}^2 \subset \mathbb{R}^3$, showing (over)-underpasses in the usual way (Fig.1a).

To every k-component g-link diagram $L \subset H_g$ we assign its *associated link:* the (k+g)-component link $\alpha(L) \subset \mathbb{R}^3$ obtained by adding g unlinked oriented circles $C_1, \ldots, C_g \subset \mathbb{R}^3$ passing through the holes of the handlebody to L (and then forgetting about the handlebody) as shown on Fig.1b. Clearly the ambient isotopy class of $\alpha(L) \subset \mathbb{R}^3$ depends only on the equivalence class of the given g-link diagram L, so that we have a well-defined map

$$\alpha : \mathcal{L}(H_g) \to \mathcal{L}(\mathbb{R}^3) \tag{1}$$

of all g-links into ordinary links (i.e. links in $\mathbb{R}^3$).

This simple observation implies the following statement.

Theorem 1. *Any ambient isotopy invariant I for links in $\mathbb{R}^3$ provides an invariant for g-links L in the handlebody H_g, namely the invariant $I(\alpha(L))$ of the associated link $\alpha(L)$.*

For example, the calculation of (say) the Homfly polynomial (see [Jo]) for the Whitehead link (Fig.2a) shows that the famous solid torus knot on Fig.2b is neither trivial, nor is it the axis $S^1 \times \{0\}$ of the solid torus $S^1 \times D^2$.

2. Alexander and Markov theorems for 1-links

Following [5], we first consider the case $g = 1$, i.e. the case of knots and links in the standard solid torus $H_1 = S^1 \times D^2 \subset \mathbb{R}^3$. In order to state the Alexander and Markov theorems for $g = 1$, we must define what the analogs of a braid and a closed braid are in this case.

These definitions will not be the ones that seem natural at first glance (closed braids do *not* "go once around" the S^1 factor of the solid torus).

Algebraically, the role of Artin's braid groupB_n is played by the *group of 1-braids* B_n^1 with generators

$$\tau, \sigma_1, \sigma_2, \ldots, \sigma_{n-1} \tag{2}$$

and relations

$$\left.\begin{aligned} \sigma_i\sigma_j &= \sigma_j\sigma_i(|i-j| \geq 2); \\ \tau\sigma_i &= \sigma_i\tau(i \geq 2); \\ \sigma_i\sigma_{i+1}\sigma_i &= \sigma_{i+1}\sigma_i\sigma_{i+1} \ (1 \leq i \leq n-2); \\ \sigma_1\tau\sigma_1\tau &= \tau\sigma_1\tau\sigma_1. \end{aligned}\right\} \tag{3}$$

This group is known in the algebraic literature as one of the Coxeter groups. It is isomorphic to a subgroup of the braid group B_{n+1} by the assignment

$$\tau \mapsto \overline{\sigma}_1, \sigma_1 \mapsto \overline{\sigma}_2, \ldots, \sigma_{n-1} \mapsto \overline{\sigma}_n,$$

where $\overline{\sigma}_1, \ldots, \overline{\sigma}_n$ are the Artin generators of B_{n+1}.

Geometrically, 1-braids may be roughly describes as ordinary braids pierced by a long knitting needle. A more formal definition of the group of geometric 1-braids may be stated as follows. Consider the pierced cylinder $P = E \times [0,1] \subset \mathbf{R}^3$, where $E = \{z \in \mathbf{C} : 0 < |z| \leq 1\}$ is the punctured disk; denote the missing segment $\{0\} \times [0,1]$ by A; choose n points on the top and n points on the bottom of P:

$$p_i^\epsilon = (i/(n+1), \epsilon), \qquad \epsilon = 0, 1, \; i = 1, \ldots, n$$

and call *strand* any polygonal line in P joining one of the points p_i^0 to one of the points p_j^1 and intersecting each layer $E \times \{t\}$, $t \in [0,1]$ at exactly one point; we call any set of n nonintersecting strands in P a *1-braid* (Fig.3).

Two braids are considered *equivalent* if one can be taken to the other by a finite sequence of triangular moves (see Sec.1) within P, each of the intermediate strands being a strand disjoint from the other strands (of course the triangles [a,b,c] are not allowed to intersect the other strands or the missing segment A). When we wish to distinguish a specific 1-braid $\beta \subset P$ from its equivalence class (also called 1-braid), we call it a *1-braid diagram* and picture it in the expected way (as in Fig.3).

A 1-braid diagram $\beta \subset P$ is called *proper* if not more than one strand intersects any layer $E \times \{t\}$, $t \in [0,1]$ to the left of the missing segment A (so that all the double points lie to the right of A).

Lemma 2.1. *Any 1-braid is equivalent to a proper 1-braid.*

The proof is a simple exercise in geometric topology (see [Sol]).

Just like ordinary braids, 1-braids (with a fixed number of strands n) can be multiplied (by placing one pierced cylinder on top the other); this is clearly a group operation; the group thus obtained is generated by the elements $\tau, \sigma_1, \sigma_2, \ldots, \sigma_{n-1}$ (pictured on Fig.4) which obviously satisfy the relations (3). It can be proved that these relations suffice, i.e. the geometric group of 1-braid in n strands is isomorphic to the algebraic group of 1-braids B_n^1 presented by (2), (3) (see [Sol], [Xi]).

The *closure* operation $cl : B_n^1 \to \mathcal{L}(\mathbf{R}^3)$ assigns a 1-link $cl(\beta) \subset S^1 \times D^2$ to each 1-braid $\beta \in B_n^1$ as follows; extend the strands of β and the segment A to oriented closed curves $\overline{A}, C_1, \ldots, C_n$ as shown on Fig.5, delete a small tubular neighborhood U of the curve $\overline{A} \supset A$, identify $(\mathbf{R}^3 \cup \{\infty\}) - U$ with $S^1 \times D^2$ and take the link $C_1 \cup \cdots \cup C_n \subset S^1 \times D^2$ to be $cl(\beta)$. Note that the link $\overline{A} \cup C_1 \cup \cdots \cup C_n$ is associated link for $cl(\beta)$. Denote $\mathcal{B}^1 = \cup_n B_n^1$; we have defined a map $cl : \mathcal{B}^1 \to \mathcal{L}(S^1 \times D^2)$.

Our Alexander theorem says this map is surjective:

Theorem 2.2. *Any 1-link $L \in S^1 \times D^2$ is the closure of 1-braid.*

The proof is a straightforward application of the so-called Alexander trick (see [Sol],[So2]).

Now let us define algebraic Markov moves M_1, M_2 for the group $\mathcal{B}^1$ exactly as it is done in [Ma] or [Jo] for the ordinary braid group $\mathcal{B} = \cup B_n$: the operation M_1 is conjugation in B_n^1 $\quad(\beta \leftrightarrow \alpha\beta\alpha^{-1}, \alpha, \beta \in B_n^1)$ and M_2 is *Markov stabilization* $(\beta \leftrightarrow \beta\sigma_n^{\pm 1}, \beta \in B_n^1)$. The Markov theorem in this case is:

Theorem 2.3. *Two 1-braids have equivalent closures iff there is a finite sequence of algebraic Markov moves taking one 1-braid to the other.*

There are two proofs of this theorem. One is based on the geometric Markov moves ($\mathcal{H}$ and $\mathcal{W}$ operations in the terminology of [Bi]), the other is more algebraic and uses associated links (see [Sol]).

3. The g-braid group B_n^g

We now consider the general case $g \geq 1$ and begin by defining the analogs of braids and closed braids in a handlebody H_g of genus g.

Consider a standard embedding of H_g in $\mathbf{R}^3$ and choose a paralellipiped P (called the *tangle-box*) inside H_g (Fig.6). Roughly speaking, a *g-braid* $\beta \in B_n^g$ is the union of n strings in P which are allowed to leave the tangle box to make simple loops (in the positive or negative direction) around the handles of H_g and then return inside the tangle box (see Fig.6).

Another intuitive way of describing a g-braid is to say we take an ordinary braid and stick g knitting needles into it.

A more precise definition is the following. Introduce coordinates (x, y, z) in $\mathbf{R}^3$ so that the tangle box $P = [0,1]^3 \subset H_g \subset \mathbf{R}^3$ is the unit cube in the positive octant. Define a *strand* as a downward oriented polygonal line $\lambda \subset H_g$ such that

(i) λ starts at one of the points $(i/(n+1), 1/2, 1) = p_i^1$ and ends at one of the points $(j/(n+1), 1/2, 0) = p_j^0$, where $i, j = 1, \ldots, n$;
(ii) λ intersects each horizontal level $z = const$ within P no more than once;
(iii) λ may be entirely contained in P or may leave P to form negative or positive loops; a *loop* being an arc $\mu \subset \lambda$ contained in $H_g - \text{int}P$ and going around the (say) k-th handle so that;
(iv) μ starts at a point of the form $a = (0, 1/2, h) \in P$ and returns to P at the point $b = (0, 1/2, h - \delta)$, where $\delta > 0$ is small;
(v) if μ is *positive* loop, it goes around the k-th handle counterclockwise and its projection on the plane $z = 0$ has no double points;
(vi) if μ is *negative* loop, it goes around the k-th handle clockwise and its projection on the plane $z = 0$ has a single double point, where the branch coming back to P passes over the outgoing one.

Now define a *g-braid* $\beta \in B_n^g$ as the union of n nonintersecting strands such that

(vii) no strand (other than μ) intersects the segment $[a, b] \subset \partial P$ in situations described in condition (iv);
(viii) any loop μ' that leaves P later (i.e. at a lower point) than some other loop μ'' has no underpasses at double points formed with μ'' under projection on $\{z = 0\}$.

This long definition is actually quite natural - see Fig.7.

Two g-braids are considered *equivalent* if one can be taken to the other (staying in the class of g-braids) by a finite sequence of triangular moves within intH_g.

The definition of g-braid *multiplication* is similar to that of ordinary braids, except that we must remove the second factor (tangle-box and loops) from its handlebody before attaching it to the bottom of the first tangle-box, and then put its loops in place (around the corresponding handles of the *first* handlebody), taking care to have them going *over* all the loops of the first factor (and over-under each other in the same way they did originally); the reader should have no difficulty in formalizing this definition in the style of (i)-(viii) above; the result will be a group that we denote by B_n^g.

This group is clearly generated by the elements (see Fig.8)

$$\tau_1, \dots, \tau_g, \sigma_1, \sigma_2, \dots, \sigma_{n-1} \tag{4}$$

and satisfies the relations

$$\left.\begin{aligned} \sigma_i\sigma_j = \sigma_j\sigma_i(|i-j| \geq 2); \tau_k\sigma_i = \sigma_i\tau_k(k \geq 1, i \geq 2),\\ \sigma_i\sigma_{i+1}\sigma_i = \sigma_{i+1}\sigma_i\sigma_{i+1} \ (i = 1, \dots, n-1),\\ \tau_k\sigma_1\tau_k\sigma_1 = \sigma_1\tau_k\sigma_1\tau_k \ (k = 1, \dots, g),\\ \sigma_1\tau_{k+p}\sigma_1^{-1}\tau_k = \tau_k\sigma_1\tau_{k+p}\sigma_1^{-1}. \end{aligned}\right\} \tag{5}$$

Presumably (which means I don't have a detailed written proof of this statement) these relations suffice, i.e. *the abstract group generated by (4) with relations (5) is isomorphic to the g-braid group B_n^g described geometrically above.*

The group B_n^g is clearly isomorphic to a subgroup of the ordinary braid group B_{n+g} by the assignment:

$$\left.\begin{aligned} \sigma_1 \mapsto \overline{\sigma}_{g+1}, \sigma_2 \mapsto \overline{\sigma}_{g+2}, \dots, \sigma_{n-1} \mapsto \overline{\sigma}_{n+g-1},\\ \tau_k \mapsto \overline{\sigma}_g^{-2}, \tau_{k-1} \mapsto \overline{\sigma}_g^{-1}\overline{\sigma}_{g-1}^{-2}\overline{\sigma}_g, \dots \tau_1 \mapsto \overline{\sigma}_g^{-1}\overline{\sigma}_{g-1}^{-1} \dots \overline{\sigma}_1^{-2}\overline{\sigma}_2 \dots \overline{\sigma}_g, \end{aligned}\right\} \tag{6}$$

where $\overline{\sigma}_1 \dots \overline{\sigma}_{n+g-1}$ are the Artin generators of B_{n+g}. This isomorphism is illustrated by Fig.9.

4. Closed g-braids in handlebodies

By the *closure* of a g-braid $\beta \subset H_g$ we mean the g-link obtained by joining the point p_1^1 to $p_1^0, \dots,$ the point p_n^1 to p_n^0 by curves in $H_g - P$ with nonintersecting projection on $\{z = 0\}$ circling three quarters of the way around P on its right (see Fig.10).

A g-link will be called *proper* if it is contained in the tangle-box P (see Sec.3) except possibly for several loops (see Sec.3,(iii)-(vi)) with pairwise nonintersecting projection (on $\{z = 0\}$) without double points. The following statement is similar to Lemma 2. and is just as easy to prove.

Lemma 4.1. *Every g-link is equivalent to a proper g-link.*

This lemma and the Alexander trick suffice to prove our version of the Alexander theorem in this case:

Theorem 4.2. *Every g-link is equivalent to a closed g-braid*

Since every closed compact 3-manifold M^3 has a Heegaard splitting into a union of two handlebodies $M^3 = H_g \cup H_g'$ and any link $L \subset M^3$ is isotopic to a link L_0 contained in H_g, the previous theorem has the following consequence.

Corollary 4.3. *Every link $L \subset M^3$ in a closed compact 3-manifold M^3 is isotopic to a closed braid in a handlebody $H_g \subset M^3$, where $M^3 = H_g \cup H'_g$ is any fixed Heegaard splitting of M^3.*

The braid, of course, depends on the choice of the Heegaard splitting. This Alexander theorem for 3-manifold should be compared to R.Skora's elegant version [Sk], where the braid depends on the choice of the "axis" of the open book decomposition for M^3.

Returning to g-links and g-braids (i.e. to a fixed handlebody H_g), let us note that *algebraic Markov g-moves* for the group B_n^g can be defined exactly like Markov moves for the Artin group B_n. (This contrasts with [Sk], where in the Markov conjugation move one of factors must be "twisted" by the homeomorphism $\psi : F \to F$ defining the open book decomposition.) Using J.Birman's techniques ($\mathcal{H}$- and $\mathcal{W}$- operations) or, perhaps, Morton's simpler proof of Markov's theorem [Mo], it appears possible to prove a handlebody version of this theorem. Since the detailed proof is not yet written down, I prefer to state this as a conjecture.

Conjecture 4.4. *The Markov theorem is true for handlebodies, i.e. two closed g-braids $cl(\beta_1), cl(\beta_2), \beta_1, \beta_2 \in B_n^g$ are equivalent iff β_1 can be taken to β_2 by a finite sequence of algebraic Markov g-moves.*

5. Wishful thinking

The Markov theorem for handlebodies may seem to be useless tool, not only because we do not have the algebra needed to use it to produce a Jones type invariant for the handlebodies (to my knowledge, there is no analog of the Ocneanu trace directly related to B_n^g), but also because polynomial invariants for handlebodies may be produced almost trivially via associated links (see Sec.1 above).

However, our ultimate goal is to use V.Jones' strategy to define invariants for links in arbitrary closed 3-manifold via Heegaard splitting. More specifically, I would like to hope that the Heegaard homeomorphism $\psi : \partial H_g \to \partial H_g$ determining $M^3 = H_g \amalg_\psi H_g$ induces (whatever that means) effectively computable rules transforming invariants of links in H_g into invariants in M^3. The presence of the generators $\tau_1, \ldots, \tau_g$ (of the group B_n^g) corresponding to the handles of H_g supports this approach, and perhaps justifies a closer algebraic scrutiny of the group B_n^g and related algebras. Or, very hopefully, the inclusion $i : B_n^g \hookrightarrow B_{n+g}$ and some transformation $T_\psi : B_{n+g} \to B_{n+g}$ (involving the elements $i(\tau_k), k = 1, \ldots, g$) might yield an invariant via the (known) Ocneanu trace for B_{n+g}.

References

[Al] Alexander J.W., *The combinatorial theory of complexes*, Annals of Math **31(2)** (1930), 294–322.

[Bi] Birman J., *Braids, Links and Mapping Class Groups*, Annals of Math Studies **82** (1974), Princeton University Press.

[Jo] Jones V., *Hecke algebra representations of braid groups and link polynomials*, Annals of Math **126** (1987), 335–388.

[Ka] Kauffman L., *State models and the Jones polynomial*, Topology **26** (1987), 395–407.

[Mo] Morton H., *Threading knot diagrams*, Proc. Cambr. Phil. Soc. **99** (1986), 247–160.

[RT] Reshetikhin N. and Turaev V., *Ribbon graphs and their invariants derived from quantum groups*, Commun. Math. Phys. **127** (1990).

[Sk] Skora R., *On closed braids in 3-manifolds*, Preprint (1990).

[So1] Sossinsky A., *Links in solid tori*, Preprint (1990). (in Russian)

[So2] Sossinsky A., *Knots and braids*, Kvant 2 (1989), 6–14. (in Russian)

[TV] Turaev V. and Viro O., *State sum invariants of 3-manifolds and quantum 6j-symbols*, Preprint LOMI (1990); Topology (to appear).

[Xi] Xiang Song Lin, *Markov theorems for links in $L(p,1)$*, Preprint (1990).

[Wi] Witten E., *Quantum field theory and the Jones polynomial*, Commun. Math. Phys. **121** (1989), 351–399.

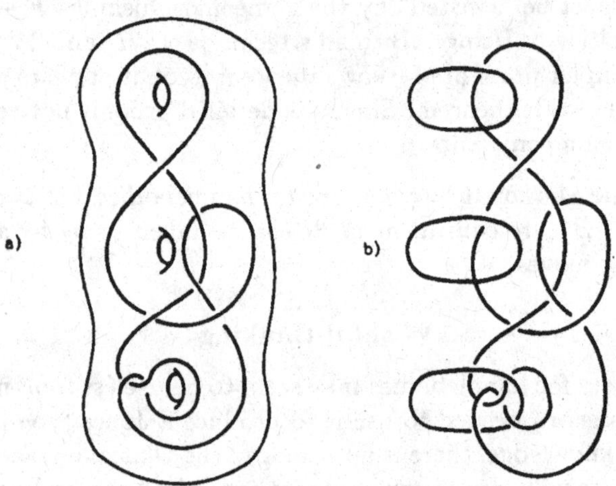

Fig.1 A 3-link and its associated link

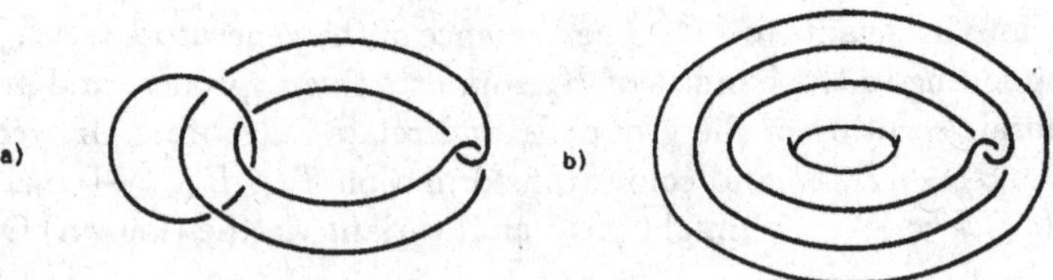

Fig.2 The Whitehead link

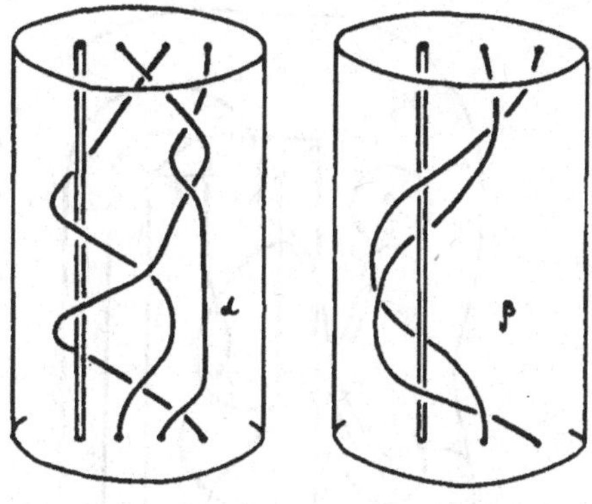

Fig.3 A(proper) 1-braid $\alpha \in B_3^1$ and a (nonproper) 1-braid $\beta \in B_2^1$

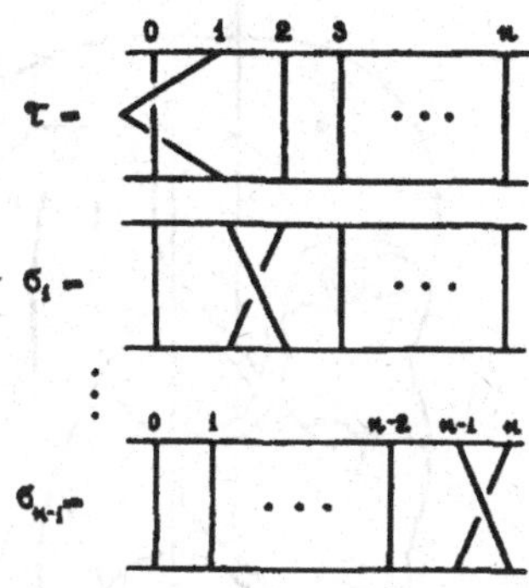

Fig.4 Generators of the group B_n^1

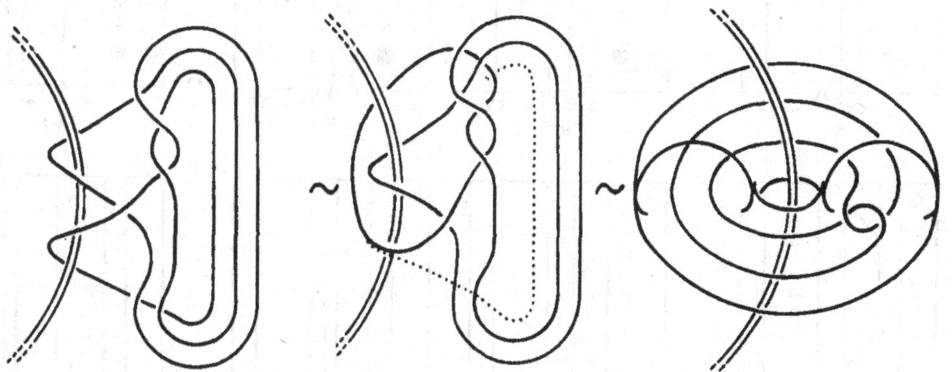

Fig.5 The closure of the 1-braid $\sigma_1\sigma_2\tau\sigma_1^{-1}\tau\sigma_1^{-2}$ is the Whitehead link

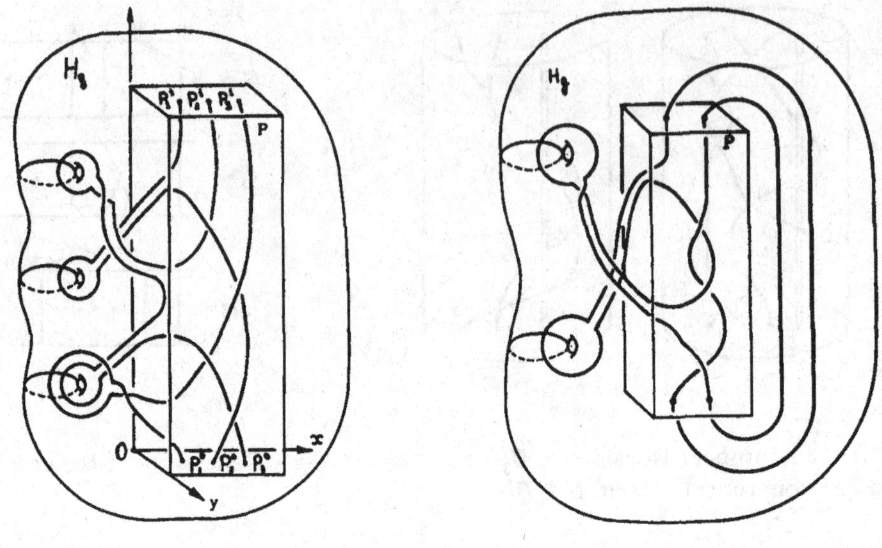

Fig.6 A 3-braid

Fig.7 The closure of the 2-braid $\tau_2\sigma_1^{-2}\tau_1^{-1}\sigma_1$

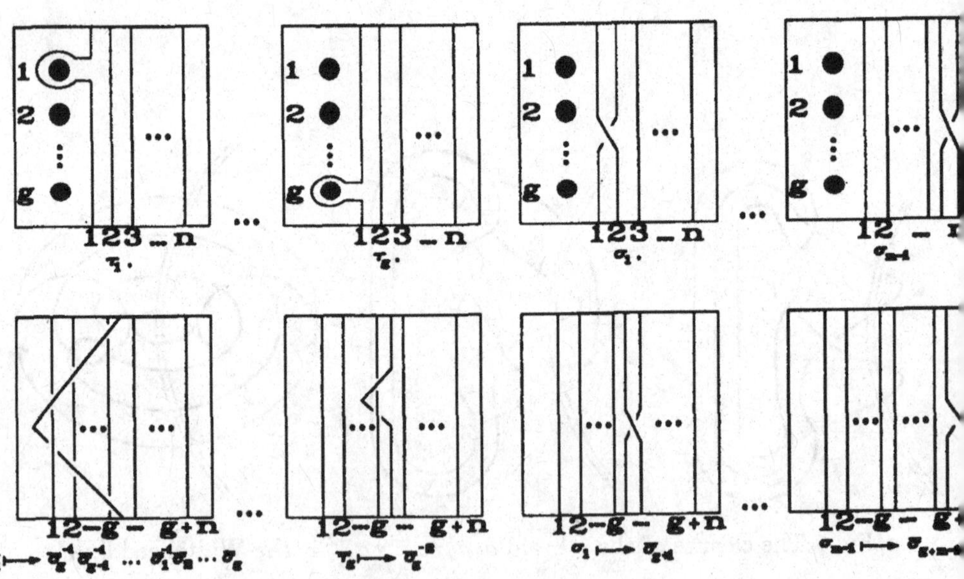

Fig.8 Generators of the group B_n^g and the monomorphism $B_n^g \to B_{n+g}$.

Moscow Institute of Electronic Machinebuilding, Moscow, B. Vuzovski, 3/12

QUANTUM INVARIANTS OF 3–MANIFOLD AND A GLIMPSE OF SHADOW TOPOLOGY

VLADIMIR TURAEV

Leningrad Branch of Steklov Mathematical Institute
and Université Louis Pasteur, Strasbourg

ABSTRACT. The paper announces explicit formulas relating two different approaches to constructing quantum invariants of 3-manifolds (see [1], [2]). A new formula for the linking number of classical knots is presented which enables one to extend the first of these approaches to ghost 3-manifold over $\mathbb{C}$.

Two parallel approaches have been developed recently to produce "quantum" invariants of 3-manifold and links in 3-manifolds on the ground of the theory of quantum groups (see [1], [2], [3], or a survey [4]). In particular, with each primitive complex root of unity Q of degree $4r$ one associates two topological invariants $\tau_Q(M) \in \mathbb{C}$, $|M|_Q \in \mathbb{R}$ of any closed oriented 3-manifold M (see [1] and [2]). On the algebraic part, the construction of these invariants is based on the theory of $U_q(sl_2(\mathbb{C}))$, $q = Q^4$. The definition of $\tau_Q(M)$ proceeds "outside" M and uses a presentation of M as the result of surgery on S^3 along a framed link in S^3. The invariant $|M|_Q$ is defined intrinsically as a state sum on a triangulation of M. In fact, $|M|_Q$ does not depend on the choice of orientation in M, being defined also for non-orientable manifolds. The following theorem establishes a relationship between these two approaches.

Theorem 1. *For any closed oriented 3-manifold M*

$$|M|_Q = |\tau_Q(M)|^2.$$

Note that

$$|\tau_Q(M)|^2 = \tau_Q(M)\overline{\tau_Q(M)} = \tau_Q(M)\tau_Q(-M).$$

Theorem 1 implies a remarkable positivity property $|M_Q| \geq 0$ which has a physical flavour in the frameworks of 3-dimensional quantum gravity. One can view $|M|_Q$ as a number of homomorphisms of $\pi_1(M)$ into the quantum group corresponding to $U_q(sl_2(\mathbb{C}))$, $q = Q^4$. More generally, for a framed colored link $L \subset M$ one defines invariants $\tau_Q(M, L)$ and $|M, L|_Q$ (see [1], [3]). The latter invariant is defined as a statistical sum on a triangulation of the exterior of L in M.

Theorem 2.

$$|M, L|_Q = \tau_Q(M, L)\overline{\tau_Q(M)}.$$

Theorem 2 shows that for links in M the invariants $\tau_Q(M, L)$, $|M, L|_Q$ are essentially equivalent. This gives an intrinsic definition of $\tau_Q(M, L)$ up to the normalizing factor $\tau_Q(-M)$. If $M = S^3$ then $\tau_Q(M) = 1$ so that

$$|S^3, L|_Q = \tau_Q(S^3, L).$$

This equality was already established in [3], producing thus a state model (on the link exterior) for the Jones polynomial. In the case of even r one refines $|M|_Q$ to an invariant $|M; \Delta|_Q$ where $\Delta \in H^1(M; \mathbf{Z}/2\mathbf{Z})$ (see [2]). The invariant $\tau_Q(M)$ may also be refined to an invariant $\tau_Q(M, \theta)$ where θ is a spin structure on M if $r = 0 \mod 4$ and $\theta \in H^1(M; \mathbf{Z}/2\mathbf{Z})$ if $r = 2 \mod 4$ (see [4]). Here

$$|M|_Q = \sum_{\Delta} |M, \Delta|_Q$$

and

$$\tau_Q(M) = \sum_{\theta} \tau_Q(M; \theta).$$

Theorem 3. *For any even* $r \geq 2$ *and any* $\Delta \in H^1(M; \mathbf{Z}/2\mathbf{Z})$

$$|M; \Delta|_Q = \sum_{\theta} \tau_Q(M; \theta)\overline{\tau_Q(M; \theta + \Delta)}.$$

Theorem 3 obviously implies Theorem 1 in the case of even r. These two theorems were conjectured in [4]. Theorems 1 – 3 may be considerably generalized to embrace the case of 3-cobordisms with links sitting inside, establishing thus a relationship between the topological quantum field theories constructed in [1], [2] and [3]. The whole approach extends to invariants derived from modular Hopf algebras (see [1]) and modular tensor categories due to Moore and Seiberg. The proofs of Theorems 1 – 3 are based on ideas of the so-called shadow topology in dimensions 3, 4. This approach was originated in [5] and will be fully presented in [6]. Here I describe one of the ideas involved: an elementary but apparently new formula for the linking number of knots in S^3.

Let $L = L_1 \cup L_2$ be an oriented 2-component link in S^3 presented by an oriented diagram D in $\mathbf{R}^2 = S^2 \setminus \infty$. This diagram splits S^2 into a finite number of disjoint open connected domains called regions of D. For any such region Y one may move along ∂Y in the direction specified by the counter-clockwise orientation in Y. Let us compute the number of jumps up along ∂Y, subtracts the number of jumps down and divide by 2. Denote the resulting number by $g(Y) \in (1/2)\mathbf{Z}$. Fix a point a of S^2 not lying on D. (A canonical choice: $a = \infty$). For $i = 1, 2$ and for any region Y of D we define $rot_i(Y, a)$ to be intersection index of an arc leading from a to any point of Y with the loop of D presenting L_i.

Theorem 4. *For any $a \in S^2 \setminus D$*

$$lk(L_1, L_2) = -\sum_Y g(Y) rot_1(Y, a) rot_2(Y, a), \tag{1}$$

where Y runs over all regions of D.

Theorem 4 directly follows from the standard computation of $lk(L_1, L_2)$ as algebraic sum of local crossing numbers.

There is a similar formula for the self-linking number l of a framed knot $L_1 \subset S^3$. Namely, one presents L_1 by a diagram D on S^2 with the framing orthogonal to S^2 and then for any $a \in S^2 \setminus D$

$$l = -\sum_Y g(Y)(rot_1(Y, a))^2 \tag{2}$$

where Y runs over regions of D. The advantage of formulas (1), (2) is that they may be directly generalized to complex shadow links (see [5]; here I will call these links *ghost links*). Recall that a complex shadow on S^2 is a finite generic collection s of loops on S^2, such that each region of s (i.e. a component of $S^2 \setminus s$) is equipped with a complex number. In terminology of [5] these numbers are "modified gleamps". In particular, each link diagram on S^2 produces a shadow with the same underlying loops and the gleams $Y \longmapsto g(Y)$. This observation enables one to define a version of Reidemeister moves for shadows. A ghost link (over $\mathbf{C}$) on S^2 is an equivalence class of complex shadows on S^2 up to these moves. One also defines framed ghost links, oriented ghost links, and ghost sublinks (see [5]).

Let L be an oriented 2-component ghost link on S^2 with components (ghost knots) L_1 , L_2. Let s be a shadow on S^2 presenting L (and formed by 2 loops). Fix a point $a \in S^2$ not lying on the loops of s, and define as above $rot_i(Y, a) \in \mathbf{Z}$ for any region Y of s and $i = 1, 2$. Put

$$r_i^a = \sum_Y g(Y) rot_i(Y, a),$$

where $g(Y)$ is the gleam of Y and $i = 1, 2$. Let $g = \sum_Y g(Y)$ be the total gleam of L.

Theorem 5. *If $g \neq 0$ then the complex number*

$$lk(L_1, L_2) = -\sum_Y g(Y) rot_1(Y, a) rot_2(Y, a) + r_1^a r_2^a g^{-1} \tag{3}$$

is an invariant of L independent of the choice of a and s.

If $g = 0$ then one may define $lk(L_1, L_2)$ by the formula (1) under the assumption $r_1^a = r_2^a = 0$. (This assumption correspondents to homological triviality of knots in $S^1 \times S^2$ necessary to define their linking numbers).

The self-linking number of a framed ghost knot L_1 is defined by the formula

$$l = -\sum_Y g(Y)(rot_1(Y, a))^2 + g^{-1}(r_1^a)^2. \tag{4}$$

The self-linking number of a framed ghost knot is defined by the formula (4).

The formulas (3), (4) being applied to shadows of knots in the lens spaces $L(n,1)$ produce the linking and self-linking numbers of the knots. The same formulas work for homological trivial ghost knots on arbitrary oriented surfaces and for genuine knots in circle bundles over the surfaces (cf. [5]).

The main application of the ghost linking numbers is the extension of the invariants of [1] to ghost 3-manifolds. A closed ghost 3-manifold is defined as an equivalence class of framed ghost links on S^2 with total gleam 1 up to the shadow version of the Kirby-Fenn-Rourke moves (cf. [5]). The matrix of the linking and self-linking numbers of the components of a framed ghost link is symmetric. Its signature may be combined with the Jones-type invariants of the ghost links introduced in [5] to produce Witten type invariants of ghost manifolds. More generally, one may define ghost 3-cobordisms and construct corresponding ghost topological quantum field theories.

Remarks. 1. The corank of the square matrix mentioned above is, of course, the first Betti number of the ghost 3-manifold. Do other classical invariants, like π_1, $\overline{\mu}$-invariants the Alexander polynomial extend to ghost manifolds and links? The theory seems to be more rich if one uses other groups of coefficients (say $(1/n)\mathbb{Z}$) instead of $\mathbb{C}$.

2. Theorem 1 was recently announced by K. Walker, [7].

References

[1] N.Y.Reshetikhin,V.G.Turaev, *Invariants of 3-manifolds via link polynomials and quantum groups*, Invent. Math. **103** (1991), 547.

[2] V.Turaev and O.Viro, *State sum invariants of 3-manifolds and quantum 6j-symbols*, Topology (to appear).

[3] V.Turaev, *Quantum invariants of links and 3-valent graphs in 3-manifolds*, Preprint (1990).

[4] V.Turaev, *State sum models in low-dimensional topology*, Proc. ICM-90, Kyoto (to appear).

[5] V.Turaev, *Shadow links and IRF-models of statistical mechanics*, Publ. Inst. Rech. Math. Av (1990), Strasbourg; J. Diff. Geom. (to appear).

[6] V.Turaev, *Topology of shadows*, In preparation.

[7] K.Walker, *On Witten's 3-manifolds invariants*, Preprint (November 1990), Strasbourg.

C.N.R.S. – U.R.A. Université Louis Pasteur, Dep. de Mathématique, 7, rue Rene Descartes 67084, Strasbourg, Cedex, France

MOVES OF TRIANGULATIONS OF A PL–MANIFOLD

O. VIRO

Leningrad Branch of Steklov Mathematical Institute

). Introduction. The paper is devoted to one of the key questions on triangulations: ιow to transform a triangulation of a PL-manifold to any other its triangulation. Re-:ently this question became actual again because of the new progress in constructing :ombinatorial invariants of 3-manifolds. In particular, this work has been motivated by ooking for invariance conditions for state sums constructed on triangulations, cf. [TV].

There is a classic answer to this question. It had been given by J. W. Alexander [Al] n 1930. The answer given in this paper seems to be simpler and more convenient in ιpplications.

The Alexander elementary transformations of triangulations were subdivisions. May ɔe it is related to the fact that the notion of equivalence of triangulations is based on he notion of subdivision. Now the Alexander elementary transformations are called *tar subdivisions*. Strictly speaking, they constitute infinite series.

The transformations of this work are elementary replacements. They appear to be omewhat similar to the Morse modifications of manifolds. In particular there are :xactly $n+1$ transformations for triangulations of n-dimensional manifolds. They are elated rather to elementary collapsings of $(n+1)$-manifolds than to subdivisions of ι-manifolds.

In [TV, Section A.2.5] these transformations and their generalizations for singular tri-ιngulations were introduced and it was announced that any two singular triangulations ɔf a PL-manifold can be transformed one to another by a series of such transformations. n this talk I announce this statement for non-singular (usual) triangulations and discuss ι new point of view on the transformations providing another prove. The statement for ingular triangulations is an easy corollary of the non-singular case. This proof seems ɔaser and more traditional than the proof which we meant in [TV].

ι. Two-dimensional case. To begin with, consider the most visualizable case — the :ase of surfaces. There are 3 transformations, they are shown in Fig. 1. The third of hem is inverse to the first.

On the first glance the second transformation has nothing common with the others. Iowever the three-dimensional view presented in Fig. 2 suggests that the transforma-ions of Fig. 1 are exactly what happens with the triangulation of the boundary of a riangulated 3-manifold under the operations inverse to elementary simplicial collapsings ɔf it, i.e. under gluing a tetrahedron along one, two or three faces respectively.

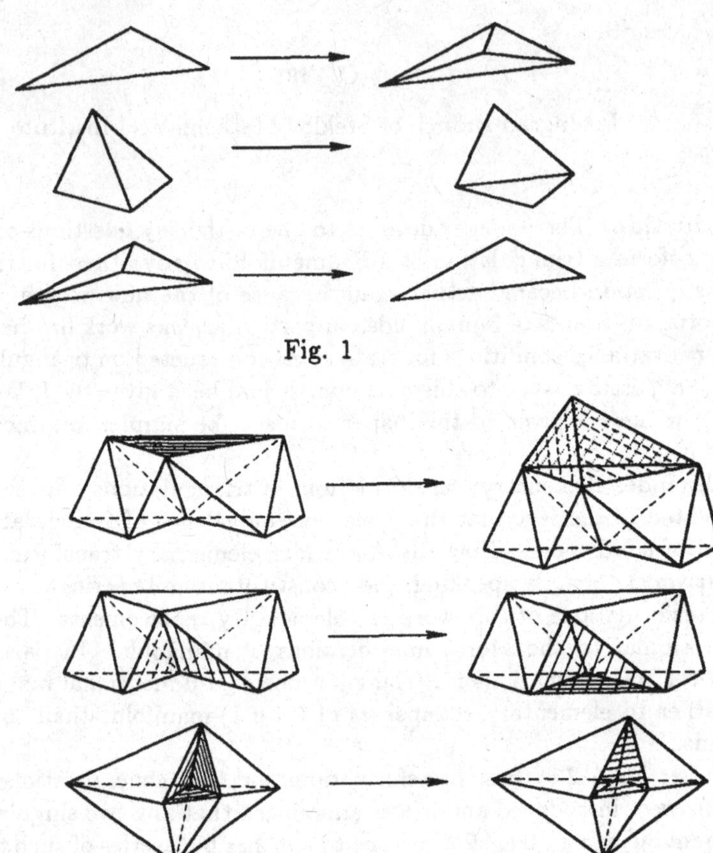

Fig. 1

Fig. 2

2. Structure of the star of a simplex. To prepare a definition of the transformation in high dimensions, remind well known facts on stars and links. A reader familiar wit classic notions related to triangulations can omit this subsection.

The *(closed) star* of a simplex σ in a triangulated space is the union of all close simplexes containing σ, it is denoted by $St(\sigma)$. The *link* of a simplex σ is the union c all closed simplexes contained in $St(\sigma)$ which do not intersect σ, it is denoted by $lk(\sigma$

The *join* $X * Y$ of spaces X and Y is defined to be the quotient space of $X \times Y \times [0,$ obtained by contracting subsets $pt \times Y \times 0$ and $X \times pt \times 1$. It can be thought of a the union of straight segments joining X and Y such that any two of these segment intersect at most at one point, which can be only an end point for both segments. Sinc the join of two simplexes is a simplex, triangulations of X and Y define a triangulatio of $X * Y$.

Obviously,

$$St(\sigma) = lk(\sigma) * \sigma,$$

Therefore, the boundary $\partial St(\sigma)$ of the star $St(\sigma)$ is the join of $lk(\sigma)$ and the boundar

$\partial(\sigma)$ of the simplex σ.

In the case of PL-manifold, the link is PL-homeomorphic to the boundary of a simplex of the complementary dimension: if our simplex σ has dimension p, then the link $lk(\sigma)$ is PL-homeomorphic to the boundary of q-dimensional simplex, where $p+q=n$ (this is nothing but a definition of PL-manifold).

3. Internal description of the transformations. Consider the case when the link of a p-simplex σ in an n-dimensional manifold *is isomorphic (as a triangulated space)* to the boundary of q-simplex ($p+q=n$). Then the boundary $\partial St(\sigma)$ is isomorphic (also as a triangulated space) to the join of the boundaries of p-dimensional and q-dimensional simplexes. Note that in the boundary of $St(\sigma)$ they are involved in absolutely similar ways.

Replace the star of our simplex with the join of q-simplex and the boundary of p-simplex. This is the transformation of index p.

It is clear that the inverse transformation to the transformation of index p is the transformation of index $q=n-p$.

In the special case $p=n$ the transformation above is the usual star subdivision centered at a simplex of the high dimension. In the other cases it is not a subdivision.

4. External description. Let X be a triangulated PL-manifold of dimension n. Take a triangulated PL-manifold Y such that X is a component of ∂Y, say $Y=X\times I$ where $I=[0,1]$.

Let T be an $(n+1)$-simplex and S the union of any number $<n+2$ of its n-faces. Since these faces make the star of their intersection in ∂T, S is a PL-ball of dimension n.

Glue T to Y by a simplicial (with respect to the triangulations under consideration) embedding $S\to X$. The result, say Y', is PL-homeomorphic to Y (it collapses to Y). Therefore the component X' of the boundary of Y' containing X is PL-homeomorphic to X. The natural triangulation of X' coincides with the triangulation of X outside the image of S, and so X' as a *triangulated space* can be obtain from X by replacing S by $\partial T\setminus \operatorname{Int} S$.

This operation coincides with the transformation described in the previous section. To prove that, it is enough to show that:

(1) the join of p-simplex with the boundary of q-simplex can be embedded as a triangulated space into the boundary of $(p+q+1)$-simplex so that the closure of the complement of the image is isomorphic to the join of q-simplex with the boundary of p-simplex and
(2) the union of any number $<n+1$ of n-faces of an $(n+1)$-simplex T is the star of some simplex σ in the boundary of T, it is homeomorphic to the join of σ and the boundary of some simplex $\tau\subset\partial T$, and the closure of its complement in ∂T is the join of τ and $\partial\sigma$.

The first follows from the fact that $(p+q+1)$-simplex is isomorphic to the join of p-simplex and q-simplex. In the second assertion for σ one can take the intersection of the n-faces and for τ the simplex spanned on the vertexes of T which are not contained in σ.

I propose to call the transformation of index p discussed in the previous and this sections a ***simplex move of index*** p.

5. Three-dimensional case. In the case $n = 3$ there are 4 simplex moves. They form two pairs of transformations inverse to each other. One pair consists of the star subdivision centered at an internal point of a 3-simplex and the inverse transformation. See Fig. 3. The other pair shown in Fig. 4. It exchanges a pair of tetrahedra meeting in a common face by a triple of tetrahedra with common edge.

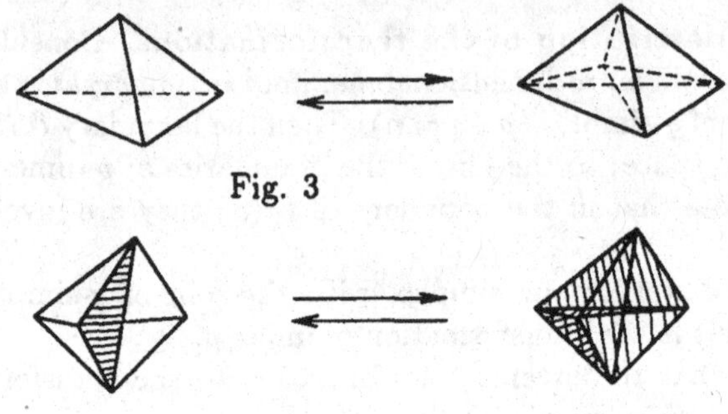

Fig. 3

Fig. 4

6. Four-dimensional case. The four-dimensional case seems to be the most interesting due to potential constructions of state sum invariants of 4-manifolds. In this case there are essentially three simplex moves (two pairs of moves inverse to each other and one move inverse to itself). The first one is just a star subdivision centered at 4-simplex. The simplex move of index 3 replaces a couple of 4-simplexes meeting in a tetrahedron by four 4-simplexes which meets in a new edge connecting the non-common vertices of the original 4-simplexes and makes the star of this edge. The simplex move of index 2 reglue a star of 2-simplex. This star should consist of three 4-simplexes.

7. Main Theorem. *Any two triangulations of a PL-manifold can be transformed to each other by a sequence of simplex moves. If these triangulations coincide on some subpolyhedron of the manifold, such moves can be found changing triangulations only outside the subpolyhedron.*

Idea of proof. Take the product $X \times I$ of the manifold X under consideration by the interval $I = [0, 1]$ and triangulate it so that the triangulations induced on $X \times 0$ and $X \times 1$ would be the given two triangulations. It is possible to do this in such a way that there exists an order on the set of high dimensional simplexes of $X \times I$ such that adjoinment of them to $X \times 0$ in this order induces a desired sequence of simplex moves.

8. Dualization and generalizations. First, remind some folklore notions concerning stratifications of PL-manifolds (it is borrowed mainly from [TV, Appendix 2]). By a *stratification* of a PL-manifold X we mean a partition of X on disjoint parts (which are called *strata*) with the following properties:

(1) each stratum is a submanifold of either $Int\, X$ or ∂X;
(2) as a manifold each stratum has empty boundary;
(3) the closure of each stratum is a subpolyhedron of X which is the union of a finite number of strata.

We consider only stratifications which satisfy an additional property of "local triviality" along strata. An arbitrary stratification can be canonically subdivided to satisfy this property. A stratification of a manifold X is said to be *locally trivial* if each point of any stratum S has a neighborhood U in X such that there exists a homeomorphism $U \to V \times \mathbf{R}^q$, where V is $\mathbf{R}^p$ or $\mathbf{R}^p_+$, mapping $U \cap S$ onto $pt \times \mathbf{R}^q$ and the intersection of U with any stratum onto $C \times \mathbf{R}^q$, where C is a submanifold of V. These C's constitute a stratification of V.

If each stratum is homeomorphic to an Euclidean space, then the stratification is said to be *cellular*.

A locally trivial stratification of X is said to be *simple*, if for each point of X the stratification of V mentioned above is homeomorphic to the cone over the standard stratification (by faces) of the boundary of the p-dimensional simplex, in the case $V = \mathbf{R}^p$, and , in the case $V = \mathbf{R}^p_+$, to the same stratification, but with one p-dimensional stratum removed.

The most classical stratifications of manifolds are triangulations. They are locally trivial cellular stratifications, but in general they are not simple (in the sense specified above).

Another classical set of locally trivial cellular stratifications are stratifications dual to triangulations of manifolds without boundary, i.e. partitions of manifolds on barycentric stars of simplexes of triangulations. These are simple stratifications.

In the case of a manifold with non-empty boundary the corresponding simple stratifications consist of the intersections of the barycentric stars with the interior and the boundary of the manifold.

An important property of triangulations is that each triangulated space can be described up to strata preserving homeomorphisms in a purely combinatorial way. For this it is sufficient to describe the set of its vertices and the collection of finite subsets of this set, which are the sets of vertices of simplexes. The only restriction on a collection of the subsets is that together with each subset it should contain all the subsets of it.

Thus triangulations provide a method of combinatorial description of manifolds. But usually triangulations have so many simplexes that it is hard to use them in practice when some manifold is to be described. The natural way to avoid this difficulty is to generalize triangulations. The most usual generalization is the notion of CW-complex, but it leads to a loss of the combinatorial character of the theory. The following notion lies between the notions of triangulation and CW-complex.

Let X be a space. A family of continuous maps $\varphi_\alpha : T^{d_\alpha} \to X$, $\alpha \in A$ of standard simplexes T^{d_α}, $d_\alpha \in \mathbf{Z}_+$, is called a *singular triangulation* of X, if

(1) all $\varphi_\alpha|_{IntT^{d_\alpha}}$ are embeddings,
(2) $\varphi_\alpha(IntT^{d_\alpha})$ are open cells of some CW-decomposition of X,
(3) for any face F of T^{d_α} the restriction $\varphi_\alpha|_F$ can be obtained from some φ_β by composition with a linear isomorphism $F \to T^{d_\beta}$.

Note that replacing $\varphi_\alpha|_{IntT^{d_\alpha}}$ by φ_α in (1) and incorporating the condition that the intersection of any two simplexes is their common face convert this definition into a definition of triangulation.

A singular triangulation also can be completely described (up to strata preserving homeomorphism) in a discrete combinatorial way. For this, it is sufficient to describe the set of its vertices and the collection of maps of some finite sets to this set. These maps are the restrictions of φ_α to sets of vertices of T^{d_α}. The only restriction on a collection of the maps is that together with each map it should contain all the restrictions of this map.

The construction of the barycentric star stratification can be generalized in an obvious way to a construction which assigns to any cellular stratification of a manifold a dual stratification defined up to ambient isotopy. An application of this construction to a singular triangulation of a PL-manifold gives a simple cellular stratification. Conversely an application of this construction to any simple cellular stratification gives a singular triangulation. Thus the construction yields a 1-1 correspondence between singular triangulations and simple cellular stratifications.

For a given manifold one can find usually singular triangulations and simple cellular stratifications which are considerably smaller than triangulations. For example it is easy to prove that any closed connected manifold has a singular triangulation with only one vertex. A closed orientable surface of genus g has singular triangulations with $4g-2$ triangles.

Now generalize the simplex moves defined above to the case of singular triangulations. The generalized simplex move of index p can be applied, iff the closure of the barycentric star of one of the p-simplexes is strata preserving homeomorphic to the canonically triangulated q-simplex with $q = n-p$, where n is the dimension of the manifold. Here the baricentric star appeared instead of link since the link is not related to structure of the star as closely as the baricentric star (contrary to the case of non-singular triangulations where they are isomorphic). Indeed, for any simplex S of a singular stratification there is a natural strata preserving map of the join of the closure of S and the boundary of its barycentric star onto the closed star of S, this map is identity on S and 1-1 on the complement of the boundary of the barycentric star. In the situation under consideration the generalized simplex move of index p replaces the stratification of the star of the p-simplex by the image under this map of the triangulation of the join above presented as the join of the barycentric star with the boundary of the p-simplex.

The simplest (however slightly implicit) description of the move of a simple cellular triangulation corresponding to the index p simplex move of the dual singular triangulation is the following one (given in [TV]): there should exists a simple cellular stratification of $X \times I$, where X is the manifold under consideration, such that

(1) on $X \times 0$ it coincides with the initial stratification,
(2) on $X \times 1$ it coincides with the resulting stratification,
(3) the natural projection $X \times I \to I$ restricted to any stratum of $X \times I$ has no critical points,
(4) there is only one 0-stratum in $X \times (0, 1)$.
(5) $p+1$ strata of dimension 1 join this 0-stratum with $X \times 1$ and the others $n-p$ with $X \times 0$.

These conditions mean that in the sense of the Goresky-MacPherson stratified Morse theory the natural projection of $X \times [0,1] \to [0,1]$ is a Morse function with only one critical point, and at this point it has index $n-p$ in the sense of Khovansky. Thus one can consider our move as a kind of stratified Morse modification of index $n-p$.

References

[Al] J. W. Alexander, *The combinatorial theory of complexes*, Ann. Math. **(2) 31 (1930)**, 294–322.
[TV] V. G. Turaev, O. Ya. Viro, *State sum invariants of 3-manifolds and quantum 6j- symbols*, Topology (to appear).

LOMI, Fontanka 27, Leningrad 191011, USSR

YANG–BAXTER RELATION, EXACTLY SOLVABLE MODELS AND LINK POLYNOMIALS

MIKI WADATI[1], TETSUO DEGUCHI[1] AND YASUHIRO AKUTSU[2]

[1]Department of Physics, Faculty of Science,
University of Tokyo
[2]Department of Physics, Faculty of Science,
Osaka University

ABSTRACT. Presented is a general theory to construct link polynomials, topological invariants for knots and links, from exactly solvable (integrable) models. Representations of the braid group and the Markov traces on the representations are made through the general theory which is based on fundamental properties of exactly solvable models. Various examples including Alexander, Jones, Kauffman and a hierarchy of link polynomials are explicitly shown.

1. Introduction

The Yang-Baxter relation[1,2] is a sufficient condition for the solvability of models n statistical physics and quantum field theory. More precisely, to each solvable model ve can associate a family of commuting transfer matrices which are generators of an nfinite number of conserved quantities, and can show that the Yang-Baxter relation is sufficient condition for commutativity of the transfer matrices. It is a right occasion o recall that universality and importance of the Yang-Baxter relation became clear hrough the development of the quantum soliton theory. The classical soliton system as an infinite number of conserved quantities and is proved to be a completely integrable ystem. A method for proving these properties is called inverse scattering method. Its xtension to quantum systems is the quantum inverse scattering method [3–7]. By this levelopment, we get a unified viewpoint on exactly solvable models in (1+1)-dimensional ield theory and in 2-dimensional classical statistical mechanics.

Recently, the Yang-Baxter relation has been found to be a key to several important levelopments in mathematical physics. Among them, various link polynomials [8–12] nd their extensions have been obtained from exactly solvable models through a method ased on a general theory [13–27]. The purpose of this paper is to summarize the general nethod for construction of those link polynomials from exactly solvable models. Several roblems in physics such as path integrals, fractional statistics and quantum gravity are elated to the braid group [28–33]. In particular, it is interesting that solvable models nd conformal field theories share many mathematical features in common [34–41].

The paper consists of the following. In Sec.2, factorized S-matrices, vertex models nd IRF models are introduced. Then, a method to make the representations of the raid group is given. In Sec.3, link polynomials are constructed by algebraic and graphic approaches. The crossing symmetry is used for the graphical calculation of the link polynomials. In Sec.4, applications to several models are shown. Link polynomials obtained in our approach include Alexander, Jones, Kauffman and new link polynomials. In Sec.5, link polynomials include Alexander and Jones polynomials are constructed from solvable models with graded symmetry. The last section is devoted to concluding remarks.

2. Exactly solvable models and braids

Let us first introduce factorized S-matrices[42]. We write the amplitude of the scattering process: $i \to k$, $j \to l$ as $S_{jl}^{ik}(u)$ (Fig.1(a)), where u is the rapidity difference of incoming (outgoing) particles. In general, the "charge" or "spin" variables i, j, k and l of $S_{jl}^{ik}(u)$ take vector values (weight vectors). The factorized S-matrices represent the elastic scattering of particles where only the exchanges of momenta and the phase shifts occur. The rapidity difference of the scattering particles can be depicted by the angle in the diagram. When $S_{jl}^{ik}(u)$ is non-zero only the case $i + j = k + l$, we say that the model has "charge conversation" property [13,21,22].

The Yang-Baxter relation for the S-matrices reads as

$$\sum_{abc} S_{cr}^{bq}(u) S_{kc}^{ap}(u+v) S_{jb}^{ia}(v) = \sum_{abc} S_{bq}^{ap}(u) S_{cr}^{ia}(u+v) S_{kc}^{jb}(u). \tag{1}$$

This relation is often referred to as the factorization equation [1,42].

In two-dimensional statistical mechanics [2,22], we may consider two types of solvable models, vertex models and IRF models. The models is called to be exactly solvable (or simply, solvable) when the Boltzmann weight satisfy the Yang-Baxter relation. We introduce vertex models. The Boltzmann weight $w(i, j, k, l; u)$ of a vertex model is defined for a configuration $\{i, j, k, l\}$ round a vertex. The parameter u is called spectral parameter which controls the anisotropy (and strength) of the interactions for the model.

For vertex models the Yang-Baxter relation is given by

$$\begin{aligned} &\sum_{abc} w(b, c, q, r; u) w(a, k, p, c; u+v) w(i, j, a, b; v) \\ &\sum_{abc} w(a, b, p, q; v) w(i, c, a, r; u+v) w(j, k, b, c; u). \end{aligned} \tag{2}$$

It is known that factorized S-matrices are mathematically equivalent to corresponding solvable vertex models. Explicitly, we identity $S_{jl}^{ik}(u) = w(i, j, k, l; u)$

We introduce IRF models. The Boltzmann weight of an IRF model $w(a, b, c, d; u)$ is defined on a contribution $\{a, b, c, d\}$ round a face (Fig.1(b)). IRF models have constraints on the configurations. By $b \sim a$ we denote that the "spin" b is admissible to the "spin" a under the constraint of the model. If the conditions $b \sim a, a \sim d, b \sim c$ and $c \sim d$ are all satisfied, then the configuration $\{a, b, c, d\}$ in Fig.1(b) is called to be allowed. The Boltzmann weights for not-allowed configurations are set to be 0. For IRF models the Yang-Baxter relation is written as

$$\begin{aligned} &\sum_{c} w(b, d, c, a; u) w(d, e, f, c; u+v) w(c, f, g, a; v) \\ &\sum_{c} w(d, e, c, b; v) w(b, c, g, a; u+v) w(c, e, f, g; u). \end{aligned} \tag{3}$$

The IRF configuration a, b, c, d in Fig.1(b) corresponds to the vertex configuration in Fig.1(a) by $i = a - d, j = b - a, k = b - c$ and $l = c - d$.

The Boltzmann weights for the vertex model and the IRF model satisfy the following basic relations in addition to the Yang-Baxter relation [13,17,18,21,22].

(1) standard initial condition

$$S_{jl}^{ik}(u = 0) = \delta_{il}\delta_{jl}, \qquad w(a, b, c, dd; u = 0) = \delta_{ac}, \tag{4}$$

where δ_{ij} is the Kronecker delta.

(2) inversion relation (unitary condition)

$$\sum_{mp} S_{pl}^{mk}(u) S_{jm}^{ip}(-u) = \rho(u)\rho(-u)\delta_{il}\delta_{jk}, \tag{5}$$

$$\sum_{e} w(e, c, d, a; u) w(b, c, e, a; -u) = \rho(u)\rho(-u)\delta_{bd} \tag{6}$$

where $\rho(u)$ is a model-dependent function.

(3) Second inversion relation (second unitary condition)

$$\sum_{pm} S_{pl}^{im}(\lambda - u) S_{mj}^{kp}(\lambda + u) \cdot \left(\frac{r(m)r(p)}{r(i)r(j)r(k)r(l)} \right)^{1/2} = \rho(u)\rho(-u)\delta_{ij}\delta_{kl}, \tag{7}$$

$$\sum_{e} w(c, e, a, b; \lambda - u) w(a, e, c, d; \lambda + u) \frac{\psi(e)\psi(b)}{\psi(a)\psi(c)} = \rho(u)\rho(-u)\delta_{bd}. \tag{8}$$

We call the parameter λ crossing parameter, and $\{r(i)\}$ and $\{\psi(a)\}$ crossing multipliers.

(4) crossing symmetry (Fig.2)

$$S_{jl}^{ik}(u) = S_{ki}^{j\bar{l}}(\lambda - u) \left(\frac{r(i)r(l)}{r(j)r(k)} \right)^{1/2}, \tag{9}$$

$$w(a, b, c, d; u) = w(b, c, d, a; \lambda - u) \left(\frac{\psi(e)\psi(b)}{\psi(a)\psi(c)} \right)^{1/2}. \tag{10}$$

Here, we have used the notation $\bar{j}$ for the "antiparticle" of j. We assume that $r(\bar{j}) = 1/r(j)$. Note that the second inversion relation and the crossing symmetry define the crossing multipliers.

We shall see the basic relations and the Yang-Baxter relation are related to the local moves on link diagrams, known as the Reidemeister moves in knot theory.

In order to connect exactly solvable models to the braid group we introduce Yang-Baxter operator $X_i(u)$ [13,17,21,22]. For vertex models (factorized S-matrices) we define Yang-Baxter operator by

$$X_i(u) = \sum_{abcd} S_{da}^{cb}(u) I^{(1)} \otimes \cdots \otimes e_{ac}^{(i)} \otimes e_{bd}^{(i+1)} \otimes I^{(i+2)} \otimes \cdots \otimes I^{(n)}. \tag{11}$$

Here $I^{(i)}$ denotes the identity matrix and e_{ab} a matrix such that $(e_{ab})_{jk} = \delta_{ja}\delta_{kb}$. The Yang-Baxter operators $\{X_i(u)\}$ satisfy the following relations (Yang-Baxter algebra),

$$X_i(u)X_{i+1}(u+v)X_i(v) = X_{i+1}(u)X_i(u+v)X_{i+1}(v), \tag{12}$$

$$X_i(u)X_j(v) = X_j(v)X_i(u), \ |i-j| \geq 2. \tag{13}$$

In terms of the Yang-Baxter operators, the Yang-Baxter relations for factorized S-matrices, vertex models and IRF models are in the same form.

We recall the definition of the braid group [43]. The braid group B_n is defined by a set of generators, $b_1, \ldots, b_{n-1}$ which satisfy

$$\begin{aligned} b_i b_{i+1} b_i &= b_{i+1} b_i b_{i+1}, \\ b_i b_j &= b_j b_i, \ |i-j| \geq 2. \end{aligned} \tag{14}$$

The operation b_i and the inverse b_i^{-1} are best understood by the graphs (Fig.3).

It is known that any oriented link can be expressed by a closed braid. The equivalent braids expressing the same link are mutually transformed by a finite sequence of two types of operations, Markov moves I and II. The Markov trace $\phi(\cdot)$ is the linear functional on the representation of the braid group which have the following properties (the Markov properties):

I

$$\phi(AB) = \phi(BA), \qquad A, B \in B_n, \tag{15}$$

II

$$\phi(Ab_n) = \tau\phi(A), \ \ \phi(Ab_n^{-1}) = \bar{\tau}\phi(A), \ \ A \in B_n, \ \ b_n \in B_{n+1}, \tag{16}$$

where

$$\tau = \phi(b_i), \ \ \bar{\tau} = \phi(b_i^{-1}), \ \text{ for all } \ i. \tag{17}$$

From the Markov trace we obtain a link polynomial $\alpha(\cdot)$ as [13,21,22]

$$\alpha(A) = (\tau\bar{\tau})^{-\frac{n-1}{2}}\left(\frac{\bar{\tau}}{\tau}\right)^{\frac{1}{2}e(A)}\phi(A), \ \ A \in B_n. \tag{18}$$

Here $e(A)$ is the exponent sum of b_i's in the braid A, which is equivalent to the writhe of the link diagram. (cf. 34)) It is easy to show that $\alpha(\cdot)$ defined by (18) is invariant under the Markov moves.

We now relate the Yang-Baxter operator with representation of the braid generator (hereafter we call the latter *braid operator*). The braid operator $G(+)_i$, the inverse operator $G(-)_i$ and the identity I are given by [13]

$$G(\pm)_i = \lim_{u\to\infty} X_i(\pm u)/\rho(\pm u), \tag{19}$$

$$I = X_i(0). \tag{20}$$

The limit $u \to \infty$ (more precisely, an infinite limit in a certain direction in the complex u-plane) requires that factorized S-matrices (the Boltzmann weights) be parametrized by hyperbolic or trigonometric functions. In statistical mechanics, it implies that the model is at the criticality. Hereafter we write the matrix elements of the braid operator as

$$G^{ab}_{cd}(\pm) = \lim_{u\to\infty} S^{cb}_{da}(\pm u)/\rho(\pm u). \tag{21}$$

Then we can express the braid operator (19) constructed from the Yang-Baxter operator as

$$G(\pm)_i = \sum_{abcd} G^{ab}_{cd}(\pm) I^{(1)} \otimes \cdots \otimes e^{(i)}_{ac} \otimes e^{(i+1)}_{bd} \otimes I^{(i+2)} \otimes \cdots \otimes I^{(n)}. \tag{22}$$

We can also construct braid operators for IRF models by (19). Thus corresponding to an exactly solvable model, we obtain a representation of the braid group by using the formula (19).

3. Construction of link polynomials

3.1 The Markov trace. We shall obtain link polynomials by constructing the Markov trace on the representations of the braid group derived from the solvable models. For factorized S-matrices and vertex models, the Markov trace takes the following form [13,21,22]

$$\begin{aligned} \phi(A) &= \frac{\check{T}r(H(n)A)}{\check{T}r(H(n))}, \qquad A \in B_n, \\ [H(n)]^{a_1 a_2 \cdots a_n}_{b_1 b_2 \cdots b_n} &= \prod_{j=1}^{n} r^2(a_j)\delta^{a_j}_{b_j}, \end{aligned} \tag{23}$$

where δ^a_b is the Kronecker delta. For the models with the crossing symmetry (and the second inversion relation), $r(p)$ is nothing but the crossing multiplier of the model. The trace $\phi(\cdot)$ defined in (23) is the Markov trace since we can prove the Markov property I by the "charge conservation" property and the Markov property II by the following conditions:

$$\sum_b G^{ab}_{ab}(\pm) r^2(b) = \xi(\pm) \text{ (independent of } a). \tag{24}$$

The τ-factors are related to $\xi(\pm)$ as $\bar{\tau}/\tau = \xi(-)/\xi(+)$.

We can prove the extended Markov property [17,19,21,22],

$$\sum_b X^{ab}_{ab}(u) h(b) = H(u;\eta)\rho(u) \text{ (independent of } a), \tag{25}$$

where $H(u;\eta)$ is called characteristic function. This relation is an extension of (24) into the case of finite spectral parameter.

For IRF models we introduce a "constrained trace" $\tilde{T}r(A)$ [17,19,21,22]:

$$\tilde{T}r(A) = \tilde{\sum_{l_1 l_2 \cdots l_n}} A^{l_0 l_1 \cdots l_n}_{l_0 l_1 \cdots l_n} \frac{\psi(l_n)}{\psi l_0}, \ (l_0 \ : \ fixed) \tag{26}$$

where symbol $\sim$ on $\sum$ represents the summation over admissible multi-indices $\{l_i : l_{i+1} \sim l_i\}$ for $i = 0, \cdots, n-1$ with l_0 being fixed. Then the Markov trace $\phi(\cdot)$ is written as

$$\phi(A) = \frac{\tilde{Tr}(A)}{\tilde{Tr}(I(n))}, \qquad A \in B_n, \tag{27}$$

where $I(n)$ is the "identity" operator for n strings. We can prove the extended Markov property also for IRF models [17,19,21,22]. To summarize, the extended Markov property (and the charge conservation condition for vertex models) is sufficient for the existence of the Markov trace. This completes the algebraic construction of link polynomials from exactly solvable models.

3.2 Graphical calculation. The crossing symmetry is significant in algebraic and graphical aspects of the knot theory. For solvable models with the crossing symmetry the Yang-Baxter operator with $u = \lambda$ becomes the Temperley-Lieb operator [18]. In fact, by setting (Fig. 4)

$$E_i = X_i(\lambda), \tag{28}$$

we find that the operators $\{E_i\}$ satisfy the following relations (Temperley-Lieb algebra) [44]

$$\begin{aligned} E_i E_{i\pm1} E_i &= E_i, \qquad E_i^2 = q^{\frac{1}{2}} E_i, \\ E_i E_j &= E_j E_i, \quad |i-j| \geq 2, \end{aligned} \tag{29}$$

where the quantity $q^{1/2}$ is related to the crossing multipliers $r(a)$ (or $\psi(i)$) by [13,17,18]

$$q^{\frac{1}{2}} = \sum_j r^2((j), \text{ for } S\text{-matrix (vertex model)}, \tag{30}$$

$$= \sum_{b\sim a} \frac{\psi(b)}{\psi(a)}, \text{ for IRF model.} \tag{31}$$

In (31) the summation is over all states b allowable to a.

Let us consider the graphical meaning of the relations (29). From the crossing symmetry and the standard initial condition we have (Fig.4) [18,21]

$$\begin{aligned} S_{da}^{cb} &= \left(\frac{r(a)r(c)}{r(b)r(d)}\right)^{\frac{1}{2}} S_{b\bar{c}}^{d\bar{a}}(0) \\ & r(a)\delta(a,\bar{b}) \cdot r(c)\delta(c < \bar{d}), \end{aligned} \tag{32}$$

where $\delta(a,c) = \delta_{ac}$ is the Kronecker delta. We regard the elements $r(c)\delta(c,\bar{d})$ and $r(a)\delta(a,\bar{b})$ as the weights for the pair-annihilation diagram and the pair-creation diagram, respectively (Fig.5). Then, the Yang-Baxter operator at $u = \lambda$ is depicted as the monoid diagram, which gives a graphical explanation of the Temperley-Lieb algebra. This interpretation is consistent with a fact that the energy at $u = \lambda$ is related to the pair-creation energy.

For IRF models, the weights $\{\psi(a)/\psi(b)\}^{\frac{1}{2}}$ and $\{\psi(c)/\psi(b)\}^{\frac{1}{2}}$ correspond to the pair-annihilation and pair-creation diagrams, respectively (Fig.5).

We can formulate link polynomials with the crossing symmetry directly on link diagrams. Link diagram $\hat{L}$ is a 2-dimensional projection of a link L. The $w(\hat{L})$ is the sum of signs for all crossings C_i is the sum of signs for all crossings C_i in the link diagram (Fig.6):

$$w(\hat{L}) = \sum_{C_i} \epsilon(C_i). \tag{33}$$

We calculate "statistical sum" $Tr(\hat{L})$ on the diagram $\hat{L}$ by the rules given in Fig.5. The link polynomial for the link L is expressed as

$$\alpha(L) = c^{-w(\hat{L})} \frac{Tr(\hat{L})}{Tr(\hat{K}_0)}. \tag{34}$$

Here $\hat{K}_0$ is the trivial knot diagram (a loop) and the constant c is defined by a relation

$$G_i E_i = c E_i, \tag{35}$$

or by (cf.(24))

$$c = \left(\frac{\xi(-)}{\xi(+)} \right)^{\frac{1}{2}}. \tag{36}$$

It is not difficult to show that $\alpha(L)$ is invariant under the Reidemeister moves (Fig.7), and therefore $\alpha(L)$ is a topological invariant of the link L. Thus we have shown that the link polynomials constructed from solvable models with the crossing symmetry are also graphically formulated. The monoid diagram and the weights for the creation and annihilation diagrams were used by L.H. Kauffman [45] for the Bracket polynomial which gives a graphical calculation of the Jones polynomial. We have derived monoid operators from the crossing symmetry of solvable models by a general formula (28).

We have a remark. The graphical formulation applied to closed braids yields the Markov trace (Fig.8). For the link polynomials made from the models with the crossing symmetry, the formulation based on the Markov trace is thus equivalent to the graphical formulation.

It is interesting that the link diagrams are considered as the Feynman diagrams for the high energy processes of charged particles and the link polynomials as the scattering amplitudes. At the lowest point in the diagram there occurs a pair creation and at the highest point a pair annihilation. It is also interesting that, if we regard the link diagrams as distorted 2-dimensional lattices, the link polynomials are considered as the partition functions.

4. Various Examples

4.1 N-state vertex model. From the N-state vertex models, a hierarchy of link polynomials is obtained by the general method presented in Sec.3 and Sec.4 [13]. The model corresponds to the factorized S-matrices with spin $s = (N-1)/2$ particles [46,47].

Using the N-state vertex model(asymmetrized by the symmetry breaking transformation), we get the braid operator which satisfies an N-order relation [13]:

$$(G_i - C_1)(G_i - C_2) \cdots (G_i - C_N) = 0, \tag{37}$$

where for $j = 1, 2, \cdots, N$

$$G_j = (-1)^{(j+N)} t^{\frac{1}{2}N(N-1)-\frac{1}{2}j(j-1)}, \qquad t = e^{2\lambda}. \tag{38}$$

We call a relation for G_i such as relation (37) reduction relation of the braid operator The crossing multiplier for the asymmetrized N-state vertex model is

$$r(k) = e^{-\lambda k} = t^{-k/2}, \qquad k = -s, -s+1, \cdots, s, \tag{39}$$

where $s = (N-1)/2$.

The extended Markov property [17,22] is satisfied with the characteristic functio given as

$$H(u;\lambda) = \frac{\sinh(N\lambda - u)}{\sinh(\lambda - u)}. \tag{40}$$

The constant τ and $\bar{\tau}$ are

$$\tau = 1/(1 + t + \cdots + t^{N-1}), \tag{41}$$

$$\bar{\tau} = t^{N-1}/(1 + t + \cdots + t^{N-1}). \tag{42}$$

It is remarkable [13,21,22] that there exists an infinite sequence of link polynomials corre sponding to the N-state vertex models ($N = 2, 3, 4, 5, \cdots$). The $N = 2$ case correspond to the Jones polynomial [9]. In the $N \geq 3$ cases we have new link polynomials. Fron the reduction relation, we obtain the skein relations (the Alexander-Conway relations for the link polynomials:

$$\alpha(L_+) = (1-t)t^{\frac{1}{2}}\alpha(L_0) + t^2\alpha(L_-), \ (N = 2) \tag{43}$$

$$\begin{aligned}\alpha(L_{2+}) &= t(1 - t^2 + t^3)\alpha(L_+) + (t^4 - t^5 + t^7)\alpha(L_0) - \\ &\quad - t^8\alpha(L_-), \ (N = 3)\end{aligned} \tag{44}$$

$$\begin{aligned}\alpha(L_{3+}) &= t^{3/2}(1 - t^3 + t^5 - t^6)\alpha(L_{2+}) + t^6(1 - t^2 + t^3 + t^5 - t^6 + t^8)\alpha(L_+) - \\ &\quad - t^{25/2}(-1 + t - t^3 + t^6)\alpha(L_0) - t^{20}\alpha(L_-), \ (N = 4).\end{aligned} \tag{45}$$

In (43), by L_+, L_0 and L_- we have denoted links which have the configuration of b_i, b and b_i^{-1}, at an intersection. Similarly, L_{2+}, L_+, L_0 and L_- in (44) and L_{3+}, L_{2+}, L_+, L and L_- in (45) should be understood. For general N, the skein relation is of N-t degree relating links $L_{(N-1)+}, \cdots, L_0$ and L_{-1}.

4.2 Graph state IRF model. We can construct solvablle IRF models correspondin to arbitrary graphs in any dimensions [48,17]. We may express the constraint of th model by a graph. In the graph each point represents the spin state. When a spin c i admissible to d then the point for c is connected by a line to the point for d. For AD type graphs the models are called ADE models [49]. There also exist solvable mode with elliptic parameterization for extended Dynkin diagrams [48,49,50].

Let us construct the graph state IRF models [17]. We solve the eigenvalue equatio for the graph

$$\sum_{b \sim a} \psi(b) = \Lambda \psi(a), \tag{46}$$

where the summation is over all spin state b admissible to a. Using the eigenvectors $\{\psi(a)\}$, we construct the Temperley-Lieb operator

$$[E_i]_{k_1\cdots k_n}^{p_1\cdots p_n} = \prod_{j=0}^{i-1} \delta_{k_i-1}^{k_i+1} \frac{\psi(p_i)\psi(k_i)}{\psi(p_{i-1}} \prod_{j=i+1}^{n} \delta_{k_j}^{p_j}. \tag{47}$$

And then we have the Yang-Baxter operator

$$X_i(u) = \frac{\sinh(\lambda-u)}{\sinh(\lambda)}\left(I + \frac{\sinh(u)}{\sinh(\lambda-u)}E_i\right). \tag{48}$$

It is essential that the eigenvectors $\{\psi(a)\}$ are the crossing multipliers of the model. We obtain braid operator by taking the limit $u \to \infty$ and the Markov trace on the braid group representation by using the crossing multipliers. The link polynomial satisfies the skein relation of second degree.

We can consider vertex models corresponding to the graph state IRF models under the Wu-Kadanoff-Wegner transformation and the base-point-infinity limit [24]. From these vertex and IRF models we have multi-variable braid matrices [24].

4.3 ABCD IRF models. The IRF model corresponding to affine Lie algebra $A_{m-1}^{(1)}(B_m^{(1)}, C_m^{(1)}, D_m^{(1)})$ is called $A_{m-1}^{(1)}(B_m^{(1)}, C_m^{(1)}, D_m^{(1)})$ model [51]. The crossing parameter λ and the sign factor σ are summarized as

$$\lambda = m\omega/2 \qquad \sigma = 1 \qquad \text{for } A_{m-1}^{(1)}, \tag{49}$$

$$\lambda = (2m-1)\omega/2 \qquad \sigma = 1 \qquad \text{for } B_m^{(1)}, \tag{50}$$

$$\lambda = (m+1)\omega \qquad \sigma = -1 \qquad \text{for } C_m^{(1)}, \tag{51}$$

$$\lambda = (m-1)\omega \qquad \sigma = 1 \qquad \text{for } D_m^{(1)}, \tag{52}$$

where ω is a parameter. The reduction relations are [19,21,22]

$$(G_i - 1)(G_i + \gamma^2) = 0 \text{ for } A_{m-1}^{(1)}, \tag{53}$$

$$(G_i - 1)(G_i - \beta)(G_i + \gamma^2) = 0 \text{ for } B_{m-1}^{(1)}, C_m^{(1)}, D_m^{(1)}, \tag{54}$$

with

$$\gamma = e^{-i\omega} \text{for } A_{m-1}^{(1)}, B_m^{(1)}, C_m^{(1)} \text{ and } D_m^{(1)} \tag{55}$$

$$\beta = \sigma e^{-i[2\lambda+\omega(1+\sigma)]} \text{for } B_m^{(1)}, C_m^{(1)} \text{ and } D_m^{(1)}. \tag{56}$$

The extended Markov property is proved and the characteristic functions are calculated as

$$H(u) = \frac{\sin m\omega - u}{\sin\omega - u} \text{ for } A_{m-1}^{(1)} \tag{57}$$

$$H(u) = \frac{\sigma \sin 2\lambda - u}{\sin\omega - u} \text{ for } B_m^{(1)} C_m^{(1)} \text{ and } D_m^{(1)}. \tag{58}$$

(The explicit forms of the crossing multipliers are given in [19].) Using the reduction relations and the Markov traces, we obtain the (generalized) skein relations:

$$\alpha(L_+) = (1-t)t^{(m-1)/2}\alpha(L_0) + t^m\alpha(L_-), \qquad \text{for } A^{(1)}_{m-1}, \tag{59}$$

$$\begin{aligned}\alpha(L_{2+}) =&(1-t+\beta)e^{-i(2\lambda+\omega(\sigma-1))}\cdot\alpha(L_+)\\ &+(t+\beta(t-1))e^{-2i(2\lambda+\omega(\sigma-1))}\cdot\alpha(L_0)\\ &-t\beta e^{-3i(2\lambda+\omega(\sigma-1))}\cdot\alpha(L_-), \qquad \text{for } B^{(1)}_m, C^{(1)}_m \text{ and } D^{(1)}_m\end{aligned} \tag{60}$$

where

$$t = e^{-2i\omega}. \tag{61}$$

For $A^{(1)}_{m-1}$ model, the Alexander polynomial is obtained by the limit $m \to 0$, while $m = 2$ corresponds to the Jones polynomial.

Link polynomials thus obtained are one-variable invariants for each fixed m. It is noted that m is independent of t. We now have two variables t and m. The link polynomial constructed from $A^{(1)}_{m-1}$ model corresponds to the two-variable extension [10] of the Jones polynomial. The link polynomial from $A^{(1)}_{m-1}$ models corresponds to the Kauffman polynomial [12]. We thus have explicit realizations of the Kauffman polynomial and the two-variable extension of the Jones polynomial (HOMFLY polynomial). The braid matrices given by Turaev [52,53] correspond to the vertex-model analog of the braid matrices constructed from $A^{(1)}_{m-1}, B^{(1)}_m, C^{(1)}_m, D^{(1)}_m$ IRF models. From the IRF models we can construct braid matrices and the Markov trace for the vertex model by the Wu-Kadanoff-Wegner transformation and the base-point-infinity limit [17]. For example, from A-type IRF models we derive the multi-state vertex models [54] related to $SU(n)$. The Markov traces [19] for the IRF models lead to those [52,53] for the vertex models.

5. Vertex models with graded symmetry and link polynomials

5.1 Vertex models associated with $gl(M|N)$. We shall explain construction of link polynomials from vertex models with grade symmetry. [25,26] We consider a family of solvable vertex models obtained by C.L.Schultz [55] through a direct calculation. We find that these models are associated with Lie superalgebras $gl(M|N)$ [26]. The graded Yang-Baxter relation was introduced by Kulish and Sklyanin [56]. We can show that rational solutions for the above models are equivalent to solutions of the graded Yang-Baxter relation associated with $gl(M|N)$ [56].

Let us introduce a set of signs $\{\epsilon_i\}$

$$\epsilon_i = 1 \text{ or } -1, \text{ for } i = 1, \ldots, M+N. \tag{62}$$

The sign ϵ_i represents the "parity" of the edge state i. We also introduce "grade" $p(i) \in \{0,1\}$ of the edge state i as $\epsilon_i = (-1)^{p(i)}$. The number of positive (resp. negative) signs is given by M (resp. N). Thus we have defined the graded symmetry. For any set of signs $\{\epsilon_i\}$ we have a solution of the Yang-Baxter relation. The Boltzmann weights are given as follows:

$$\begin{aligned} w(a,a,a,a;u) &= \sinh\eta - \epsilon_a u/\sinh\eta, \\ w(a,b,b,a;u) &= \begin{cases} \exp(-u), & \text{for } a<b, \\ \exp(u), & \text{for } a>b, \end{cases} \\ w(a,b,a,b;u) &= \pm\sinh u/\sinh\eta, \text{ for } a\neq b. \end{aligned} \tag{63}$$

Here η is a parameter and the edge variables a and b take values $1,2,\cdots,M+N$. The models have the charge conservation property: $w(a,b,c,d;u)=0$, unless $a+b=c+d$.

Non-zero elements of the braid matrices are given in the following [25,26]:

$$\begin{aligned} G^{aa}_{aa}(+) &= \begin{cases} 1 & \text{for} \quad \epsilon_a = 1, \\ -t & \text{for} \quad \epsilon_a = -1, \end{cases} \\ G^{ab}_{ab}(+) &= \begin{cases} 0 & \text{for} \quad a<b, \\ 1-t & \text{for} \quad a>b, \end{cases} \\ G^{ab}_{ba}(+) &= \mp t^{1/2} \qquad for \quad a\neq b. \end{aligned} \tag{64}$$

Here a variable t is defined by $t=\exp(2\eta)$. We obtain 2^{M+N} different representations depending on the choice of the signs $\{\epsilon_a\}$ [26]. Note that by replacing t with t^{-1} and multiplying the braid matrix by $-t$, we have an equivalent representation.

Each representation has only two eigenvalues 1 and $-t$. The braid matrices satisfy the Hecke algebra relations. Thus we have seen that the Hecke algebra also appears in the braid matrices associated with the Lie superalgebra $gl(M|N)$ [25,26].

By taking the limit $\eta\to 0$ we get the graded permutation operator from the representation of the braid group (64). In this sense, the braid operator is q-analogue of the graded permutation operator [26].

.2 Link polynomials. Through the general theory we construct the Markov trace on the representations derived in the previous section. For any grading $\{\epsilon_i\}$ the Markov trace is given by

$$\begin{aligned} \phi(A) &= \frac{Tr(H(n)A)}{Tr(H(n))}, \qquad A\in B_n, \\ [H(n)]^{a_1a_2\cdots a_n}_{b_1b_2\cdots b_n} &= \prod_{j=1}^{n} h(j)\delta^a_b, \end{aligned} \tag{65}$$

Here the diagonal matrix h is

$$h(j)=\epsilon_j \exp\{\eta(\sum_{k=1}^{j-1} 2\epsilon_k+\epsilon_j-M+N)\}, \text{ for } j=1,\cdots,M+N. \tag{66}$$

In the limit $\eta\to 0$, the trace with matrix h reduces to the supertrace $str(A)=\sum_i \epsilon_i A_{ii}$ [26]. We can prove the extended Markov property [17,19,21,22],

$$\sum_b X^{ab}_{ab}(u)h(b) = H(u;\eta)\rho(u) \text{ (independent of } a), \tag{67}$$

where characteristic function $H(u;\eta)$ is given by

$$H(u;\eta) = \frac{\sinh(M-N)\eta - u}{\sinh\eta - u}, \tag{68}$$

This is a generalization of the characteristic function for the $A_{M-1}(sl(M))$ model [19,21,22].

It has been pointed out in [26] that $Tr(H(n))$ in (65) becomes zero when $M = n$ and that an alternative form of the Markov trace can be used for calculations.

The link polynomial obtained from the vertex model associated with $gl(M|N)$ has the skein relation:

$$\alpha(L_+) = t^{p/2}(1-t)\alpha(L_0) + t^{p+1}\alpha(L_-). \tag{69}$$

where

$$p = M - N - 1. \tag{70}$$

Since the skein relation is of second degree, the link polynomial is calculable only by the relation. We now have a hierarchy of link polynomials which depends on the number $p = M-N-1$. It is interesting that as far as p is common we have the same link polynomia [25,26]. To repeat, from different models related to $gl(M|N)$ with $p = M-N-1$ we obtain the same link polynomial. Note that the hierarchy includes the case $p = 0$ where $\bar{\tau}/\tau = 1$

The HOMFLY polynomial [10] is characterized by the second degree skein relation:

$$\alpha(L_+) = \omega^{1/2}(1-t)\alpha(L_0) + \omega t\alpha(L_-). \tag{71}$$

Here t and ω are independent (continuous) variables. We see that the link polynomial constructed from the $gl(M|N)$ type vertex models correspond to the cases $\omega = t^p$, $p \in$ of the HOMFLY polynomial. Based on the Markov traces we thus obtain a hierarch of link polynomials corresponding to the HOMFLY polynomial [25,26].

The link for $p = -1$ is the Alexander polynomial [8]. The case $p = 1$ corresponds t the Jones polynomial [9]. Therefore we have a number of braid matrices with differen sizes which lead to the Alexander polynomial and the Jones polynomial [25,26].

6. Concluding Remarks

We have shown that various link polynomials are systematically constructed fro exactly solvable models. The Yang-Baxter relation, which is a sufficient condition fc the solvability of the models, plays a central role in the theory. Due to the limited space we have omitted the following discussions.

(1) The existence and properties of the link polynomials [13] constructed from th N-state vertex model [47] can be proved also by the construction of compo ite models (fusion method) in terms of the Temperley-Lieb algebra and th graphical formulation derived from the crossing symmetry [18]. Note that th combination of the crossing symmetry and the Temperley-Lieb algebra chara terized the link polynomials.

(2) We can construct two-variable link invariants [15,16,21,22]. Those invarian may be regarded as two-variable extension of the link polynomials constructe

from A type composite vertex and IRF models. In the papers [15,16] an algorithm for calculation of the two-variable link invariants for any links has been established, and some examples have been given.

(3) For any combinations of braid matrices which have the Markov traces we can construct multivariable link polynomials with higher skein relations [27]. If there are braid matrices with the Markov traces, we obtain a composite (hybrid-type) braid matrix and a composite Markov trace from them, and therefore a link polynomial [27]. Thus we have a variety of multivariable link polynomials.

Acknowledgements. One of the author (M.W.) thanks L.D.Faddeev, P.P.Kulish, E.K.Sklyanin and A.G.Izergin for comments of this work and hospitality during his stay in the Euler International Mathematical Institute, Leningrad. He also thanks H.Morton,K.Millet and L.H.Kauffman for keen interest in our work.

References

[1] C.N.Yang, Phys.Rev.Lett. **19** (1967), 1312.
[2] R.J.Baxter, Ann.of Phys. **70** (1972), 323.
R.J.Baxter, *Exactly Solved Models in Statistical Mechanics*, Academic Press, 1982.
[3] L.D.Faddeev, Sov.Sci.Rev. **C1** (1980), 107.
[4] P.P.Kulish and E.K.Sklyanin, Lect. Notes Phys. **151** (1982), 61.
[5] H.B.Thacker, Rev.Mod.Phys. **53** (1981), 253.
[6] M.Wadati, Dynamical Problems in Soliton Systems (S.Takeno, eds.), Springer, Berlin, 1985, p. 68.
[7] L.A.Takhtadzhan and L.D.Faddeev, Russian Math.Surveys **34** (1979), 11.
[8] J.W.Alexander, Trans.Amer.Math.Soc. **30** (1928), 275.
[9] V.F.R.Jones, Bull.Amer.Math.Soc. **12** (1985), 103.
[10] P.Freyd,D.Yetter,J.Hoste,W.B.R.Liskorish,K.Millett and A.Ocneanu, Bull.Amer.Math.Soc. **12** (1985), 239.
[11] J.H.Przytycki and K.P.Traczyk, Kobe Univer. Math. J. **4** (1987), 115.
[12] L.H.Kauffman, *On Knots*, Princeton University Press, 1987; Trans.Amer.Math.Soc. **318** (1990), 417.
[13] Y.Akutsu and M.Wadati, J.Phys.Soc.Jpn. **56** (1987), 839; 3039.
[14] Y.Akutsu, T.Deguchi and M.Wadati, J.Phys.Soc.Jpn. **56** (1987), 3464.
[15] Y.Akutsu and M.Wadati, Commun.Math.Phys. **117** (1988), 243.
[16] T.Deguchi,Y.Akutsu and M.Wadati, J.Phys.Soc.Jpn. **57** (1988), 757.
[17] Y.Akutsu, T.Deguchi and M.Wadati, J.Phys.Soc.Jpn. **57** (1988), 1173.
[18] T.Deguchi,M.Wadati and Y.Akutsu, J.Phys.Soc.Jpn. **57** (1988), 1905.
[19] T.Deguchi,M.Wadati and Y.Akutsu, J.Phys.Soc.Jpn. **57** (1988), 2921.
[20] M.Wadati and Y.Akutsu, Progr.Theor.Phys.Suppl. **94** (1988), 1.
[21] Y.Akutsu, T.Deguchi and M.Wadati, Braid Group, Knot Theory and Statistical Mechanics (C.N.Yang and M.L.Ge, eds.), World Scientific, 1989, p. 151.
[22] M.Wadati,T.Deguchi and Y.Akutsu, Phys.Reports **180** (1989), 427.
T.Deguchi,M.Wadati and Y.Akutsu, Adv.Stud. in Pure Math. **19** (1989), 193.
M.Wadati,Y.Akutsu and T.Deguchi, *Link Polynomials and Exactly Solvable Models*, Lecture Notes Math., Springer-Verlag, 1990, p. 111.
Y.Akutsu, *Link Polynomials, Linking Numbers and Exactly Solvable Models*, KEK Report, vol. 89-22, 1990, p. 45.
[23] M.Wadati,T.Deguchi and Y.Akutsu, Nonlinear Evolution Equations,Integrability and Spectral Methods(A.Fordy, ed.), Manchester University Press, 1991.
[24] T.Deguchi, Int.J.Mod.Phys. **A5** (1990), 2195.
[25] T.Deguchi, J.Phys.Soc.Jpn. **58** (1990), 3441.
[26] T.Deguchi and Y.Akutsu, J.Phys.A:Math.Gen. **23** (1990), 1861.
[27] T.Deguchi, J.Phys.Soc.Jpn. **59** (1990), 1119.
[28] Y.S.Wu, Phys.Rev.Lett. **52** (1984), 2103.

[29] G.W.Semenoff, Phys.Rev.Lett. **61** (1988), 517.
[30] A.M.Polyakov, Mod.Phys.Lett. **3A** (1988), 325.
[31] C.Rovelli and L.Smolin, Phys.Rev.Lett. **61** (1988), 1155.
[32] E.Witten, Commun.Math.Phys. **121** (1989), 351.
[33] A.Kuniba,Y.Akutsu and M.Wadati, J.Phys.Soc.Jpn. **55** (1986), 3285.
[34] A.Tsuchiya and Y.Kanie, Adv.Stud. in Pure Math **16** (1988), 297.
[35] T.Kohno, Ann.Inst.Fourier **37** (1987), 139.
[36] E.Verlinde, Nucl.Phys. **B300[FS22]** (1988), 360.
[37] G.Moore and N.Seiberg, Phys.Lett. **B212** (1988), 451; Commun.Math.Phys. **123** (1989), 177.
[38] J.Fröhlich,, *Statistics of fields, the Yang-Baxter equation*, Nonpertubative Quantum Field Theory (G.'t Hooft et.al., eds.), Plenum Pub., 1988, p. 71.
G.Felder,J.Fröhlich and G.Keller, Commun.Math.Phys. **124** (1989), 417,647.
[39] K.H.Rehren and B.Schroer, Nucl.Phys. **B312** (1989), 715.
[40] H.C.Lee,M.-L.Ge,M.Couture and Y.S.Wu, Int.J.Mod.Phys. **A,4** (1989), 2333.
[41] M.Wadati,Y.Yamada and T.Deguchi, J.Phys.Soc.Jpn. **58** (1989), 1153.
[42] A.B.Zamolodchikov and A.B.Zamolodchikov, Ann.of Phys. **120** (1979), 253.
M.Karowsi,H.J.Thun,T.T.Truong and P.H.Weisz, Phys.Lett. **67B** (1977), 321.
K.Sogo,M.Uchimani,A.Nakamura and M.Wadati, Prog.Theor.Phys. **66** (1981), 1284.
[43] J.S.Birman, *Braids, Links and Mapping Class Groups*, Princeton University Press, 1974.
[44] H.N.V.Temperley and E.H.Lieb, Proc.Roy.Soc. **A322** (1971), 251.
[45] L.H.Kauffman, Contemp.Math. **78** (1988), 263; ibid **96** (1989), 221.
[46] A.B.Zamolodchikov and V.A.Fateev, Sov.J.Nucl.Phys. **32** (1980), 293.
[47] K.Sogo,Y.Akutsu and T.Abe, Prog.Theor.Phys. **70** (1983), 730,739.
[48] Y.Akutsu,A.Kuniba and M.Wadati, J.Phys.Soc.Jpn. **55** (1986), 1486.
[49] V.Pasquier, J.Phys.A:Math.Gen. **20** (1987), L217,221.
[50] A.Kuniba and T.Yajima, J.Phys.A:Math.Gen. **21** (1988), 519; J.Stat.Phys. **50 no. 3/4** (1988), 829.
[51] M.Jimbo,T.Miwa and M.Okado, Commun.Math.Phys. **116** (1988), 353.
[52] V.G.Turaev, Invent.Math. **92** (1988), 527.
[53] N.Yu.Reshetikhin, *LOMI preprints E-4-87,E-17-87*, Leningrad, 1988.
[54] I.V.Cherednik, Theor.Math.Phys. **43** (1980), 356.
O.Babelon,H.J.deVega and C.M.Viallet, Nucl.Phys. **B190** (1981), 542.
J.H.H.Perk and C.L.Schultz, Phys.Lett. **84A** (1981), 407.
[55] C.L.Schultz, Phys.Rev.Lett. **46** (1981), 629.
[56] P.P.Kulish and E.K.Sklyanin, J.Sov.Math. **19** (1981), 1596.

Figure caption

Fig.1(a) vertex configuration (scattering process)$\{i,j,k,l\}$,
(b) IRF configuration a, b, c, d.
Fig.2 Crossing symmetry.
Fig.3 Operations b_i and b_i^{-1}.
Fig.4 Scattering with $u = \lambda$ correspond to annihilation-creation process.
Fig.5 Elements of link diagram. Left (right) four graphs are for vertex (IRF) model
(1) pair-annihilation diagram: $r(c)\delta(c,\bar{d}); (\psi(a)/\psi(b))^{1/2}$.
(2) pair-creation diagram: $r(a)\delta(a,\bar{b}); (\psi(a)/\psi(b))^{1/2}$.
(3) braid diagram with $\epsilon = -1 : G^{ab}_{cd}(+); G(a,b,c,d;+)$.
(4) braid diagram with $\epsilon = 1 : G^{ab}_{cd}(+); G(a,b,c,d;-)$.
Fig.6 Sign $\epsilon(C)$ for a crossing.
Fig.7 Reidemeister moves.
Fig.8 Equivalence of the Markov trace (left) and the graphical calculation (right). Black circles denote crossing multipliers.

Fig.1

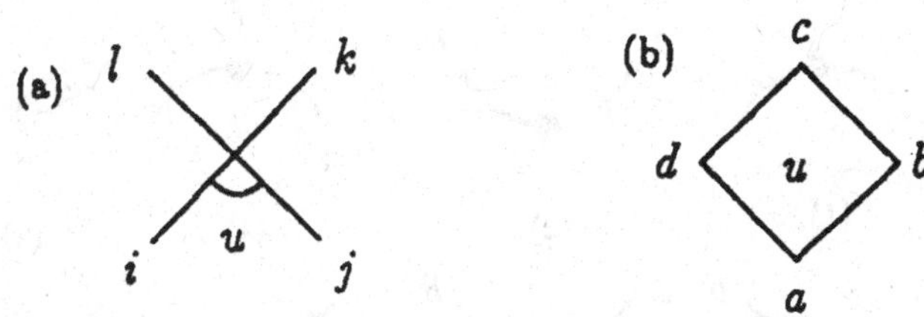

Fig.2

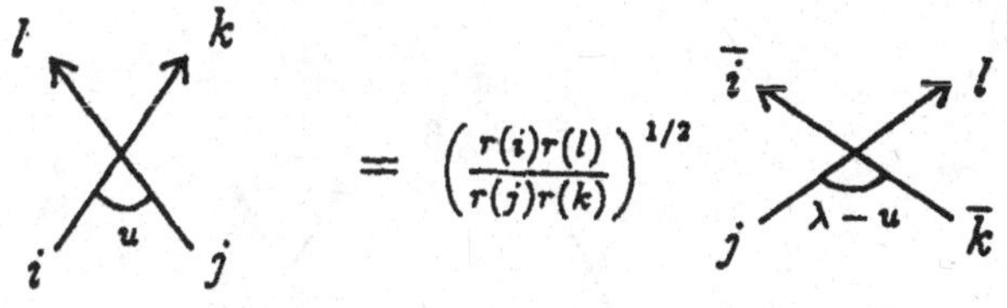

$$l \quad k \qquad\qquad \bar{i} \quad l$$

$$u \quad = \left(\frac{r(i)r(l)}{r(j)r(k)}\right)^{1/2} \quad \lambda - u$$

$$i \quad j \qquad\qquad j \quad \bar{k}$$

Fig.3

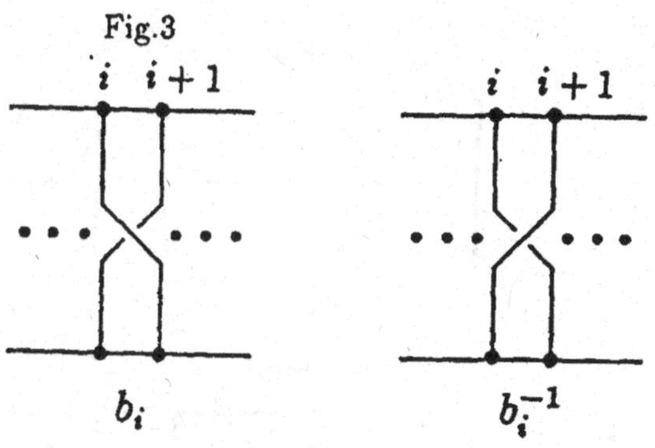

Fig.4

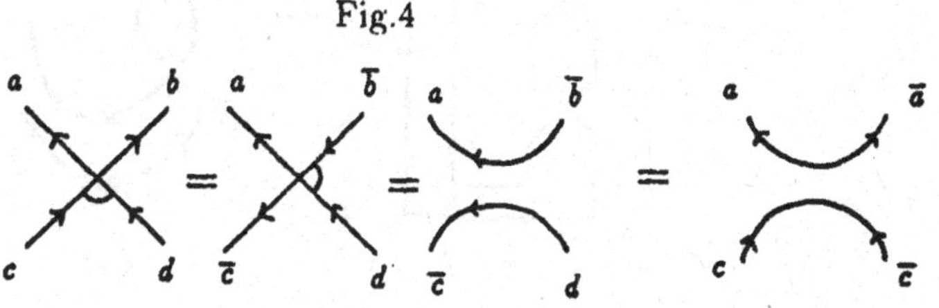

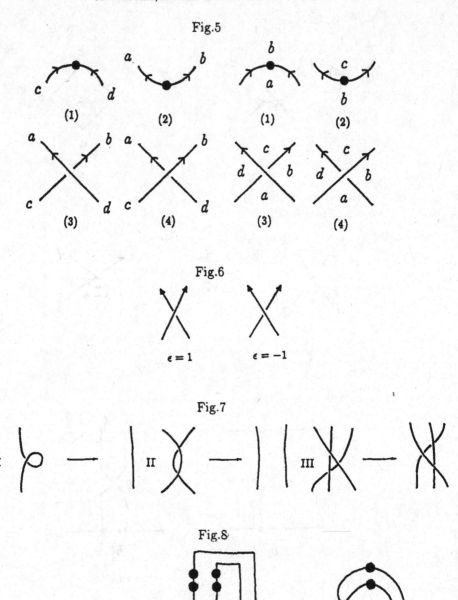

Fig.5

Fig.6

$\epsilon = 1$ $\epsilon = -1$

Fig.7

Fig.8

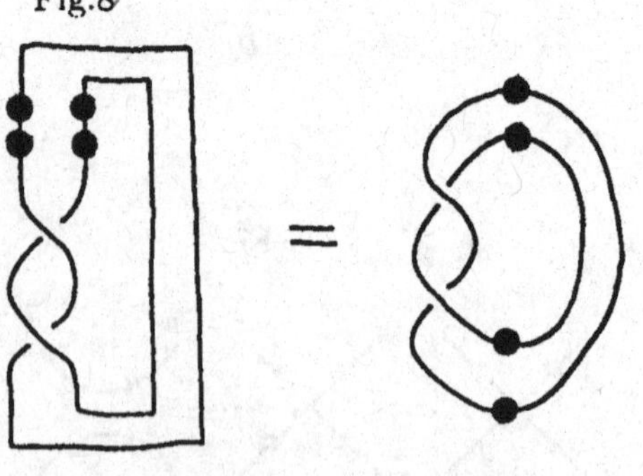

DEPARTMENT OF PHYSICS, FACULTY OF SCIENCE, UNIVERSITY OF TOKYO, HONGO 7-3-1, BUNKYO-KU, TOKYO 113, JAPAN

TRIANGULARITY OF TRANSITION MATRICES FOR GENERALIZED HALL–LITTLEWOOD POLYNOMIALS

S.V.KEROV

We shall mainly use notations from [1]. Let $\mathcal{A} = \oplus\mathcal{A}_n$ be the graded $\mathbb{C}$-algebra of symmetric polynomials in infinite number of variables $x = (x_1, x_2, \dots)$ and let $\mathcal{P}_n$ be the set of all partitions $\lambda = (\lambda_1, \dots, \lambda_r)$ of $n = \lambda_1 + \dots + \lambda_r$. The monomial symmetric functions $m_\lambda(x) = x_1^{\lambda_1} x_2^{\lambda_2} \dots +$ symmetric terms, $\lambda \in \mathcal{P}_n$, form basis of $\mathcal{A}_n$ as well as the power sum functions $p_\lambda = p_{\lambda_1} p_{\lambda_2} \dots$, $p_n = x_1^n + x_2^n + \dots$.Denote by $\ll$ the row lexicographic ordering and by $\leq$ the dominance (natural) partial order on $\mathcal{P}_n$. For both orderings (1^n) is the first and (n) is the last partition.

Fix a sequence $w = (w_1, w_2, \dots) \in \mathbb{C}^\infty$ and let $w_\lambda = w_1^{\lambda_1} w_2^{\lambda_2} \dots$ and $z_\lambda = 1^{r_1} r_1! 2^{r_2} r_2! \dots$ where $\lambda = (1^{r_1}, 2^{r_2}, \dots)$. Consider the bilinear form

$$(p_\rho, p_\sigma)_w = \delta_{\rho\sigma} z_\rho w_\rho \tag{1}$$

on $\mathcal{A}$. We introduce generalized Hall-Littlewood polynomials $P_\lambda = P_\lambda(x; w)$ by means of orthogonalization process applied to of the basis $\{m_\lambda\}$ with respect to the form $(.,.)_w$ and lexicographic ordering:

$$P_\lambda(x, w) = m_\lambda(x) + \sum_{\mu < \lambda} u_{\lambda\mu}(w) m_\mu(x) \,, \tag{2}$$

$$(P_\lambda, P_\mu)_w = 0 \quad \text{for } \lambda \neq \mu \tag{3}$$

Ordinary Hall-Littlewood polynomials correspond to $w_n = (1 - t^n)^{-1}$, $n = 1, 2, \dots$. In particular, for $t = 0$ we get the Schur functions $s_\lambda(x) = P_\lambda(x; 1, 1, \dots)$.

In [2] Macdonald has proved that for

$$w_n = (1 - q^n)/(1 - t^n), \quad n = 1, 2, \dots \tag{4}$$

the transition coefficients $u_{\lambda\mu}(w)$ in (2) are unitriangular with respect to the dominance order (not only lexicographical order):

$$u_{\lambda\mu}(w) = 0 \quad \text{unless} \quad \mu \leq \lambda. \tag{5}$$

Let us choose $\theta \in \mathbb{R}^*$ and $m = 1, 2, \dots$ and let $r = e^{2\pi i/m}$ be the m-th root of unity. Considering the limit $q, t \to r$ while $q - r = (t - r) \cdot \theta$, one can check that (5) is also valid in cases

$$w_n = \theta \quad \text{for } n \equiv 0 \bmod m, \quad w_n = 1 \text{ otherwise.} \tag{6}$$

In the cases (4), (6) GHL-polynomials P_λ can be interpreted as spherical functions for classical symmetric spaces over real or p-adic fields [3] and for quantum q-analogs thereof [4]. In [5,6] these two cases were described axiomatically. It is shown there that (up to an evident normalization $w_n \to w_n \cdot a^n$, $a \in \mathbf{C}^*$) solutions (4), (6) are characterized by the equations

$$(f_k - f_{n+1})(f_{n-k+1} - f_{n+1}) = (f_1 - f_{n+1})(f_n - f_{n+1}) \tag{7}$$

for all $1 \leq k \leq n < \infty$. Here $f_n = b_n/b_{n-1}$ and $b_n = \sum 1/z_\rho w_\rho$, where the sum ranges over all partitions ρ of n.

Problem. *For what sequences $w = (w_1, w_2, \dots)$ do the GHL-polynomials $P_\lambda(x; w)$ satisfy triangularity condition (5)? Are there any examples besides (4) and (6)?*

References

[1] I.G.Macdonald, *Symmetric functions and Hall polynomials*, Oxford, 1979.
[2] I.G.Macdonald, *Symmetric functions (2)*, Preprint.
[3] I.G.Macdonald, *Orthogonal polynomials, associated with root systems*, Preprint.
[4] T.H.Koornwinder, *Orthogonal polynomials in connection with quantum groups*, Orthogonal polynomials: Theory and Practice (ed. by Neval), 257–292.
[5] S.V.Kerov, *Hall-Littlewood functions and orthogonal polynomials*, Funkz. Anal. Pril. 25 no. 1 (1991). (in Russian)
[6] S.V.Kerov, *Generalized Hall-Littlewood symmetric functions and orthogonal polynomials*, Adv. Sov Math. (1991) (to appear).

AN OPEN PROBLEM IN QUANTUM GROUPS

SHAHN MAJID

An elementary problem concerns what could be called the Yang Baxter variety, YB_n, where n is a natural number. This is the cubic variety in $\mathbb{C}P^{n^4-1}$ defined by the equation $R_{12}R_{13}R_{23} = R_{23}R_{13}R_{12}$ for $R \in M_n(\mathbb{C}) \otimes M_n(\mathbb{C})$($R$'s of different scale considered equal). We are interested in R invertible. I ask: what does YB_n look like?

In other words I want to take a geometric view of the moduli space of solutions of the quantum Yang-Baxter equation (QYBE). What is its geometric structure? We can of course study this from an algebraic-geometric point of view, and maybe this is the best. For the moment though, I want to take an even more naive point of view, namely just as some subspace in $\mathbb{C}P^{n^4-1}$ or even, working with the representatives, in $\mathbb{C}P^{n^4-1}$. We already know a bit about its structure. It is singular at the identity $R = 1$, where the linearization of the QYBE gives the Classical Yang-Baxter equations (CYBE). A number of lines or "conical singularities" come out from this, indexed, roughly speaking, by Dynkin diagrams (the q-deformations of the identity that give rise to the quantum analogs of the classical Lie groups). On the other hand there are definitely other non-standard points on the variety, for example the solutions connected with the Alexander-Conway knot polynomial[6][5]. This leads to a non-standard quantum group computed in [4].

Indeed, it is well-known that every point of the Yang-Baxter variety gives rise to a bialgebra $A(R)$ as defined by Faddeev, Reshetikhin and Takhtajan in [2]. In fact, this $A(R)$ always has a quotient, $\breve{A}(R)$, which is a Hopf algebra (in a certain weak sense) and the dual of quasitriangular Hopf algebra $\breve{U}(R)$ (again in a certain weak sense)[7]. Of course, for the standard R-matrices we just get the results of [2]: the point about [7] is that we really get a "bundle" of quantum groups (with quasitriangular structures) defined in a uniform way over the whole variety.

So it is interesting to know the structure of YB_n, and we propose here a geometrical view of it.. For example, is it path connected? In some recent work[3] it is shown that it is NOT. It is found that the point $R = \tau$ is isolated. That is, if $R = \tau + \hbar r$ solves the QYBE to lower order in $\hbar$ then $R = \lambda\tau$ for some λ, i.e. the same point in YB_n. The rigidity of the point $R = \tau$ is also partially understood from the point of view of the deformation theory of Gerstenhaber and Schack applied to this "bundle"[3]. This is because $A(\tau) = TM_n$ (the tensor algebra on M_n with matrix coproduct) has no non-trivial deformations[3]. Hence, if a one parameter family R_λ through τ in YB_n did exist then we could consider the bialgebras $A(R_\lambda)$ and by the deformation theory this would have to be isomorphic to $A(\tau)$. Further details are in [3].

On the other hand there are plenty of solutions around the twist map, namely those of[1] of the form $R = \tau = aB^{-1t} \otimes B$ for a non-degenerate bilinear form B_{ij}. Here $a + a^{-1} + TrB^{-1t}B = 0$. At least for $n = 2$, these solutions *are* path connected to the identity[1]. So you see that YB_n has a rich structure. What is the structure then of YB_n? Even for $n = 2$? I propose here that this can be attacked quite naively as simply solving a non-linear differential equation, perhaps along the following lines.

Step 1. Understand the linearized problem (the R-CYBE) at any point R of YB_n. Preferably this should be done as uniformly as possible in terms of arbitrary R. The solution space of this defines the tangent space $T_R YB_n$. A conjecture for this is that the R-CYBE could be expressed uniformly in terms of R-Lie algebras, i.e. Lie algebras in the (braided) category generated by R (or comodules of $A(R)$). I don't know how these should be defined exactly when R is not triangular, but see my talk on braided groups.

Step 2. Knowing the tangent space, find some natural vector fields $v = v(R)$, again uniformly for generic R (i.e. defined for general R without reference to a specific R).

Step 3. Exponentiate such v to generate flows in YB_n, i.e. solve $\frac{d}{dt}R(t) = v(R(t))$.

In this way, one can hope to map out the structure of YB_n. In practice, of course we may not be able to solve Steps 1 or 2 generically but instead iterate Steps 1 and 2 alternatively. Thus, pick an R_0, compute $T_{R_0} YB_n$, pick $v(R_0)$ in this. Move to $R_1 = R_0 + \hbar v(R_0)$ (obeys QYBE to some accuracy for $\hbar$ sufficiently small), compute $T_{R_1} YB_n$ etc. In this way we might also attempt to generate paths in YB_n.

References

[1] M.Dubois-Violette and G.Launer, *The quantum group of a non-degenerate bilinear form*, Phys Lett.B. (1990).

[2] L.D.Faddeev,N.Yu.Reshetikhin and L.A.Takhtajan, *Quantization of Lie groups and Lie algebras* Algebraic Analysis,Vol.I (M.Kashivara and T.Kawai, eds.), Academic Press, 1988, pp. 129–139.

[3] M.Gerstehaber,S.Majid and S.D.Schack, *In preparation.*

[4] M.Jing,M-L.Ge and Y-S.Wu, *A new quantum group associated with "non-standard" braid group representation*, *Preprint*, 1990.

[5] L.Kauffman, *Knots,abstract tensors and the Yang-Baxter equations*, Knots,Topology and Quantum Field Theory, World Scientific, 1989.

[6] H.C.Lee,M.Couture and N.C.Schmeig, *Connected knot polynomials*, *Preprint*, 1988.

[7] S.Majid, *More examples of bicrossproduct and double cross product Hopf algebras*, Isr.J.Math. **72** (1990), 133–148.

TWO PROBLEMS IN QUANTIZED ALGEBRAS OF FUNCTIONS

YA.S.SOIBELMAN

1. Let $C(K)_q$ be the algebra of functions which are continuous on compact quantum group K (see [1] for definition). It is well known from [1] that there is an imbedding of $C(K)_q$ into the algebra of continuous operator-functions on the maximal torus $T \subset K$. Describe the image. The answer is known for $K = SU(2)$ (see [2]). There is such a description for odd-dimensional quantum spheres (see [3]).

2. Let g be a finite dimensional complex Lie algebra, $C[g^*]$ be an algebra of polynomial functions on dual space g^*. Let us equip $C[g^*]$ with Poisson brackets by using the Lie-Kirillov formulas. Quantization of this Hopf-Poisson algebra is known: it is the universal enveloping algebra $U(g)$. Therefore the usual method of orbit (Konstant-Kirillov) gives rise to the relation between the representation theory of the algebra of functions on the quantum group g^* and symplectic leaves in the Poisson-Lie group g^* (these are coadjoint orbits in our case).

Problem. *How to generalize this to the case of more general quantum groups?*

I mean the generalization of the method of orbits. The correspondence between representations and leaves was investigated in [1,2] for compact quantum groups.

References

[1] Ya.Soibelman, *Algebra of functions on compact quantum group and its representations.*, Algebra Anal **2** no. 1 (1990), 190–212. (in Russian)

[2] L.Vaksman, Ya.Soibelman, *Algebra of functions on quantum group SU(2)*, Funkz. Anal. Pril. **22** no. 3 (1988), 1–14. (in Russian)

[3] L.Vaksman, Ya.Soibelman, *Algebra of functions on quantum group SU(n+1) and odd-dimensional quantum spheres*, Algebra Anal. **2** no. 5 (1990), 101–120. (in Russian)

UNSOLVED PROBLEMS

EARL J. TAFT

Let A be the Hopf algebra $k[x]$, k a field, with x primitive. The Hopf algebra dual A^0 is the space of linearly recursive sequences (B.Peterson and E.J.Taft, Aequationes Math. 20 (1980), 1-17). We gave an algorithm for diagonalizing a linearly recursive sequence in terms of the finite-dimensional subcoalgebra it generates. The product is that of divided-power series. Now let $W_1 = \operatorname{Der} k[x]$, the Witt algebra. (Take characteristic $k \neq 2$). The Lie algebra W_1 has basis $e_i = x^{i+1}\frac{d}{dx}$ for $i \geq -1$, with $[e_i, e_j] = (j-i)e_{i+j}$. W.Michaelis has recently shown that W_1 is a Lie bialgebra, with Lie coalgebra structure given by $\delta(e_n) = n(e_n \wedge e_{-1}) + (n+1)(e_0 \wedge e_{n-1})$, where $a \wedge b = a \otimes b - b \otimes a$ in $W_1 \otimes W_1$. This seems to be related to the fact that $e_0 \wedge e_{-1}$ satisfies the classical Yang-Baxter property in $W_1 \otimes W_1 \subseteq U(W_1) \otimes U(W_1)$. W_1 is a locally finite Lie coalgebra. Now W_1^0, the Lie coalgebra dual to the Lie algebra W_1, is also a Lie subalgebra of the convolution Lie algebra W_1^*, and W_1^0 is a Lie bialgebra. W.Nichols has recently shown that W_1^0 is the space of linearly recursive sequences.

Problem 1. *Is there an algorithm to compute $\delta(f)$ in $W_1^0 \otimes W_1^0$ for f in W_1^0? This is easy if $f = e_n^*$ in the dual basis, i.e., for the finite sequences. W_1^0 is not locally finite as a Lie coalgebra, so this may not be analogous to Δf in $A^0 \otimes A^0$.*

Problem 2. *Is there any relation between the (commutative, cocommutative) Hopf algebra A^0 and the Lie bialgebra W_1^0 (both of which are the space of linearly recursive sequences)? Since everything starts from $k[x]$, there should be some relation between the two structures.*

Problem 3. *Is there a quantum deformation of any of these structures - A, A^0, W_1 and W_1^0?*

ON CLASSIFICATION OF Z–GRADED LIE ALGEBRAS OF CONSTANT GROWTH WHICH HAVE ALGEBRA $\mathbb{C}[h]$ AS CARTAN SUBALGEBRA

A. M. VERSHIK

Let us recall the main definitions (see our papers with M. V. Saveliev, Commun. Math. Phys. **126**, no.3 (1989);Phys. Lett.A. **143**, no.3 (1990)).

Let $\mathfrak{g}$ be a **Z**-graded Lie algebra $\mathfrak{g} = \bigoplus_{i\in\mathbb{Z}} \mathfrak{g}_i$, and $\mathfrak{g}_0$ is an abelian subalgebra which is called Cartan subalgebra. Contrary to the definition of Kac-Moody algebras we do not suppose that the dimension of $\mathfrak{g}_0$ is finite, but we supply $\mathfrak{g}_0$ with the structure of a commutative associative algebra E over $\mathbb{C}$. By definition the root system is the spectrum of this algebra. Moreover, we suppose that $\mathfrak{g}_{+1}, \mathfrak{g}_{-1}$ are isomorphic to $\mathfrak{g}_0$ as linear spaces. We denote the elements of $\mathfrak{g}_0, \mathfrak{g}_{+1}, \mathfrak{g}_{-1}$ by $x_i(\phi)$ where $\phi \in E,\ i = -1, 0, +1$.

We consider the "local part" $\mathfrak{g}_{-1} \oplus \mathfrak{g}_0 \oplus \mathfrak{g}_{+1}$ and suppose that this part generates the Lie algebra $\mathfrak{g}$. Now we shall formulate our axioms: (1) $[x_0(\phi), x_0(\psi)] = 0, \forall \phi, \psi \in E$; (2) $[x_0(\phi), x_{\pm 1}(\psi)] = \pm x_{\pm 1}(K\phi \cdot \psi)$ where multiplication is in E and K is a linear operator in E (the Cartan operator); (3)$[x_{+1}(\phi), x_{-1}(\psi)] = x_0(\phi \cdot \psi)$; (4) $\forall i \in \mathbf{Z}, \forall x \in \mathfrak{g}_i, x \neq 0, \exists y \in \mathfrak{g}_\varepsilon, \varepsilon = -\operatorname{sgn} i$ and $[x, y] \neq 0$. Let $\mathfrak{g}(E, K)$ be quotient of freely generated by local part Lie algebra factorizing by maximal nilpotent ideal having trivial intersection with $\mathfrak{g}_0$. If $\mathfrak{g}$ has also a property: $\mathfrak{g}_i \cong E, i \in \mathbf{Z}$ (linear isomorphism), we shall say that $\mathfrak{g}(E, K)$ has a constant growth.

Now we return to our problem: which graded Lie algebras have the algebra of polynomials as Cartan subalgebra? Let us suppose that $E = \mathbb{C}[h]$ is the algebra of all complex polynomials of one variable.

Theorem. *If $(Kf)(h) = f(h+1) - 2f(h) + f(h-1)$ then $\mathfrak{g}(\mathbb{C}[h], K)$ is isomorphic as a **Z**-graded Lie algebra to Lie algebra $<< \partial, L_x >>$ where ∂ is differentiation and L_x is multiplication by x , $[\partial, L_x] = 1$.*

This example is not unique; there are many other Cartan operators which generate algebras of constant growth. The following series was considered by B.Feigin and E.Frenkel. Let $U = U(sl(2))$ be the universal enveloping algebra of $sl(2, \mathbf{R})$ and Δ its Casimir element.Let us consider U as a Lie algebra with gradation generated by $\deg h = 0, \deg f = -1, \deg e = +1$.

Theorem. *If I_c is the ideal generated by $\Delta - c1$ then the **Z** graded Lie algebra U/I_c is of the type $(\mathbb{C}[h], K_c)$ for some K_c, $c \in \mathbf{R}$.*

Is this list complete?

List of Participants

First workshop "Quantum Groups, Deformation Theory and Representation Theory"

(15-28 October 1990)

Biedenharn L.C.	Durham, USA	Noumi M.	Tokyo, Japan
Celeghini E.	Florence, Italy	Olshanskii G.I.	Moscow, USSR
Drinfeld V.G.	Kharkov, USSR	Polivanov M.K.	Moscow, USSR
Faddeev L.D.	Leningrad, USSR	Retakh V.S.	Moscow, USSR
Feigin B.L.	Moscow, USSR	Rubtsov V.N.	Moscow, USSR
Gerstenhaber M.	Philadelphia, USA	Saveliev M.V.	Protvino, USSR
Giachetti R.	Florence, Italy	Schack S.T.	Buffallo N.Y., USA
Gurevich D.I.	Moscow, USSR	Semenov-Tian	
Karasev M.V.	Moscow, USSR	-Shansky M.A.	Leningrad, USSR
Kerov S.V.	Leningrad, USSR	Sklyanin E.K.	Leningrad, USSR
Kirillov A.N.	Leningrad, USSR	Soibelman Ya.	Rostov-na-Donu, USSR
Klimyk A.U.	Kiev, USSR	Stasheff J.D.	Chapel Hill, USA
Korogodsky L.I.	Rostov-na-Donu, USSR	Taft E.J.	New Brunswick, USA
Kulish P.P.	Leningrad, USSR	Tarlini M.	Florence, Italy
Lyubashenko V.	Kiev, USSR	Ueno K.	Tokyo, Japan
Majid S.	Cambridge, UK	Vershik A.M.	Leningrad, USSR
Nazarov M.L.	Moscow, USSR	Woronowicz S.	Warsaw, Poland

Second workshop "Quantum Groups, Symmetries of Dynamical Systems and Conformal Field Theory"

(12-25 November 1990)

Alekseev A.Yu.	Leningrad, USSR	Date E.	Osaka, Japan
Babelon O.	Paris, France	Faddeev L.D.	Leningrad, USSR
Bernard D.	Saclay, France	Fock V.V.	Moscow, USSR
Cherednik I.V.	Moscow, USSR	Gervais J.-L.	Paris, France
Chekhov L.	Moscow,USSR	Izergin A.G.	Leningrad,USSR

Kashaev R.M.	Protvino, USSR	Schrader R.	Berlin, Germany
Kulish P.P.	Leningrad, USSR	Semenoff G.	Vancouver, Canada
Leites D.	Stockholm, Sweden	Semikhatov A.	Moscow, USSR
Lukierski J.	Wroclaw, Poland	Sergeev S.M.	Protvino, USSR
Nakanishi T.	Nagoya, Japan	Sklyanin E.K.	Leningrad, USSR
Niemi A.J.	Helsinki, Finland	Smirnov F.A.	Leningrad, USSR
Nijhoff F.W.	Potsdam, USA	Sorace E.	Florence, Italy
Pogrebkov A.K.	Moscow,USSR	Takebe T.	Tokyo, Japan
Reyman A.G.	Leningrad, USSR	Tarasov V.O.	Leningrad, USSR
Rittenberg V.	Bonn, Germany	Todorov I.T.	Sofia, Bulgaria
Sasaki R.	Kyoto, Japan	Volkov A.Yu.	Leningrad, USSR

Third workshop "Quantum Groups, Low-Dimensional Topology and Link Invariants"

(3-16 December 1990)

Belokolos E.	Kiev, USSR	Morton H.R.	Liverpool, UK
Breen L.S.	Paris, France	Murakami J.	Osaka, Japan
Drobotukhina Yu.	Leningrad, USSR	Piunikhin S.	Moscow, USSR
Faddeev L.D.	Leningrad, USSR	Postnikov M.M.	Moscow, USSR
Falqui G.	Trieste, Italy	Radul A.O.	Moscow, USSR
Ge M.-L.	Tianjin, P.R.China	Schüler A.	Leipzig, Germany
Gusarov M.	Leningrad, USSR	Sossinsky A.B.	Moscow, USSR
Kauffman L.H.	Chicago,USA	Thang L.T.Q	Hanoi, Vietnam
Khovanov M.	Moscow, USSR	Vaksman L.L.	Rostov-na-Donu, USSR
Kerov S.V.	Leningrad, USSR	Vassil'ev V.	Moscow, USSR
Kirillov A.N.	Leningrad, USSR	Viro O.Ya.	Leningrad, USSR
Kohno T.	Fukuoka, Japan	Wadati M.	Tokyo, Japan
Kulish P.P.	Leningrad, USSR	Zakrzewski S.	Warsaw, Poland
Lyubashenko V.	Kiev, USSR	Zavialov O.	Moscow, USSR
Millett K.	Santa Barbara, USA		

Lecture Notes in Mathematics

For information about Vols. 1–1312
please contact your bookseller or Springer-Verlag

Vol. 1313: F. Colonius. Optimal Periodic Control. VI, 177 pages. 1988.

Vol. 1314: A. Futaki. Kähler-Einstein Metrics and Integral Invariants. IV, 140 pages. 1988.

Vol. 1315: R.A. McCoy, I. Ntantu, Topological Properties of Spaces of Continuous, Functions. IV, 124 pages. 1988.

Vol. 1316: H. Korezlioglu, A.S. Ustunel (Eds.), Stochastic Analysis and Related Topics. Proceedings, 1986. V, 371 pages. 1988.

Vol. 1317: J. Lindenstrauss, V.D. Milman (Eds.), Geometric Aspects of Functional Analysis. Seminar, 1986–87. VII, 289 pages. 1988.

Vol. 1318: Y. Felix (Ed.), Algebraic Topology – Rational Homotopy. Proceedings, 1986. VIII, 245 pages. 1988.

Vol. 1319: M. Vuorinen, Conformal Geometry and Quasiregular Mappings. XIX, 209 pages. 1988.

Vol. 1320: H. Jürgensen, G. Lallement, H.J. Weinert (Eds.), Semigroups, Theory and Applications. Proceedings, 1986. X, 416 pages. 1988.

Vol. 1321: J. Azéma, P.A. Meyer, M. Yor (Eds.), Séminaire de Probabilités XXII. Proceedings. IV, 600 pages. 1988.

Vol. 1322: M. Métivier, S. Watanabe (Eds.), Stochastic Analysis. Proceedings, 1987. VII, 197 pages. 1988.

Vol. 1323: D.R. Anderson, H.J. Munkholm, Boundedly Controlled Topology. XII, 309 pages. 1988.

Vol. 1324: F. Cardoso, D.G. de Figueiredo, R. Iório, O. Lopes (Eds.), Partial Differential Equations. Proceedings, 1986. VIII, 433 pages. 1988.

Vol. 1325: A. Truman, I.M. Davies (Eds.), Stochastic Mechanics and Stochastic Processes. Proceedings, 1986. V, 220 pages. 1988.

Vol. 1326: P.S. Landweber (Ed.), Elliptic Curves and Modular Forms in Algebraic Topology. Proceedings, 1986. V, 224 pages. 1988.

Vol. 1327: W. Bruns, U. Vetter, Determinantal Rings. VII, 236 pages. 1988.

Vol. 1328: J.L. Bueso, P. Jara, B. Torrecillas (Eds.), Ring Theory. Proceedings, 1986. IX, 331 pages. 1988.

Vol. 1329: M. Alfaro, J.S. Dehesa, F.J. Marcellan, J.L. Rubio de Francia, J. Vinuesa (Eds.): Orthogonal Polynomials and their Applications. Proceedings, 1986. XV, 334 pages. 1988.

Vol. 1330: A. Ambrosetti, F. Gori, R. Lucchetti (Eds.), Mathematical Economics. Montecatini Terme 1986. Seminar. VII, 137 pages. 1988.

Vol. 1331: R. Bamón, R. Labarca, J. Palis Jr. (Eds.), Dynamical Systems, Valparaiso 1986. Proceedings. VI, 250 pages. 1988.

Vol. 1332: E. Odell, H. Rosenthal (Eds.), Functional Analysis. Proceedings. 1986–87. V, 202 pages. 1988.

Vol. 1333: A.S. Kechris, D.A. Martin, J.R. Steel (Eds.), Cabal Seminar 81–85. Proceedings, 1981–85. V, 224 pages. 1988.

Vol. 1334: Yu.G. Borisovich, Yu.E. Gliklikh (Eds.), Global Analysis – Studies and Applications III. V, 331 pages. 1988.

Vol. 1335: F. Guillén, V. Navarro Aznar, P. Pascual-Gainza, F. Puerta, Hyperrésolutions cubiques et descente cohomologique. XII, 192 pages. 1988.

Vol. 1336: B. Helffer, Semi-Classical Analysis for the Schrödinger Operator and Applications. V, 107 pages. 1988.

Vol. 1337: E. Sernesi (Ed.), Theory of Moduli. Seminar, 1985. VIII, 232 pages. 1988.

Vol. 1338: A.B. Mingarelli, S.G. Halvorsen. Non-Oscillation Domains of Differential Equations with Two Parameters. XI, 109 pages. 1988.

Vol. 1339: T. Sunada (Ed.), Geometry and Analysis of Manifolds. Proceedings, 1987. IX, 277 pages. 1988.

Vol. 1340: S. Hildebrandt, D.S. Kinderlehrer, M. Miranda (Eds.), Calculus of Variations and Partial Differential Equations. Proceedings, 1986. IX, 301 pages. 1988.

Vol. 1341: M. Dauge, Elliptic Boundary Value Problems on Corner Domains. VIII, 259 pages. 1988.

Vol. 1342: J.C. Alexander (Ed.), Dynamical Systems. Proceedings, 1986–87. VIII, 726 pages. 1988.

Vol. 1343: H. Ulrich, Fixed Point Theory of Parametrized Equivariant Maps. VII, 147 pages. 1988.

Vol. 1344: J. Král, J. Lukes, J. Netuka, J. Vesely´ (Eds.), Potential Theory – Surveys and Problems. Proceedings, 1987. VIII, 271 pages. 1988.

Vol. 1345: X. Gomez-Mont, J. Seade, A. Verjovski (Eds.), Holomorphic Dynamics. Proceedings, 1986. VII. 321 pages. 1988.

Vol. 1346: O.Ya. Viro (Ed.), Topology and Geometry – Rohlin Seminar. XI, 581 pages. 1988.

Vol. 1347: C. Preston, Iterates of Piecewise Monotone Mappings on an Interval. V, 166 pages. 1988.

Vol. 1348: F. Borceux (Ed.), Categorical Algebra and its Applications. Proceedings, 1987. VIII, 375 pages. 1988.

Vol. 1349: E. Novak, Deterministic and Stochastic Error Bounds in Numerical Analysis. V, 113 pages. 1988.

Vol. 1350: U. Koschorke (Ed.), Differential Topology Proceedings, 1987, VI, 269 pages. 1988.

Vol. 1351: I. Laine, S. Rickman, T. Sorvali (Eds.), Complex Analysis, Joensuu 1987. Proceedings. XV, 378 pages. 1988.

Vol. 1352: L.L. Avramov, K.B. Tchakerian (Eds.), Algebra – Some Current Trends. Proceedings. 1986. IX, 240 Seiten. 1988.

Vol. 1353: R.S. Palais, Ch.-l. Teng, Critical Point Theory and Submanifold Geometry. X, 272 pages. 1988.

Vol. 1354: A. Gómez, F. Guerra, M.A. Jiménez, G. López (Eds.), Approximation and Optimization. Proceedings, 1987. VI, 280 pages. 1988.

Vol. 1355: J. Bokowski, B. Sturmfels, Computational Synthetic Geometry. V, 168 pages. 1989.

Vol. 1356: H. Volkmer, Multiparameter Eigenvalue Problems and Expansion Theorems. VI, 157 pages. 1988.

Vol. 1357: S. Hildebrandt, R. Leis (Eds.), Partial Differential Equations and Calculus of Variations. VI, 423 pages. 1988.

Vol. 1358: D. Mumford, The Red Book of Varieties and Schemes. V, 309 pages. 1988.

Vol. 1359: P. Eymard, J.-P. Pier (Eds.) Harmonic Analysis. Proceedings, 1987. VIII, 287 pages. 1988.

Vol. 1360: G. Anderson, C. Greengard (Eds.), Vortex Methods. Proceedings, 1987. V, 141 pages. 1988.

Vol. 1361: T. tom Dieck (Ed.), Algebraic Topology and Transformation Groups. Proceedings. 1987. VI, 298 pages. 1988.

Vol. 1362: P. Diaconis, D. Elworthy, H. Föllmer, E. Nelson, G.C. Papanicolaou, S.R.S. Varadhan. École d´ Été de Probabilités de Saint-Flour XV–XVII. 1985–87 Editor: P.L. Hennequin. V, 459 pages. 1988.

Vol. 1363: P.G. Casazza, T.J. Shura, Tsirelson´s Space. VIII, 204 pages. 1988.

Vol. 1364: R.R. Phelps, Convex Functions, Monotone Operators and Differentiability. IX, 115 pages. 1989.

Vol. 1365: M. Giaquinta (Ed.), Topics in Calculus of Variations. Seminar, 1987. X, 196 pages. 1989.

Vol. 1366: N. Levitt, Grassmannians and Gauss Maps in PL-Topology. V, 203 pages. 1989.

Vol. 1367: M. Knebusch, Weakly Semialgebraic Spaces. XX, 376 pages. 1989.

Vol. 1368: R. Hübl, Traces of Differential Forms and Hochschild Homology. III, 111 pages. 1989.

Vol. 1369: B. Jiang, Ch.-K. Peng, Z. Hou (Eds.), Differential Geometry and Topology. Proceedings, 1986–87. VI, 366 pages. 1989.

Vol. 1370: G. Carlsson, R.L. Cohen, H.R. Miller, D.C. Ravenel (Eds.), Algebraic Topology. Proceedings, 1986. IX, 456 pages. 1989.

Vol. 1371: S. Glaz, Commutative Coherent Rings. XI, 347 pages. 1989.

Vol. 1372: J. Azéma, P.A. Meyer, M. Yor (Eds.), Séminaire de Probabilités XXIII. Proceedings. IV, 583 pages. 1989.

Vol. 1373: G. Benkart, J.M. Osborn (Eds.), Lie Algebras. Madison 1987. Proceedings. V, 145 pages. 1989.

Vol. 1374: R.C. Kirby, The Topology of 4-Manifolds. VI, 108 pages. 1989.

Vol. 1375: K. Kawakubo (Ed.), Transformation Groups. Proceedings, 1987. VIII, 394 pages, 1989.

Vol. 1376: J. Lindenstrauss, V.D. Milman (Eds.), Geometric Aspects of Functional Analysis. Seminar (GAFA) 1987–88. VII, 288 pages. 1989.

Vol. 1377: J.F. Pierce, Singularity Theory, Rod Theory, and Symmetry-Breaking Loads. IV, 177 pages. 1989.

Vol. 1378: R.S. Rumely, Capacity Theory on Algebraic Curves. III, 437 pages. 1989.

Vol. 1379: H. Heyer (Ed.), Probability Measures on Groups IX. Proceedings, 1988. VIII, 437 pages. 1989.

Vol. 1380: H.P. Schlickewei, E. Wirsing (Eds.), Number Theory, Ulm 1987. Proceedings. V, 266 pages. 1989.

Vol. 1381: J.-O. Strömberg, A. Torchinsky, Weighted Hardy Spaces. V, 193 pages. 1989.

Vol. 1382: H. Reiter, Metaplectic Groups and Segal Algebras. XI, 128 pages. 1989.

Vol. 1383: D.V. Chudnovsky, G.V. Chudnovsky, H. Cohn, M.B. Nathanson (Eds.), Number Theory, New York 1985–88. Seminar. V, 256 pages. 1989.

Vol. 1384: J. Garcia-Cuerva (Ed.), Harmonic Analysis and Partial Differential Equations. Proceedings, 1987. VII, 213 pages. 1989.

Vol. 1385: A.M. Anile, Y. Choquet-Bruhat (Eds.), Relativistic Fluid Dynamics. Seminar, 1987. V, 308 pages. 1989.

Vol. 1386: A. Bellen, C.W. Gear, E. Russo (Eds.), Numerical Methods for Ordinary Differential Equations. Proceedings, 1987. VII, 136 pages. 1989.

Vol. 1387: M. Petkovi´c, Iterative Methods for Simultaneous Inclusion of Polynomial Zeros. X, 263 pages. 1989.

Vol. 1388: J. Shinoda, T.A. Slaman, T. Tugué (Eds.), Mathematical Logic and Applications. Proceedings, 1987. V, 223 pages. 1989.

Vol. 1000: Second Edition. H. Hopf, Differential Geometry in the Large. VII, 184 pages. 1989.

Vol. 1389: E. Ballico, C. Ciliberto (Eds.), Algebraic Curves and Projective Geometry. Proceedings, 1988. V, 288 pages. 1989.

Vol. 1390: G. Da Prato, L. Tubaro (Eds.), Stochastic Partial Differential Equations and Applications II. Proceedings, 1988. VI, 258 pages. 1989.

Vol. 1391: S. Cambanis, A. Weron (Eds.), Probability Theory on Vector Spaces IV. Proceedings, 1987. VIII, 424 pages. 1989.

Vol. 1392: R. Silhol, Real Algebraic Surfaces. X, 215 pages. 1989.

Vol. 1393: N. Bouleau, D. Feyel, F. Hirsch, G. Mokobodzki (Eds.), Séminaire de Théorie du Potentiel Paris, No. 9. Proceedings. VI, 265 pages. 1989.

Vol. 1394: T.L. Gill, W.W. Zachary (Eds.), Nonlinear Semigroups, Partial Differential Equations and Attractors. Proceedings, 1987. IX, 233 pages. 1989.

Vol. 1395: K. Alladi (Ed.), Number Theory, Madras 1987. Proceedings. VII, 234 pages. 1989.

Vol. 1396: L. Accardi, W. von Waldenfels (Eds.), Quantum Probability and Applications IV. Proceedings, 1987. VI, 355 pages. 1989.

Vol. 1397: P.R. Turner (Ed.), Numerical Analysis and Parallel Processing. Seminar, 1987. VI, 264 pages. 1989.

Vol. 1398: A.C. Kim, B.H. Neumann (Eds.), Groups – Korea 1988. Proceedings. V, 189 pages. 1989.

Vol. 1399: W.-P. Barth, H. Lange (Eds.), Arithmetic of Complex Manifolds. Proceedings, 1988. V, 171 pages. 1989.

Vol. 1400: U. Jannsen. Mixed Motives and Algebraic K-Theory. XIII, 246 pages. 1990.

Vol. 1401: J. Steprans, S. Watson (Eds.), Set Theory and its Applications. Proceedings, 1987. V, 227 pages. 1989.

Vol. 1402: C. Carasso, P. Charrier, B. Hanouzet, J.-L. Joly (Eds.), Nonlinear Hyperbolic Problems. Proceedings, 1988. V, 249 pages. 1989.

Vol. 1403: B. Simeone (Ed.), Combinatorial Optimization. Seminar, 1986. V, 314 pages. 1989.

Vol. 1404: M.-P. Malliavin (Ed.), Séminaire d'Algèbre Paul Dubreil et Marie-Paul Malliavin. Proceedings, 1987–1988. IV, 410 pages. 1989.

Vol. 1405: S. Dolecki (Ed.), Optimization. Proceedings, 1988. V, 223 pages. 1989. Vol. 1406: L. Jacobsen (Ed.), Analytic Theory of Continued Fractions III. Proceedings, 1988. VI, 142 pages. 1989.

Vol. 1407: W. Pohlers, Proof Theory. VI, 213 pages. 1989.

Vol. 1408: W. Lück, Transformation Groups and Algebraic K-Theory. XII, 443 pages. 1989.

Vol. 1409: E. Hairer, Ch. Lubich, M. Roche. The Numerical Solution of Differential-Algebraic Systems by Runge-Kutta Methods. VII, 139 pages. 1989.

Vol. 1410: F.J. Carreras, O. Gil-Medrano, A.M. Naveira (Eds.), Differential Geometry. Proceedings, 1988. V, 308 pages. 1989.

Vol. 1411: B. Jiang (Ed.), Topological Fixed Point Theory and Applications. Proceedings. 1988. VI, 203 pages. 1989.

Vol. 1412: V.V. Kalashnikov, V.M. Zolotarev (Eds.), Stability Problems for Stochastic Models. Proceedings, 1987. X, 380 pages. 1989.

Vol. 1413: S. Wright, Uniqueness of the Injective III_1 Factor. III, 108 pages. 1989.

Vol. 1414: E. Ramirez de Arellano (Ed.), Algebraic Geometry and Complex Analysis. Proceedings, 1987. VI, 180 pages. 1989.

Vol. 1415: M. Langevin, M. Waldschmidt (Eds.), Cinquante Ans de Polynômes. Fifty Years of Polynomials. Proceedings, 1988. IX, 235 pages.1990.

Vol. 1416: C. Albert (Ed.), Géométrie Symplectique et Mécanique. Proceedings, 1988. V, 289 pages. 1990.

Vol. 1417: A.J. Sommese, A. Biancofiore, E.L. Livorni (Eds.), Algebraic Geometry. Proceedings, 1988. V, 320 pages. 1990.

Vol. 1418: M. Mimura (Ed.), Homotopy Theory and Related Topics. Proceedings, 1988. V, 241 pages. 1990.

Vol. 1419: P.S. Bullen, P.Y. Lee, J.L. Mawhin, P. Muldowney, W.F. Pfeffer (Eds.), New Integrals. Proceedings, 1988. V, 202 pages. 1990.

Vol. 1420: M. Galbiati, A. Tognoli (Eds.), Real Analytic Geometry. Proceedings, 1988. IV, 366 pages. 1990.

Vol. 1421: H.A. Biagioni, A Nonlinear Theory of Generalized Functions, XII, 214 pages. 1990.

Vol. 1422: V. Villani (Ed.), Complex Geometry and Analysis. Proceedings, 1988. V, 109 pages. 1990.

Vol. 1423: S.O. Kochman, Stable Homotopy Groups of Spheres: A Computer-Assisted Approach. VIII, 330 pages. 1990.

Vol. 1424: F.E. Burstall, J.H. Rawnsley, Twistor Theory for Riemannian Symmetric Spaces. III, 112 pages. 1990.

Vol. 1425: R.A. Piccinini (Ed.), Groups of Self-Equivalences and Related Topics. Proceedings, 1988. V, 214 pages. 1990.

Vol. 1426: J. Azéma, P.A. Meyer, M. Yor (Eds.), Séminaire de Probabilités XXIV, 1988/89. V, 490 pages. 1990.

Vol. 1427: A. Ancona, D. Geman, N. Ikeda, École d'Eté de Probabilités de Saint Flour XVIII, 1988. Ed.: P.L. Hennequin. VII, 330 pages. 1990.

Vol. 1428: K. Erdmann, Blocks of Tame Representation Type and Related Algebras. XV. 312 pages. 1990.

Vol. 1429: S. Homer, A. Nerode, R.A. Platek, G.E. Sacks, A. Scedrov, Logic and Computer Science. Seminar, 1988. Editor: P. Odifreddi. V, 162 pages. 1990.

Vol. 1430: W. Bruns, A. Simis (Eds.), Commutative Algebra. Proceedings. 1988. V, 160 pages. 1990.

Vol. 1431: J.G. Heywood, K. Masuda, R. Rautmann, V.A. Solonnikov (Eds.), The Navier-Stokes Equations – Theory and Numerical Methods. Proceedings, 1988. VII, 238 pages. 1990.

Vol. 1432: K. Ambos-Spies, G.H. Müller, G.E. Sacks (Eds.), Recursion Theory Week. Proceedings, 1989. VI, 393 pages. 1990.

Vol. 1433: S. Lang, W. Cherry, Topics in Nevanlinna Theory. II, 174 pages.1990.

Vol. 1434: K. Nagasaka, E. Fouvry (Eds.), Analytic Number Theory. Proceedings, 1988. VI, 218 pages. 1990.

Vol. 1435: St. Ruscheweyh, E.B. Saff, L.C. Salinas, R.S. Varga (Eds.), Computational Methods and Function Theory. Proceedings, 1989. VI, 211 pages. 1990.

Vol. 1436: S. Xambó-Descamps (Ed.), Enumerative Geometry. Proceedings, 1987. V, 303 pages. 1990.

Vol. 1437: H. Inassaridze (Ed.), K-theory and Homological Algebra. Seminar, 1987–88. V, 313 pages. 1990.

Vol. 1438: P.G. Lemarié (Ed.) Les Ondelettes en 1989. Seminar. IV, 212 pages. 1990.

Vol. 1439: E. Bujalance, J.J. Etayo, J.M. Gamboa, G. Gromadzki. Automorphism Groups of Compact Bordered Klein Surfaces: A Combinatorial Approach. XIII, 201 pages. 1990.

Vol. 1440: P. Latiolais (Ed.), Topology and Combinatorial Groups Theory. Seminar, 1985–1988. VI, 207 pages. 1990.

Vol. 1441: M. Coornaert, T. Delzant, A. Papadopoulos. Géométrie et théorie des groupes. X, 165 pages. 1990.

Vol. 1442: L. Accardi, M. von Waldenfels (Eds.), Quantum Probability and Applications V. Proceedings, 1988. VI, 413 pages. 1990.

Vol. 1443: K.H. Dovermann, R. Schultz, Equivariant Surgery Theories and Their Periodicity Properties. VI, 227 pages. 1990.

Vol. 1444: H. Korezlioglu, A.S. Ustunel (Eds.), Stochastic Analysis and Related Topics VI. Proceedings, 1988. V, 268 pages. 1990.

Vol. 1445: F. Schulz, Regularity Theory for Quasilinear Elliptic Systems and – Monge Ampère Equations in Two Dimensions. XV, 123 pages. 1990.

Vol. 1446: Methods of Nonconvex Analysis. Seminar, 1989. Editor: A. Cellina. V, 206 pages. 1990.

Vol. 1447: J.-G. Labesse, J. Schwermer (Eds), Cohomology of Arithmetic Groups and Automorphic Forms. Proceedings, 1989. V, 358 pages. 1990.

Vol. 1448: S.K. Jain, S.R. López-Permouth (Eds.), Non-Commutative Ring Theory. Proceedings, 1989. V, 166 pages. 1990.

Vol. 1449: W. Odyniec, G. Lewicki, Minimal Projections in Banach Spaces. VIII, 168 pages. 1990.

Vol. 1450: H. Fujita, T. Ikebe, S.T. Kuroda (Eds.), Functional-Analytic Methods for Partial Differential Equations. Proceedings, 1989. VII, 252 pages. 1990.

Vol. 1451: L. Alvarez-Gaumé, E. Arbarello, C. De Concini, N.J. Hitchin, Global Geometry and Mathematical Physics. Montecatini Terme 1988. Seminar. Editors: M. Francaviglia, F. Gherardelli. IX, 197 pages. 1990.

Vol. 1452: E. Hlawka, R.F. Tichy (Eds.), Number-Theoretic Analysis. Seminar, 1988–89. V, 220 pages. 1990.

Vol. 1453: Yu.G. Borisovich, Yu.E. Gliklikh (Eds.), Global Analysis – Studies and Applications IV. V, 320 pages. 1990.

Vol. 1454: F. Baldassari, S. Bosch, B. Dwork (Eds.), p-adic Analysis. Proceedings, 1989. V, 382 pages. 1990.

Vol. 1455: J.-P. Françoise, R. Roussarie (Eds.), Bifurcations of Planar Vector Fields. Proceedings, 1989. VI, 396 pages. 1990.

Vol. 1456: L.G. Kovács (Ed.), Groups – Canberra 1989. Proceedings. XII, 198 pages. 1990.

Vol. 1457: O. Axelsson, L.Yu. Kolotilina (Eds.), Preconditioned Conjugate Gradient Methods. Proceedings, 1989. V, 196 pages. 1990.

Vol. 1458: R. Schaaf, Global Solution Branches of Two Point Boundary Value Problems. XIX, 141 pages. 1990.

Vol. 1459: D. Tiba, Optimal Control of Nonsmooth Distributed Parameter Systems. VII, 159 pages. 1990.

Vol. 1460: G. Toscani, V. Boffi, S. Rionero (Eds.), Mathematical Aspects of Fluid Plasma Dynamics. Proceedings, 1988. V, 221 pages. 1991.

Vol. 1461: R. Gorenflo, S. Vessella, Abel Integral Equations. VII, 215 pages. 1991.

Vol. 1462: D. Mond, J. Montaldi (Eds.), Singularity Theory and its Applications. Warwick 1989, Part I. VIII, 405 pages. 1991.

Vol. 1463: R. Roberts, I. Stewart (Eds.), Singularity Theory and its Applications. Warwick 1989, Part II. VIII, 322 pages. 1991.

Vol. 1464: D. L. Burkholder, E. Pardoux, A. Sznitman, Ecole d'Eté de Probabilités de Saint- Flour XIX-1989. Editor: P. L. Hennequin. VI, 256 pages. 1991.

Vol. 1465: G. David, Wavelets and Singular Integrals on Curves and Surfaces. X, 107 pages. 1991.

Vol. 1466: W. Banaszczyk, Additive Subgroups of Topological Vector Spaces. VII, 178 pages. 1991.

Vol. 1467: W. M. Schmidt, Diophantine Approximations and Diophantine Equations. VIII, 217 pages. 1991.

Vol. 1468: J. Noguchi, T. Ohsawa (Eds.), Prospects in Complex Geometry. Proceedings, 1989. VII, 421 pages. 1991.

Vol. 1469: J. Lindenstrauss, V. D. Milman (Eds.), Geometric Aspects of Functional Analysis. Seminar 1989-90. XI, 191 pages. 1991.

Vol. 1470: E. Odell, H. Rosenthal (Eds.), Functional Analysis. Proceedings, 1987-89. VII, 199 pages. 1991.

Vol. 1471: A. A. Panchishkin, Non-Archimedean L-Functions of Siegel and Hilbert Modular Forms. VII, 157 pages. 1991.

Vol. 1472: T. T. Nielsen, Bose Algebras: The Complex and Real Wave Representations. V, 132 pages. 1991.

Vol. 1473: Y. Hino, S. Murakami, T. Naito, Functional Differential Equations with Infinite Delay. X, 317 pages. 1991.

Vol. 1474: S. Jackowski, B. Oliver, K. Pawałowski (Eds.), Algebraic Topology, Poznań 1989. Proceedings. VIII, 397 pages. 1991.

Vol. 1475: S. Busenberg, M. Martelli (Eds.), Delay Differential Equations and Dynamical Systems. Proceedings, 1990. VIII, 249 pages. 1991.

Vol. 1476: M. Bekkali, Topics in Set Theory. VII, 120 pages. 1991.

Vol. 1477: R. Jajte, Strong Limit Theorems in Noncommutative L_2-Spaces. X, 113 pages. 1991.

Vol. 1478: M.-P. Malliavin (Ed.), Topics in Invariant Theory. Seminar 1989-1990. VI, 272 pages. 1991.

Vol. 1479: S. Bloch, I. Dolgachev, W. Fulton (Eds.), Algebraic Geometry. Proceedings, 1989. VII, 300 pages. 1991.

Vol. 1480: F. Dumortier, R. Roussarie, J. Sotomayor, H. Żoładek, Bifurcations of Planar Vector Fields: Nilpotent Singularities and Abelian Integrals. VIII, 226 pages. 1991.

Vol. 1481: D. Ferus, U. Pinkall, U. Simon, B. Wegner (Eds.), Global Differential Geometry and Global Analysis. Proceedings, 1991. VIII, 283 pages. 1991.

Vol. 1482: J. Chabrowski, The Dirichlet Problem with L^2-Boundary Data for Elliptic Linear Equations. VI, 173 pages. 1991.

Vol. 1483: E. Reithmeier, Periodic Solutions of Nonlinear Dynamical Systems. VI, 171 pages. 1991.

Vol. 1484: H. Delfs, Homology of Locally Semialgebraic Spaces. IX, 136 pages. 1991.

Vol. 1485: J. Azéma, P. A. Meyer, M. Yor (Eds.), Séminaire de Probabilités XXV. VIII, 440 pages. 1991.

Vol. 1486: L. Arnold, H. Crauel, J.-P. Eckmann (Eds.), Lyapunov Exponents. Proceedings, 1990. VIII, 365 pages. 1991.

Vol. 1487: E. Freitag, Singular Modular Forms and Theta Relations. VI, 172 pages. 1991.

Vol. 1488: A. Carboni, M. C. Pedicchio, G. Rosolini (Eds.), Category Theory. Proceedings, 1990. VII, 494 pages. 1991.

Vol. 1489: A. Mielke, Hamiltonian and Lagrangian Flows on Center Manifolds. X, 140 pages. 1991.

Vol. 1490: K. Metsch, Linear Spaces with Few Lines. XIII, 196 pages. 1991.

Vol. 1491: E. Lluis-Puebla, J.-L. Loday, H. Gillet, C. Soulé, V. Snaith, Higher Algebraic K-Theory: an overview. IX, 164 pages. 1992.

Vol. 1492: K. R. Wicks, Fractals and Hyperspaces. VIII, 168 pages. 1991.

Vol. 1493: E. Benoît (Ed.), Dynamic Bifurcations. Proceedings, Luminy 1990. VII, 219 pages. 1991.

Vol. 1494: M.-T. Cheng, X.-W. Zhou, D.-G. Deng (Eds.), Harmonic Analysis. Proceedings, 1988. IX, 226 pages. 1991.

Vol. 1495: J. M. Bony, G. Grubb, L. Hörmander, H. Komatsu, J. Sjöstrand, Microlocal Analysis and Applications. Montecatini Terme, 1989. Editors: L. Cattabriga, L. Rodino. VII, 349 pages. 1991.

Vol. 1496: C. Foias, B. Francis, J. W. Helton, H. Kwakernaak, J. B. Pearson, H_∞-Control Theory. Como, 1990. Editors: E. Mosca, L. Pandolfi. VII, 336 pages. 1991.

Vol. 1497: G. T. Herman, A. K. Louis, F. Natterer (Eds.), Mathematical Methods in Tomography. Proceedings 1990. X, 268 pages. 1991.

Vol. 1498: R. Lang, Spectral Theory of Random Schrödinger Operators. X, 125 pages. 1991.

Vol. 1499: K. Taira, Boundary Value Problems and Markov Processes. IX, 132 pages. 1991.

Vol. 1500: J. P. Serre, Lie Algebras and Lie Groups. VII, 168 pages. 1992.

Vol. 1501: A. De Masi, E. Presutti, Mathematical Methods for Hydrodynamic Limits. IX, 196 pages. 1991.

Vol. 1502: C. Simpson, Asymptotic Behavior of Monodromy. V, 139 pages. 1991.

Vol. 1503: S. Shokranian, The Selberg-Arthur Trace Formula (Lectures by J. Arthur). VII, 97 pages. 1991.

Vol. 1504: J. Cheeger, M. Gromov, C. Okonek, P. Pansu, Geometric Topology: Recent Developments. Editors: P. de Bartolomeis, F. Tricerri. VII, 197 pages. 1991.

Vol. 1505: K. Kajitani, T. Nishitani, The Hyperbolic Cauchy Problem. VII, 168 pages. 1991.

Vol. 1506: A. Buium, Differential Algebraic Groups of Finite Dimension. XV, 145 pages. 1992.

Vol. 1507: K. Hulek, T. Peternell, M. Schneider, F.-O. Schreyer (Eds.), Complex Algebraic Varieties. Proceedings, 1990. VII, 179 pages. 1992.

Vol. 1508: M. Vuorinen (Ed.), Quasiconformal Space Mappings. A Collection of Surveys 1960-1990.IX, 148 pages. 1992.

Vol. 1509: J. Aguadé, M. Castellet, F. R. Cohen (Eds.), Algebraic Topology - Homotopy and Group Cohomology. Proceedings, 1990. X, 330 pages. 1992.

Vol. 1510: P. P. Kulish (Ed.), Quantum Groups. Proceedings, 1990. XII, 398 pages. 1992.